ADVANCES AND CHALLENGES IN PHARMACEUTICAL TECHNOLOGY

ADVANCES AND CHALLENGES IN PHARMACEUTICAL TECHNOLOGY

Materials, Process Development and Drug Delivery Strategies

Edited by

AMIT KUMAR NAYAK
Department of Pharmaceutics, Seemanta Institute of Pharmaceutical Sciences, Mayurbhanj, Odisha, India

KUNAL PAL
Department of Biotechnology and Medical Engineering, National Institute of Technology Rourkela, Rourkela, Odisha, India

INDRANIL BANERJEE
Department of Bioscience and Bioengineering, Indian Institute of Technology Jodhpur, Jodhpur, Rajasthan, India

SAMARENDRA MAJI
Department of Chemistry, Faculty of Engineering and Technology, SRM Institute of Science and Technology, SRM Nagar, Kattankulathur, Tamil Nadu, India

UPENDRANATH NANDA
Department of Pharmaceutics, Seemanta Institute of Pharmaceutical Sciences, Mayurbhanj, Odisha, India

ELSEVIER

ACADEMIC PRESS
An imprint of Elsevier

Academic Press is an imprint of Elsevier
125 London Wall, London EC2Y 5AS, United Kingdom
525 B Street, Suite 1650, San Diego, CA 92101, United States
50 Hampshire Street, 5th Floor, Cambridge, MA 02139, United States
The Boulevard, Langford Lane, Kidlington, Oxford OX5 1GB, United Kingdom

Library of Congress Cataloging-in-Publication Data
A catalog record for this book is available from the Library of Congress

British Library Cataloguing-in-Publication Data
A catalogue record for this book is available from the British Library

ISBN 978-0-12-820043-8

For information on all Academic Press publications
visit our website at https://www.elsevier.com/books-and-journals

Publisher: Andre Gerhard Wolff
Acquisitions Editor: Erin Hill-Parks
Editorial Project Manager: Billie Jean Fernandez
Production Project Manager: Omer Mukthar
Cover Designer: Greg Harris

Typeset by SPi Global, India

Contents

Contributors

Mohd Afzal
Department of Chemistry, College of Science, King Saud University, Riyadh, Saudi Arabia

Syed Anees Ahmed
Department of Pharmacology, Hygia Institute of Pharmaceutical Education and Research, Lucknow, Uttar Pradesh, India

Saad Alkahtani
Department of Zoology, College of Science, King Saud University, Riyadh, Saudi Arabia

Feras M. Al-Marshad
Department of Internal Medicine, College of Medicine, Shaqra University, Shaqra, Saudi Arabia

Arfat Anis
SABIC Polymer Research Center, Department of Chemical Engineering, King Saud University, Riyadh, Saudi Arabia

Mohammed Tahir Ansari
School of Pharmacy, University of Nottingham Malaysia, Jalan Broga, Semenyih, Kajang, Selangor, Malaysia

Pablo Ricardo Arantes
Laboratory of Genetic Toxicology, Federal University of Health Sciences of Porto Alegre—UFCSPA, Porto Alegre, Rio Grande do Sul, Brazil

Dariana Aristizabal Bedoya
Chemistry Science Faculty, Organic Chemistry Department, National University of Cordoba; Institute for Research and Development in Process Engineering and Applied Chemistry, Science and Technology National Council, Córdoba, Argentine

Mabilly Cox Holanda de Barros Dias
Athlone Institute of Technology, Athlone, County Westmeath, Ireland

Ranjit Barua
Centre for Healthcare Science and Technology, Indian Institute of Engineering Science and Technology, Shibpur, Howrah, India

Ananya Barui
Centre for Healthcare Science and Technology, Indian Institute of Engineering Science and Technology, Shibpur, Howrah, India

Pooja Das Bidla
Pharmaceutics Research Projects Laboratory, Department of Pharmaceutical Sciences, Dr. Harisingh Gour Vishwavidyalaya, Sagar, MP, India

Sourav Chakraborty
Scottish Church College, Kolkata, India

Tamalika Chakraborty
Department of Pharmaceutics, Guru Nanak Institute of Pharmaceutical Science and Technology, Kolkata, India

Hongbo Cheng
School of Pharmacy, Shenyang Pharmaceutical University, Shenyang, People's Republic of China

Amit Roy Chowdhury
Centre for Healthcare Science and Technology, Indian Institute of Engineering Science and Technology, Shibpur, Howrah, India

Zhixiang Cui
School of Pharmacy, Shenyang Pharmaceutical University, Shenyang, People's Republic of China

Sabya Sachi Das
Department of Pharmaceutical Sciences and Technology, Birla Institute of Technology, Mesra, Ranchi, Jharkhand, India

Shalini Dasgupta
Centre for Healthcare Science and Technology, Indian Institute of Engineering Science and Technology, Shibpur, Howrah, India

Pallab Datta
Centre for Healthcare Science and Technology, Indian Institute of Engineering Science and Technology, Shibpur, Howrah, India

Sudipto Datta
Centre for Healthcare Science and Technology, Indian Institute of Engineering Science and Technology, Shibpur, Howrah, India

N.G. Devi
Department of Biomedical Engineering, Manipal Institute of Technology, MAHE, Manipal, India

Somali Dhal
Department of Biotechnology and Medical Engineering, National Institute of Technology Rourkela, Rourkela, Odisha, India

Francisco N. Figueroa
Chemistry Science Faculty, Organic Chemistry Department, National University of Cordoba; Institute for Research and Development in Process Engineering and Applied Chemistry, Science and Technology National Council, Córdoba, Argentine

Md Saquib Hasnain
Department of Pharmacy, Shri Venkateshwara University, Gajraula, Uttar Pradesh, India

Jeferson Gustavo Henn
Laboratory of Genetic Toxicology, Federal University of Health Sciences of Porto Alegre—UFCSPA, Porto Alegre, Rio Grande do Sul, Brazil

Ankit Jain
Department of Materials Engineering, Indian Institute of Science, Bangalore, Karnataka, India

Sanjay K. Jain
Pharmaceutics Research Projects Laboratory, Department of Pharmaceutical Sciences, Dr. Harisingh Gour Vishwavidyalaya, Sagar, MP, India

Chandi Charan Kandar
Institute of Pharmacy, Jalpaiguri, West Bengal, India

Jiyauddin Khan
School of Pharmacy, Management and Science University, Selangor Darul Ehsan, Malaysia

Aisha Khatoon
Molecular Cell Biology Laboratory, Integral Information and Research Centre-4 (IIRC-4), Department of Biosciences, Faculty of Science, Integral University, Lucknow, Uttar Pradesh, India

Sidra Khatoon
Department of Physics, University of Lucknow, Lucknow, Uttar Pradesh, India

Doman Kim
Department of International Agricultural Technology & Institute of Green BioScience and Technology, Seoul National University, Seoul, Republic of Korea

Micaela A. Macchione
Chemistry Science Faculty, Organic Chemistry Department, National University of Cordoba; Institute for Research and Development in Process Engineering and Applied Chemistry, Science and Technology National Council; Academic Pedagogical Institute of Human Sciences, National University of Villa María, Villa María, Córdoba, Argentine

Samarendra Maji
Department of Chemistry, Faculty of Engineering and Technology, SRM Institute of Science and Technology, SRM Nagar, Kattankulathur, Tamil Nadu, India

Shirui Mao
School of Pharmacy, Shenyang Pharmaceutical University, Shenyang, People's Republic of China

Biswaranjan Mohanty
Department of Pharmaceutics, Institute of Pharmacy and Technology, Salipur, Cuttack, Odisha, India

Dinara Jaqueline Moura
Laboratory of Genetic Toxicology, Federal University of Health Sciences of Porto Alegre—UFCSPA, Porto Alegre, Rio Grande do Sul, Brazil

Amit Kumar Nayak
Department of Pharmaceutics, Seemanta Institute of Pharmaceutical Sciences, Mayurbhanj, Odisha, India

Michael Nugent
Athlone Institute of Technology, Athlone, County Westmeath, Ireland

Kunal Pal
Department of Biotechnology and Medical Engineering, National Institute of Technology Rourkela, Rourkela, Odisha, India

Pritish Kumar Panda
Pharmaceutics Research Projects Laboratory, Department of Pharmaceutical Sciences, Dr. Harisingh Gour Vishwavidyalaya, Sagar, MP, India

Kamla Pathak
Faculty of Pharmacy, Uttar Pradesh University of Medical Sciences, Etawah, Uttar Pradesh, India

Dilshad Qureshi
Department of Biotechnology and Medical Engineering, National Institute of Technology Rourkela, Rourkela, Odisha, India

Sarjana Raikwar
Pharmaceutics Research Projects Laboratory, Department of Pharmaceutical Sciences, Dr. Harisingh Gour Vishwavidyalaya, Sagar, MP, India

Fiona Concy Rodrigues
Department of Biomedical Engineering, Manipal Institute of Technology, MAHE, Manipal, India

Sumana Roy
Department of Pharmaceutics, Guru Nanak Institute of Pharmaceutical Science and Technology, Kolkata, India

Ram Kumar Sahu
Department of Pharmaceutical Science, Assam University (A Central University), Silchar, Assam, India

Shivani Saraf
Pharmaceutics Research Projects Laboratory, Department of Pharmaceutical Sciences, Dr. Harisingh Gour Vishwavidyalaya, Sagar, MP, India

Preetam Sarkar
Department of Food Process Engineering, National Institute of Technology, Rourkela, India

Pavel Sengupta
Centre for Healthcare Science and Technology, Indian Institute of Engineering Science and Technology, Shibpur, Howrah, India

Luiza Steffens
Laboratory of Genetic Toxicology, Federal University of Health Sciences of Porto Alegre—UFCSPA, Porto Alegre, Rio Grande do Sul, Brazil

Miriam C. Strumia
Chemistry Science Faculty, Organic Chemistry Department, National University of Cordoba; Institute for Research and Development in Process Engineering and Applied Chemistry, Science and Technology National Council, Córdoba, Argentine

Mohammad Tabish
Department of Pharmacology, College of Medicine, Shaqra University, Shaqra, Saudi Arabia

Goutam Thakur
Department of Biomedical Engineering, Manipal Institute of Technology, MAHE, Manipal, India

Ankita Tiwari
Pharmaceutics Research Projects Laboratory, Department of Pharmaceutical Sciences, Dr. Harisingh Gour Vishwavidyalaya, Sagar, MP, India

Amit Verma
Pharmaceutics Research Projects Laboratory, Department of Pharmaceutical Sciences, Dr. Harisingh Gour Vishwavidyalaya, Sagar, MP, India

Xin Zhang
School of Pharmacy, Shenyang Pharmaceutical University, Shenyang, People's Republic of China

CHAPTER 1

Natural polymers as useful pharmaceutical excipients

Chandi Charan Kandar[a], Md Saquib Hasnain[b], and Amit Kumar Nayak[c]
[a]Institute of Pharmacy, Jalpaiguri, West Bengal, India
[b]Department of Pharmacy, Shri Venkateshwara University, Gajraula, Uttar Pradesh, India
[c]Department of Pharmaceutics, Seemanta Institute of Pharmaceutical Sciences, Mayurbhanj, Odisha, India

1 Introduction

Polymers are macromolecules comprised of a large number of repeating units, which are chemically inter-linked [1, 2]. These long chain macromolecules are of high molecular weights. The basic difference between simple compounds having low molecular weight and polymers emerges from the spatial disposition and dynamics of long chain polymers, which exhibit unpredictable interaction among the long chains [3]. On the basis of various criteria, such as molecular weight distribution, crystallinity and average molecular weight, the properties of polymers vary from each other [1, 3]. Besides these above-mentioned parameters, processing methodologies and processing conditions (which can be monitored during syntheses of the polymers) also influence the properties and functions of the polymers. In this regard, previous knowledge is essential to ensure the reproducible effects exerted by the polymers for significant biomedical applications [4–9]. Polymers may be of synthetic, semi-synthetic or naturally-originated.

Recently, natural polymers are extensively being used in place of synthetic polymers for their extremely good biodegradability, inexpensive and sustainable production, large abundances of raw materials from the natural resources, *etc.* [10–14]. Among all varieties of polymers, natural polymers are well recognized as widely used industrial excipients and have increased the interest in the subject of exhaustive research because of their production sustainability, biodegradability and biosafety [15–18]. Natural polymers are very familiar and more useful materials due to their easy availability, biocompatibility, biodegradability and environment-friendly. In recent years, numerous natural polymers are being produced from various natural origins, *e.g.*, plants, animals, algal, microbes, *etc.* [19–21]. Natural polymers are both physico-chemically and structurally diversified materials due to presence of various functional groups in their structures, which may

Advances and Challenges in Pharmaceutical Technology
https://doi.org/10.1016/B978-0-12-820043-8.00012-8

be of effective sites for the chemical functionalizations (modifications) [6, 22–27]. These also possess many impressive biomaterial properties, like aqueous solubility, high swelling capacity, biodegradability, biocompatibility, *etc.* [10–12]. Natural polymers are capable to form various three-dimensional inter-linked molecular networking structures, which are well known as "gels." On the molecular structures of polymers, ionic strength, pH and temperature, the gel strengths of these natural polymers are reported to be dependent [7, 20, 28]. These above mentioned properties of natural polymers have led their utilization as most important category of pharmaceutical excipients to formulate different types of dosage forms [29, 30].

Recently, the investigation of these natural polymers as potential pharmaceutical excipients for the formulations of several dosage forms has been proceeding in consistence manner in terms of diversity and property [12, 29]. The use of these excipients in pharmaceutical applications including drug delivery is originating from their conventional auxiliary functions in formulations toward their important role to improve stability, control desired drug releasing, to enhance the target specificity and finally, to attain excellent bioavailability [9, 31–38]. Since past few decades, numerous natural polymers have already been well acquainted for the uses as thickeners, stabilizers, granulating agents, gel-formers, mucoadhesives, binders, disintegrants, emulsifying agents, suspending agents, film-formers, matrix-forming agents, release-retarding agents, enteric resistants, *etc.*, in a variety of pharmaceutical dosage forms [28, 32–48]. Natural polymers possess the additional advantage that they contain extracellular substances, which can be used as ligands to bind with the cell receptors and due to this characteristic feature, they are used in targeted drug delivery systems [10, 21]. The capability of some natural polymers related to their degradation by the microorganisms together with enzymatic activities can be employed as the potential advantages in some instances; for example colon targeting drug delivery systems [9, 49]. In this context, the current chapter discusses about various natural polymers and their uses as pharmaceutical applications in different pharmaceutical formulations.

2 Classifications of natural polymers

The sources of natural polymers include plants, microbes, algae, fungi and animals. On the basis of sources, natural polymers can be classified as [8, 9]:

(i) Plant originated natural polymers (for examples: pectin, cellulose, starch, gum acacia, gum tragacanth, glucomannan, locust bean gum, cashew gum, tamarind gum, sterculia gum, okra gum, guar gum, psyllium polysaccharide, *etc.*)

(ii) Microbial originated natural polymers (for examples: curdlan, gellan gum, xanthan gum, *etc.*)

(iii) Algal originated natural polymers (for examples: alginate, carrageenan, agar, *etc.*)

(iv) Fungal originated natural polymers (for examples: pullulan, scleroglucan, *etc.*)

(v) Animal originated natural polymers (for examples: gelatin, albumin, *etc.*)

In addition, on the basis of their molecular structure and properties, different natural polymers can be classified as [9, 10, 21]:

(i) Natural polysaccharides (for examples: cellulose, hemicellulose, lignin, chitin, chitosan, starch, alginate, gellan gum, pectin, gum acacia, gum tragacanth, guar gum, *etc.*)

(ii) Proteins-based natural polymers (for examples: albumin, collagen, gelatin, silk fibroin, fibrin, *etc.*)

3 Some important natural polymers and their use as pharmaceutical excipients

3.1 Cellulose

An important naturally occurring homopolysaccharide is cellulose, which is the basic structural material of plant cell-walls. It is obtained from different renewable sources like plant fibers (*i.e.*, jute, linen, wood fibers, hemp, and cotton, *etc.*) [50]. Cellulose is also produced from some bacteria. It contains β (1,4) glycosidic linkage among D-(+)-β-glucose, *i.e.*, polymer of anhydro-β-glucose [30]. The molecular structure of cellulose is shown in Fig. 1. Its linear unbranched polysaccharide chain and many parallel cellulose molecules help to form crystalline microfibrils [8]. Since these crystalline

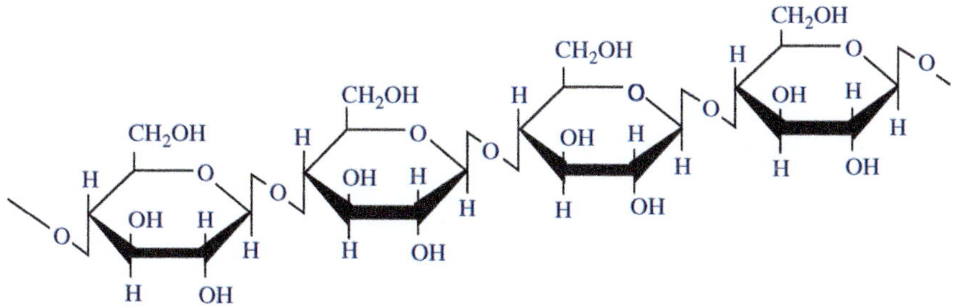

Fig. 1 Molecular structure of cellulose.

cellulose microfibrils are mechanically very strong, well resistant to the enzymatic attacks and are aligned with each other, they provide structure of the plant as well as bacterial cell walls [51]. Powdered cellulose can be prepared by mechanically disintegration from those cellulose having fibrous materials like cotton and wood. Powdered cellulose has proved its usefulness as filler in pharmaceutical tablet preparations [8]. Very good quality powdered cellulose is produced by chemically treatment with HCl, which is known as microcrystalline cellulose and it is more chosen over the powdered cellulose for the reason that it is very free-flowing cellulose material having non-fibrous natured particles. It is also used as diluent and binder in pharmaceutical tablets for both direct compression as well as granulation methods [52]. *In vivo*, cellulose is poorly biodegradable; however, it can be made hydrolysable by changing its higher order structure [30]. The ease in which it can be converted to various derivatives makes it a smart raw material for many biomedical applications. Different cellulose derivatives can be synthesized by various functionalization processes, such as esterification, carboxymethylation, graft copolymerization and cross-linking reactions [53]. Functionalizations of cellulose yield different cellulose derivatives (semi-synthetic): hydroxylpropyl methyl cellulose (HPMC) and carboxylmethyl cellulose (CMC), while the esterification of cellulose produces various useful semi-synthetic cellulose derivatives (for examples: cellulose nitrate, cellulose acetate and cellulose acetate phthalate, *etc.*) [53, 54]. The cellulose derivatives are used for the preparation of different membrane controlled drug releasing dosage forms (for examples: enteric coating on tablets and granules, semi-permeable membranes in osmotic pumps, *etc.*). Some recent researches on cellulose and its derivatives as pharmaceutical excipients are presented in Table 1.

3.2 Pectins

Pectins are non-starch, inexpensive, non-toxic, water soluble, natural complex biopolysaccharides occurred in the plant cell-walls [66]. These are industrially produced from various kinds of plant-based resources, such as citrus peels, apple pomaces, sugar beet roots, *etc.*, has already been employed as thickening agents, stabilizers, emulsifying agents and gelling agents [66]. These are principally linearly structured biopolymers of α (1,4) glycosidic linkage of D-galacturonic acid residues. In the molecular structure of pectin, the galacturonic acid main chains also possess the occasional rhamnose groups that disrupts the arrangement of chain helix and α-L-rhamnopyranose by α-(1–2) linkage (Fig. 2) [9, 67]. The carboxylic acid

Table 1 Recent researches on cellulose and its derivatives as pharmaceutical excipients.

Pharmaceutical formulations	Uses as excipients	References
Bisoprolol fumarate matrix tablets made of alginate, HPMC and microcrystalline cellulose	HPMC: Matrix-former, sustained releasing agent Microcrystalline cellulose: Disintegrating agent	Malakar et al. (2012) [55]
In situ cross-linked alginate matrix tablets for sustained salbutamol sulfate prepared using microcrystalline cellulose	Disintegrating agent	Malakar et al. (2014) [56]
Ethyl cellulose microparticles of metformin HCl	Matrix-former, sustained releasing agent, encapsulating agent	Maji et al. (2012) [57]
Gastroretentive hydrodynamically balanced system of ofloxacin	Buoyant (floating) imparting agent	Nayak et al. (2013) [58]
Floating capsules containing alginate-HPMC based beads of salbutamol sulfate	Buoyant (floating) imparting agent, matrix-former, sustained releasing agent	Malakar et al. (2014) [59]
Gliclazide-loaded alginate-methyl cellulose mucoadhesive microcapsules	Matrix-former, sustained releasing agent, mucoadhesive agent	Pal and Nayak (2011) [60]
Alginate-HPMC and alginate-sodium CMC buccal patches of atenolol	Sustained releasing agent, mucoadhesive agent	Rath Adhikari et al. (2010) [61]
Alginate/HPMC-based *in situ* gelling ophthalmic system for gatifloxacin	Gelling agent, sustained releasing agent, mucoadhesive agent	Liu et al. (2006) [62]
CMC/graphene oxide bio-nanocomposite hydrogel beads as anticancer drug release carrier	Matrix-former, sustained releasing agent	Monireh and Hassan (2017) [63]
Bacterial cellulose and gelatin-based hydrogel composites for drug delivery	Matrix-former, sustained releasing agent	Treesuppharat et al. (2017) [64]
Wood hemicellulose/chitosan-based semi-interpenetrating network hydrogels for drug release	Matrix-former, sustained releasing agent	Karaaslan et al. (2010) [65]

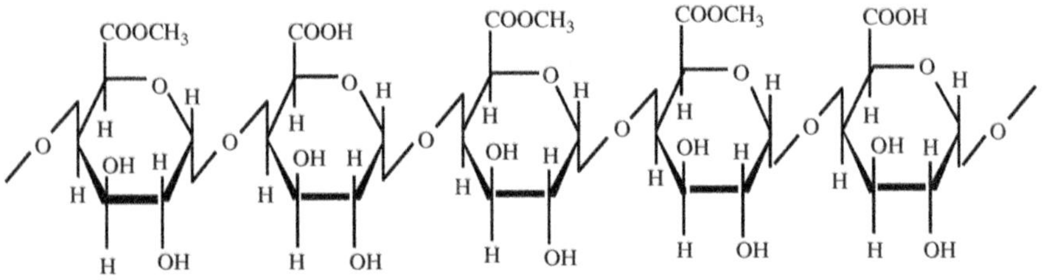

Fig. 2 Molecular structure of pectin.

groups of galacturonic acid residues of pectin-backbone are mostly esterified by methoxy group. In general, the characterization of pectin molecules is determined by the degree of esterification and also, by its degree of amidation in some instances. Both the characterization parameters (*i.e.*, degree of esterification as well as degree of amidation) are attributed by the percentages of esterified and amidated carboxyl groups present in the pectin molecules, respectively [68]. On the basis of degree of methoxylation (%), pectins are generally categorized into two different categories: (i) high methoxy pectins (50–80%) and (ii) low methoxy pectins (25–50%) and [69]. The degree of esterification exerts a major impact on various important characteristics of pectins, such as aqueous solubility, good swelling capacity and excellent gel-producing capability [67]. On the adding up of high sucrose quantity (greater than 50%) in an acidic medium, high methoxy pectins produce gels. Usually, low methoxy pectins can form gelled structures by the so-called ionotropic cross-linking interaction of low methoxy pectins with divalent metal cations (*e.g.*, Ca^{2+}, Zn^{2+}, *etc.*) and the formed ionotropically gelled pectinate matrices are already being researched and employed as potential sustained drug releasing carriers [67, 70]. The ionotropic cross-linking interaction in-between the carboxyl groups of low methoxy pectins and divalent metal cations (as ionotropic cross-linking agents) induce the formation of the so-called "Egg-Box" model-structure of ionotropic gelation. Although, to some extent, it differs from the "Egg-Box" model-structure formed for the cross-linking gelation of sodium alginate [66]. Since many years, pectins have widely been employed as useful pharmaceutical excipients in the numerous dosage formulations, such as emulsions, suspensions, tablets, multiparticulates (for examples: granules, beads, microparticles, nanoparticles, *etc.*), gels, hydrogels, films, scaffolds, *etc.* [70–72]. Pectins have already been employed as matrix-forming and/or sustained releasing excipients in the formulations of many pharmaceutical matrix tablets [73]. Recently, low methoxy pectin has been exploited for the formulations of numerous ionotropically gelled pectinate microparticles and beads for sustained releasing and floating multiple-unit systems [67, 74–78]. Due to water soluble character, pectins are unable for the protection of drug loading during the passage of pectins-based dosage forms through the stomach and small intestine [77, 78]. Recent years, various pectinate-based systems already have been designed for colonic drug delivery/targeting [79, 80]. Some recent researches on pectins as pharmaceutical excipients are represented in Table 2.

Table 2 Recent researches on pectins as pharmaceutical excipients.

Pharmaceutical formulations	Uses as excipients	References
Calcium pectinate-fenugreek seed mucilage mucoadhesive beads of metformin HCl	Matrix-former, sustained releasing agent, mucoadhesive agent, encapsulating agent	Nayak et al. (2013) [81]
Calcium pectinate-ispaghula mucilage mucoadhesive beads of metformin HCl	Matrix-former, sustained releasing agent, mucoadhesive agent, encapsulating agent	Nayak et al. (2014) [75]
Jackfruit seed starch-pectinate mucoadhesive beads of metformin HCl	Matrix-former, sustained releasing agent, mucoadhesive agent, encapsulating agent	Nayak et al. (2014) [76]
Aceclofenac-loaded pectinate-poly (vinyl pyrrolidone) beads	Matrix-former, sustained releasing agent, encapsulating agent	Nayak et al. (2013) [78]
Mucoadhesive-floating zinc-pectinate-sterculia gum interpenetrating polymer network beads of ziprasidone HCl	Matrix-former, sustained releasing agent, buoyant (floating) imparting agent, encapsulating agent	Bera et al. (2015) [74]
Gellan gum/pectin beads for the targeted colonic delivery of resveratrol	Matrix-former, sustained releasing agent, colon targeting agent, encapsulating agent	Prezotti et al. (2018) [80]
Pectin beads loaded with chitosan-iron microspheres for colonic specific delivery of ciprofloxacin	Matrix-former, sustained releasing agent, colon targeting agent, encapsulating agent	Reynaud et al. (2015) [82]

Continued

Table 2 Recent researches on pectins as pharmaceutical excipients—cont'd

Pharmaceutical formulations	Uses as excipients	References
Pectin-metronidazole prodrug bearing microspheres for colon targeting	Matrix-former, sustained releasing agent, colon targeting agent, encapsulating agent	Vaidya et al. (2015) [83]
Pectin coated vitamin C liposomes	Coating agent	Zhou et al. (2018) [84]
Pectin-coated nanoliposomal delivery of phloridzin	Coating agent	Haghighi et al. (2018) [85]
Phosphate crosslinked pectin based dual responsive hydrogel networks and nanocomposites for drug delivery	Matrix-former, sustained releasing agent	Eswaramma et al. (2017) [86]

3.3 Gum Arabica

Gum Arabica is obtained from *Acacia nilotica* plant of family Leguminosae and it belongs to polysaccharide category of biopolymers [86]. Currently, gum Arabica is mostly produced from 2 commonly known acacia plant species: *Acacia seyal* and *Acacia Senegal* [87]. It is a biocompatible as well as biodegradable gummy polysaccharide [88]. Being a pharmaceutical biopolymer, gum Arabica exhibits different kind of functional physical characteristics, such as aqueous solubility, high swelling ability, good stability in different pHs as well as excellent gel-producing capability [86, 89]. Gum Arabica is a little acidic heteropolysaccharide. It exists as the combination of various metal salts (*e.g.*, potassium, calcium, and magnesium) of polysaccharidic acids, where a main chain of $(1 \rightarrow 3)$-β-D-galactopyranosyl units and the side chains of D-galactopyranosyl, L-rhamnopyranosyl, L-arabinofuranosyl and D-glucopyranosyl uronic acid units are present (Fig. 3) [89, 90]. Since many decades, gum Arabica has been exploited as stabilizer, thickening agent, emulsifier, suspending agent, *etc.*, in many food products [87, 91]. In various pharmaceutical as well as cosmeceutical formulations, gum Arabica has widely been used as a potential excipient material [87, 88]. Gum Arabica has already been exploited as binding agents and disintegrating agents in the formulations tablets for a variety of drugs [92, 93]. Some researches uses on gum Arabica as pharmaceutical excipients are presented in Table 3.

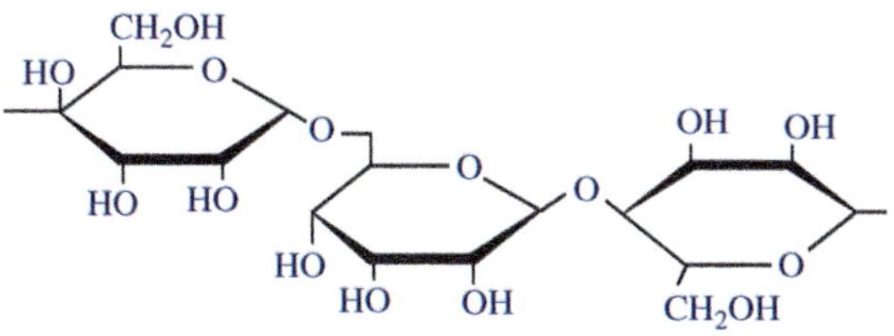

Fig. 3 Molecular structure of gum Arabica.

Table 3 Recent researches on gum Arabica as pharmaceutical excipients.

Pharmaceutical formulations	Uses as excipients	References
Film coated immediate release oral tablet formulation	Film coating former	Alkarib et al. (2016) [94]
Calcium alginate-gum Arabic beads of glibenclamide	Matrix-former, sustained releasing agent	Nayak et al. (2012) [90]
Insulin-loaded nanoparticles using chitosan and gum Arabic	Encapsulating material and release retardant	Avadi et al. (2010) [88]
Monolithic osmotic tablet system made of gum Arabic	Osmotic agent, suspending agent, and expanding agent	Lu et al. (2003) [95]
Self-gelling primaquine-gum Arabic conjugate	Injectable gelling agent	Nishi and Jayakrishnan (2007) [96]
Isoniazid-encapsulated chitosan-gum Arabic particles for sustained release formulation	Encapsulating material and release retardant	Kadare et al. (2014) [97]
Nanoemulsion-based vitamin E delivery systems using gum Arabic	Emulsifying agent	Ozturk et al. (2015) [98]
Potassium diclofenac-loaded pH-responsive hydrogel based on gum arabic conjugates	Matrix-former, sustained releasing agent	Reis et al. (2016) [99]
Bioinspired alginate-gum arabic hydrogel for controlled drug release in chronic wound healing	Matrix-former, sustained releasing agent	Li et al. (2017) [100]

3.4 Gum tragacanth

Gum tragacanth is another one natural polysaccharide candidate derived from the plants of various species of Iranian, Turkish, Indian, Afghanistan and Russian leguminous family of the *Astragalus* genus, mainly from *A. gummifer*, *A. tragacantha* and *A. microcephalus* Willd [101]. Astragalus plants

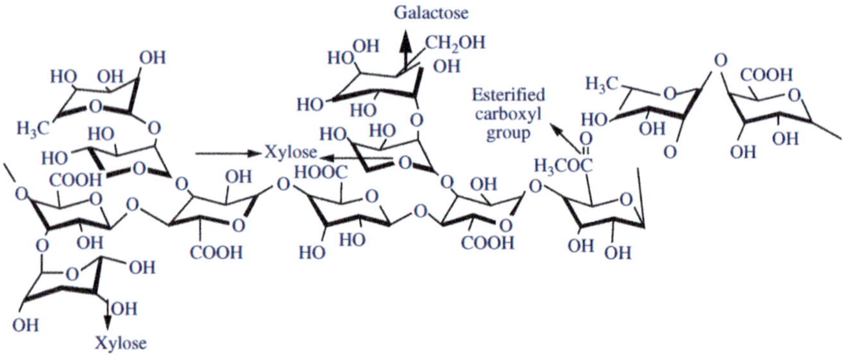

Fig. 4 Molecular structure of gum tragacanth.

exudate sap to safeguard the wounds of the plant to tissue injury as a defense mechanism and the dried product of sap is called exudate gum. The injured plant part is covered by the exudates, which becomes hard into flakes at one time on exposure to air and sunlight [101, 102]. The highly complex polysaccharidic molecular structure of gum tragacanth comprises acidic natured heterogeneous and branched proteoglycan having high molecular weight of 840 kDa, approximately and the shape is elongated of 4500 × 19 Å (Fig. 4) [102]. Gum tragacanth exhibits two separate fractions: one is 60–70% compound called as bassorin and another is 20–30% known as tragacanthin. Tragacanthin is water-soluble and further fractioned into ethanolic insoluble part, tragacanthic acid and ethanolic soluble part, arabinogalactan [103]. Both bassorin and tragacanthic acid are water insoluble in nature, but they can form viscous gel. Both fractions of tragacanth comprise little quantities of methoxyl groups and protein natured substance [101–103]. The protein natured substance present in the aqueous soluble fractions in high amounts [102]. Since several past decades, gum tragacanth has extensively been employed as suspending agent, stabilizer, thickener, emulsifying agent, *etc.*, in many foods and formulations of pharmaceuticals [52, 102]. It also has been used as widely accepted pharmaceutical excipients in numerous pharmaceutical tablets as matrix-former and sustained drug releasing agents [104]. It has experienced its wide applications in the development of several drug releasing hydrogels with its chemical as well as physical modifications [105–107]. Some recent researches on gum tragacanth as pharmaceutical excipients are presented in Table 4.

Table 4 Recent researches on gum tragacanth as pharmaceutical excipients.

Pharmaceutical formulations	Uses as excipients	References
Sustained release matrix tablets of diclofenac sodium prepared with gum tragacanth	Matrix-former, sustained releasing agent	Shahiwala and Zarar (2018) [108]
Gum tragacanth-based nanogel as a superparamagnetic molecularly imprinted polymer for quercetin release	Sustained releasing agent, gelling agent	Hemmati et al. (2016) [105]
Thermo/pH sensitive drug delivery systems based on tragacanth gum	Matrix-former, sustained releasing agent, gelling agent	Hemmati and Ghaemy (2016) [109]
Gum tragacanth-based pH responsive hydrogels for drug delivery	Matrix-former, sustained releasing agent, gelling agent	Singh and Sharma (2014) [106]
Crosslinked poly(vinyl pyrrolidone)-acrylic acid-gum tragacanth hydrogels for drug delivery	Matrix-former, sustained releasing agent, gelling agent	Singh and Sharma (2017) [107]
Oral gum tragacanth-acrylic acid amphotericin B carrier for systemic candidiasis	Matrix-former, sustained releasing agent, gelling agent	Mohamed et al. (2018) [110]
Lecithin-gum tragacanth mucoadhesive hybrid nano-carrier system of amphotericin B	Matrix-former, sustained releasing agent, gelling agent, mucoadhesive agent	Jabri et al. (2018)[111]
PLGA-gum tragacanth electrospun nanofibers containing tetracycline HCl for periodontal regeneration	Matrix-former, sustained releasing agent, gelling agent, mucoadhesive agent	Ranjbar-Mohammadi et al. (2016) [112]
Nanohydrogels made of gum tragacanth for drug delivery	Matrix-former, sustained releasing agent, gelling agent	Sadat Hosseini et al. (2016) [113]
Curcumin-loaded gum tragacanth-poly(ε-caprolactone) electrospun nanofibers	Matrix-former, sustained releasing agent, gelling agent, mucoadhesive agent	Ranjbar-Mohammadi et al. (2016) [114]

3.5 Guar gum

Guar gum is an example of plant derived polymers of non-ionic polysaccharide group, which is extracted from the seeds of the plant—*Cyamopsis tetragonoloba* (family: Leguminosae) [115]. The important characteristic and favorable properties of guar gum are physicochemical stability, aqueous solubility, high swelling ability, biodegradability and biocompatibility [44, 115–117]. The molecular structure of guar gum comprises linear polysaccharidic chains of (1→4)-β-D-mannopyranosyl residues and branching occurs with α-D-galactopyranosyl residues interconnected via the (1→6) linkages (Fig. 5) [44, 115, 116]. Guar gum has already been utilized as thickener, stabilizer, emulsifier, gel-producing agent, mucoadhesive agent, *etc.*, in many pharmaceutical dosage forms [115]. In addition, it has also been employed for the preparation of different controlled releasing dosage forms for oral administrations (such as tablets, beads, microparticles, pellets, *etc.*) due to its excellent gelling network structure [118]. Recent years, physically as well as chemically functionalized derivatives of guar gum have been synthesized to control the extensive swelling of native guar gum in different pH milieu and these guar gum derivatives have already been used to formulate many drug delivery systems of improved properties [115, 116]. The guar gum is reported for the enzymatic degradation by the enzymes produced by colonic microorganisms at the colonic fluids and owing to such characteristic property, guar gum has been widely employed for the development of many orally administrable colon-targeted drug delivery dosage forms [115, 116, 118, 119]. Some recent researches on gum tragacanth as pharmaceutical excipients are presented in Table 5.

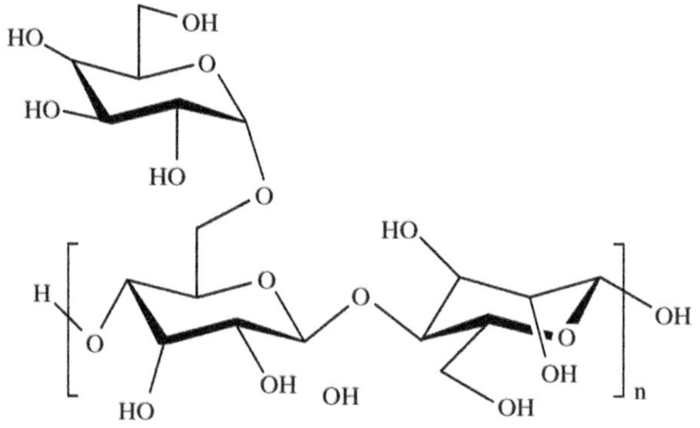

Fig. 5 Molecular structure of guar gum.

Table 5 Recent researches on guar gum as pharmaceutical excipients.

Pharmaceutical formulations	Uses as excipients	References
Carboxymethyl guar gum nanoparticles for drug delivery	Matrix-former, sustained releasing agent	Dodi et al. (2016) [120]
Semi-interpenetrating hydrogels from carboxymethyl guar gum and gelatin for ciprofloxacin release	Matrix-former, sustained releasing agent	Ghosh et al. (2018) [121]
pH-responsive guar gum hydrogels for intestinal delivery of dexamethasone	Matrix-former, sustained releasing agent, gelling agent	Das and Subuddhi (2015) [122]
PLGA nanoparticles contained mucoadhesive guar-gum buccal films for delivery of antihypertensive peptide	Film-former, matrix-former, sustained releasing agent, gelling agent, mucoadhesive agent	Castro et al. (2018) [123]
Ketorolac tromethamine colon targeted guar gum compression coated tablets	Matrix-former, sustained releasing agent, colon targeting agent	Vemula and Katkum (2015) [119]
Guar gum succinate-sodium alginate pH-sensitive beads for colon-specific drug delivery	Matrix-former, sustained releasing agent, colon targeting agent	Seeli et al. (2016) [116]
Guar gum oleate-graft-poly (methacrylic acid) hydrogel as a colon-specific controlled drug delivery	Matrix-former, sustained releasing agent, colon targeting agent	Sathya Seeli et al. (2017) [124]
Guar gum microspheres of 5-aminosalicylic acid for colon targeting	Matrix-former, sustained releasing agent, colon targeting agent	Koushik et al. (2016) [125]
Mesoporous silica nanoparticle based enzyme responsive system for colon specific drug delivery through guar gum capping	Matrix-former, sustained releasing agent, colon targeting agent	Kumar et al. (2017) [117]
Acrylic acid grafted guar gum-nanosilica membranes for transdermal diclofenac delivery	Film-former, matrix-former, sustained releasing agent	Giri et al. (2013) [126]

3.6 Locust bean gum

Locust bean gum is also known as carob bean gum [127]. It is a plant originated seed gum derived from the carob tree (*Ceratonia siliqua*) seeds [127, 128]. The extracted locust bean gum is available in powder form and it is of whitish to yellowish-white in color. It is one of the branched biopolysaccharides, which is non-ionic in nature [127]. The molecular structure of locust bean gum comprises a galactomannan type heteropolysaccharide containing galactose as well as mannose residues interconnected via the glycosidic linkage in a ratio of 1:4 [127, 128]. It consists of (1, 4)-linked β-D-mannopyranose skeleton connected to α-D-galactose and the branching occurs at 6-position of mannopyranose (Fig. 6) [128]. Locust bean gum is less water soluble and for this, it is required to apply mild heat to prepare aqueous solutions [129]. Even at the lower concentrations of locust bean gum, it has the capability to prepare more viscous natured aqueous solutions, which remain approximately unchanged by the further incorporations of salts, alteration of temperature and pH [130]. Locust bean gum is generally familiar as biocompatible, biodegradable non-teratogenic and non-mutagenic biopolymer [129]. It has been utilized as potential pharmaceutical excipients in many pharmaceutical dosage systems [127–129]. Locust bean gum is generally employed in the formulations of various drugs releasing tablets for oral administrations [131, 132]. As locust bean gum has already been reported for its improved biomucoadhesive potential, it has recently been employed as biomucoadhesive polymeric excipient in various mucoadhesive drug delivery systems [133]. Some recent researches on locust bean gum as pharmaceutical excipients are presented in Table 6.

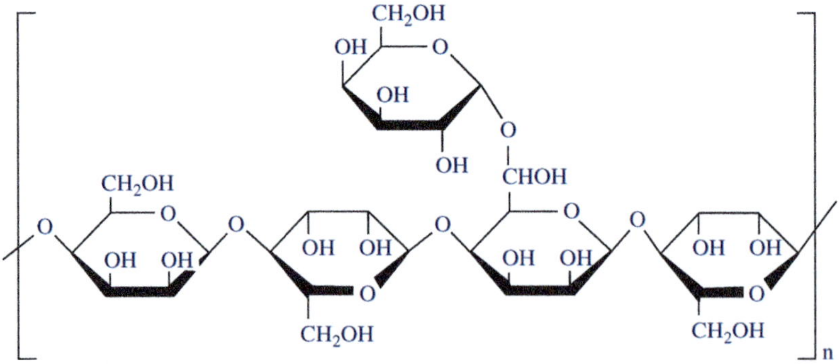

Fig. 6 Molecular structure of locust bean gum.

Table 6 Recent researches on locust bean gum as pharmaceutical excipients.

Pharmaceutical formulations	Uses as excipients	References
Nimesulide tablets	Superdisintegrant	Malik et al. (2011) [131]
Matrix tablets made of interpolymeric network of locust bean gum-poly (vinyl alcohol)	Matrix former, release retardant	Kaity et al. (2013) [132]
Alginate-locust bean gum mucoadhesive beads of aceclofenac	Matrix former, release retardant	Prajapati et al. (2014) [133]
Alginate-locust bean gum beads of gliclazide	Matrix former, release retardant	Prajapati et al. (2013) [134]
Alginate-locust bean gum interpolymeric network microspheres of aceclofenac	Matrix former, release retardant	Jana et al. (2015) [135]
Locust bean gum and xanthan gum based microparticles of celecoxib	Matrix-former, sustained releasing agent	Sharma et al. (2016) [136]
Chitosan—locust bean gum interpenetrating polymeric network nanocomposites for delivery of aceclofenac	Matrix-former, sustained releasing agent	Jana et al. (2017) [137]
Alginate beads of captopril using locust bean gum	Matrix-former, sustained releasing agent	Pawar et al. (2015) [138]
Modified locust bean-alginate interpenetrating beads for oral glipizide delivery	Matrix-former, sustained releasing agent	Dey et al. (2015) [139]
Locust bean gum-alginate interpenetrating polymeric network of capecitabine	Matrix-former, sustained releasing agent	Upadhyay et al. (2018) [140]

3.7 Alginates

Alginates are the marine-derived algal biopolymeric group composed of alginic acid salts, which are the naturally obtained polysaccharides commercially extracted from different types of brown marine algae, like, *Laminaria hyperborea*, *Ascophyllum nodosum*, *Macrocystis pyrifera*, *etc.* [20, 141]. Chemically, alginate molecules consists of the linear blocks of (1→4)-linkage between monomers of β-D-mannuronic acid (M unit) and α-L-guluronic acid (G unit) (Fig. 7). These linear blocks of alginate molecular structure are characterized as anionic copolymer comprised of α-L-guluronate (G unit) as well as β-D-mannuronate (M unit) arranged in asymmetrical arrangement of various proportions of G-G, M-G and M-M units, having

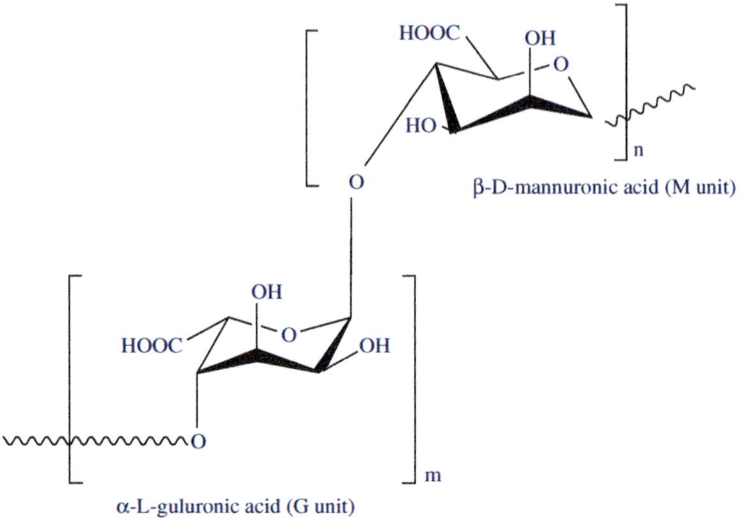

Fig. 7 Molecular structure of alginate.

1, 4-glycosidic linkages among these [142]. In nature, they are available as a mixture of salts of different metal cations occurred in the seawater, *e.g.*, as Na^{+}, Mg^{2+}, Ba^{2+}, Sr^{2+}, *etc.* [20]. Sodium alginate, the monovalent sodium salt of alginic acid is used in the formulation of various pharmaceutical products as raw materials. Characteristically, sodium alginate undergo ionotropic cross-linking interaction based gelation by the influence of some di-/tri-valent metal cations, such as Ca^{2+}, Ba^{2+}, Cu^{2+}, Pb^{2+}, Zn^{2+}, Cd^{2+}, Al^{3+}, Fe^{3+}, *etc.* [20, 141]. The ionotropic cross-linking interaction in-between monovalent sodium ions in the alginate structure with the cross-linking di-/tri-valent metal cations causes ionotropic gelation of sodium alginate, which is attributed by the fact of the stacking of guluronic groups of alginate structure to create typical "egg-box" structure. The junction zones are formed at the ionotropic cross-linking sites by the di-/tri-valent metal cations causing inter-polysaccharide binding [143, 144]. The two polymer chains come together because the di-/tri-valent metal cations compete with the mono-valent sodium ions of sodium alginate molecules. Ionotropic cross-linking di-/tri-valent metal ions are accommodated in the interstices of two polyuronate chains comprising a close ion-pair interactions with carboxylate ions of sodium alginate molecules and adequate co-ordination by other electro-negative oxygen atoms [20]. Because of their most favorable physico-chemical characteristics like solubility, viscosity, cross-linking, sol-gel transformation, *etc.*, and developed biological properties like immunogenicity, biocompatibility, bioadhesion *etc.*, alginates are

Table 7 Recent researches on alginates as pharmaceutical excipients.

Pharmaceutical formulations	Uses as excipients	References
In situ cross-linked alginate matrix tablets for sustained salbutamol sulfate release	Matrix-former, sustained releasing agent, gelling agent	Malakar et al. (2014) [56]
Zinc alginate-carboxymethyl cashew gum microbeads for isoxsuprine HCl release	Matrix-former, sustained releasing agent, encapsulating agent	Das et al. (2014) [145]
Linum usitatissimum polysaccharide-alginate mucoadhesive beads of diclofenac sodium	Matrix-former, sustained releasing agent, mucoadhesive agent, encapsulating agent	Hasnain et al. (2018) [48]
Floating capsules containing alginate-based beads of salbutamol sulfate	Matrix-former, sustained releasing agent	Malakar et al. (2014) [59]
Modified starch (cationized)-alginate beads containing aceclofenac	Matrix-former, sustained releasing agent, encapsulating agent	Malakar et al. (2014) [146]
Metformin HCl-loaded tamarind seed polysaccharide-alginate beads	Matrix-former, sustained releasing agent, mucoadhesive agent, encapsulating agent	Nayak et al. (2016) [148]
Alginate-nanohydroxyapatite composite matrices of diclofenac sodium	Matrix-former, sustained releasing agent, encapsulating agent	Hasnain et al. (2016) [149]
Okra gum-alginate mucoadhesive beads of glibenclamide	Matrix-former, sustained releasing agent, mucoadhesive agent, encapsulating agent	Sinha et al. (2015) [143]
Alginate hydrogel core-shell systems for combination delivery of ranitidine HCl and aceclofenac	Matrix-former, sustained releasing agent, encapsulating agent	Jana et al. (2015) [150]
Testosterone-loaded alginate nanocapsules	Sustained releasing agent, encapsulating agent	Jana et al. (2015) [151]

commonly used as very useful water soluble excipients in many pharmaceutical dosage forms, such as emulsions, gels, capsules, tablets, buccal patches, particulates (like beads, microparticles, nanoparticles, *etc.*) [45, 46]. Recently, various tailor-made alginate materials are being synthesized and exploited to formulate various improved and smart drug delivery systems [145–147]. Some recent researches on alginates as pharmaceutical excipients are presented in Table 7.

3.8 Chitosan

Chitosan is one of marine derived biopolymer, which is derived from chitin by deacetylation reaction [152]. From nature crustacean shells or the cell wall of fungi, chitin is obtained. Chitin is also another example of marine derived biopolymer and used as raw material in the chitosan production [26, 153]. Chemically, chitosan molecules comprise an aminopolysaccharide structure of α-1, 4-linked 2-amino-2-deoxy-α-D-glucose (N-acetyl glucosamine) [24] (Fig. 8). Being a cationic polymer in the neutral or alkaline pHs, chitosan molecules contain free amino groups and is reported as an aqueous

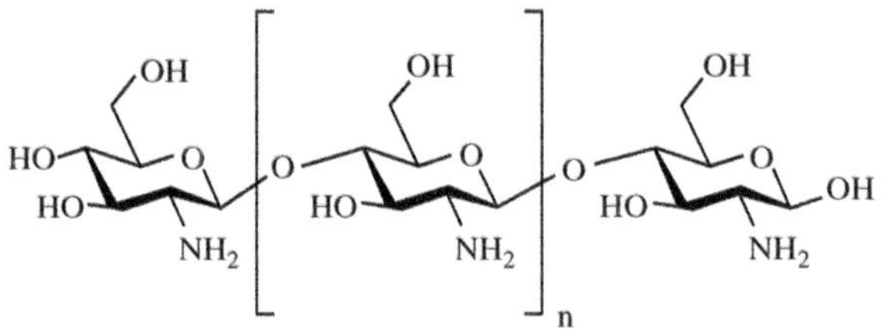

Fig. 8 Molecular structure of chitosan.

Table 8 Recent researches on chitosan as pharmaceutical excipients.

Pharmaceutical formulations	Uses as excipients	References
Chitosan-hydroxypropyl methylcellulose matrices as carriers for hydrodynamically balanced capsules of moxifloxacin HCl	Matrix-former, sustained releasing agent	Verma et al. (2017) [154]
Carbopol gel containing chitosan-egg albumin nanoparticles for transdermal aceclofenac delivery	Sustained releasing agent, encapsulating agent	Jana et al. (2014) [26]
Chitosan-egg albumin nanoparticles for oral drug delivery	Sustained releasing agent, encapsulating agent	Jana et al. (2013) [27]
Photoresponsive chitosan conjugated prodrug nanocarrier of 5-fluorouracil	Matrix-former, sustained releasing agent, photoresponsive agent	Horo et al. (2019) [155]
Tyrosol and ferulic acid encapsulated chitosan-gelatin films	Film-former, matrix-former, sustained releasing agent	Benbettaïeb et al. (2016) [156]

Table 8 Recent researches on chitosan as pharmaceutical excipients—cont'd

Pharmaceutical formulations	Uses as excipients	References
Aceclofenac-loaded chitosan-tamarind seed polysaccharide interpenetrating polymeric network microparticles	Matrix-former, sustained releasing agent, encapsulating agent	Jana et al. (2013) [153]
Magnetic stimuli-responsive chitosan-based drug delivery biocomposite for multiple triggered release	Matrix-former, sustained releasing agent, stimuli responsive agent	Harris et al. (2017) [157]
pH-responsive chitosan/polymer core-shell composite nanoparticles for drug delivery	Matrix-former, sustained releasing agent, pH responsive agent, encapsulating agent	Huang et al. (2017) [158]
On-chip made paclitaxel-loaded chitosan nanoparticles for cancer therapeutics	Matrix-former, sustained releasing agent, encapsulating agent	Majedi et al. (2014) [159]
Polysorbate 80 coated crosslinked chitosan nanoparticles of ropinirole HCl for brain targeting	Matrix-former, sustained releasing agent, gelling agent, encapsulating agent	Ray et al. (2018) [152]

insoluble polymer [154]. In the lower pH (acidic milieu), amino groups of chitosan molecules experience protonation and thus, make it soluble in the aqueous medium [152]. The aqueous solubility of chitosan is mainly due to the presence of free amino group as well as N-acetyl groups [26, 27]. Chitosan, being the cationic natured biopolysaccharide containing many free amino groups, is capable of cross-linking [24, 153]. During past few decades, chitosan is being used as drug carrier material in a variety of pharmaceutical dosage forms for the delivery of numerous drugs due to its biocompatibility and biodegradability nature [26, 27]. But, the extensive applications of chitosan as a potential biopolymeric excipient in the formulations of various drug delivery carriers is restricted by reason of its fast degradation or dissolution in the lower pH of the stomach (acidic environment), which restricts the controlled drug releasing from various chitosan made oral drug delivery systems [16, 24]. To minimize this demerit of chitosan as excipient, various functionalized (chemical modified) chitosan materials are being synthesized and used to formulate many controlled drug releasing carrier systems [16]. Some recent researches on chitosan as pharmaceutical excipients are presented in Table 8.

3.9 Gellan gum

Gellan gum is a microbial biopolysaccharide, which is obtained as a fermentation product by a microorganism—Pseudomonas *elodea* [39, 70]. Chemically, two types of gellan gum are available: (i) one is native form of gellan gum, which contains higher number of acyl groups, and (ii) another form contains lower number acyl groups, also known as deacetylated gellan gum [160]. Both the forms of gellan gum comprise an analogous linear tetrasaccharide structural feature, which is composed of (1 → 4)-L-rhamnose-α (1 → 3)-D-glucose-β-(1 → 4)-D-glucoronic acid-β-(1 → 4)-D-glucose as the repeating sugar units of in a molar ratio of 2:1:1 [39, 161–164]. The native gellan gum structure possesses 2 acyl forms: high acyl gellan gum and low acyl gellan gum (Fig. 9) [160]. This dissimilarity in substitutions between 2 form gellan gum (*i.e.*, native and deacetylated forms) presents a potential disparity in the gelling characteristics of gellan gum [162]. Gellan gum is being used as biopolymeric pharmaceutical excipients in different drug delivery dosage forms, such as tablets, gels, hydrogels, beads, microparticles, nanoparticles, films, *etc.* [39]. Recent years, in many reported researches, ionotropically gelled gellan gum beads have been developed by using deacetylated gellan gum via the ionotropic cross-linking gelation process in the presence of di-/tri-valent metal cations [70, 161–164]. Some recent researches on gellan gum as pharmaceutical excipients are presented in Table 9.

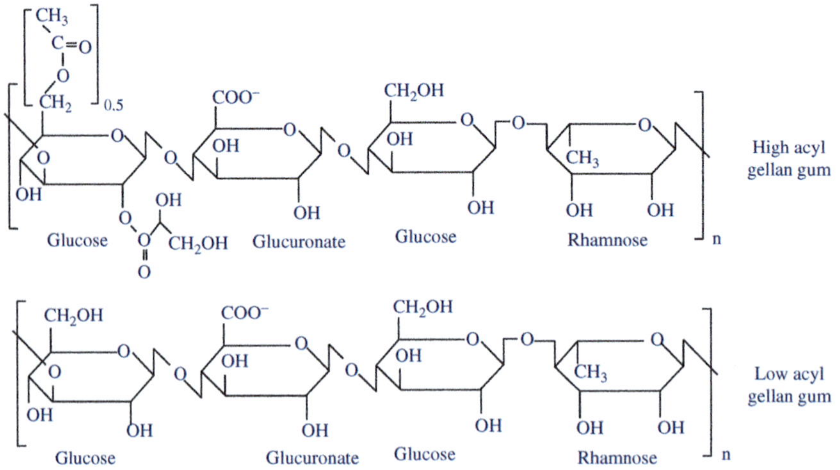

Fig. 9 Molecular structure of gellan gum.

Table 9 Recent researches on gellan gum as pharmaceutical excipients.

Pharmaceutical formulations	Uses as excipients	References
Aceclofenac-loaded unsaturated esterified alginate/gellan gum microspheres	Matrix-former, sustained releasing agent, encapsulating agent	Jana et al. (2013) [160]
Artocarpus heterophyllus L. seed starch-gellan gum mucoadhesive beads of metformin HCl	Matrix-former, sustained releasing agent, encapsulating agent	Nayak et al. (2014) [161]
Tamarind seed polysaccharide-gellan gummucoadhesive beads of metformin HCl	Matrix-former, sustained releasing agent, encapsulating agent	Nayak et al. (2014) [162]
Ispaghula mucilage-gellan gum mucoadhesive beads of metformin HCl	Matrix-former, sustained releasing agent, encapsulating agent	Nayak et al. (2014) [163]
Trigonella foenum-graecum L. seed mucilage-gellan gum mucoadhesive beads of metformin HCl	Matrix-former, sustained releasing agent, encapsulating agent	Nayak and Pal (2014) [164]
Gellan gum-based mucoadhesive microspheres of almotriptan for nasal administration	Sustained releasing agent, encapsulating agent, mucoadhesive agent	Abbas and Marihal (2014) [165]
Gellan gum fluid gels for topical administration of diclofenac	Gelling agent, sustained releasing agent, mucoadhesive agent	Mahdi et al. (2016) [166]
Chitosan-ibuprofen-gellan ternary nanogel for controlled transdermal delivery of ibuprofen	Gelling agent, sustained releasing agent	Abioye et al. (2015) [167]
Gellan nanohydrogels of piroxicam for cutaneous administration	Gelling agent, sustained releasing agent	Musazzi et al. (2018) [168]
Injectable gellan gum-based nanoparticles-loaded system of vancomycin	Matrix-former, sustained releasing agent, gelling agent	Posadowska et al. (2018) [169]

3.10 Tamarind gum

Tamarind gum is obtained from the seed endosperm of the plant: *Tamarindus indica* L. (*imli*, in Hindi; "Indian date") belonging to the Leguminosae family [31, 170]. It is occurred as a cell-wall storage unit in tamarind seed endosperm and is isolated from the tamarind seed kernel powder using various established methods [170–172]. The tamarind seed kernel powder was prepared from the mature and harvested tamarind seeds via a series of steps, such as selection of raw seeds, removal of seed coat, seed kernel separation, milling, grinding and then, sieving [31]. Tamarind gum is a water soluble biopolysaccharide and regarded as a galactoxyloglucan. The molecular structure of tamarind gum comprises of (1→4)-β-D-glucan backbone substituted with the side chains of α-D-xylopyranose and β-D-galactopyranosyl (1→2)-α-D-xylopyranose linked (1→6) to glucose residues, where glucose, xylose and galactose units are in a ratio of 2.8:2.25:1.0 (Fig. 10) [6, 31, 42]. It is aqueous soluble, hydrophilic, stability in low pH (acidic environment), biodegradable, biocompatible and nonirritant in nature [32]. Tamarind gum is employed as thickener, stabilizer, emulsifier, suspending agent, tablet binder, gel-producing agent, film-producing agent, mucoadhesive, matrix-former and release retardant agent in various dosage forms [6, 31, 42, 148, 170–173]. It is also employed as matrix-former in the formulation of sustained drug releasing matrix tablets for different kinds of drugs [31, 42, 174]. Additionally, because of the excellent hydrophilic as well as biomucoadhesive properties, tamarind gum is being used in the formulations of various types of biomucoadhesive systems for drug delivery [67, 162]. Recently, several modifications of tamarind gum have been attempted and evaluated for their applications as pharmaceutical excipients in many kinds of dosage forms [6, 153]. Some recent researches on tamarind gum as pharmaceutical excipients are presented in Table 10.

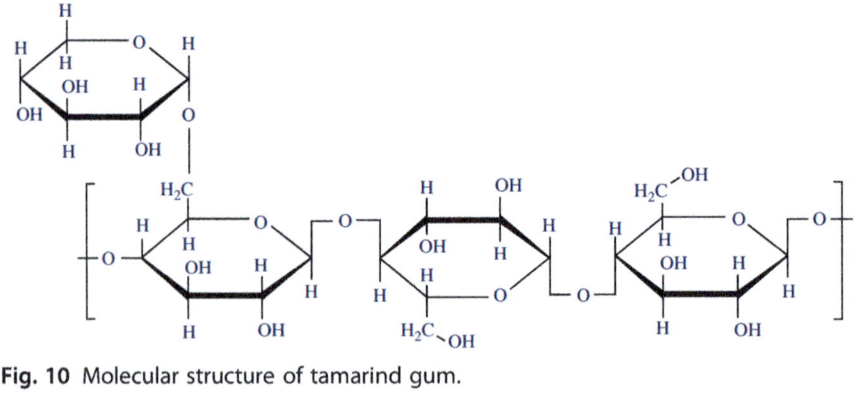

Fig. 10 Molecular structure of tamarind gum.

Table 10 Recent researches on tamarind gum as pharmaceutical excipients.

Pharmaceutical formulations	Uses as excipients	References
Graft-modified tamarind gum-based matrix tablets of aspirin	Matrix-former, sustained releasing agent	Ghosh and Pal (2013) [174]
Tamarind glucan-based mucoadhesive buccal films of rizatriptan benzoate	Film-former, matrix-former, sustained releasing agent, mucoadhesive agent	Avachat et al. (2013) [175]
Calcium pectinate-tamarind polysaccharide mucoadhesive beads of metformin HCl	Matrix-former, sustained releasing agent, mucoadhesive agent, encapsulating agent	Nayak et al. (2014) [67]
Aceclofenac-loaded chitosan-tamarind polysaccharide interpenetrating polymeric network microparticles	Matrix-former, sustained releasing agent, encapsulating agent	Jana et al. (2013) [153]
Tamarind polysaccharide-gellan mucoadhesive beads of metformin HCl	Matrix-former, sustained releasing agent, mucoadhesive agent, encapsulating agent	Nayak et al. (2014) [162]
Alginate gel-coated oil-entrapped alginate-tamarind gum-magnesium stearate buoyant beads of risperidone	Sustained releasing agent, encapsulating agent, mucoadhesive agent	Bera et al. (2015) [176]
Metformin HCl-loaded tamarind polysaccharide-alginate beads	Matrix-former, sustained releasing agent, mucoadhesive agent, encapsulating agent	Nayak et al. (2016) [148]
Tropicamide-loaded tamarind seed xyloglucan nanoaggregates for ophthalmic delivery	Gelling agent, sustained releasing agent, encapsulating agent	Dilbaghi et al. (2013) [177]
Chronotherapeutic drug delivery from colon targeted directly compressed propranolol HCl matrix tablets made of tamarind gum, chitosan and okra gum controlled release	Matrix-former, sustained releasing agent, colon targeting agent	Newton et al. (2015) [178]
Metronidazole loaded carboxymethyl tamarind kernel polysaccharide-polyvinyl alcohol cryogels	Matrix-former, sustained releasing agent, gelling agent	Meenakshi and Ahuja (2015) [179]

3.11 Xanthan gum

Xanthan gum, a natural biopolymer, is derived by a fermentation product of extracellular polysaccharide in presence of sugars like sucrose, glucose with the help of a bacteria namely *Xanthomonas campestris* [180]. The xanthan gum molecular structure is composed of pentasaccharidic subunits and also contains of D-glucosyl, D-mannosyl, and D-glucuronyl acid residues maintaining the ratio of 2:2:1 with different proportions of O-acetyl and pyruvyl residues (Fig. 11) [181]. Xanthan gum molecules are linear (*i.e.*, unbranched) and also exhibit a conformational transition for the thermal induction [180, 181]. The water solubility characteristic of xanthan gum is excellent [181]. It exhibits good biocompatibility and excellent biodegradability [180]. It is not affected by the pH alterations, temperature changes and wider range of ionic strength. Xanthan gum is being used in many applications in the food as well as pharmaceutical industries [182]. As potential pharmaceutical excipient, xanthan gum has been used in the formulations of many drug delivery systems, such as oral, buccal, topical, *etc.*, drug delivery systems [104, 182–184]. Some recent researches on xanthan gum as pharmaceutical excipients are presented in Table 11.

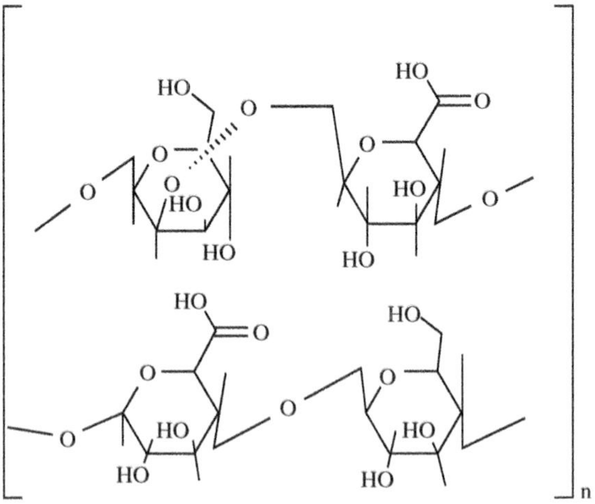

Fig. 11 Molecular structure of xanthan gum.

Table 11 Recent researches on xanthan gum as pharmaceutical excipients.

Pharmaceutical formulations	Uses as excipients	References
Modified xanthan gum-chitosan based hydrogel for ciprofloxacin delivery	Matrix-former, sustained releasing agent	Hanna and Saad (2019) [185]
Co-polymer micelles and xanthan-O-carboxymethyl hydrogel particles	Matrix-former, sustained releasing agent	Maiti and Mukherjee (2014) [186]
Diltiazem HCl from cross-linked interpenetrating polymer network beads of pectin and modified xanthan gum	Matrix-former, sustained releasing agent	Giri et al. (2013) [183]
Liranaftate-loaded xanthan gum based hydrogel for topical delivery	Matrix-former, sustained releasing agent, encapsulating agent	Mishra et al. (2018) [187]
Hydrazone-linkage-based self-healing and injectable xanthan-poly(ethylene glycol) hydrogels	Matrix-former, sustained releasing agent, gelling agent	Sharma et al. (2018) [188]
Xanthan gum coated mucoadhesive liposomes for efficient nose to brain delivery of curcumin	Coating agent, mucoadhesive agent	Samudre et al. (2015) [189]
Xanthan gum-based bilayered mucoadhesive buccal patches of zolmitriptan	Film-former, matrix-former, sustained releasing agent, mucoadhesive agent	Shiledar et al. (2014) [184]
Xanthan gum-locust bean gum based microparticles of celecoxib	Matrix-former, sustained releasing agent	Sharma et al. (2016) [136]
Modified xanthan gum for buccal delivery in treating sialorrhea	Film-former, matrix-former, sustained releasing agent, mucoadhesive agent	Laffleur and Michalek (2017) [190]

3.12 Sterculia gum

Sterculia gum is also called as karaya gum. It is a plant originated water soluble polysaccharide possessing medicinal importance [40]. Sterculia gum is obtained from the exudates of the plant—*Sterculia urens* (family: sterculiaceae) [191, 192]. The exudates are obtained by means of peeling or cutting the tree bark of *Sterculia urens*, which is treated as crude raw material for the

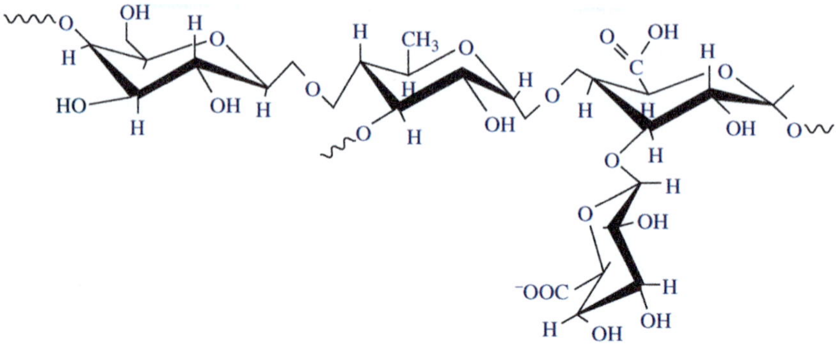

Fig. 12 Molecular structure of sterculia gum.

production of sterculia gum. It is a partially acetylated biopolysaccharide containing 3 dissimilar chains (Fig. 12) [192]. The 1st chain of sterculia gum molecules comprises approximately 50% of the total polysaccharide made of 4 galacturonic acid residues as repeating units, L-rhamnose residues at the reducing end and β-D-galactose branch. The 2nd chain of sterculia gum molecules comprises approximately 17% of the total polysaccharide made of an oligorhamnan containing D-galactose residues and D-galacturonic acid branch. The 3rd chain of sterculia gum molecules comprises approximately 30% of the total polysaccharide made of D-glucuronic acid residues comprising galactose, rhamnose and uronic acid residue [40, 191, 192]. Sterculia gum has many distinguishable properties, like high viscous rheology, excellent acidic stability, good swelling ability, biodegradability, biocompatibility, nonallergic, nonteratogenic, *etc.* [171, 191]. Sterculia gum also possesses antimicrobial property [192]. During past few years, sterculia gum is utilized as an important pharmaceutical excipient material in tablets, microparticles, beads, hydrogels, buccoadhesive drug delivery systems, *etc.* [40, 171, 191, 192]. Some recent researches on sterculia gum as pharmaceutical excipients are presented in Table 12.

3.13 Agar agar

Agar-agar (also called as agar) is a jelly-type biopolymeric material, derived from the red algae, *Rhodophyta phyllum* [201]. The polymeric structure of agar-agar consists of galactose, 3,6-anhydrogalactose. [202]. In addition, an inorganic sulfate is present in the agar-agar molecular structure, which is bonded to the carbohydrate. It is made up of two components, *i.e.*, agarose and agaropectin [201, 202]. Agarose is a linear polysaccharide. Agaropectin

Table 12 Recent researches on sterculia gum as pharmaceutical excipients.

Pharmaceutical formulations	Uses as excipients	References
Mucoadhesive-floating zinc-pectinate-sterculia gum interpenetrating polymer network beads of ziprasidone HCl	Matrix-former, sustained releasing agent, mucoadhesive agent, encapsulating agent	Bera et al. (2015) [74]
Alginate-sterculia gum gel-coated oil-entrapped alginate beads of risperidone	Matrix-former, sustained releasing agent, mucoadhesive agent, encapsulating agent	Bera et al. (2015) [193]
Oil-entrapped sterculia gum-alginate buoyant systems of aceclofenac	Matrix-former, sustained releasing agent	Guru et al. (2013) [194]
Sterculia crosslinked PVA and PVA-poly(AAm)hydrogel wound dressings for slow drug delivery	Matrix-former, sustained releasing agent, encapsulating agent	Singh and Pal (2012) [195]
Barium ions crosslinked alginate and sterculia gum-based gastroretentive floating drug delivery	Matrix-former, sustained releasing agent, encapsulating agent	Singh and Chauhan (2011) [196]
Sterculia gum-based double potential antidiarrheal drug delivery system	Matrix-former, sustained releasing agent, encapsulating agent	Singh and Sharma (2011) [197]
Sterculia gum-based interpenetrating polymer network microparticles containing repaglinide	Matrix-former, sustained releasing agent, encapsulating agent	Kulkarni et al. (2014) [198]
Sterculia gum-based osmotic core tablet for colon-specific drug delivery of azathioprine	Matrix-former, sustained releasing agent, colon targeting agent	Nath and Nath (2013) [199]
Sterculia gum-based tablet coated with chitosan/Eudragit RLPO mixed blend polymers for possible colonic drug delivery	Matrix-former, sustained releasing agent, colon targeting agent	Nath and Nath (2013) [200]

is the blend of smaller molecules. The aqueous solubility of agar-agar increases with the increasing of temperature. Most common examples of agar-agar are nutrient and non-nutrient agar, chocolate agar, blood agar, pharmaceutical agar, sabouraud agar, trypticase soy agar or tryptone soy agar, neomycin agar, *etc.* [201, 202]. Due to its so many advantageous characteristics, such as easy availability, inexpensiveness, cold-setting ability, nontoxicity, biodegradability, *etc.*, pharmaceutical agar is commonly utilized as excipient in many pharmaceutical products like tablets, pills, granules, microparticles, nanoparticles, *etc.* [202, 203]. Recently, agar-agar has been used to formulate hydrogels for controlling the drug release [203].

3.14 Gelatin

Gelatin is a naturally occurring biodegradable and multifunctional protein type biopolymer possessing thermoreversible nature [204]. Gelatin is a mixture of proteins and peptides [205]. It is derived by partial hydrolysis of collagen and the collagen for the production of gelatin is obtained from the animal bones and skin trimmings [204, 205]. Gelatin is rich source of proline, hydroxyproline, and glycine in its polymeric chain structure (Fig. 13). On the basis of the hydrolysis mechanism, gelatin is classified into two types: type A and type B. Acid hydrolyzed gelatin is known as type A, while base hydrolyzed gelatin is known as type B [204]. Due to its thermoreversible characteristics, gelatin is regarded as a special type of biopolymer showing the sol-gel transformation property. It has commonly been used as thickener, stabilizer, and emulsifier. Pharmaceutically, it is employed as constituent material in the formula of hard as well as soft gelatin capsules [204]. Due to its high swelling property, gelatin has been used in drug delivery area for its hydrogel producing capability [206]. Along with these properties, gelatin is capable to produce films. It has already been employed in the formulations of microparticles and nanoparticles the delivery of various drugs, proteins, enzymes, genes, *etc.* [206–209]. Some recent researches on gelatin as pharmaceutical excipients are presented in Table 13.

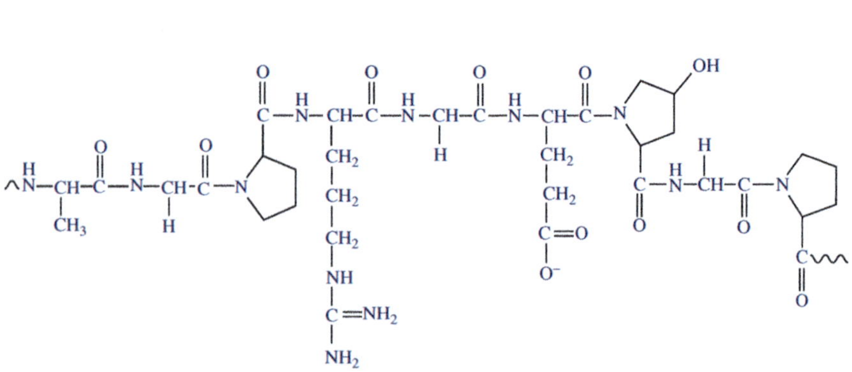

Fig. 13 Molecular structure of gelatin.

Table 13 Recent researches on gelatin as pharmaceutical excipients.

Pharmaceutical formulations	Uses as excipients	References
Bacterial cellulose and gelatin-based hydrogel composites for drug delivery	Matrix-former, sustained releasing agent, gelling agent, encapsulating agent	Treesuppharat et al. (2017) [64]
Semi-interpenetrating hydrogels from carboxymethyl guar gum and gelatin for ciprofloxacin	Matrix-former, sustained releasing agent, encapsulating agent	Ghosh et al. (2018) [121]
Controlled release of tyrosol and ferulic acid encapsulated in chitosan-gelatin films	Film-former, matrix-former, sustained releasing agent	Benbettaïeb et al. (2016) [156]
Modulated release from liposomes entrapped in chitosan-gelatin hydrogels	Release modifying agent, gelling agent	Ciobanu et al. (2014) [206]
Mucoadhesive buccal tablets based on chitosan-gelatin microparticles for delivery of propranolol HCl	Matrix-former, sustained releasing agent, mucoadhesive agent	Abruzzo et al. (2015) [207]
Moxifloxacin-loaded gelatin nanoparticles for ocular delivery	Matrix-former, sustained releasing agent, encapsulating agent, mucoadhesive agent	Mahor et al. (2016) [208]
Isoniazid-gelatin conjugate microparticles containing rifampicin for the treatment of tuberculosis	Sustained releasing agent, encapsulating agent	Manca et al. (2013) [209]
Stavudine-loaded gelatin liposomes for HIV therapy	Mucoadhesive agent	Nayak et al. (2017) [210]
Chitosan-polylacticacid-polyethylene glycol-gelatin nanoparticles for controlled drug delivery	Sustained releasing agent, gelling agent, encapsulating agent	Rajan and Raj (2013) [211]
EGFR-targeted gelatin nanoparticles for systemic administration of gemcitabine in an orthotopic pancreatic cancer model	Gelling agent, encapsulating agent	Singh et al. (2016) [212]

3.15 Albumin

Albumin is also an important naturally obtained protein polymer and is also known as the most profuse blood protein [213]. It serves as a reservoir and is also connected with the transport of plethora of substances like nutrients, hormones, metals and toxins. Albumin is produced mainly in liver hepatocytes

[214]. Albumin is available in three types: bovine serum albumin, human serum albumin and ovalbumin [213]. Monomeric phosphoglycoprotein, ovalbumin finds immense applications in food and pharmaceutical products owing to its easy accessibility, inexpensive, emulsion stabilization as well as pH and temperature responsive behavior [214]. Bovine serum albumin is widely utilized in drug delivery applications owing to its extraordinary ligand binding property [215, 216]. The most abundant hydrophilic plasma protein, human serum albumin is another important component in different drug delivery system owing to its inertness and biodegradability [214]. Recently, albumins have been developed for the formulation of nanoparticles used in different aspects of drug delivery [26, 27, 216, 217]. Some current researches on albumin as pharmaceutical excipients are presented in Table 14.

Table 14 Recent researches on albumin as pharmaceutical excipients.

Pharmaceutical formulations	Uses as excipients	References
Albumin vehicles for pulmonary drug delivery	Encapsulating agent	Woods et al. (2015) [215]
Glutathione-conjugated asiatic acid-loaded bovine serum albumin nanoparticles for brain-targeted drug delivery	Encapsulating agent, targeting agent	Raval et al. (2015) [216]
Albumin nanoparticles for the delivery of gabapentin	Encapsulating agent	Wilson et al. (2014) [217]
Carbopol gel containing chitosan-egg albumin nanoparticles for transdermal aceclofenac delivery	Matrix-former, encapsulating agent	Jana et al. (2014) [26]
Chitosan-egg albumin nanoparticles of alprazolam for oral delivery	Matrix-former, encapsulating agent	Jana et al. (2013) [27]
Lapatinib-loaded human serum albumin nanoparticles	Encapsulating agent	Wan et al. (2016) [218]
Acyclovir-loaded albumin nanoparticles	Encapsulating agent, targeting agent	Suwannoi et al. (2017) [219]
Cabazitaxel-loaded human serum albumin nanoparticles	Encapsulating agent	Qu et al. (2016) [220]
Albumin nanoparticles of albendazole in ovarian cancer	Matrix-former, encapsulating agent	Noorani et al. (2015) [221]
Gemcitabine conjugated albumin nanoparticles	Encapsulating agent, targeting agent	Kushwah et al. (2017) [222]

4 Disadvantages of natural polymer

Apart from advantages in natural polymer, some disadvantages are also associated with the uses as pharmaceutical applications:

(i) Microbial contamination during production from their natural sources when exposed to the external environment, uncontrolled hydration rate because of differences in availability and the presence of different species.

(ii) Batch to batch variation in percentage of components resulting from difference resources and resource regions.

(iii) The rate of production is dependent upon the environmental issues and several additional issues, it can't be altered.

(iv) Potential impurities may also result in unwanted immune responses. In particular, heavy metal contaminations that often related with the herbal polymeric excipients are also recognized as potential drawback.

(v) Some of the natural polymers like starches entail higher sensitivity to acidic attack; however, functionalizations or modification has been demonstrated to impart resistances to acids.

5 Conclusion

Natural polymers have already gained significant consideration in last few decades by reason of their remarkable uses in the field of pharmaceutical sciences. The above-discussed natural polymers are of market importance. The main reasons behind it are that various natural polymers are of low cost, readily available, biodegradable and biocompatible. Their broad range of uses covers from the food industries to the pharmaceutical industries. In the pharmaceutical industry and researches, natural polymers are mainly utilized in the many formulations of pharmaceutical drug delivery dosage forms, such as oral, topical, brain, buccal, *etc.*, drug deliveries are highly demanding. Due to a number of distinguishable characteristics of natural polymers like enhanced viscosity, higher acidic stability and good swelling ability, biodegradability, biocompatibility, nonallergic, *etc.*, they are recognized as potential biopolymeric pharmaceutical excipients to formulate many kinds of dosage forms like tablets, capsules, pills, pellets, spheroids, beads, microparticles, nanoparticles, films, gels, hydrogels, scaffolds, *etc.* The future outlook for the explorations and exploitations of new natural polymers is extremely promising.

References

[1] M.Y. Kariduraganavar, A.A. Kittur, R.R. Kamble, Polymer synthesis and processing, in: S.G. Kumbar, C.T. Laurencin, M. Deng (Eds.), Natural and Synthetic Biomedical Polymers, Elsevier Inc, USA, 2014, pp. 1–32.

[2] T.G. Volova, E.I. Shishatskaya, A.J. Sinskey, Degradable Polymers: Production, Properties and Applications, Nova Science Pub., Inc, NY, USA, 2013.

[3] M. Parthasarathy, S. Sethuraman, Hierarchical characterization of biomedical polymers, in: S.G. Kumbar, C.T. Laurencin, M. Deng (Eds.), Natural and Synthetic Biomedical Polymers, Elsevier Inc, USA, 2014, pp. 32–42.

[4] L. Nair, C. Laurencin, Biodegradable polymers as biomaterials, Prog. Polym. Sci. 32 (2007) 762–798.

[5] M.S. Hasnain, A.K. Nayak, Recent progress in responsive polymer-based drug delivery systems, in: A.S.H. Makhlouf, N.Y. Abu-Thabit (Eds.), Stimuli Responsive Polymeric Nanocarriers for Drug Delivery Applications: Vol. 2: Advanced Nanocarriers for Therapeutics, Elsevier Ltd., 2019, pp. 569–595.

[6] A.K. Nayak, D. Pal, Functionalization of tamarind gum for drug delivery, in: V.K. Thakur, M.K. Thakur (Eds.), Functional Biopolymers, Springer International Publishing, Switzerland, 2018, pp. 35–56.

[7] A.K. Nayak, B. Das, Introduction to polymeric gels, in: K. Pal, I. Bannerjee (Eds.), Polymeric Gels Characterization, Properties and Biomedical Applications, Elsevier Ltd, 2018, pp. 3–27.

[8] I. Ogaji, E. Nep, J.D. Audu-Peter, Advances in natural polymers as pharmaceutical excipients, Pharm. Anal. Acta 3 (2011) 1–16.

[9] A. Nayak, O. Olatunji, D.B. Das, G. Vladisavljevic, Pharmaceutical applications of natural polymers, in: O. Olatunji (Ed.), Natural Polymers, Springer International Publishing, Switzerland, 2016, pp. 229–313.

[10] P.B. Malafaya, G.A. Silva, R.L. Reis, Natural-origin polymers as carriers and scaffolds for biomolecules and cell delivery in tissue engineering applications, Adv. Drug Deliv. Rev. 59 (2007) 207–233.

[11] D. Pal, A.K. Nayak, Interpenetrating polymer networks (IPNs): natural polymeric blends for drug delivery, in: M. Mishra (Ed.), Encyclopedia of Biomedical Polymers and Polymeric Biomaterials, vol. VI, Taylor & Francis Group, USA, 2015, pp. 4120–4130.

[12] M.S. Hasnain, A.K. Nayak, R. Singh, F. Ahmad, Emerging trends of natural-based polymeric systems for drug delivery in tissue engineering applications, Sci. J. UBU 1 (2) (2010) 1–13.

[13] A.K. Nayak, D. Pal, D.R. Pany, B. Mohanty, Evaluation of *Spinacia oleracea* L. leaves mucilage as innovative suspending agent, J. Adv. Pharm. Technol. Res. 1 (2010) 338–341.

[14] A.K. Nayak, D. Pal, Natural starches-blended ionotropically-gelled micrparticles/beads for sustained drug release, in: V.K. Thakur, M.K. Thakur, M.R. Kessler (Eds.), Handbook of Composites from Renewable Materials, Nanocomposites: Advanced Applications, vol. 8, WILEY-Scrivener, USA, 2017, pp. 527–560.

[15] Y. Yu, M. Shena, Q. Song, J. Xie, Biological activities and pharmaceutical applications of polysaccharide from natural resources: a review, Carbohydr. Polym. 183 (2018) 91–101.

[16] M.S. Hasnain, A.K. Nayak, Chitosan as responsive polymer for drug delivery applications, in: A.S.H. Makhlouf, N.Y. Abu-Thabit (Eds.), Stimuli Responsive Polymeric Nanocarriers for Drug Delivery Applications, Volume 1, Types and Triggers, Elsevier Ltd., 2018, pp. 581–605.

[17] M.S. Hasnain, S.A. Ahmad, N. Chaudhary, M.N. Hoda, A.K. Nayak, Biodegradable polymer matrix nanocomposites for bone tissue engineering, in: Inamuddin, A.M. Asiri,

A. Mohammad (Eds.), Applications of Nanocomposite Materials in Orthopedics, Elsevier Inc., 2019, pp. 1–37.
[18] A.K. Nayak, S.A. Ahmed, M. Tabish, M.S. Hasnain, Natural polysaccharides in tissue engineering application, in: M.S. Hasnain, A.K. Nayak (Eds.), Natural Polysaccharides in Drug Delivery and Biomedical Applications, Academic Press, Elsevier Inc., 2019, pp. 531–548.
[19] S.M.M. Hasnain, M.S. Hasnain, A.K. Nayak, Natural polysaccharides: Sources and extraction methodologies, in: M.S. Hasnain, A.K. Nayak (Eds.), Natural Polysaccharides in Drug Delivery and Biomedical Applications, Academic Press, Elsevier Inc, 2019, pp. 1–14.
[20] D. Pal, A.K. Nayak, Alginates, blends and microspheres: controlled drug delivery, in: M. Mishra (Ed.), Encyclopedia of Biomedical Polymers and Polymeric Biomaterials, vol. I, Taylor & Francis Group, USA, 2015, pp. 89–98.
[21] J.F. Mano, G.A. Silva, H.S. Azevedo, P.B. Malafaya, R.A. Sousa, S.S. Silva, L.F. Boesel, J.M. Oliveira, T.C. Santos, A.P. Marques, N.M. Neves, R.L. Reis, Natural origin biodegradable systems in tissue engineering and regenerative medicine: present status and some moving trends, J. R. Soc. Interface 4 (2007) 999–1030.
[22] S.O. Sen, A.K. Nayak, P. Devbhuti, A. Ghosh, Modification of gums by periodate oxidation: a natural cross-linker, Int. J. Pharm. Pharm. Sci. (2019) 11: 1–6.
[23] A.K. Nayak, H. Bera, M.S. Hasnain, D. Pal, Synthesis and characterization of graft-copolymers of plant polysaccharides, in: V.K. Thakur (Ed.), Biopolymer Grafting, Synthesis and Properties, Elsevier Inc, 2018, pp. 1–62.
[24] A.K. Nayak, D. Pal, Chitosan-based interpenetrating polymeric network systems for sustained drug release, in: A. Tiwari, H.K. Patra, J.-W. Choi (Eds.), Advanced Theranostics Materials, WILEY-Scrivener, USA, 2015, pp. 183–208.
[25] D. Pal, A.K. Nayak, S. Saha, Interpenetrating polymer network hydrogels of chitosan: applications in controlling drug release, in: M.I.H. Mondal (Ed.), Cellulose-Based Superabsorbent Hydrogels, Springer, Cham, 2019, pp. 1727–1768.
[26] S. Jana, S. Manna, A.K. Nayak, K.K. Sen, S.K. Basu, Carbopol gel containing chitosan-egg albumin nanoparticles for transdermal aceclofenac delivery, Colloids Surf. B Biointerfaces 114 (2014) 36–44.
[27] S. Jana, N. Maji, A.K. Nayak, K.K. Sen, S.K. Basu, Development of chitosan-based nanoparticles through inter-polymeric complexation for oral drug delivery, Carbohydr. Polym. 98 (2013) 870–876.
[28] E.M. Elmowafy, G.A.S. Awad, S. Mansour, A.E.-H.A. El-Shamy, Ionotropically emulsion gelled polysaccharide beads: preparation, *in vitro* and *in vivo* evaluation, Carbohydr. Polym. 75 (2009) 135–142.
[29] S. Jana, Gandhi, K.K. Sen, S.K. Basu, Natural polymers and their application in drug delivery and biomedical field, J. Pharm. Sci. Technol. 1 (2011) 16–27.
[30] A.K. Nayak, D. Pal, Natural polysaccharides for drug delivery in tissue engineering, Everymans Sci. XLVI (2012) 347–352.
[31] A.K. Nayak, D. Pal, Tamarind seed polysaccharide: an emerging excipient for pharmaceutical use, Indian J. Pharm. Educ. Res. 51 (2017) S136–S146.
[32] A.K. Nayak, D. Pal, K. Santra, Screening of polysaccharides from tamarind, fenugreek and jackfruit seeds as pharmaceutical excipients, Int. J. Biol. Macromol. 79 (2015) 756–760.
[33] A.K. Nayak, D. Pal, J. Pradhan, T. Ghorai, The potential of *Trigonella foenum-graecum* L. seed mucilage as suspending agent, Indian J. Pharm. Educ. Res. 46 (2012) 312–317.
[34] A.K. Nayak, M.T. Ansari, D. Pal, M.S. Hasnain, Hyaluronic acid (hyaluronan): pharmaceutical applications, in: A.K. Nayak, M.S. Hasnain, D. Pal (Eds.), Natural Polymers for Pharmaceutical Applications, Vol. III: Animal Derived Polymers, Apple Academic Press, USA, 2019, pp. 1–32.

[35] D. Pal, A.K. Nayak, S. Saha, M.S. Hasnain, Pharmaceutical applications of chondroitin, in: A.K. Nayak, M.S. Hasnain, D. Pal (Eds.), Natural Polymers for Pharmaceutical Applications, Vol. III: Animal Derived Polymers, Apple Academic Press, USA, 2019, pp. 117–132.
[36] A.K. Nayak, H. Bera, *In situ* polysaccharide-based gels for topical drug delivery applications, in: S. Maiti, S. Jana (Eds.), Polysaccharide Carriers for Drug Delivery, Elsevier Inc, 2019, pp. 615–638.
[37] H. Bera, S. Maiti, S. Saha, A.K. Nayak, Biopolymers-based gastroretentive buoyant systems for therapeutic management of Helicobacter pylori infection, in: S. Maiti, S. Jana (Eds.), Polysaccharide Carriers for Drug Delivery, Elsevier Inc, 2019, pp. 713–736.
[38] P. Saini, N. Sharma, Natural polymers used in fast disintegrating tablets: a review, Int. J. Drug Dev. Res. 4 (2012) 18–27.
[39] M. Milivojevic, I. Pajic-Lijakovic, B. Bugarski, A.K. Nayak, M.S. Hasnain, Gellan gum in drug delivery applications, in: M.S. Hasnain, A.K. Nayak (Eds.), Natural Polysaccharides in Drug Delivery and Biomedical Applications, Academic Press, Elsevier Inc, 2019, pp. 145–186.
[40] H. Bera, Y.F. Abbasi, M.S. Hasnain, A.K. Nayak, Sterculia gum in drug delivery applications, in: M.S. Hasnain, A.K. Nayak (Eds.), Natural Polysaccharides in Drug Delivery and Biomedical Applications, Academic Press, Elsevier Inc, 2019, pp. 223–247.
[41] A.K. Nayak, M.T. Ansari, F. Sami, H. Bera, M.S. Hasnain, Cashew gum in drug delivery applications, in: M.S. Hasnain, A.K. Nayak (Eds.), Natural Polysaccharides in Drug Delivery and Biomedical Applications, Academic Press, Elsevier Inc, 2019, pp. 263–283.
[42] S. Dey, B.C. Nandy, J.N. De, M.S. Hasnain, A.K. Nayak, Tamarind gum in drug delivery applications, in: M.S. Hasnain, A.K. Nayak (Eds.), Natural Polysaccharides in Drug Delivery and Biomedical Applications, Academic Press, Elsevier Inc, 2019, pp. 285–306.
[43] A. Samanta, A. De, M.S. Hasnain, H. Bera, A.K. Nayak, Gum odina as pharmaceutical excipient, in: M.S. Hasnain, A.K. Nayak (Eds.), Natural Polysaccharides in Drug Delivery and Biomedical Applications, Academic Press, Elsevier Inc, 2019, pp. 327–337.
[44] S. Jana, S. Maiti, S. Jana, K.K. Sen, A.K. Nayak, Guar gum in drug delivery applications, in: M.S. Hasnain, A.K. Nayak (Eds.), Natural Polysaccharides in Drug Delivery and Biomedical Applications, Academic Press, Elsevier Inc, 2019, pp. 187–201.
[45] D. Pal, S. Saha, A.K. Nayak, M.S. Hasnain, Marine-derived polysaccharides: pharmaceutical applications, in: A.K. Nayak, M.S. Hasnain, D. Pal (Eds.), Natural Polymers for Pharmaceutical Applications, Vol. II: Marine and Microbiologically Derived Polymers, Apple Academic Press, USA, 2019, pp. 1–36.
[46] A.K. Nayak, M.G. Laskar, M.S. Hasnain, D. Pal, Pharmaceutical applications of alginates, in: A.K. Nayak, M.S. Hasnain, D. Pal (Eds.), Natural Polymers for Pharmaceutical Applications, Vol. II: Marine and Microbiologically Derived Polymers, Apple Academic Press, USA, 2019, pp. 37–70.
[47] M. Ahuja, A. Kumar, P. Yadav, K. Singh, *Mimosa pudica* seed mucilage: isolation, characterization and evaluation as tablet disintegrant and binder, Int. J. Biol. Macromol. 57 (2013) 105–110.
[48] M.S. Hasnain, P. Rishishwar, S. Rishishwar, S. Ali, A.K. Nayak, Isolation and characterization of *Linum usitatisimum* polysaccharide to prepare mucoadhesive beads of diclofenac sodium, Int. J. Biol. Macromol. 116 (2018) 162–172.
[49] V.R. Sinha, R. Kumria, Polysaccharides in colon-specific drug delivery, Int. J. Pharm. 224 (2001) 19–38.

[50] D. Pal, A.K. Nayak, S. Saha, Cellulose-based hydrogels: present and future, in: M.S. Akhtar, M.K. Swamy, U.R. Sinnaih (Eds.), Natural Bio-active Compounds, Vol. 1: Production and Applications, Springer Nature Singapore Pte Ltd., 2019, pp. 285–332.
[51] D.J. Cosgrove, Growth of the plant cell wall, Nat. Rev. Mol. Cell Biol. 6 (2005) 850–861.
[52] A.H. Kibbe, Handbook of Pharmaceutical Excipients, The Pharmaceutical Press, Washington, DC, 2000.
[53] S. Sajjad Keshipour, A. Maleki, Modification of cellulose, in: M.S. Akhtar, M.K. Swamy, U.R. Sinnaih (Eds.), Natural Bio-active Compounds, Vol. 1: Production and Applications, Springer Nature Singapore Pte Ltd., 2019, pp. 435–486.
[54] D.N.S. Hon, Cellulose and Its Derivatives: Structures, Reactions and Medical Uses, Marcel Dekker, New York, USA, 1996.
[55] J. Malakar, A.K. Nayak, S. Goswami, Use of response surface methodology in the formulation and optimization of bisoprolol fumarate matrix tablets for sustained drug release, ISRN Pharm. (2012), 730628.
[56] J. Malakar, K. Das, A.K. Nayak, *In situ* cross-linked matrix tablets for sustained salbutamol sulfate release—formulation development by statistical optimization, Polim. Med. 44 (2014) 221–230.
[57] R. Maji, B. Das, A.K. Nayak, S. Ray, Ethyl cellulose microparticles containing metformin HCl by emulsification-solvent evaporation technique: effect of formulation variables, ISRN Polym. Sci. (2012), 801827.
[58] A.K. Nayak, B. Das, R. Maji, Gastroretentive hydrodynamically balanced system of ofloxacin: formulation and *in vitro* evaluation, Saudi Pharm. J. 21 (2013) 113–117.
[59] J. Malakar, P. Dutta, S.D. Purokayastha, S. Dey, A.K. Nayak, Floating capsules containing alginate-based beads of salbutamol sulfate: *in vitro-in vivo* evaluations, Int. J. Biol. Macromol. 64 (2014) 181–189.
[60] D. Pal, A.K. Nayak, Development, optimization and anti-diabetic activity of gliclazide-loaded alginate-methyl cellulose mucoadhesive microcapsules, AAPS PharmSciTech 12 (2011) 1431–1441.
[61] N. Rath Adhikari, B.S. Nayak, A.K. Nayak, B. Mohanty, Formulation and evaluation of buccal patches for delivery of atenolol, AAPS PharmSciTech 11 (2010) 1034–1044.
[62] Z. Liu, J. Li, S. Nie, H. Liu, P. Ding, W. Pan, Study of an alginate/HPMC-based in situ gelling ophthalmic delivery system for gatifloxacin, Int. J. Pharm. 315 (2006) 12–17.
[63] R. Monireh, N. Hassan, Carboxymethyl cellulose/graphene oxide bio-nanocomposite hydrogel beads as anticancer drug carrier agent, Carbohydr. Polym. 168 (2017) 320–326.
[64] W. Treesuppharat, P. Rojanapanthu, C. Siangsanoh, H. Manuspiya, S. Ummartyotin, Synthesis and characterization of bacterial cellulose and gelatin-based hydrogel composites for drug delivery systems, Biotechnol. Rep. 15 (2017) 84–91.
[65] A.M. Karaaslan, M.A. Tshabalala, G. Buschle-Diller, Wood hemicellulose/chitosan-based semi-interpenetrating network hydrogels: mechanical, swelling and controlled drug release properties, Bioresources 5 (2010) 1036–1054.
[66] F. Munarin, M.C. Tanzi, P. Petrini, Advances in biomedical applications of pectin gels, Int. J. Biol. Macromol. 51 (2012) 681–689.
[67] A.K. Nayak, D. Pal, K. Santra, Development of calcium pectinate-tamarind seed polysaccharide mucoadhesive beads containing metformin HCl, Carbohydr. Polym. 101 (2014) 220–230.
[68] A. Assifoui, O. Chambin, P. Cayot, Drug release from calcium and zinc pectinate beads: impact of dissolution medium composition, Carbohydr. Polym. 85 (2011) 388–393.

[69] J.-S. Lee, E.-J. Kim, D. Chung, H.G. Lee, Characteristics and antioxidant activity of catechin-loaded calcium pectinate gel beads prepared by internal gelation, Colloids Surf. B Biointerfaces 74 (2009) 17–22.
[70] A.K. Nayak, D. Pal, Plant-derived polymers: Ionically gelled sustained drug release systems, in: M. Mishra (Ed.), Encyclopedia of Biomedical Polymers and Polymeric Biomaterials, vol. VIII, Taylor & Francis Group, USA, 2016, pp. 6002–6017.
[71] T. Marras-Marquez, J. Peña, M. Veiga-Ochoa, Robust and versatile pectin-based drug delivery systems, Int. J. Pharm. 479 (2) (2015) 265–276.
[72] C.T. Musabayane, O. Munjeri, T.P. Matavire, Transdermal delivery of chloroquine by amidated pectin hydrogel matrix patch in the rat, Ren. Fail. 25 (2003) 525–534.
[73] T. Ugurlu, M. Turkoglu, U.S. Gurer, B.G. Akarsu, Colonic delivery of compression coated nisin tablets using pectin/HPMC polymer mixture, Eur. J. Pharm. Biopharm. 67 (2007) 202–210.
[74] H. Bera, S. Boddupalli, A.K. Nayak, Mucoadhesive-floating zinc-pectinate-sterculia gum interpenetrating polymer network beads encapsulating ziprasidone HCl, Carbohydr. Polym. 131 (2015) 108–118.
[75] A.K. Nayak, D. Pal, K. Santra, Development of pectinate-ispagula mucilage mucoadhesive beads of metformin HCl by central composite design, Int. J. Biol. Macromol. 66 (2014) 203–221.
[76] A.K. Nayak, D. Pal, Blends of jackfruit seed starch-pectin in the development of mucoadhesive beads containing metformin HCl, Int. J. Biol. Macromol. 62 (2013) 137–145.
[77] P. Sriamornsak, S. Sungthongjeen, S. Puttipipatkhachorn, Use of pectin as a carrier for intragastric floating drug delivery: carbonate salt contained beads, Carbohydr. Polym. 67 (2007) 436–445.
[78] A.K. Nayak, S. Kalia, M.S. Hasnain, Optimization of aceclofenac-loaded pectinate-poly (vinyl pyrrolidone) beads by response surface methodology, Int. J. Biol. Macromol. 62 (2013) 194–202.
[79] W. Zhang, K.M. Mahuta, B.A. Mikulski, J.N. Harvestine, J.Z. Crouse, J.C. Lee, M.G. Kaltchev, C.S. Tritt, Novel pectin-based carriers for colonic drug delivery, Pharm. Dev. Technol. 21 (2016) 127–130.
[80] F.G. Prezotti, F.I. Boni, N.N. Ferreira, S.P. Campana-Filho, A. Almeida, T. Vasconcelos, M.P.D. Gremião, B.S.F. Cury, B. Sarmento, Gellan gum/pectin beads are safe and efficient for the targeted colonic delivery of resveratrol, Polymers 10 (2018) 50.
[81] A.K. Nayak, D. Pal, S. Das, Calcium pectinate-fenugreek seed mucilage mucoadhesive beads for controlled delivery of metformin HCl, Carbohydr. Polym. 96 (2013) 349–357.
[82] F. Reynaud, N. Tsapis, S.S. Guterres, A.R. Pohlmann, E. Fattal, Pectin beads loaded with chitosan–iron microspheres for specific colonic adsorption of ciprofloxacin, J. Drug Deliv. Sci. Technol. 30 (2015) 494–500.
[83] A. Vaidya, S. Jain, R.K. Agrawal, S.K. Jain, Pectin–metronidazole prodrug bearing microspheres for colon targeting, J. Saudi Chem. Soc. 19 (3) (2015) 257–264.
[84] W. Zhou, W. Liu, L. Zou, W. Liu, C. Liu, R. Liang, J. Chen, Storage stability and skin permeation of vitamin C liposomes improved by pectin coating, Colloids Surf. B Biointerfaces 117 (2014) 330–337.
[85] S. Eswaramma, N.S. Reddy, K.K. Rao, Phosphate crosslinked pectin based dual responsive hydrogel networks and nanocomposites: development, swelling dynamics and drug release characteristics, Int. J. Biol. Macromol. 103 (2017) 1162–1172.
[86] E. Dauqan, A. Abdullah, Utilization of gum Arabic for industries and human health, Am. J. Appl. Sci. 10 (2013) 1270–1279.
[87] S. Patel, A. Goyal, Applications of natural polymer gum Arabic: a review, Int. J. Food Prop. 18 (2015) 986–998.

[88] M.R. Avadi, A.M.M. Sadeghi, N. Mohammadpour, S. Abedin, F. Atyabi, R. Dinarvand, M. Rafiee-Tehrani, Preparation and characterization of insulin nanoparticles using chitosan and Arabic gum with ionotropic gelation method, Nanomed.: Nanotechnol., Biol. Med. 6 (2010) 58–63.
[89] A.K. Nayak, M.S. Hasnain, Gum Arabic based multiple units for oral drug delivery, in: A.K. Nayak, M.S. Hasnain (Eds.), Plant Polysaccharides-Based Multiple-Unit Systems for Oral Drug Delivery, Springer, Singapore, 2019, pp. 25–30.
[90] A.K. Nayak, B. Das, R. Maji, Calcium alginate/gum Arabic beads containing glibenclamide: development and *in vitro* characterization, Int. J. Biol. Macromol. 51 (2012) 1070–1078.
[91] M.P. Yadav, J.M. Igartuburu, Y. Yan, E.A. Nothnagel, Chemical investigation of the structural basis of the emulsifying activity of gum arabic, Food Hydrocoll. 21 (2007) 297–308.
[92] A.K. Mistry, C.D. Nagda, D.C. Nagda, B.C. Dixit, R.B. Dixit, Formulation and *in vitro* evaluation of ofloxacin tablets using natural gums as binders, Sci. Pharm. 82 (2014) 441–448.
[93] A.K. Jha, D. Chetia, Development of natural gum based fast disintegrating tablets of glipizide, Asian J. Pharm. 6 (2012) 282–288.
[94] S.Y. Alkarib, D.E. Mohamed Elhassan, A.O. Nur, Evaluation of gum Arabic as a film coating former for immediate release oral tablet formulation, World J. Pharm. Pharm. Sci. 5 (2016) 32–41.
[95] E.X. Lu, Z.Q. Jiang, Q.Z. Zhang, X.G. Jiang, A water-insoluble drug monolithic osmotic tablet system utilizing gum arabic as an osmotic, suspending, and expanding agent, J. Control. Release 92 (2003) 375–382.
[96] K.K. Nishi, A. Jayakrishnan, Self-gelling primaquine-gum arabic conjugate: an injectable controlled delivery system for primaquine, Biomacromolecules 8 (2007) 84–90.
[97] P. Kadare, P. Maposa, A. Dube, C.C. Maponga, Encapsulation of isoniazid in chitosan-gum arabic and poly (lactic-co-glycolic acid) PVA particles to provide a sustained release formulation, J. Pharm. Pharmacol. S (1) (2014) 1–6.
[98] B. Ozturk, S. Argin, M. Ozilgen, D.J. McClements, Formation and stabilization of nanoemulsion-based vitamin E delivery systems using natural biopolymers: whey protein isolate and gum arabic, Food Chem. 188 (2015) 256–263.
[99] A.V. Reis, T.A. Moia, D.L.A. Sitta, M.R. Mauricio, E.T. Tenorio-Neto, M.R. Guilherme, A.F. Rubira, E.C. Muniz, Sustained release of potassium diclofenac from a pH-responsive hydrogel based on gum arabic conjugates into simulated intestinal fluid, J. Appl. Polym. Sci. 133 (2016).
[100] M. Li, H. Li, X. Li, H. Zhu, Z. Xu, L. Liu, J. Ma, M. Zhang, A bioinspired alginate-gum arabic hydrogel with micro-/nanoscale structures for controlled drug release in chronic wound healing, ACS Appl. Mater. Interfaces 9 (2017) 22160–22175.
[101] B. Sima, A.M. Mohammad, Z. Azizollaah, Physicochemical and rheological characterization of gum tragacanth exudates from six species of Iranian Astragalus, Food Biophys. 5 (2010) 57–71.
[102] D. Verbeken, S. Dierckx, K. Dewettinck, Exudate gums: occurrence, production, and applications, Appl. Microbiol. Biotechnol. 63 (2003) 10–21.
[103] G.O. Aspinall, J. Baillie, Gum tragacanth. Part I fractionation of the gum and the structure of tragacanthic acid, J. Chem. Soc. (1963) 1702–1714.
[104] A. Rasul, M. Iqbal, G. Murtaza, M.K. Waqas, M. Hanif, S.A. Khan, N.S. Bhatti, Design, development and in-vitro evaluation of metoprolol tartrate tablets containing xanthan-tragacanth, Acta Pol. Pharm. 67 (2010) 517–522.
[105] K. Hemmati, A. Masoumi, M. Ghaemy, Tragacanth gum-based nanogel as a superparamagnetic molecularly imprinted polymer for quercetin recognition and controlled release, Carbohydr. Polym. 136 (2016) 630–640.

[106] B. Singh, V. Sharma, Influence of polymer network parameters of tragacanth gum-based pH responsive hydrogels on drug delivery, Carbohydr. Polym. 101 (2014) 928–940.
[107] B. Singh, V. Sharma, Crosslinking of poly(vinylpyrrolidone)/acrylic acid with tragacanth gum for hydrogels formation for use in drug delivery applications, Carbohydr. Polym. 157 (2017) 185–195.
[108] A. Shahiwala, A. Zarar, Pharmaceutical product lead optimization for better in vivo bioequivalence performance: a case study of diclofenac sodium extended release matrix tablets, Curr. Drug Deliv. 15 (2018) 705–715.
[109] K. Hemmati, M. Ghaemy, Synthesis of new thermo/pH sensitive drug delivery systems based on tragacanth gum polysaccharide, Int. J. Biol. Macromol. 87 (2016) 415–425.
[110] H.A. Mohamed, R.R. Radwan, A.I. Raafat, A.E. Ali, Antifungal activity of oral (tragacanth/acrylic acid) amphotericin B carrier for systemic candidiasis: *in vitro* and *in vivo* study, Drug Deliv. Transl. Res. 8 (2018) 191–203.
[111] T. Jabri, M. Imran, Shafiullah, K. Rao, I. Ali, M. Arfan, M.R. Shah, Fabrication of lecithin-gum tragacanth muco-adhesive hybrid nano-carrier system for *in-vivo* performance of amphotericin B, Carbohydr. Polym. 194 (2018) 89–96.
[112] M. Ranjbar-Mohammadi, M. Zamani, M.P. Prabhakaran, S.H. Bahrami, S. Ramakrishna, Electrospinning of PLGA/gum tragacanth nanofibers containing tetracycline hydrochloride for periodontal regeneration, Mater. Sci. Eng. C 58 (2016) 521–531.
[113] M. Sadat Hosseini, K. Hemmati, M. Ghaemy, Synthesis of nanohydrogels based on tragacanth gum biopolymer and investigation of swelling and drug delivery, Int. J. Biol. Macromol. 82 (2016) 806–815.
[114] M. Ranjbar-Mohammadi, S. Rabbani, S.H. Bahrami, M.T. Joghataei, F. Moayer, Antibacterial performance and in vivo diabetic wound healing of curcumin loaded gum tragacanth/poly(epsilon-caprolactone) electrospun nanofibers, Mater. Sci. Eng. C 69 (2016) 1183–1191.
[115] M. Prabaharan, Prospective of guar gum and its derivatives as controlled drug delivery systems, Int. J. Biol. Macromol. 49 (2011) 117–124.
[116] D.S. Seeli, S. Dhivya, N. Selvamurugan, M. Prabaharan, Guar gum succinate-sodium alginate beads as a pH-sensitive carrier for colon-specific drug delivery, Int. J. Biol. Macromol. 91 (2016) 45–50.
[117] B. Kumar, S.G. Kulanthaivel, A. Mondal, S. Mishra, B. Banerjee, A. Bhaumik, I. Banerjee, S. Giri, Mesoporous silica nanoparticle based enzyme responsive system for colon specific drug delivery through guar gum capping, Colloids Surf. B Biointerfaces 150 (2017) 352–361.
[118] T.M. Aminabhavi, M.N. Nadagouda, S.D. Joshi, U.A. More, Guar gum as platform for the oral controlled release of therapeutics, Expert Opin. Drug Deliv. 11 (2014) 753–766.
[119] S.K. Vemula, R. Katkum, Formulation, development and pharmacokinetics of ketorolac tromethamine colon targeted guar gum compression coated tablets, Anal. Chem. Lett. 5 (2015) 149–161.
[120] G. Dodi, A. Pala, E. Barbu, D. Peptanariu, D. Hritcu, M.I. Popa, B.I. Tamba, Carboxymethyl guar gum nanoparticles for drug delivery applications: preparation and preliminary in-vitro investigations, Mater. Sci. Eng. C 63 (2016) 628–636.
[121] S.K. Ghosh, A. Das, A. Basu, A. Halder, S. Das, S. Basu, M.F. Abdullah, A. Mukherjee, S. Kundu, Semi-interpenetrating hydrogels from carboxymethyl guar gum and gelatin for ciprofloxacin sustained release, Int. J. Biol. Macromol. 120 (2018) 1823–1833.
[122] S. Das, U. Subuddhi, pH-responsive guar gum hydrogels for controlled delivery of dexamethasone to the intestine, Int. J. Biol. Macromol. 79 (2015) 856–863.

[123] P.M. Castro, P. Baptist, A.R. Madureira, B. Sarmento, M.E. Pintado, Combination of PLGA nanoparticles with mucoadhesive guar-gum films for buccal delivery of antihypertensive peptide, Int. J. Pharm. 547 (2018) 593–601.
[124] D. Sathya Seeli, M. Prabaharan, Guar gum oleate-graft-poly(methacrylic acid) hydrogel as acolon-specific controlled drug delivery carrier, Carbohydr. Polym. 158 (2017) 51–57.
[125] D. Kaushik, K. Sharma, S. Sardana, Colon targeting guar gum microspheres of 5-aminosalicylic acid: evaluation of various process variables, characterization and *in-vitro* drug release, Indian J. Pharm. Educ. Res. 50 (2016) 106–114.
[126] A. Giri, T. Bhunia, S.R. Mishra, L. Goswami, A.B. Panda, S. Pald, A. Bandyopadhyay, Acrylic acid grafted guar gum–nanosilica membranes for transdermal diclofenac delivery, Carbohydr. Polym. 91 (2) (2013) 492–501.
[127] A.K. Nayak, M.S. Hasnain, Locust bean gum based multiple units for oral drug delivery, in: A.K. Nayak, M.S. Hasnain (Eds.), Plant Polysaccharides-Based Multiple-Unit Systems for Oral Drug Delivery, Springer, Singapore, 2019, pp. 61–66.
[128] M.S. Hasnain, A.K. Nayak, M.T. Ansari, D. Pal, Pharmaceutical applications of locust bean gum, in: A.K. Nayak, M.S. Hasnain, D. Pal (Eds.), Natural Polymers for Pharmaceutical Applications, Volume I: Plant Derived Polymers, Apple Academic Press, USA, 2019, pp. 139–162.
[129] V.D. Prajapati, G.K. Jani, N.G. Moradiya, N.P. Randeria, B.J. Nagar, Locust bean gum: a versatile biopolymer, Carbohydr. Polym. 94 (2013) 814–821.
[130] V. Mathur, N. Mathur, Fenugreek and other less known legume galactomannan polysaccharides: scope for developments, J. Sci. Ind. Res. 64 (2005) 475–481.
[131] K. Malik, G. Arora, I. Singh, Locust bean gum as superdisintegrant—formulation and evaluation of nimesulide, Polym. Med. 41 (2011) 17–28.
[132] S. Kaity, J. Isaac, A. Ghosh, Interpenetrating polymer network of locust bean gum-poly (vinyl alcohol) for controlled release drug delivery, Carbohydr. Polym. 94 (2013) 456–467.
[133] V.D. Prajapati, G.K. Jani, N.G. Moradiya, N.P. Randeria, P.M. Maheriya, B.J. Nagar, Locust bean gum in the development of sustained release mucoadhesive macromolecules of aceclofenac, Carbohydr. Polym. 113 (2014) 138–148.
[134] V.D. Prajapati, K.H. Mashuru, H.K. Solanki, G.K. Jani, Development of modified release gliclazide biological macromolecules using natural biodegradable polymers, Int. J. Biol. Macromol. 55 (2013) 6–14.
[135] S. Jana, A. Gandhi, S. Sheet, K.K. Sen, Metal ion-induced alginate–locust bean gum IPN microspheres for sustained oral delivery, Int. J. Biol. Macromol. 72 (2015) 47–53.
[136] N. Sharma, R.D. Deshpande, D. Sharma, R.K. Sharma, Development of locust bean gum and xanthan gum based biodegradable microparticles of celecoxib using a central composite design and its evaluation, Ind. Crop. Prod. 82 (2016) 161–170.
[137] S. Jana, K.K. Sen, Chitosan—locust bean gum interpenetrating polymeric network nanocomposites for delivery of aceclofenac, Int. J. Biol. Macromol. 102 (2017) 878–884.
[138] H.A. Pawar, K. Lalitha, K. Ruckmani, Alginate beads of captopril using galactomannan containing *Senna tora* gum, guar gum and locust bean gum, Int. J. Biol. Macromol. 76 (2015) 119–131.
[139] P. Dey, B. Sa, S. Maiti, Impact of gelation period on modified locust bean-alginate interpenetrating beads for oral glipizide delivery, Int. J. Biol. Macromol. 76 (2015) 176–180.
[140] M. Upadhyay, S.K.R. Adena, H. Vardhan, S. Pandey, B. Mishra, Development and optimization of locust bean gum and sodium alginate interpenetrating polymeric network of capecitabine, Drug Dev. Ind. Pharm. 44 (2018) 511–521.

[141] D. Pal, A.K. Nayak, Plant polysaccharides-blended ionotropically-gelled alginate multiple-unit systems for sustained drug release, in: V.K. Thakur, M.K. Thakur, M.R. Kessler (Eds.), Handbook of Composites from Renewable Materials, Volume 6, Polymeric Composites, WILEY-Scrivener, USA, 2017, pp. 399–400.

[142] M.S. Hasnain, A.K. Nayak, Inamuddin, A.M. Asiri, A. Mohammad, Alginate-inorganic composite particles as sustained drug delivery matrices, in: Applications of Nanocomposite Materials in Drug Delivery, Elsevier Inc., 2018, pp. 39–74.

[143] P. Sinha, U. Ubaidulla, A.K. Nayak, Okra (*Hibiscus esculentus*) gum-alginate blend mucoadhesive beads for controlled glibenclamide release, Int. J. Biol. Macromol. 72 (2015) 1069–1075.

[144] P. Sinha, U. Ubaidulla, M.S. Hasnain, A.K. Nayak, B. Rama, Alginate-okra gum blend beads of diclofenac sodium from aqueous template using $ZnSO_4$ as a cross-linker, Int. J. Biol. Macromol. 79 (2015) 555–563.

[145] B. Das, S. Dutta, A.K. Nayak, U. Nanda, Zinc alginate-carboxymethyl cashew gum microbeads for prolonged drug release: development and optimization, Int. J. Biol. Macromol. 70 (2014) 505–515.

[146] J. Malakar, A.K. Nayak, A. Das, Modified starch (cationized)-alginate beads containing aceclofenac: formulation optimization using central composite design, Starch-Starke 65 (2013) 603–612.

[147] R. Tripathi, B. Mishra, Development and evaluation of sodium alginate-polyacrylamide graft-co-polymer-based stomach targeted hydrogels of famotidine, AAPS PharmSciTech 13 (2012) 1091–1102.

[148] A.K. Nayak, D. Pal, K. Santra, Swelling and drug release behavior of metformin HCl-loaded tamarind seed polysaccharide-alginate beads, Int. J. Biol. Macromol. 82 (2016) 1023–1027.

[149] M.S. Hasnain, A.K. Nayak, M. Singh, M. Tabish, M.T. Ansari, T.J. Ara, Alginate-based bipolymeric-nanobioceramic composite matrices for sustained drug release, Int. J. Biol. Macromol. 83 (2016) 71–77.

[150] S. Jana, A. Samanta, A.K. Nayak, K.K. Sen, S. Jana, Novel alginate hydrogel core–shell systems for combination delivery of ranitidine HCl and aceclofenac, Int. J. Biol. Macromol. 74 (2015) 85–92.

[151] S. Jana, A. Gangopadhaya, B.B. Bhowmik, A.K. Nayak, A. Mukhrjee, Pharmacokinetic evaluation of testosterone-loaded nanocapsules in rats, Int. J. Biol. Macromol. 72 (2015) 28–30.

[152] S. Ray, P. Sinha, B. Laha, S. Maiti, U.K. Bhattacharyya, A.K. Nayak, Polysorbate 80 coated crosslinked chitosan nanoparticles of ropinirole hydrochloride for brain targeting, J. Drug Deliv. Sci. Technol. 48 (2018) 21–29.

[153] S. Jana, A. Saha, A.K. Nayak, K.K. Sen, S.K. Basu, Aceclofenac-loaded chitosan-tamarind seed polysaccharide interpenetrating polymeric network microparticles, Colloids Surf. B Biointerfaces 105 (2013) 303–309.

[154] A. Verma, J. Dubey, N. Verma, A.K. Nayak, Chitosan-hydroxypropyl methylcellulose matrices as carriers for hydrodynamically balanced capsules of moxifloxacin HCl, Curr. Drug Deliv. 14 (2017) 83–90.

[155] H. Horo, S. Das, B. Mandal, L.M. Kundu, Development of a photoresponsive chitosan conjugated prodrug nano-carrier for controlled delivery of antitumor drug 5-fluorouracil, Int. J. Biol. Macromol. 121 (2019) 1070–1076.

[156] N. Benbettaïeb, A. Assifaoui, T. Karbowiak, F. Debeaufort, O. Chambin, Controlled release of tyrosol and ferulic acid encapsulated in chitosan–gelatin films after electron beam irradiation, Radiat. Phys. Chem. 118 (2016) 81–86.

[157] M. Harris, H. Ahmed, B. Barr, D. LeVine, L. Pace, A. Mohapatra, B. Morshed, J.D. Bumgardner, J.A. Jennings, Magnetic stimuli-responsive chitosan-based drug delivery biocomposite for multiple triggered release, Int. J. Biol. Macromol. 104 (2017) 1407–1414.

[158] W.F. Huang, C.P. Tsui, C.Y. Tang, M. Yang, G. Linxia, Surface charge switchable and pH-responsive chitosan/polymer core-shell composite nanoparticles for drug delivery application, Compos. Part B 121 (2017) 83–91.

[159] F.S. Majedi, M.M. Hasani-Sadrabadi, J.J. VanDersarl, N. Mokarram, S. Hojjati-Emami, E. Dashtimoghadam, S. Bonakdar, M.A. Shokrgozar, A. Bertsch, P. Renaud, On-chip fabrication of paclitaxel-loaded chitosan nanoparticles for cancer therapeutics, Adv. Funct. Mater. 24 (2014) 432–441.

[160] S. Jana, A. Das, A.K. Nayak, K.K. Sen, S.K. Basu, Aceclofenac-loaded unsaturated esterified alginate/gellan gum microspheres: *in vitro* and *in vivo* assessment, Int. J. Biol. Macromol. 57 (2013) 129–137.

[161] A.K. Nayak, D. Pal, K. Santra, *Artocarpus heterophyllus* L. seed starch-blended gellan gum mucoadhesive beads of metformin HCl, Int. J. Biol. Macromol. 65 (2014) 329–339.

[162] A.K. Nayak, D. Pal, K. Santra, Tamarind seed polysaccharide-gellan mucoadhesive beads for controlled release of metformin HCl, Carbohydr. Polym. 103 (2014) 154–163.

[163] A.K. Nayak, D. Pal, K. Santra, Ispaghula mucilage-gellan mucoadhesive beads of metformin HCl: development by response surface methodology, Carbohydr. Polym. 107 (2014) 41–50.

[164] A.K. Nayak, D. Pal, *Trigonella foenum-graecum* L. seed mucilage-gellan mucoadhesive beads for controlled release of metformin HCl, Carbohydr. Polym. 107 (2014) 31–40.

[165] Z. Abbas, S. Marihal, Gellan gum-based mucoadhesive microspheres of almotriptan for nasal administration: formulation optimization using factorial design, characterization, and *in vitro* evaluation, J. Pharm. Bioallied Sci. 6 (2014) 267–277.

[166] M.H. Mahdi, B.R. Conway, T. Mills, A.M. Smith, Gellan gum fluid gels for topical administration of diclofenac, Int. J. Pharm. 515 (2016) 535–542.

[167] A.O. Abioye, S. Issah, A.T. Kola-Mustapha, *Ex vivo* skin permeation and retention studies on chitosan-ibuprofen-gellan ternary nanogel prepared by *in situ* ionic gelation technique—a tool for controlled transdermal delivery of ibuprofen, Int. J. Pharm. 490 (2015) 112–130.

[168] U.M. Musazzi, C. Cencetti, S. Franzé, N. Zoratto, C. Di Meo, P. Procacci, et al., Gellan nanohydrogels: novel nanodelivery systems for cutaneous administration of piroxicam, Mol. Pharm. 15 (2018) 1028–1036.

[169] U. Posadowska, M. Brzychczy-Wloch, E. Pamula, Injectable gellan gum-based nanoparticles-loaded system for the local delivery of vancomycin in osteomyelitis treatment, J. Mater. Sci. Mater. Med. 27 (1) (2015) 9.

[170] A.K. Nayak, Tamarind seed polysaccharide-based multiple-unit systems for sustained drug release, in: S. Kalia, L. Averous (Eds.), Biodegradable and Bio-based Polymers: Environmental and Biomedical Applications, WILEY-Scrivener, USA, 2016, pp. 471–494.

[171] A.K. Nayak, M.S. Hasnain, D. Pal, Gelled microparticles/beads of sterculia gum and tamarind gum for sustained drug release, in: V.K. Thakur, M.K. Thakur (Eds.), Handbook of Springer on Polymeric Gel, Springer International Publishing, Switzerland, 2018, pp. 361–414.

[172] A.K. Nayak, S.S. Nanda, D.K. Yi, M.S. Hasnain, D. Pal, Pharmaceutical applications of tamarind gum, in: A.K. Nayak, M.S. Hasnain, D. Pal (Eds.), Natural Polymers for Pharmaceutical Applications, Volume I: Plant Derived Polymers, Apple Academic Press, USA, 2019, pp. 1–20.

[173] A.K. Nayak, M.S. Hasnain, Tamarind polysaccharide based multiple units for oral drug delivery, in: A.K. Nayak, M.S. Hasnain (Eds.), Plant Polysaccharides-Based Multiple-Unit Systems for Oral Drug Delivery, Springer, Singapore, 2019, pp. 31–59.

[174] S. Ghosh, S. Pal, Modified tamarind kernel polysaccharide: a novel matrix for control release of aspirin, Int. J. Biol. Macromol. 58 (2013) 296–300.

[175] A. Avachat, K.N. Gujar, K.V. Wagh, Development and evaluation of tamarind seed glucan-based mucoadhesive buccal films of rizatriptan benzoate, Carbohydr. Polym. 91 (2013) 537–542.
[176] H. Bera, S. Boddupalli, S. Nandikonda, S. Kumar, A.K. Nayak, Alginate gel-coated oil-entrapped alginate–tamarind gum–magnesium stearate buoyant beads of risperidone, Int. J. Biol. Macromol. 78 (2015) 102–111.
[177] N. Dilbaghi, H. Kaur, M. Ahuja, S. Kumar, Evaluation of tropicamide-loaded tamarind seed xyloglucan nanoaggregates for ophthalmic delivery, Carbohydr. Polym. 94 (2013) 286–291.
[178] A.M.J. Newton, V.L. Indana, J. Kumar, Chronotherapeutic drug delivery of tamarind gum, chitosan and okra gum controlled release colon targeted directly compressed propranolol HCl matrix tablets and *in vitro* evaluation, Int. J. Biol. Macromol. 79 (2015) 290–299.
[179] A.M. Meenakshi, Metronidazole loaded carboxymethyl tamarind kernel polysaccharide-polyvinyl alcohol cryogels: preparation and characterization, Int. J. Biol. Macromol. 72 (2015) 931–938.
[180] J.L. Candia, W.D. Deckwer, Xanthan gum, in: M.C. Flickinger, S.W. Drew (Eds.), Encyclopaedia of Bioprocess Technology: Fermentation, Biocatalysis, and Bioseparation, Wiley, New York, 1999, pp. 2695–2711.
[181] A. Kumar, K.M. Rao, S.S. Han, Application of xanthan gum as polysaccharide in tissue engineering: a review, Carbohydr. Polym. 180 (2018) 128–144.
[182] A. Alhalmi, N. Alzubaidi, M. Altowairi, M. Almoiliqy, B. Sharma, Xanthan gum; its biopharmaceutical applications: an overview, World J. Pharm. Pharm. Sci. 7 (2018) 1536–1548.
[183] T.K. Giri, C. Choudhary, A. Alexander, Ajazuddin, H. Badwaik, M. Tripathy, D.K. Tripathi, Sustained release of diltiazem hydrochloride from cross-linked biodegradable IPN hydrogel beads of pectin and modified xanthan gum, Indian J. Pharm. Sci. 75 (2013) 619–627.
[184] R.R. Shiledar, A.A. Tagalpallewar, C.R. Kokare, Formulation and in vitro evaluation of xanthan gum-based bilayered mucoadhesive buccal patches of zolmitriptan, Carbohydr. Polym. 101 (2014) 1234–1242.
[185] D.H. Hanna, G.R. Saad, Encapsulation of ciprofloxacin within modified xanthan gum- chitosan based hydrogel for drug delivery, Bioorg. Chem. 84 (2019) 115–124.
[186] S. Maiti, S. Mukherjee, Controlled drug delivery attributes of co-polymer micelles and xanthan-O-carboxymethyl hydrogel particles, Int. J. Biol. Macromol. 70 (2014) 37–43.
[187] B. Mishra, S.K. Sahoo, S. Sahoo, Liranaftate loaded xanthan gum based hydrogel for topical delivery: physical properties and *ex-vivo* permeability, Int. J. Biol. Macromol. 107 (2018) 1717–1723.
[188] P.K. Sharma, S. Taneja, Y. Singh, Hydrazone-linkage-based self-healing and injectable xanthan-poly(ethylene glycol) hydrogels for controlled drug release and 3D cell culture, ACS Appl. Mater. Interfaces 10 (2018) 30936–30945.
[189] S. Samudre, A. Tekade, K. Thorve, A. Jamodkar, G. Parashar, N. Chaudhari, Xanthan gum coated mucoadhesive liposomes for efficient nose to brain delivery of curcumin, Drug Deliv. Lett. 5 (3) (2015) 201–207.
[190] F. Laffleur, M. Michalek, Modified xanthan gum for buccal delivery—a promising approach in treating sialorrhea, Int. J. Biol. Macromol. 102 (2017) 1250–1256.
[191] A.K. Nayak, M.S. Hasnain, Sterculia gum based multiple units for oral drug delivery, in: A.K. Nayak, M.S. Hasnain (Eds.), Plant Polysaccharides-Based Multiple-Unit Systems for Oral Drug Delivery, Springer, Singapore, 2019, pp. 67–82.

[192] A.K. Nayak, D. Pal, Sterculia gum-based hydrogels for drug delivery applications, in: S. Kalia (Ed.), Polymeric Hydrogels as Smart Biomaterials, Springer Series on Polymer and Composite Materials, Springer International Publishing, Switzerland, 2016, pp. 105–151.
[193] H. Bera, S.G. Kandukuri, A.K. Nayak, S. Boddupalli, Alginate-sterculia gum gel-coated oil-entrapped alginate beads for gastroretentive risperidone delivery, Carbohydr. Polym. 120 (2015) 74–84.
[194] P.R. Guru, A.K. Nayak, R.K. Sahu, Oil-entrapped sterculia gum-alginate buoyant systems of aceclofenac: development and *in vitro* evaluation, Colloids Surf. B Biointerfaces 104 (2013) 268–275.
[195] B. Singh, L. Pal, Sterculia crosslinked PVA and PVA-poly(AAm)hydrogel wound dressings for slow drug delivery: mechanical, mucoadhesive, biocompatible and permeability properties, J. Mech. Behav. Biomed. Mater. 9 (2012) 9–21.
[196] B. Singh, D. Chauhan, Barium ions crosslinked alginate and sterculia gum-based gastrortentive floating drug delivery system for use in peptic ulcers, Int. J. Polym. Mater. 60 (2011) 684–705.
[197] B. Singh, N. Sharma, Design of sterculia gum based double potential antidiarrheal drug delivery system, Colloids Surf. B Biointerfaces 82 (2011) 325–332.
[198] R.V. Kulkarni, F.S. Patel, H.M. Nanjappaiah, A.A. Naikawadi, *In vitro* and *in vivo* evaluation of novel interpenetrating polymer network microparticles containing repaglinide, Int. J. Biol. Macromol. 69 (2014) 514–522.
[199] B. Nath, L.K. Nath, Studies on stercuia gum formulations in the form of osmotic core tablet for colon-specific drug delivery of azathioprine, PDA J. Pharm. Sci. Technol. 67 (2013) 172–184.
[200] B. Nath, L.K. Nath, Design, development, and optimization of sterculia gum-based tablet coated with chitosan/eudragit RLPO mixed blend polymers for possible colonic drug delivery, J. Pharm. (2013), 546324.
[201] R. Armisén, F. Galatas, Handbook of Hydrocolloids, second ed., Woodhead Publishing, 2009, pp. 82–107.
[202] P. Laurienzo, Marine polysaccharides in pharmaceutical applications: an overview, Mar. Drugs 8 (2010) 2435.
[203] Z.-C. Yin, Y.-L. Wang, K. Wang, A pH-responsive composite hydrogel beads based on agar and alginate for oral drug delivery, J. Drug Deliv. Sci. Technol. 43 (2018) 12–18.
[204] M. Foox, M. Zilberman, Drug delivery from gelatin-based systems, Expert Opin. Drug Deliv. 12 (9) (2015) 1547–1563.
[205] M. Xu, L. Wei, Y. Xiao, H. Bi, H. Yang, Y. Du, Physicochemical and functional properties of gelatin extracted from yak skin, Int. J. Biol. Macromol. 95 (2017) 1246–1253.
[206] B.C. Ciobanu, A.N. Cadinoiu, M. Popa, J. Desbrieres, C.A. Peptu, Modulated release from liposomes entrapped in chitosan/gelatin hydrogels, Mater. Sci. Eng. C 43 (2014) 383–391.
[207] A. Abruzzo, T. Cerchiara, F. Bigucci, M.C. Gallucci, B. Luppi, Mucoadhesive buccal tablets based on chitosan/gelatin microparticles for delivery of propranolol hydrochloride, J. Pharm. Sci. 104 (2015) 4365–4372.
[208] A. Mahor, S.K. Prajapati, A. Verma, R. Gupta, A.K. Iyer, P. Kesharwani, Moxifloxacin loaded gelatin nanoparticles for ocular delivery: formulation and in-vitro, in-vivo evaluation, J. Colloid Interface Sci. 483 (2016) 132–138.
[209] M.L. Manca, R. Cassano, D. Valenti, S. Trombino, T. Ferrarelli, N. Picci, A.M. Fadda, M. Manconi, Isoniazid-gelatin conjugate microparticles containing rifampicin for the treatment of tuberculosis, J. Pharm. Pharmacol. 65 (2013) 1302–1311.

[210] D. Nayak, A. Boxi, S. Ashe, N.C. Thathapudi, B. Nayak, Stavudine loaded gelatin liposomes for HIV therapy: preparation, characterization and in vitro cytotoxic evaluation, Mater. Sci. Eng. C 73 (2017) 406–416.
[211] M. Rajan, V. Raj, Formation and characterization of chitosan-polylacticacid-polyethylene glycol-gelatin nanoparticles: a novel biosystem for controlled drug delivery, Carbohydr. Polym. 98 (2013) 951–958.
[212] A. Singh, J. Xu, G. Mattheolabakis, M. Amiji, EGFR-targeted gelatin nanoparticles for systemic administration of gemcitabine in an orthotopic pancreatic cancer model, Nanomed. Nanotechnol. Biol. Med. 12 (2016) 589–600.
[213] J. Rozga, T. Piątek, P. Małkowski, Human albumin: old, new, and emerging applications, Ann. Transplant. 18 (2013) 205–217.
[214] G.J. Quinlan, G.S. Martin, T.W. Evans, Albumin: biochemical properties and therapeutic potential, Hepatology 41 (2005) 1211–1219.
[215] A. Woods, A. Patel, D. Spina, Y. Riffo-Vasquez, A. Babin-Morgan, R.T. de Rosales, K. Sunassee, S. Clark, H. Collins, K. Bruce, L.A. Dailey, B. Forbes, *In vivo* biocompatibility, clearance, and biodistribution of albumin vehicles for pulmonary drug delivery, J. Control. Release 210 (1–9) (2015).
[216] N. Raval, T. Mistry, N. Acharya, S. Acharya, Development of glutathione-conjugated asiatic acid-loaded bovine serum albumin nanoparticles for brain-targeted drug delivery, J. Pharm. Pharmacol. 67 (2015) 1503–1511.
[217] B. Wilson, Y. Lavanya, S.R.B. Priyadarshini, M. Ramasamy, J.L. Jenita, Albumin nanoparticles for the delivery of gabapentin: preparation, characterization and pharmacodynamic studies, Int. J. Pharm. 473 (2014) 73–79.
[218] X. Wan, X. Zheng, X. Pang, Z. Pang, J. Zhao, Z. Zhang, T. Jiang, W. Xu, Q. Zhang, X. Jiang, Lapatinib-loaded human serum albumin nanoparticles for the prevention and treatment of triple-negative breast cancer metastasis to the brain, Oncotarget 7 (2016) 34038–34051.
[219] P. Suwannoi, M. Chomnawang, N. Sarisuta, S. Reichl, C.C. Müller-Goymann, Development of acyclovir-loaded albumin nanoparticles and improvement of acyclovir permeation across human corneal epithelial T cells, J. Ocular Pharmacol. Ther. 33 (2017) 743–752.
[220] N. Qu, R.J. Lee, Y. Sun, G. Cai, J. Wang, M. Wang, J. Lu, Q. Meng, L. Teng, D. Wang, L. Teng, Cabazitaxel-loaded human serum albumin nanoparticles as a therapeutic agent against prostate cancer, Int. J. Nanomedicine 11 (2016) 3451–3459.
[221] L. Noorani, M. Stenzel, R. Liang, M.H. Pourgholami, D.L. Morris, Albumin nanoparticles increase the anticancer efficacy of albendazole in ovarian cancer xenograft model, J. Nanobiotechnol. 13 (2015) 25.
[222] V. Kushwah, A.K. Agrawal, C.P. Dora, D. Mallinson, D.A. Lamprou, R.C. Gupta, S. Jain, Novel gemcitabine conjugated albumin nanoparticles: a potential strategy to enhance drug efficacy in pancreatic cancer treatment, Pharm. Res. 34 (2017) 2295–2311.

CHAPTER 2

Synthetic and semi-synthetic polymers for pharmaceutical applications

Micaela A. Macchione[a,b,c], Dariana Aristizabal Bedoya[a,b], Francisco N. Figueroa[a,b], and Miriam C. Strumia[a,b]
[a]Chemistry Science Faculty, Organic Chemistry Department, National University of Cordoba, Córdoba, Argentine
[b]Institute for Research and Development in Process Engineering and Applied Chemistry, Science and Technology National Council, Córdoba, Argentine
[c]Academic Pedagogical Institute of Human Sciences, National University of Villa María, Villa María, Córdoba, Argentine

1 Introduction

The pharmaceutical industries dedicated to the production of drugs perform one of the most important tasks in the area of health and medicine, since they fulfill roles in the diagnosis, therapy and prevention of diseases both in humans and animals. In addition, they are also used for the preservation of food and cosmetics [1]. Therefore, the use of drugs is strictly controlled by the Federal Office of Food, Medicine and Cosmetics.

Polymers represent an important component of several pharmaceutical dosage forms. In fact, it is already well-known that the formulation and clinical performance of pharmaceutical dosage forms are dependent on the properties of the polymeric materials used in their formulation. Despite the innumerable applications that polymers have in different technological areas, they represent a relatively small percentage of global sales worldwide [2, 3].

The main advantage of the use of polymers is their ability to provide formulations with a wide variety of physical and chemical characteristics [1, 4]. Accordingly, it is possible to obtain the desired properties by varying the physicochemical features of the polymeric material such as linear, grafted, cross-linked structure, its molecular weight and polydispersity, the monomer structure (aliphatic, aromatic, polar, non-polar or with specific functional groups) and the possibility of copolymerization that can give rise to polymers with different configuration (blocks or random), or also mixture with other polymers.

Advances and Challenges in Pharmaceutical Technology
https://doi.org/10.1016/B978-0-12-820043-8.00005-0

In general, polymers have properties derived from their macrostructure, mainly from the interactions between chains depending on the nature of the monomers that compose them or their molecular mass. These properties have been extensively exploited for pharmaceutical applications, such as:

- Their capacity to form films enables the use as coating for distasteful drug in tablets.
- Their adhesion properties give the capacity of binding the components of a formulation, for example in the preparation of tablets, or bioadhesivity for patches preparation.
- Their viscoelastic behavior and the capacity of thickening allow the use of these materials as a rheology modifier, flow control and viscosity agents in pharmaceutical liquids, suspensions and emulsions.
- The possibility of controlling the hydrophilic/hydrophobic balance enables to manage the solubility in both organic and aqueous solutions.
- The ease of gelation or gel formation, in which the presence of physical or chemical crosslinks between the adjacent polymers allows the formation of porous matrices. The porous nature of the matrices allows the incorporation of hydrophilic drugs and modification of the release characteristics, improving drug stability and bioavailability for administration.
- The presence of anionic or cationic functional groups in the chemical structure of polymers confers the pH-dependent solubility or behavior to the final system. This feature makes possible to provide pharmaceutical formulations with a specific therapeutic action according to the pH of the environment.
- The huge variety of possibilities for the adaptation of polymers to the preparation of carrier formulations so that the drug reaches the target tissue which is generally inaccessible.
- The specific barrier properties of the polymeric films ensure the protection when used as packaging.

Therefore, pharmaceutical applications of polymers range from solid dosage forms (tablets, capsules), dispersed systems (suspensions, emulsions, creams, ointments), implants, particle systems (micro and nano capsules, micro and nanoparticles) and transdermal patches, where the previously mentioned properties give advantages to the dosage forms.

The new polymeric materials can be designed to meet the requirements of drug administration such as better dissolution, bioadhesion, swelling, favorable pinocytic parameters and minimal toxic, irritant or immunogenic effects [4].

In the following sections of this chapter, the most used synthetic methodologies for obtaining synthetic and semi-synthetic polymers with some representative examples of their pharmaceutical applications will be presented. Besides, there will be a brief presentation of the advances and future in the synthetic methodologies of polymer systems potentially used in pharmaceutical technology.

2 Synthetic and semi-synthetic polymers

The polymers can be classified into three types: natural, synthetic or semi-synthetic. The huge advantage of synthetic polymers is the easiness of preparing large quantities with countless possibilities of modifications to achieve the desired properties [1].

2.1 Synthetic polymers

As it was mentioned previously, a wide range of polymers are used as components of conventional pharmaceutical formulations. Mainly synthetic and semi-synthetic pharmaceutical polymers including polyethylene glycol, polyvinyl alcohol, polyethylene oxide, polyvinyl pyrrolidone, polyacrylate or polymethacrylate esters, cellulose based polymers, and many plastics and rubbers. Table 1 represents those synthetic polymers that are most commonly used in pharmacy, although it is recognized that there will be other polymers not included in this table that may also have pharmaceutical applications [2].

2.1.1 Methodologies of synthesis

In this section, the main chemical routes for the development of synthetic and semi-synthetic polymers will be reviewed.

It is well known that the process in which reactive monomers are bonded by covalent interactions producing chains or three-dimensional networks is called polymerization. In the synthesis of synthetic polymers for pharmaceutical applications, different polymerization mechanisms can be involved such as step-growth polymerization (a), chain-growth polymerization (b), and ring-opening polymerization (c) (Fig. 1).

The following section discusses the most frequently used methods for the most well-known synthetic polymer classes used in pharmaceutical applications.

Table 1 Synthetic polymers more used in pharmaceutical industries.

Classification	Subclassification	Structure	Properties	Applications
Vinyl polymers	Poly(methacrylates)	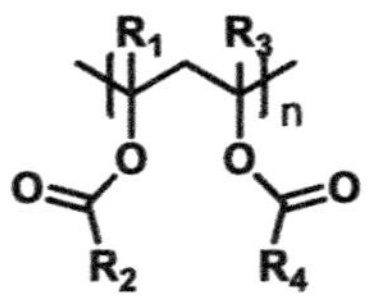	Anionic or cationic nature, non-toxic, non-irritant	Film coatings for solid dosage forms, binders for both aqueous and organic wet granulation process and as viscosity modifiers in some topical formulations, matrix in drug delivery systems
	Poly(vinylalcohol)	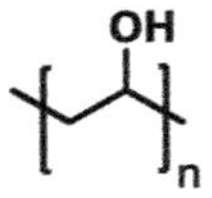	Non-toxic, non-irritant	Lubrication, viscosity modifying agent and sustained release formulations
	Poly(vinylpyrrolidone)	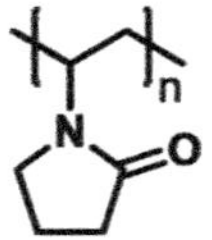	Hydrophilic, non-ionic, high solubility in water and alcohol	Binder of granules, coating, solubilizer in oral and parenteral formulations, viscosity modifying agent in topical formulations
	Poly(acrylic acid)	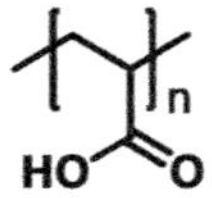	Anionic nature, the viscosity is dependent on pH	Viscosity modifying agent, binder in the formulation of granules, matrix for sustained drug release and controlled drug delivery at the site of application, bioadhesive
Polyesters	Poly(lactides) and derived	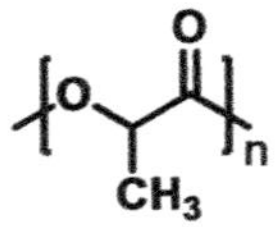	Organic acid that may be found as either the L(+) or D(−) stereoisomer and could be presented as block or random copolymer	
	Poly(caprolactone)	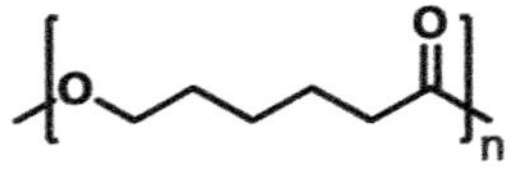	Semi crystaline, biodegradable	Matrix for controlled-release systems in a range of geometries and medical device biomaterial

Polyanhydrides		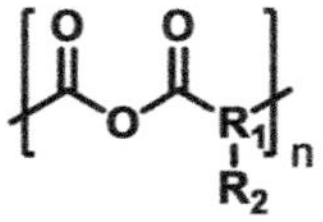	Hydrophobic polymers with hydrolytically labile anhydride linkages	Film or fibers forming, topical formulations to aid retention of the dosage form
Polyalcohols	Poly(ethyleneglycol)	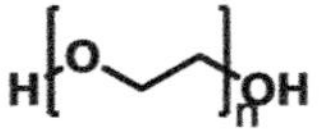	Non-immunogenic and non-toxic, amphiphilic polymers, soluble in aqueous and organic solvents	Increasing agent of solubility of poor soluble drug, parenteral and solid dosage form preparations, grafting agent of surfaces of medical devices or colloidal dosage to prevent deposition of protein or bacterial surface growth or to minimize their recognition by the immune system
	Poly(ethylene oxide) (PEO)	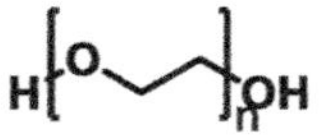	Soluble in aqueous and some organic solvents and is completely and rapidly eliminated from the body	Binder and a thickening agent. Depending of the molecular weight, it can be used to formulate immediate or sustained release formulations
Silicones		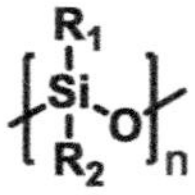	Highly dependent on the organic group attached to the silicon atom and can exist in the form of low viscosity oil to gels, rubbers and solid resins, inert and highly biocompatible	Medical devices such as cardiac valves, hydrocephalus shunts, intravaginal controlled drug delivery implants and intraocular lenses; esthetic and reconstructive prostheses

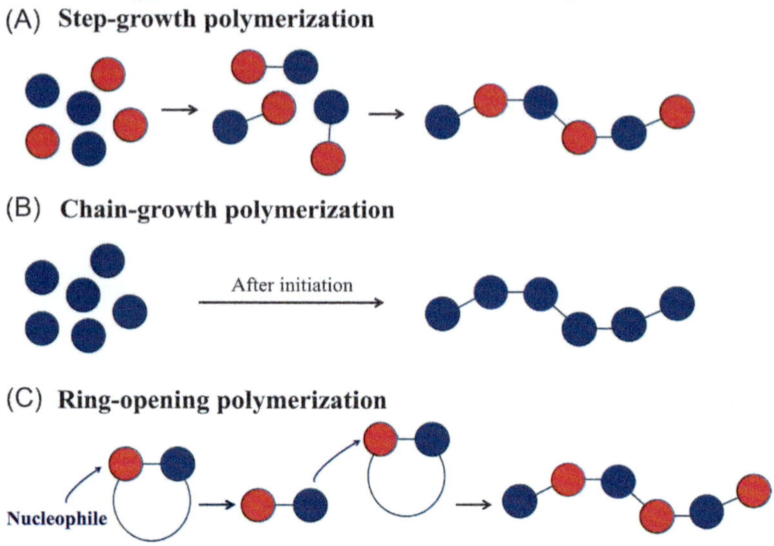

Fig. 1 Scheme of mechanisms of polymerization.

Step-growth polymerization

In the step-growth polymerization, also called condensation polymerization, the intermolecular condensation of the functional groups of monomers takes place by step [5]. Simultaneously, small molecules such as ammonia, water, HCl or methanol are eliminated. In this case, monomers should have functional groups capable of leading direct chemical reactions. Linear polymers can be produced from monomers with two reactive sites. Additionally, monomers with more reactive sites lead to cross-linked structures. Examples of polymers with pharmaceutical interest produced using this synthetic strategy are polysiloxanes such as poly(dimethylsiloxane) (PDMS), which is a material approved by the FDA to be used in foods (except milk), cosmetic and pharmaceutical products. This polymer has excellent biocompatibility, low toxicity, optical transparency, gas permeability, high thermal stability, elasticity and low production costs [1].

Chain-growth polymerization

In chain-growth polymerization (or addition polymerization), the growth of chains consists in the incorporation of a sequence of monomers without the elimination of small molecules [5]. Firstly, the formation of an active center (free radical or ion) takes place, which is the chain initiation step. Consequently, addition polymerization can be anionic, cationic and radical.

After initiation, monomer molecules are added to the polymer chain, a process called chain propagation. The termination step means the end of the chain propagation by the disappearance of the active center.

Free-radical polymerization is the conventional technique used to develop polymeric materials [6]. In this case, radicals should be firstly produced by the decomposition of the initiator (primary radicals). Then, radicals are transferred from the initiator molecules to the monomer units present. Monomers that can take part in this polymerization are alkenes or aldehydes/ketones with carbon-carbon or carbon-oxygen double bond, respectively [6, 7]. Radical adds to one atom of the double bond and produces a new active radical on the other atom. In the propagation step, the polymer chain grows by addition or atom transfer reactions. Finally, the termination step takes place either by recombination or disproportionation. Some examples of polymers synthesized by this chemical route are poly(*N*-isopropylacrylamide), poly(vinyl pyrrolidone), poly(meth)acrylates, polyacrylonitriles and poly(vinyl acetate). Then, poly(vinyl alcohol) can be obtained by partial or full hydrolyzation of poly(vinyl acetate) since the vinyl alcohol monomer is not enough stable for its polymerization.

"Living" polymerizations are special kind of chain-growth polymerization, they can be anionic, cationic or free-radical. In these mechanisms, the termination step is eliminated or suppressed [8]. For the case of living free-radical polymerization, the use of the term "reversible deactivation radical polymerization" is recommended by IUPAC. Here, the propagating radicals are deactivated in a reversible way. Consequently, active radicals and dormant species establish a new equilibrium. Living polymerization is desirable because it offers precision and control of molecular weights, tacticity and crystallinity. Examples of types of reversible-deactivation radical polymerization are atom transfer radical polymerization (ATRP), reversible addition-fragmentation chain-transfer polymerization (RAFT) and nitroxide-mediated polymerization (NMP). Materials synthesized by reversible-deactivation radical polymerization are still under research. For example, there are many reports of the synthesis of poly(*N*-isopropylacrylamide) (PNIPAAm) by several controlled radical polymerization techniques including ATRP and RAFT polymerization [9], however, PNIPAAm-based materials have not yet been approved by FDA.

Ring-opening polymerization

Ring-opening polymerization is another synthetic route in which a reactive center (ionic, cationic, or radical) can open a ring of cyclic monomer to form the polymer chains [10]. The driving force is the release of the ring strain [11, 12]. Ring structure and size influence the driving force and so the ability of polymerization [13].

Ring-opening polymerization has some similarities in kinetics and mechanisms with chain and step polymerization. It shares the steps of initiation, propagation, and termination with chain polymerization. On the other hand, the monomer addition is one by one similar as step polymerization [5, 10]. Nevertheless, monomer and larger sized species do not generally react themselves or with each other as it happens in step polymerization [10]. In addition, the initiation and subsequent steps can take place at similar speeds and, therefore, are not chain reactions from the kinetic point of view [5].

Many commercially important polymers are synthesized by this mechanism including polyesters from cyclic ester (lactones) such as poly(caprolactone), poly(glycolic acid), poly(lactic acid) (PLA), poly(ethylene glycol) (PEG).

PEG, which is also called poly(ethylene oxide), is very often applied in pharmaceutical applications. This polymer can be synthesized by anionic or cationic ring-opening polymerization of ethylene oxide being the first one the most used mechanism because it results in polymer chains with controlled molecular weight and monodisperse distribution [1, 14]. For the initiation step, the SN2 nucleophilic substitution of alkoxide of alkali metals is effectively used to produce new alkoxide species which are able to react with monomer molecules giving polyether chains. Termination agents with labile hydrogen such as alcohols and water are usually employed to produce hydroxyl end-groups [15]. On the other hand, cationic ring-opening polymerization can also be used for the synthesis of PEG. However, it leads to broad molecular weight distribution because of backbiting reactions [14], and consequently, is less used than anionic mechanism.

Ring-opening polymerization can also proceed as living polymerization.

Synthetic copolymers

In addition to synthetic homopolymers, it is possible to synthesize different copolymers from the same mechanisms of polymerization described above. In a random copolymer, the probability of finding a given monomeric unit at any given site in the chain is independent of the nature of the adjacent units [16]. On the contrary, in a block copolymer, the adjacent blocks are constitutionally different [16], two or more components (blocks) of different polymers are chemically attached to each other. The properties of these polymers depend on the copolymer sequence distribution, the chemical nature of the blocks, the average molecular weight and its distribution of the blocks and the copolymer. Living polymerization techniques can be particularly powerful tools for the synthesis of block copolymers by sequential addition of monomers [17].

There are many cases of copolymers used in pharmaceutical applications (approved by FDA) for example [1]:

- Matrix excipient (film former, tablet binder, tablet diluent) as Eudragit® (poly(methacrylic acid)-*co*-poly(methyl methacrylate);
- Drug-eluting stents with nonbiodegradable polymers as coating materials like poly(vinylidene-*co*-hexafluoropropylene) (PVDF-HFP);
- Marketed pH-responsive poly(meth)acrylates as protective coatings as poly(BMA-*co*-DMAEMA-*co*-MMA) (BMA: *n*-butylmethacrylate, DMAEMA: dimethylaminoethylmethacrylate, MMA: methyl methacrylate);
- The only commercialized vaginal ring (NuvaRing®) that is based on poly(ethylene-*co*-vinyl acetate).

2.2 Semi-synthetic polymers

Nature polymers are macromolecules that are isolated from natural sources [18], mainly polyesters, proteins, and polysaccharides derived from both animal and plant kingdoms [19]. Natural polymers have certain advantages over synthetic polymers since they are abundant in nature and their isolation can be relatively easy [20]. Likewise, when these polymers are in biologic fluids, they present hydrolytic and/or enzymatic degradation into non-toxic degradation products. Thus, their biodegradability, no cytotoxicity, and biocompatibility are their most remarkable features [21, 22]. All these factors make possible a large and distinct range of applications of natural polymers in the pharmaceutical industry. However, polymers of natural origin have often some limitations; they are generally poorly soluble and show poor thermal and mechanical properties. Consequently, their native structures are easily lost through thermal processes [19]. In addition, they can easily become contaminated with microbes when exposed to the environment [18]. The modification of natural polymers with synthetic polymers or synthetic chemicals is an important approach to overcome these limitations. Besides, the modification can be customized according to the requirements of the application. This results in a new kind of materials, semi-synthetic polymers that present the optimal features of both natural and synthetic polymers. If the mixture of suitable polymers is chosen, the resulting semi-synthetic polymer shows biocompatible behavior and, at the same time, presents good mechanical and thermal properties [23].

Specifically, in the pharmaceutical industry, the strategies for modification of natural polymers are carried out to design materials that show desired properties. Their biocompatibility and strength can be increased, or a stimuli-responsive behavior can be achieved. In addition, these modifications can alter some properties as solubility, hydrophobicity, viscosity, permeability, among other desired properties [24].

The main natural polymers studied for the development of semi-synthetic polymers for pharmaceutical applications are collagen, chitin, chitosan, keratin, silk and elastin, derived from animal body. Besides, those derived from plants, such as starch, cellulose and pectin [25]. For example, numerous semi-synthetic forms of cellulose have been fabricated, for instance, cellulose esters and cellulose ethers. These semi-synthetic cellulose products are extensively used as excipients in pharmaceutical preparations for many goals as binders, fillers, thickening agents, compressibility enhancers, and gelling agents [18, 26]. It is noteworthy, that only a small group of these semi-synthetic polymers has been approved by the FDA for application in the pharmaceutical industry [27–29].

2.2.1 Methodologies of synthesis

There are many employed strategies to obtain semi-synthetic polymers. Some of them are named in Table 2. In this section, we will focus on the main ones: chemical linkages, blending, grafting and crosslinking; these strategies are schematized in Fig. 2.

Chemical linkages

This is the simplest approach to obtain semi-synthetic polymers by inserting small functional groups in the natural polymer chain. Through this strategy, it is possible to attach polar or non-polar groups to the structure of the polymer and, hence, control its hydrophobicity or hydrophilicity. Consequently, the solubility of the material is drastically modified [30]. The esterification of cellulose is an example of this strategy. For this purpose, nitric acid is used in the presence of another acid like acetic acid, sulfuric acid, or phosphoric acid. This process allows to obtain cellulose with improved thermal and dissolution properties, and bioactivity. Cellulose esters (cellulose diacetate) have been employed like hemodialysis membranes applied in the purification of blood, for patients with renal failure. These membranes were produced by Altin (company) and these presented less toxicity than synthetic polymer-based membranes [19].

Natural polymer	Synthetic polymer or synthetic chemical	Employed strategies/commercial product	Pharmaceutical applications
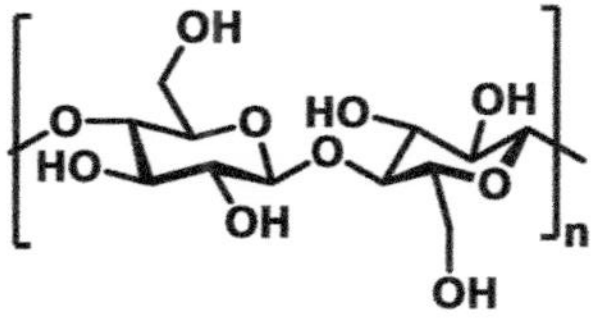 Cellulose	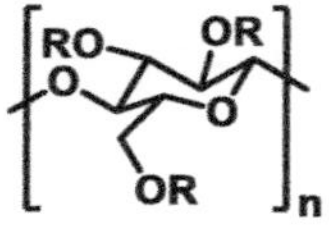 **R = H** or **CH_3** or **$CH_2CH(OH)CH_3$**	Chemical linkages/ hydroxypropyl methylcellulose (HPMC)	It is usually used as eye drops and excipient or to make capsule shells, as well as drug delivery matrix (film or gel)
	R = CH_2OCH_2 COOH Carboxymethyl cellulose	Crosslinking/ croscarmellose	These crosslinking promotes disintegrates and they can accelerate the dissolution rate of drugs in aqueous environment
	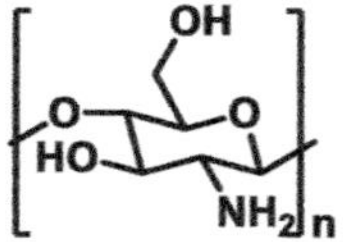 Chitosan	Blending	Wound dressing with antibacterial properties
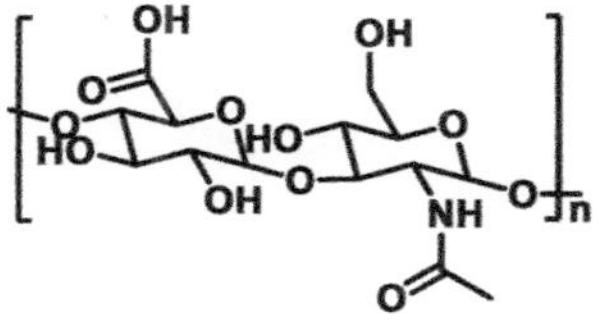 Hyaluronic acid	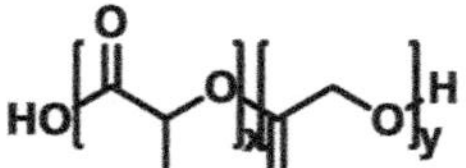 Poly(D,L-lactic acid-*co*-glycolic acid) (PLGA)	Grafting	These blending are being studied in tissue engineering applications as scaffolds. They were successful in inducing cartilage tissue formation
	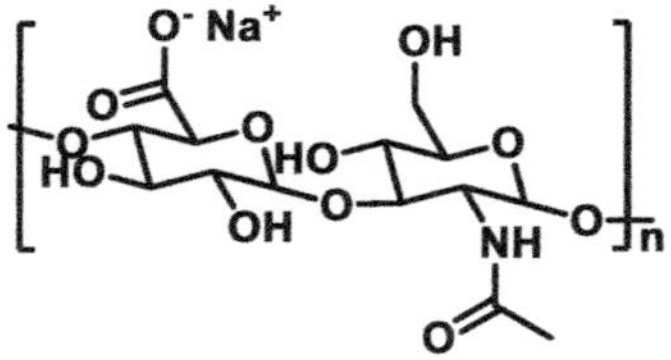	Crosslinking/ Synvisc®	Synovial fluid substitutes to reduce pain and enhance joint mobility in patients with osteoarthritis

Continued

Table 2 Some semi-synthetic polymers and their pharmaceutical applications—cont'd

Natural polymer	Synthetic polymer or synthetic chemical	Employed strategies/commercial product	Pharmaceutical applications
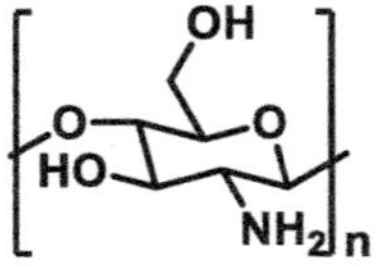 Chitosan	Hyaluronate 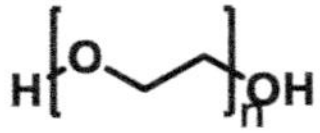	Crosslinking	These are being investigated for the development of tissue engineering scaffolds and protein delivery devices
	Poly(ethylene glycol) 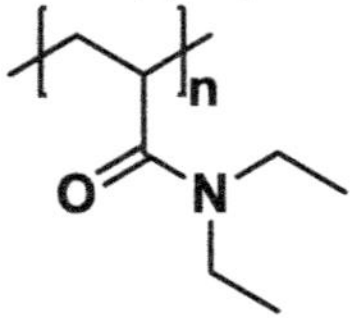	Crosslinking	Thermoresponsive interpenetrating polymer network films used as wound dressing
	Poly(*N*,*N*-diethyl-acrylamide) 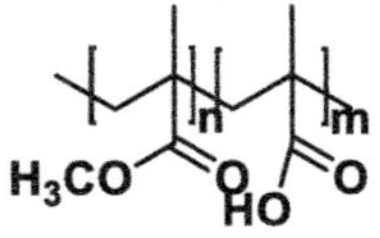	Blend	Chitosan micro-cores in which sodium diclofenac was entrapped and coated with Eudragit have a pH-dependent release profile
	Eudragit 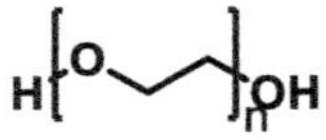	Grafting/SQZ®	Gel oral controlled release system

	Poly(ethylene glycol)		
Alginate	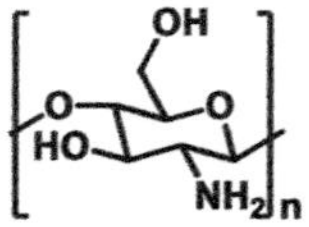 Chitosan	Blending	Excipients in tablets. The incorporation of chitosan into an alginate matrix mutually complements their modes of drug release control
	 Dodecylamine	Grafting	The formed hydrogel is expected to be more stable mechanically over long-term storage and exhibits sustained drug release property
Starch	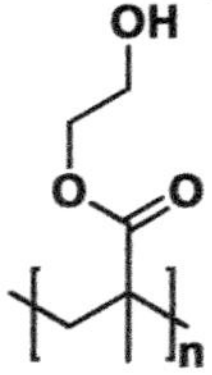 Chain-derived acrylic	Grafting	Tablets. Excipients led to enzymatic resistance and hydrophobic behavior
Collagen	Oxidized cellulose, oxidized regenerated cellulose (ORC)	Crosslinking/ Promogran®	A protease modulating matrix, for the treatment of diabetic foot ulcers
	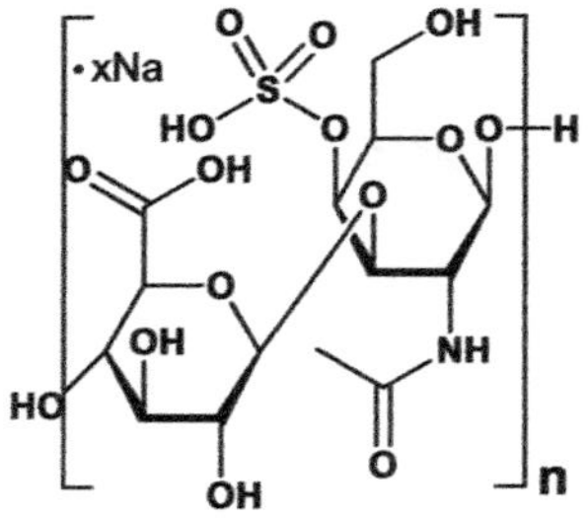 Chondroitin-6 sulfate	Crosslinking/Integra® Dermal Regeneration Template	Artificial skin substitute

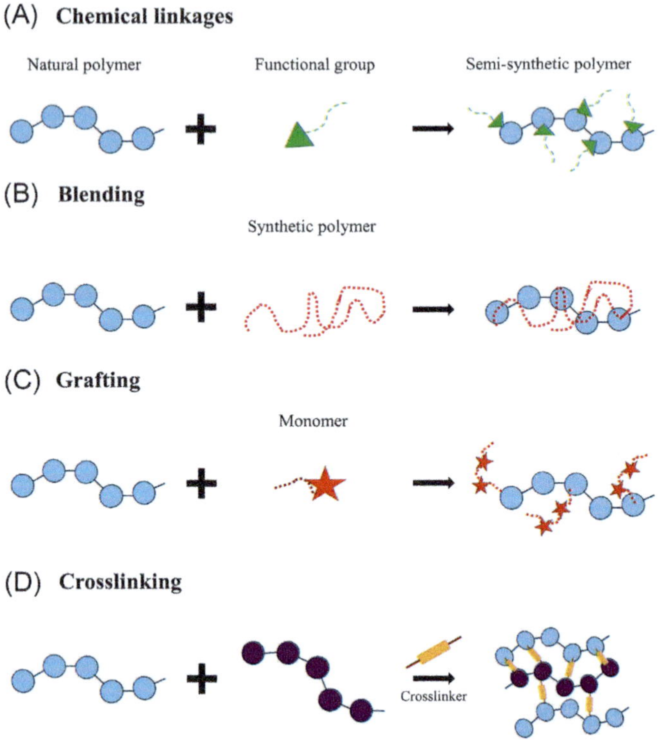

Fig. 2 Scheme of some of the employed strategies to obtain semi-synthetic polymers.

Blending

This methodology for modifying natural polymers consists in a mechanical mixture of two or more polymers, that results in a new material with improved characteristics that combine the unique features of each the individual components [31]. For example, blending methods are an alternative way of modifying hydrophilicity of the matrix without result in any major changes in the mechanical reliability [23].

In the process of polymer blending there is no break and/or formation of new bonds. Generally, it is achieved by hydrogen bonding interactions, in this case, the polymer has some functional groups that are capable of forming hydrogen bonds through dipole, pi-electrons, ionic, or charged transfer complexation [23, 32].

There are various kinds of polymer blends, but generally, they can be classified into two classes, miscible or immiscible blends, depending on the interaction between the polymers of the blend [24]. On the one hand, miscible polymer blends involve the simplest approach which is melt processing, and this can include processes such as extrusion, batch mixing and injection molding [31]. However, most polymers are thermodynamically immiscible with each other and exhibit significant phase separation due to low mixing entropy and high interfacial tension [33]. Therefore, parameters such as molecular weights, size distribution, thermal and stress-strain profiles must be controlled. as well as the control of the phase behavior and morphology becomes a key factor in determining the efficiency of polymer blend [24, 34]. On the other hand, in the case of immiscible polymer blends, generally is necessary to reduce the interfacial tension between the polymers, this is accomplished using a compatibilizer such as clays, which are additives that increase the interactional forces between their components [33].

Blending method is advantageous because it is cheaper and faster than other methods for the development of new materials. By using this method, it is possible to easily change some of the blend constituents [32]. For example, blends of collagen with other hydrophilic polymers such as polyethylene glycol or polyethylene oxide are of great interest for the development of wound-healing hydrogels or biodegradable polymeric scaffolds [35].

Grafting

Grafting is possibly the most used method to obtain semi-synthetic polymers. Modification grafting is done through covalent attachment, where one or more blocks of macromolecules are connected at different points along a polymeric backbone of another polymer that is used as the main polymeric chain [36]. In this way, the surface properties are modified while the bulk properties of the main polymer are preserved giving rise to a semi-synthetic polymer with controlled properties, such as solubility in desired solvents, water uptake, biocompatibility, chemical resistance, as well as, dye ability [20]. The main benefit of this modification strategy, beyond the ability to adjust only the surface properties of the polymer, is that it gives a stable product against desorption and establishes long-term chemical stability because of its covalent nature [24].

There are three approaches frequently followed for grafting: grafting through, grafting from, and grafting to:

(i) **Grafting to:** In this case, a preformed polymer with reactive end-group is linked with the functional groups distributed randomly along the main polymeric chain [19]. Through this approach, it is possible to get well-defined *star*-shaped polymeric structures or loosely grafted copolymers [21].

(ii) Grafting through: This methodology involves a vinyl macromonomer which is copolymerized with a low-molecular-weight comonomer. The growth of the polymeric chain is initiated in the solution with the assistance of a suitable initiator. In the propagation stage, these monomer units integrate into the backbone of a growing chain [37].

(iii) Grafting from: It is also called as surface-initiated polymerization, the growth of polymer chains occurs from initiating sites on the main polymeric backbone. The number of chains grafted to the polymeric backbone can be controlled by the number of active sites which makes it possible to get high graft density [19].

The driving force of polymer grafting is based on the generation of active sites, such as functional groups or free radicals on the polymeric backbone [23]. These active sites can be generated by employing different techniques. Those widely used include chemical, radiation, photochemical, plasma-induced, and enzymatic grafting [36]. For example, in chemical grafting, chemicals that act as initiators to produce active sites on the polymeric backbone are employed. The use of distinct redox initiators such as strong bases, Lewis acids, and metal carboxylic groups has been reported for chemical grafting [23]. Otherwise, in radiation grafting, the polymeric backbone is the initiator itself [38]. The irradiation of macromolecules can cause homolytic fission, and consequently, produces free radicals on the polymer which are used as active sites [21]. Although, in irradiation grafting, the initiator is not an essential requirement, the medium has a significant impact. For instance, peroxides may be formed on the backbone of polymers if irradiation is carried out in air [36]. Hence, both chemical initiators and the experimental conditions for grafting play a significant role in the success of grafting [23].

To date, "the grafting to" method is widely used to develop *PEGylation* process which consists in the addition of polyethylene glycol chains to other molecules or macromolecules [39]. Indeed, more than a dozen of pegylated protein therapeutics have been approved for clinical use [40]. This process produces an increase in the molecular weight of macromolecules and the *PEGylation* layer prevents proteolytic enzymes. Additionally, this process improves pharmacokinetics since it voids adsorption of serum proteins, and consequently, antibody activation [39].

Crosslinking

Crosslinking is another method to obtain semi-synthetic polymers. This method consists of interconnecting the polymer chains by crosslinkers, that is, molecules of low molecular weight with at least two reactive functional groups that allow the formation of bridges between polymeric chains [41]. This is achieved through ionic or covalent interactions to form three-dimensional networks [42]. Along these interactions, some secondary interactions like hydrogen bonds and hydrophobic interactions can also take place [43]. Sometimes, the crosslinking between polymeric chains leads to difficult degradation and reduces the availability of functional groups which causes a potential increase in cytotoxicity [42]. For this reason, the choice of crosslinker plays an important role and should be done according to the kind of polymer and the potential application of the product.

The crosslinking method can be mainly classified into two groups according to the nature of the interaction between polymer chains: chemical crosslinking and physical crosslinking. The first one is given by covalent bonds between the polymer chains either by the reaction of functional groups or by Schiff base formation. Likewise, crosslinking have been reported by reactions of condensation, addition, even crosslinking by high energy irradiation and using enzymes [44]. Otherwise, physical crosslinking can be achieved by modifying conditions like pH, ionic strength or temperature. This change produces new physicochemical interactions, like hydrogen bonding, charge interaction, hydrophobic interactions or stereocomplexation [45].

By the crosslinking method, it is possible to achieve different architectures such as films, porous and sponge-like structures or hydrogels [42]. Each of these structures have certain properties that make them particularly appropriate for diverse applications in the pharmaceutical industry. Singularly, hydrogels are hydrophilic polymeric networks that present swelling and can retain a significant amount of water into its structure [46]. They are currently the most used systems in a wide range of pharmaceutical applications such as drug delivery systems, dressings, medicated patches or applications in the medical field as scaffold for tissue engineering and implants [47]. For example, the development of polyvinyl pyrrolidone/alginate hydrogel loaded with nanosilver has been studied as wound dressings due to fluid handling capacity and its antimicrobial activity [48].

3 Applications of polymeric systems in pharmaceutical formulation and dosage forms

The great discoveries and challenges of pharmaceutical applications can be summarized according to the advance in the knowledge of those investigations carried out in this field, as an example we can named the controlled administration of drugs. This development begins in the 1960s starting with macroscopic devices (implants); after in the 1970s and 1980s, the microscopic systems (1980s) and then with the nanoscopic systems, currently in full development [49].

Doing a bit of history, the first formulation that used a synthetic polymer was known in 1994, which was designed to treatment of cancer and was clinically tested. Composed by an HPMA (*N*-(2-hydroxypropyl)methacrylamide) copolymer conjugated with doxorubicin. Later, in the decade of 2000, two new formulations became known: PEG-interferon-α (a drug with antiviral activity intended to treat chronic hepatitis C and hepatitis B) and PEG-GCSF (PEG granulocyte colony-stimulating factor). Both formulations were then introduced into the market.

The above-mentioned achievements were the first steps leading to the development of pharmaceutical formulations based on polymer interactions with different molecules, namely polymeric drugs, polymer-drug conjugates and polymer-protein conjugates, showed in Fig. 3 [49].

Today, we can cite numerous examples of research in new pharmaceutical formulations using polymers that have been successful and are already on the market or in clinical phase studies [1].

Poly(glutamic acid) (PGluA) is, for example, a biodegradable polymer that has been used in various pharmaceutical formulations giving promising results in clinical phases. One of them is the Opaxio™, which is a successful polymer-drug conjugate with a high load capacity. In this case, Taxol is used, which is already under study in Phase III and has demonstrated its efficiency against ovarian cancer. In 2008, European Medicines Agency (EMA)

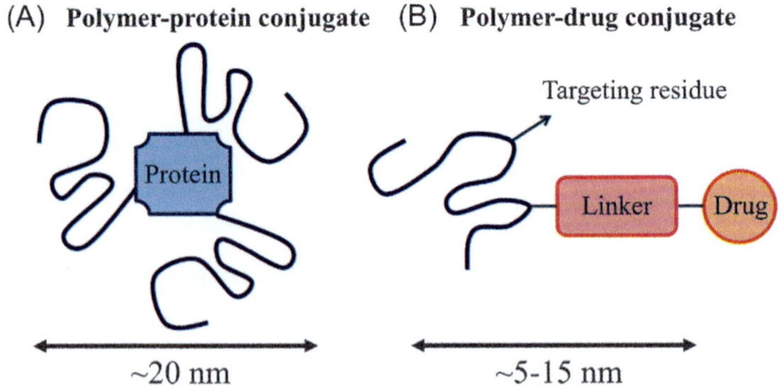

Fig. 3 Scheme of the therapeutic polymer conjugates.

announced the acceptation of Opaxio™ for first-line treatments of patients with non-small cells lung cancer. Therefore, it is probably that in the near future, it can be on the market.

Implants or insertable devices are being used for the treatment of eye diseases, and have already gone through three generations of study. The first cases consisted of insertion of eye devices that needed to be replaced or removed surgically after a certain time, such as Ocusert® or Vitrasert® [50]. An interesting advance has been the Iluvien® which is an implant that contains acetonide fluocinolone and is used for the treatment of diabetic macular edema; a rod-shaped device composed of polyimide tubes end-capped with PVA membrane. The characteristic size allows this device to remain in the cavity, even once the cargo molecule has been released completely.

Other interesting advances in clinical applications is the vascular stent. A medical device with a tubular shape employed to unblock vascular sites that affect circulatory or digestive systems. The first ones were made of metals, such as steel and alloys because the inherent capability of these metals to resist corrosion and exceptional mechanical properties. But this type of stent had to be replaced by restenosis problems. The appearance of drug-liberating stents (DES), that release antiproliferative drugs in the implanted site, was an efficient form to overcome this problem. However, released drugs themselves and polymeric coating layer may cause subsequent thrombosis and yet they remain a problem [51]. In this sense, the work realized by Chin-Quee et al. showed that the fluorinated copolymer PVDF-HFP and the phosphorylcholine polymer, both commercially available drug-eluting stents, had similar effects, giving both acute thrombotic responses [52].

4 Advances and future in pharmaceutical technology

Undoubtedly, the excellent properties of polymers in non-conventional administration forms, such as controlled-release drug delivery smart systems and specific therapeutic systems have changed the vision of the pharmaceutical industry and the interest of scientists towards this area of research. The applications mentioned above entailed new interesting perspectives for medicine, since specially designed polymers present the capability of supplying medical substances to affected tissues and cells, along with the dosage of these drugs according to a specifically controlled pharmacodynamics.

4.1 Smart polymers and nanosystems used in pharmaceutical applications

In the past few years, a wide range of smart polymers have been developed and synthesized to be employed in pharmaceutical applications, they have been called smart drug delivery systems (DDS). The word "smart" means the inherent response of the systems to small changes in the environment, or to an external stimulus, which produces changes in their properties, structure or interactions, leading to rearrangements or changes in the aggregation state [53]. The importance of DDS relies on the non-linear response triggered by small external stimuli that may produce a macroscopic, appreciable and reversible change.

These systems do not require complex sensor assemblies or sampling, or calculations to provide feedback and respond to certain stimulus [54] as the intelligent "devices" do. In that way: smart materials responsiveness is inherent to their nature.

Pharmaceutical employs polymers which respond to different stimuli: light, temperature, chemicals, biomolecules and magnetic fields [55–57]. To achieve this response, at least one monomer/component/segment with a natural response to some of this stimulus must be present in the chemical formulation/system. For example, photo-responsive polymeric systems are often supplied with photo-active groups such as, azo-groups, benzopyrans, triphenylmethane, or cinnamonyl, which can undergo a reversible structural transition when irradiated with light of specific wavelength. Smart response proceeds from groups that lead to changes in the size/shape of the polymer or the formation of ionic or zwitterionic species when irradiated [58]. Table 3 lists various stimuli and smart polymers that are most used to obtain specific responses.

In temperature sensitive polymers, the response is given by the equilibrium of the intermolecular interactions between polymer-polymer chains and polymer-solvent. These interactions, since they are reversible, can be shifted/displaced by changes in temperature. The most used polymers with this type of response are: PNIPAAm, poly-propylene oxide (PPO), PEO, PLA, proteins and polysaccharides [60,61].

pH-responsive polymers are those with acid or basic functional groups that present a reversible equilibrium/balance in aqueous solution (carboxylic, phosphoric or amine). Environmental pH changes lead to a shift in the acid/basic balance, changing the ionization degree of the species present in the polymer chains [62]. Signal reception can be originated in a physical or chemical interaction between functional groups in the polymer and signaling molecules [63].

Table 3 Various stimuli and responsive materials [59].

External stimulus	Responsive polymer
Temperature	Poloxamers
	Poly(*N*-alkylacrylamide)s
	Poly(*N*-vinylcaprolactam)s
	Cellulose
	Xyloglucan
	Chitosan
Light	Modified poly(acrylamide)s
pH	Poly(methacrylic acid)s
	Poly(vinylpyridine)s
	Poly(vinylimidazole)s
Electric field	Sulfonated polystyrenes
	Poly(thiophene)s
	Poly(ethyloxazoline)
Ultrasound	Ethylenevinylacetate

Responsive polymers interactions with signaling molecules are based upon a selective molecular recognition phenomenon which is originated in the conjugation of these responsive systems with biological molecules such as proteins, DNA, enzymes and antibodies.

These materials can be designed to achieve responsiveness to more than one stimulus at the same time, making multi-responsive devices possible [64, 65]. Wei Xiong et al. used poly (*N*-Isopropylacrylamide-*co*-acrylic acid) to synthetized nanogels with dual stimuli-responsiveness which responded to both temperature and pH [66].

The responsive mechanism of the polymer to the stimuli occurs, in general, by two mechanisms: (1) responsive polymers with sensitive groups to an external stimulus which were named above, and (2) responsive polymers which present structures susceptible to be destroyed (or formed) by an external stimulus [1].

The discovery of new diseases and the various pathophysiological conditions of other diseases which are already known but not yet cured, leads to a continuous demand for new and more efficient pharmaceutical products. Therefore, new therapeutic approaches and new designs are required, not only for drugs but also for new dosage administration forms and effective formulations. In recent years, while conventional treatment strategies have developed and improved their performance, nano-sized structures, from 50 to 500 nm [67, 68], where optimal size range is still a subject of continuous discussion and updating, have already shown that their potential properties for the pharmaceutical industry open new horizons and hopes.

This new "nano" dimension of the therapeutic systems makes possible to improve clinical trials achieving rapid and complete solubilization of pharmaceutical products for safe inoculation. These clinical trials (in dose and frequency terms) are still being optimized for a wide variety of products and therapies.

Some of the above mentioned nanosystems are represented in Fig. 4.

Nanosystems acquired a big importance in pharmaceutical applications because of their size: after intravenous administration, renal clearance cannot remove these systems because they are oversized, so they often present extended circulation time, which allows to be used as targeted delivery systems of therapeutic cargo [69]. Another important quality is that these nanosystems can be modified to become even more biocompatible and extend the circulation time in the body in order to safely reach the tissues which present a specific disease. This is achieved by, for instance, *PEGylation* of the surface of the nanocarriers, increasing half-life time and, rising consequently passive targeting via Enhanced Permeation and Retention effect (EPR), or by attaching ligands or antibodies that have specific molecular recognition to the nanocarriers surface (active targeted drug delivery).

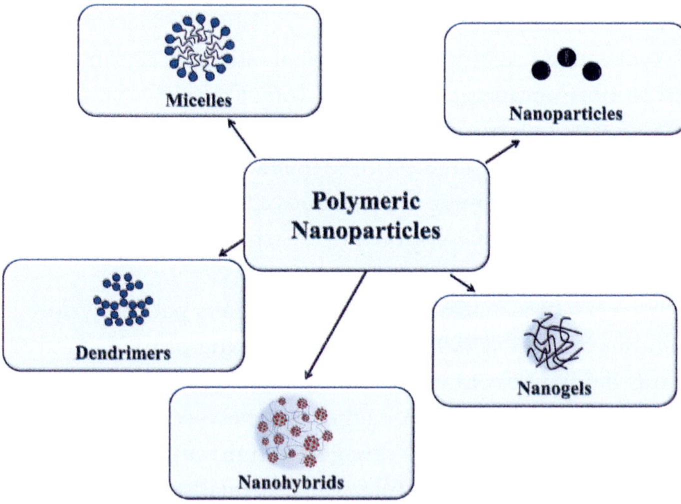

Fig. 4 Scheme of some of the most important polymeric nanosized structures for pharmaceutical applications.

The most significant advantages of nanosystems are:

(1) Size control: nanocarriers can be synthetized with strict control of the size, the surface can be chemically controlled to achieve a desire function [70].

(2) Easy synthesis: their synthesis can be straightforwardly handled by conventional techniques with slight modifications [71].

(3) High stability of encapsulated particles: nanocarriers can be formulated to retain drug molecules loaded and ensure its prolonged stability, providing a maximum therapeutic effect. Crosslinking degree can be easily regulated to control drug loading and release behavior.

(4) Maintained and controlled release: the release of the cargo molecule should occur at the targeted site, increasing therapeutic efficiency and reducing side effects.

(5) Response to stimuli: nanocarriers which present stimuli-responsiveness give the advantage of achieving a controlled release in the site of interest.

(6) Low toxicity: nanosystems should be highly biocompatible with non-toxic degradation products. Non-toxic final formulations can be obtained after synthesis or post-synthesis steps. i.e., *PEGylations* are frequently used to enhanced biocompatibility and circulation time.

(7) Multi-responsive systems: are those who can achieve, by using a combination of components, the response to two or more stimuli. Therefore, the response in the targeted site gets enhanced by achieving more specific properties.

Besides these advantages, the major and most remarkable advantages of nanocarriers that are being used in all fields of application are their nanometric size range and their surface-to-volume relation, which enhance the adsorption and/or incorporation of payloads [72]. Singh et al. developed a novel thermo-responsive nanoparticle from poly-NIPAM, to successfully load and deliver anastrozole, an aromatase enzyme inhibitor which interrupts the estrogen synthesis, and is used to treat metastasis and pre and post-menopausal breast cancer in women [73].

Sensitive nanocarriers present an outstanding quality, transitions or changes in physicochemical properties may occur, even when changes in the environment are slight. These properties, as well as the structure, can suffer significant changes triggered by external stimulus, as they can collapse into a compact form, expand in seconds or expel the absorbed solvent.

Maybe the most interesting nanostructures for medical applications are those which exhibit properties that allow diagnosis and therapy at the same time, systems called theranostics. Nanocarriers with both abilities, imaging and therapeutic properties, become quite advantageous for real-time monitoring of therapeutic efficiency. Previous research by Hsin-Cheng Chiu et al. shows an interesting example of a nanosystem with dual responsiveness [74]. They synthetized hollow hybrid nanogels with magnetic transport guidance and pH responsiveness with high drug loading efficacy. DOX-loaded hollow hybrid nanogels have shown accelerated drug release in response to pH and temperature rise as a result of a significant breakdown between the interactions that take place between the copolymer and the drug molecules.

These nanometric systems can potentially become unique candidates for therapeutic and diagnostic approaches. Therefore, synthetic procedures and the design of highly specific polymer nanomaterials have gained significant academic and industrial attention around the world. Due to this push, rapid advances are being achieved in fields like nanotechnology applied to medicine and in pharmaceutical formulations that improve the purposes of diagnosis and the possibility of specific therapeutic treatments [75].

5 Conclusions

This chapter provides a general description of the most common synthetic and semi-synthetic polymers, ranging from macro to nanoscopic size and describes some which have been used in the preparation of different pharmaceutical forms of administration. The systems described here have contributed to the improvement of different formulations for the therapy of a wide variety of human pathologies and, without a doubt, they will continue to do so in the future.

Polymers have demonstrated to be tremendously valuable for obtaining new and efficient pharmaceutical systems, a fact that is a consequence of the huge advances in chemical synthesis and in the analysis of their properties.

Different formulations used nowadays have appeared in the market after the development of patents and prototype delivery systems, but looking deeply in the history of this products we can see that the vast majority of them were the result of great basic research carried out years before. Therefore, taking into account this sequence of events and the huge number of published scientific research that may have an impact on the modern design of dosage forms, we believe that the future of the development of new pharmaceutical forms will be really successful. However, while some improvements that are quite evident, there are many developments that do not pass the clinical trial process.

Currently, the physicochemical properties of polymers, controlled release profiles and targeted behavior of nanocarriers can be modified through synthesis strategies giving a huge number of alternatives for the formulation of therapeutic agents.

With the examples of the application of different systems explained in this chapter, we have tried to demonstrate the influence of the structure/property relationships of polymers and therefore, we hope that this chapter will help scientists from different fields of research to improve and broaden their knowledge.

Acknowledgments

The authors acknowledge the support by FONCYT (PICT-2015-2477), CONICET (11220150100344CO) and SECYT-UNC (30720150100698CB, 30920150100233CB and 30720150100225CB). Dariana Aristizabal Bedoya and Francisco Figueroa thank CONICET for the fellowship awarded.

References

[1] C. Englert, J.C. Brendel, T.C. Majdanski, T. Yildirim, S. Schubert, M. Gottschaldt, N. Windhab, U.S. Schubert, Pharmapolymers in the 21st century: synthetic polymers in drug delivery applications, Prog. Polym. Sci. 87 (2018) 107–164, https://doi.org/10.1016/j.progpolymsci.2018.07.005.

[2] D. Jones, Pharmaceutical Applications of Polymers for Drug Delivery, Ismithers Rapra Publishing, 2004.

[3] E. Oledzka, M. Sobczak, Polymers in the pharmaceutical applications—natural and bioactive initiators and catalysts in the synthesis of biodegradable and bioresorbable polyesters and polycarbonates, in: E.C. Agbo (Ed.), Innovations in Biotechnology, Intech, Rijeka, Croatia, 2012, pp. 139–160.

[4] W.J. Passl, Synthetic polymers in modern pharmacy, Prog. Polym. Sci. 14 (1989) 629–677.

[5] F.W. Billmeyer, Textbook of Polymer Science, Wiley India Pvt. Limited, Michigan University, 1971.

[6] A. Rudin, P. Choi, Free radical polymerization, in: The Elements of Polymer Science & Engineering, third ed., Elsevier, 2013, pp. 341–389.

[7] G. Odian, Principles of Polymerization, fourth ed., Wiley & Sons Ltd., 2004.

[8] S. Penczek, M. Cypryk, A. Duda, P. Kubisa, S. Słomkowski, Living ring-opening polymerizations of heterocyclic monomers, Prog. Polym. Sci. 32 (2) (2007) 247–282.

[9] F. Zaera, New challenges in heterogeneous catalysis for the 21st century, Catal. Lett. 142 (2012) 501–516.

[10] W.-F. Su, Ring-opening polymerization, in: Principles of Polymer Design and Synthesis, Springer, Berlin, Heidelberg, 2013, pp. 267–299.

[11] F. Sanda, T. Endo, Radical ring-opening polymerization, J. Polym. Sci. A Polym. Chem. 39 (2001) 265–276.

[12] J.A. Song, B. Park, S. Kim, C. Kang, D. Lee, M.H. Baik, R.H. Grubbs, T.L. Choi, Living polymerization caught in the act: direct observation of an arrested intermediate in metathesis polymerization, J. Am. Chem. Soc. 141 (25) (2019) 10039–10047.
[13] S. Koltzenburg, M. Maskos, O. Nuyken, Polymer Chemistry, Springer-Verlag, Berlin, Heidelberg, 2017.
[14] R. Hoogenboom, Polyethers and polyoxazolines, in: Handbook of Ring-Opening Polymerization, Wiley-VCH Verlag GmbH & Co, 2009, pp. 141–164.
[15] A.L. Brocas, C. Mantzaridis, D. Tunc, S. Carlotti, Polyether synthesis: from activated or metal-free anionic ring-opening polymerization of epoxides to functionalization, Prog. Polym. Sci. 38 (6) (2013) 845–873.
[16] A. Jenkins, P. Kratochvil, R. Stepto, U. Suter, Glossary of basic terms in polymer (IUPAC recommendations), Pure Appl. Chem. 68 (12) (1996) 2287–2311.
[17] O.W. Webster, Living polymerization methods, Science 251 (4996) (1991) 887–893.
[18] P.K. Deb, S.F. Kokaz, S.N. Abed, A. Paradkar, R.K. Tekade, Pharmaceutical and biomedical applications of polymers, Basic Fundamentals of Drug Delivery, Elsevier Inc., 2019, pp. 203–267 (Chapter 6).
[19] A. Aravamudhan, D.M. Ramos, A.A. Nada, S.G. Kumbar, Natural polymers: polysaccharides and their derivatives for biomedical applications, Natural and Synthetic Biomedical Polymers, Elsevier Inc., 2014, pp. 67–89 (Chapter 4).
[20] N.B. Shelke, R. James, C.T. Laurencin, S.G. Kumbar, Polysaccharide biomaterials for drug delivery and regenerative engineering, Polym. Adv. Technol. 25 (2014) 448–460.
[21] S.R. Soni, A. Ghosh, Grafting onto biopolymers: application in targeted drug delivery, in: Biopolymer Grafting: Applications, Elsevier Inc., 2018, pp. 335–389 (Chapter 9).
[22] I. Mkedder, C. Travelet, A. Durand-Terrasson, S. Halila, F. Dubreuil, R. Borsali, Preparation and enzymatic hydrolysis of nanoparticles made from single xyloglucan polysaccharide chain, Carbohydr. Polym. 94 (2) (2013) 934–939.
[23] M.N. Sithole, Y.E. Choonara, L.C. Toit, P. Kumar, A review of semi-synthetic biopolymer complexes: modified polysaccharide nano-carriers for enhancement of oral drug bioavailability nano-carriers for enhancement of oral drug bioavailability, Pharm. Dev. Technol. 22 (2) (2017) 283–295, https://doi.org/10.1080/10837450.2016.1212882.
[24] V. Pillay, A. Seedat, Y.E. Choonara, L.C. du Toit, P. Kumar, V.M Ndesendo, Review article a review of polymeric refabrication techniques to modify polymer properties for biomedical and drug delivery applications, AAPS PharmSciTech 14 (2) (2013) 692–711.
[25] M.H. Othman, Polymer blends and composites from renewable resources, in: Encyclopedia of Renewable and Sustainable Materials, Elsevier Inc., 2019, pp. 179–186 (Chapter 11555).
[26] J. Shokri, K. Adibkia, Application of cellulose and cellulose derivatives in pharmaceutical industries, Cellulose-Medical, Pharmaceutical and Electronic Applications, Intech, 2013.
[27] A.K. Nayak, M.S. Hasnain, K. Pal, I. Banerjee, D. Pal, Gum-based hydrogels in drug delivery, in: Biopolymer-Based Formulations, Elsevier Inc., 2020, pp. 605–645 (Chapter 25).
[28] A.K. Nayak, H. Bera, In situ polysaccharide-based gels for topical drug delivery applications, in: Polysaccharide Carriers for Drug Delivery, Elsevier Ltd., 2019, pp. 615–638 (Chapter 21).
[29] A.K. Nayak, M.S. Hasnain, Plant polysaccharides in drug delivery applications, in: Plant Polysaccharides-Based Multiple-Unit Systems for Oral Drug Delivery, SpringerBriefs in Applied Sciences and Technology, 2019, pp. 19–23., https://doi.org/10.1007/978-981-10-6784-6_2.

[30] C.I. Alvarez Igarzabal, M. Martinelli, V. Brunetti, M.C. Strumia, Adaptation of biopolymers to specific applications, in: V.K. Thaku, A.S. Singha (Eds.), Surface Modification of Biopolymers, John Wiley & Sons, Inc., 2015, pp. 84–112.

[31] Z. Fu, H. Wang, X. Zhao, S. Horiuchi, Y. Li, Immiscible polymer blends compatibilized with reactive hybrid nanoparticles: morphologies and properties, Polymer 132 (2017) 353–361.

[32] Y. He, B. Zhu, Y. Inoue, Hydrogen bonds in polymer blends, Prog. Polym. Sci. 29 (10) (2004) 1021–1051.

[33] A. Taguet, P. Cassagnau, J.-M. Lopez-Cuesta, Structuration, selective dispersion and compatibilizing effect of (nano)fillers in polymer blends, Prog. Polym. Sci. 39 (8) (2014) 1526–1563.

[34] M. Si, T. Araki, H. Ade, A.L.D. Kilcoyne, R. Fisher, J.C. Sokolov, M.H. Rafailovich, Compatibilizing bulk polymer blends by using organoclays, Macromolecules 39 (14) (2006) 4793–4801.

[35] A. Sionkowska, Current research on the blends of natural and synthetic polymers as new biomaterials: review, Prog. Polym. Sci. 36 (9) (2011) 1254–1276.

[36] A. Bhattacharya, B.N. Misra, Grafting: a versatile means to modify polymers techniques, factors and applications, Prog. Polym. Sci. 29 (8) (2004) 767–814.

[37] M. Henze, D. Mädge, O. Prucker, J. Rühe, "Grafting through": mechanistic aspects of radical polymerization reactions with surface-attached monomers, Macromolecules 47 (9) (2014) 2929–2937.

[38] V. Costamagna, D. Wunderlin, M. Larrañaga, I. Mondragon, M. Strumia, Surface functionalization of polyolefin films via the ultraviolet-induced photografting of acrylic acid: topographical characterization and ability for binding antifungal agents, J. Appl. Polym. Sci. 102 (2006) 2254–2263.

[39] J.M. Harris, R.B. Chess, Effect of pegylation on pharmaceuticals, Nat. Rev. Drug. Discov. 2 (2003) 214–221.

[40] X. Liu, J. Sun, W. Gao, Site-selective protein modification with polymers for advanced biomedical applications, Biomaterials 178 (2018) 413–434.

[41] J. Berger, M. Reist, J.M. Mayer, O. Felt, N.A. Peppas, R. Gurny, Structure and interactions in covalently and ionically crosslinked chitosan hydrogels for biomedical applications, Eur. J. Pharm. Biopharm. 57 (1) (2004) 19–34, https://doi.org/10.1016/S0939-6411(03)00161-9.

[42] N. Reddy, R. Reddy, Q. Jiang, Crosslinking biopolymers for biomedical applications, Trends Biotechnol. 33 (6) (2015) 362–369, https://doi.org/10.1016/j.tibtech.2015.03.008.

[43] A. Muthumeenal, S.S. Pethaiah, A. Nagendran, Biopolymer composites in fuel cells, Biopolymer Composites in Electronics, Elsevier Inc., 2017, pp. 185–217 (Chapter 6).

[44] F. Ullah, M.B.H. Othman, F. Javed, Z. Ahmad, H.M. Akil, Classification, processing and application of hydrogels: a review, Mater. Sci. Eng. C 57 (2015) 414–433.

[45] S. Khunmanee, Y. Jeong, H. Park, Crosslinking method of hyaluronic-based hydrogel for biomedical applications, J. Tissue Eng. 8 (2017) 1–16.

[46] E. Ahmed, M. Hydrogel, Preparation, characterization, and applications: a review, J. Adv. Res. 6 (2) (2015) 105–121.

[47] R. Parhi, Cross-linked hydrogel for pharmaceutical applications: a review, Adv. Pharm. Bull. 7 (4) (2017) 515–530.

[48] R. Singh, D. Singh, Radiation synthesis of PVP/alginate hydrogel containing nanosilver as wound dressing, J. Mater. Sci. Mater. Med. 23 (2012) 2649–2658.

[49] K.J. Gandhi, S.V. Deshmane, K.R. Biyani, Polymers in pharmaceutical drug delivery system: a review, Int. J. Pharm. Sci. Rev. Res. 14 (2) (2012) 57–66.

[50] J. Wang, A. Jiang, M. Joshi, J. Christoforidis, Drug delivery implants in the treatment of vitreous inflammation, Mediat. Inflamm. 2013 (2013) 1–8.

[51] C.M. Han, K.S. Park, Y.K. Joung, Recent alternative approaches of vascular drug-eluting stents, J. Pharm. Investig. 48 (2018) 153–165.
[52] S.L. Chin-Quee, S.H. Hsu, K.L. Nguyen-Ehrenreich, K.T. Tai, G.M. Abraham, S.D. Pacetti, Y.F. Chan, G. Nakazawa, F.D. Kolodgie, R. Virmani, N.N. Ding, L.A. Coleman, Endothelial cell recovery, acute thrombogenicity, and monocyte adhesion and activation on fluorinated copolymer and phosphorylcholine polymer stent coatings, Biomaterials 31 (4) (2010) 648–657.
[53] M. Motornov, Y. Roiter, I. Tokarev, S. Minko, Stimuli-responsive nanoparticles, nanogels and capsules for integrated multifunctional intelligent systems, Prog. Polym. Sci. 35 (1–2) (2010) 174–211.
[54] J. Hu, Y. Zhu, H. Huang, J. Lu, Recent advances in shape-memory polymers: structure, mechanism, functionality, modeling and applications, Prog. Polym. Sci. 37 (12) (2012) 1720–1763.
[55] C. De las Heras Alarcón, S. Pennadam, C. Alexander, Stimuli-responsive polymers for biomedical applications, Chem. Soc. Rev. 34 (3) (2005) 276–285.
[56] O. Onaca, R. Enea, D.W. Hughes, W. Meier, Stimuli-responsive polymersomes as nanocarriers for drug and gene delivery, Macromol. Biosci. 9 (2) (2009) 129–139.
[57] P. Bawa, V. Pillay, Y.E. Choonara, L.C. Du Toit, Stimuli-responsive polymers and their applications in drug delivery, Biomed. Mater. 4 (2) (2009) 022001–022016.
[58] J.G. Hardy, E. Larrañeta, R.F. Donnelly, N. McGoldrick, K. Migalska, M.T.C. McCrudden, N.J. Irwin, L. Donnelly, C.P. McCoy, Hydrogel-forming microneedle arrays made from light-responsive materials for on-demand transdermal drug delivery, Mol. Pharm. 13 (3) (2016) 907–914.
[59] H. Priya James, R. John, A. Alex, K.R. Anoop, Smart polymers for the controlled delivery of drugs—a concise overview, Acta Pharm. Sin. B 4 (2) (2014) 120–127.
[60] D. Schmaljohann, D. Beyerlein, M. Nitschke, G. Werner, Thermo-reversible swelling of thin hydrogel films immobilized by low-pressure plasma, Langmuir 20 (23) (2004) 10107–10114.
[61] D. Schmaljohann, Thermo- and pH-responsive polymers in drug delivery, Adv. Drug Deliv. Rev. 58 (2006) 1655–1670.
[62] J.C. Cuggino, M. Molina, S. Wedepohl, C.I. Alvarez Igarzabal, M. Calderón, L.M. Gugliotta, Responsive nanogels for application as smart carriers in endocytic pH-triggered drug delivery systems, Eur. Polym. J. 78 (2016) 14–24.
[63] P.D. Thornton, R.J. Mart, S.J. Webb, R.V. Ulijn, Enzyme-responsive hydrogel particles for the controlled release of proteins: designing peptide actuators to match payload, Soft Matter 4 (2008) 821–827.
[64] X. Pan, J. Chen, M. Yang, J. Wu, G. He, Y. Yin, M. He, W. Xu, P. Xu, W. Cai, F. Zhang, Enzyme/pH dual-responsive polymer prodrug nanoparticles based on 10-hydroxycamptothecin-carboxymethylchitosan for enhanced drug stability and anticancer efficacy, Eur. Polym. J. 117 (2019) 372–381.
[65] S. Chen, Q. Bian, P. Wang, H. Zheng, L. Lv, Z. Dang, G. Wang, Photo, pH and redox multi-responsive nanogels for drug delivery and fluorescence cell imaging, Polym. Chem. 8 (2017) 6150–6157.
[66] W. Xiong, W. Wang, Y. Wang, Y. Zhao, H. Chen, H. Xu, H. Yang, Dual temperature/pH-sensitive drug delivery of poly(N-isopropylacrylamide-co-acrylic acid) nanogels conjugated with doxorubicin for potential application in tumor hyperthermia therapy, Colloids Surf. B Biointerfaces 84 (2) (2011) 447–453.
[67] Y.H. Bae, K. Park, Targeted drug delivery to tumors: myths, reality and possibility, J. Control. Release 153 (3) (2011) 198–205.
[68] J. Oshiro Junior, M.P. Abuçafy, E.B. Manaia, B.L. da Silva, B.G. Chiari-Andréo, L.A. Chiavacci, Drug delivery systems obtained from silica based organic-inorganic hybrids, Polymers (Basel) 8 (4) (2016) 91–105.

[69] D.M. Eckmann, R.J. Composto, A. Tsourkas, V.N.R. Muzykantov, Nanogel carrier design for targeted drug delivery, J. Mater. Chem. B 2 (46) (2014) 8085–8097.
[70] C. Gonçalves, P. Pereira, M. Gama, Self-assembled hydrogel nanoparticles for drug delivery applications, Materials (Basel) 3 (2) (2010) 1420–1460.
[71] J. Li, S. Yu, P. Yao, M. Jiang, Lysozyme—dextran core—shell nanogels prepared via a green process, Langmuir 24 (7) (2008) 3486–3492.
[72] I.N. Kurniasih, J. Keilitz, R. Haag, Dendritic nanocarriers based on hyperbranched polymers, Chem. Soc. Rev. 44 (2015) 4145–4164.
[73] N. Singh, L.A. Lyon, Au nanoparticle templated synthesis of pNIPAm nanogels, Chem. Mater. 19 (4) (2007) 719–726.
[74] W.H. Chiang, V.T. Ho, H.H. Chen, W.C. Huang, Y.F. Huang, S.C. Lin, C.S. Chern, H.C. Chiu, Superparamagnetic hollow hybrid nanogels as a potential guidable vehicle system of stimuli-mediated MR imaging and multiple cancer therapeutics, Langmuir 29 (21) (2013) 6434–6443.
[75] Y. Zheng, S. Li, Z. Weng, C. Gao, Hyperbranched polymers: advances from synthesis to applications, Chem. Soc. Rev. 44 (2015) 4091–4130.

CHAPTER 3

Grafted polysaccharides as advanced pharmaceutical excipients

Hongbo Cheng, Xin Zhang, Zhixiang Cui, and Shirui Mao
School of Pharmacy, Shenyang Pharmaceutical University, Shenyang, People's Republic of China

1 Introduction

Polysaccharides are complex carbohydrates consist of repeating units (either monosaccharides or disaccharides) linked via glycosidic bonds. They are condensation polymers with general formula Cx(H_2O)y, and the structure may be formed by linear or branched stretching. In nature, polysaccharides come from a variety of sources, including algal (such as alginate), plants (such as pectin, guar gum), microbials (such as dextran, xanthan gum), and animals (such as chitosan, chondroitin). And the type of polysaccharides can be classified based on different properties. According to the charge properties, polysaccharides can be categorized into non-charged polysaccharides (e.g. pullulan, dextran), positively charged polysaccharides (e.g. chitosan) and negatively charged polysaccharides (e.g. alginate, hyaluronic acid) [1, 2]. Chemically, polysaccharides can be classified into homopolysaccharides, which produce the same monosaccharide units after hydrolysis (e.g. cellulose, glycogen), and heteropolysaccharides, which have two or more types of monosacchride units after hydrolysis (e.g. pectin, hyaluronic acid). According to their function, polysaccharides can also be sorted as digestible polysaccharides (starch, glycogen, etc.) and indigestible polysaccharides (cellulose, chitin, etc.) [3]. Owing to their biocompatibility, biodegradability, renewability and advanced biological activity such as antimicrobial, antiviral, anticancer activities, natural polysaccharides have gained immense interest of research for pharmaceutical applications [2, 4–10].

Nowadays, the growing demand for development of diverse materials as required and the fast development of nanomedicine have generated requirements to modify the structure of these polysaccharides so as to endow them new properties and therefore broaden their applications [11–13]. Thus, the aims of this chapter are to review the chemical modification of some

Advances and Challenges in Pharmaceutical Technology
https://doi.org/10.1016/B978-0-12-820043-8.00010-4

representative polysaccharides, such as chitosan, alginate, hyaluronic acid, carrageenan, pectin, pullulan, and present their applications in pharmaceutical area, especially as drug carriers.

2 Chitosan

2.1 Characteristics of chitosan

Chitosan (CS), a natural polymer obtained from chitin, is the only cationic polysaccharide in nature [14]. It is a nontoxic, biodegradable, and biocompatible polysaccharide consisting of D-glucosamine and *N*-acetyl-D-glucosamine (Fig. 1). Chitosan can be characterized by degree of deacetylation (DD) and molecular weight (MW), which are two main parameters affecting the physiochemical properties and bioactivity of chitosan [15]. The molecular weight of commercially available chitosan ranges from 40 to 600 kD, which is soluble in aqueous acids and dimethylsulfoxide [16]. Due to its abundant primary amino groups in the backbone, chitosan has various bioactivities. For this reason, chitosan is extensively used in biomedical field as the carrier for drug and/or gene delivery [17, 18].

2.2 Grafting modification of chitosan

In order to obtain better physicochemical and biological properties, the modifications of chitosan have been widely explored. The primary amino groups and hydroxyl groups in D-glucosamine unit of chitosan are two major functional ones (Fig. 1). As the primary amino groups have stronger reactivity than that of the hydroxyl groups, they become the main sites for chitosan modification. And the hydroxyl groups of chitosan can also be modified when the primary amino groups are protected [19]. Chitosan can be modified to increase either its hydrophilicity or hydrophobicity based on the specific requirement (Fig. 1).

In order to increase chitosan' solubility in neutral and alkaline environment and enlarge its application for drug delivery, the hydrophilic modification, such as quaternization, thiolation, PEGylation, and succinylation, are widely developed (Table 1).

Hydrophobic modification can endow chitosan many additional functions, which can be achieved by inducing hydrophobic moieties, such as saturated and unsaturated acyl groups, into the chitosan chain (Table 1). And this modification endows chitosan amphiphilic property, which makes

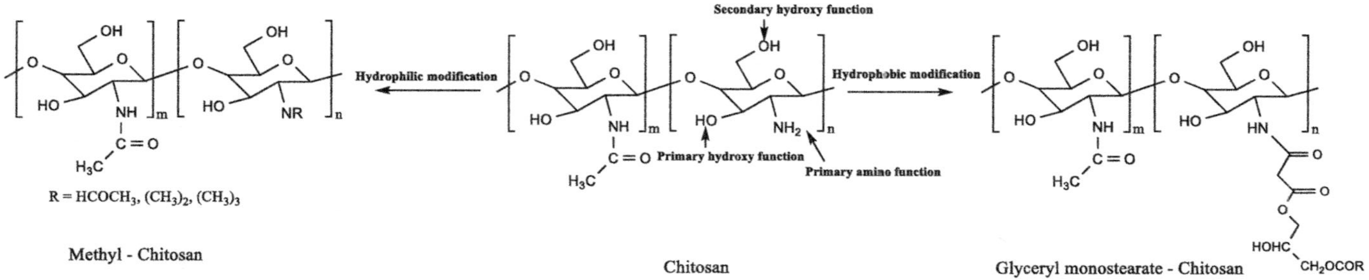

Fig. 1 Chemical structure of chitosan and its derivatives.

Table 1 Various strategies to modify chitosan.

Modification	Materials	Characteristics	References
Hydrophilic modification	Methyl/ethyl groups	Increase solubility, enhance mucosa adhesiveness and permeation ability	[20, 21]
	PEG	Improve stability, prolong therapeutic effects	[22, 23]
	Thiol groups	Improve mucoadhesion and permeation capability, inhibit the intestinal P-gp	[24–26]
	Succinyl groups	Excellent absorption enhancer, pH sensitivity and mucoadhesion	[27–29]
Hydrophobic modification	Saturated (C_8-C_{18}) and unsaturated acyl groups	Modified solubility, amphiphilic property, self-assembly behavior	[30–32]
	Sterol group	Increase the cellular absorption of drug	[33]

chitosan an ideal nanocarrier for water-insoluble drugs delivery by self-assembling into nanoparticles in aqueous phase.

2.3 Pharmaceutical applications of grafted chitosan derivatives

As novel carrier materials, chitosan derivatives have been widely explored for various drug delivery in the form of nanoparticles, microparticles, coating membranes and hydrogels, etc.

2.3.1 Delivery of poorly water soluble drugs

With appropriate modification, chitosan derivatives modified with hydrophobic groups can self-assemble into nanoparticles, with hydrophobic drugs encapsulated in the core. Thus, hydrophobically modified chitosan derivatives can be designed to deliver poorly water soluble drugs. Moreover, the hydrophilic backbone of chitosan can form the shell to protect drugs against the adverse influence of gastrointestinal environment [33, 34]. Representative studies on using grafted-chitosan for the delivery of poorly soluble small molecule drugs are presented in Table 2.

Table 2 Representatives of delivery of poorly soluble small molecule drugs by grafted-chitosan derivatives.

Polymers	Model drugs	Formulations	Results	References
N-Octyl-*O*-sulfate chitosan	Paclitaxel	Polymeric micelles	Compared with the commercial product, the PTX-loaded micelles showed a 5.7- and 3.5-fold higher apparent volume of distribution and total body clearance, respectively	[35]
PEGylated stearic acid-chitosan	Doxorubicin	Polymeric micelles	Presented excellent DOX encapsulation efficiency and loading capacity; Compared with PEGylated chitosan, PEGylated stearic acid-chitosan micelles showed enhanced accumulation in tumor tissue	[36]
Ureido-modified chitosan	Amoxicillin	Nanoparticles	Favorable gastric retention property (>6 h), which is beneficial for the oral drug delivery to eradicate *Helicobacter pylori* infection in stomach	[37]
Poly(acrylic acid-*co*-acrylamide)-chitosan	Curcumin	Nanoparticles	Curcumin showed a faster release at pH 5.4 relative to that at pH 7.4	[38]
Glyceryl monooleate-chitosan	Hydroxycamptothecin (HCPT)	Nanomicelles	After oral delivery, HCPT nanomicelles produced a prolonged circulation time in vivo (72 h) and showed a 2.7-fold increase in bioavailability relative to the HCPT injection	[39]

2.3.2 Delivery of proteins and peptides

Due to their high safety, potency and specificity, proteins/peptides have developed rapidly in recent years. However, due to their large molecular weight and poor stability, the approved biological drugs are mainly administrated by injection in clinic. Therefore, in order to improve patients' compliance, non-invasive administration routes are highly desired. Chitosan, as a cationic polysaccharide, is a commonly used nanocarrier to delivery peptides and proteins through non-invasive administration [18, 40]. And chitosan derivatives have been widely explored for better protection of fragile protein and peptides, controlled delivery at target sites and improved permeation. Table 3 summarizes representative studies of chitosan derivatives based protein and peptide delivery system.

2.3.3 Gene delivery

Gene therapy has shown promising therapeutic effect in clinic. However, due to the lack of safe and effective gene vectors, their clinical application is often restricted. With strong gene binding ability and flexible chemical modifiability, chitosan, as a promising non-viral vector, has been widely developed for gene delivery [17]. The challenges of chitosan as a gene delivery vector are its poor solubility at neutral and alkaline pH, low transfection efficiency, lack of cell specificity and low escape capacity from endosomes into the cytoplasm. To overcome above disadvantages, a series of chitosan derivatives have been synthesized and investigated for gene delivery, some representatives are presented in Table 4.

2.3.4 Vaccines delivery

In the field of immunotherapy, vaccines are playing a more and more important role in disease prevention and treatment. In the process of delivery, efficient vaccine adjuvants are generally required to improve the immunogenicity of vaccines. And due to the good biosafety, chitosan-based immune adjuvants have been widely developed for vaccines delivery (Table 5).

3 Alginate

3.1 Properties of alginate

Alginate is a natural anionic polysaccharide typically originated from brown seaweed. It is a linear polymer composed of β-D-mannuronopyranosyl (M) and α-l-guluronopyranosyl (G) units (Fig. 2) [60]. The molecular weight of commercially available sodium alginates ranges from 32 to

Table 3 Representatives of grafted-chitosan-based systems for protein and peptide delivery.

Polymers	Model drugs	Formulations	Results	References
Glyceryl monostearate-chitosan	Enoxaparin	Nanocomplexes	The oral bioavailability of enoxaparin was enhanced to 12.96%	[31]
Trimethyl-chitosan	Insulin	Polyelectrolyte complexes (PEC)	PEC improved insulin stability in simulated intestinal fluid and high temperature	[41]
Glyceryl monostearate-chitosan	Insulin	Insulin existed in molecular form	Revealed enhanced and prolonged hypoglycemic effect	[11]
mPEG-chitosan-glyceryl monocaprylate	Insulin	Polyelectrolyte nanocomplexes	The pharmacological availability was up to 5.9% with prolonged therapeutic effect	[23]
Quaternary-ammoniumpalmitoyl glycol chitosan	Leucine(5)-enkephalin	Nanoparticles	Showed a 1.67-fold increase in brain exposure and produced 38% maximum possible effect responders	[42]
PEGylated chitosan	Human parathyroid hormone 1-34	Nanoparticles	Showed a 4-fold increase in oral bioavailability than the PTH solution	[43]
Chitosan-PASP-PEG	Bovine serum albumin	Polyelectrolyte nanoparticles	Showed a sustained drug release for more than 24 h	[44]
Oleoyl-carboxymethyl chitosan	Bovine serum albumin	Nanoparticles	The protein was released in a sustained manner in PBS and the permeation via the mucous layer was enhanced	[45]

Table 4 Representative examples of chitosan derivatives mediated gene delivery.

Polymers	Genes	Formulations	Results	References
PEG-chitosan	DNA	Nanocomplexes	Complexes formed with 1.9% PEGylated chitosan remained stable in PBS and showed increased transfection efficacy compared to non-PEGylated chitosans	[46]
Trimethyl chitosan	DNA	Microcomplexes	Showed a 8.5-fold increase in cellular uptake relative to the chitosan polyplexes and produced a 678-fold increased transfection efficiency in NIH/3T3 cells	[47]
PEI-chitosan	DNA	Nanoparticles	Showed 141-, 58-, and 3-fold higher transfection efficiency than that of naked DNA, PEI and chitosan in liver, respectively	[48]
Arginine-chitosan	pDNA	Nanoparticles	Compared with the unmodified chitosan, the arginine-chitosan nanoparticles showed a 3.4–5.4-fold increased cellular uptake and 13–30-fold greater transfection efficiency in HEK 293 cells	[49]
Stearic acid-grafted chitosan	DNA	Nanoparticles	Improved the uptake of DNA by cells and DNA was easier to detach from the carrier	[50]
Thiolate chitosan	pSEAP	Nanoparticles	Increased pSEAP mucoadhesion and penetration in cells	[51]
Deoxycholic acid-modified chitosan	siRNA	Nanoparticles mixed with PLGA	The gene silencing effect was maintained at least up to 72 h, proving their prolonged effect	[52]
Folic acid-poly(ethylene glycol)-chitosan oligosaccharide lactate	siRNA	Nanoparticles	Showed strong targeting to tumor while little uptake by liver	[53]

Table 5 Representative examples of chitosan derivatives mediated vaccines delivery.

Polymers	Vaccines	Formulations	Results	References
N-Trimethyl chitosan	Ovalbumin	Nanoparticles	Showed a 1000-fold higher serum IgG levels than the un-adjuvanted OVA after nasal administration	[54]
Mannosylated chitosan	Anti-GRP DNA vaccine	Nanoparticles	Showed a 1.42-fold increase in antitumor activity relative to the pGRP group	[55]
Phosphorylated chitosan (PCS)	Ovalbumin	PCS aqueous solutions containing ovalbumin	After intramuscular injection, the antigen was controlled release from PCS gel network and produced a prolonged stimuli to the immune system	[56]
Glycol chitosan (GC)	Hepatitis B surface antigen	Nanoparticles	GC NPs elicited a relatively stronger immune response than the CS NPs after nasal administration	[57]
CKSTHPLSC-chitosan	BmpB vaccine	Microparticles	Increased M-cell targeting and transcytosis ability, improved secretory IgA responses in the mucosal tissue and produced systemic IgG antibody responses	[58]

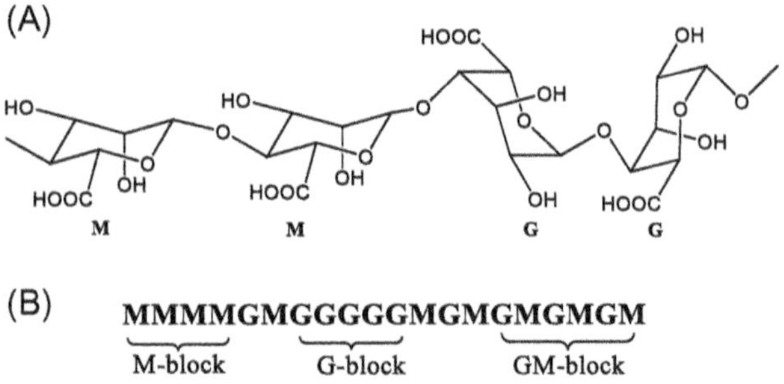

Fig. 2 Typical alginate structure: (A) chain conformation, (B) block distribution [59].

400 KDa, and required molecular weight can be easily obtained via oxidative degradation [61]. Alginates are hydrophilic, and the carboxylate ion in alginate backbone enhances its solubility in water at pH $\geq$5. Meanwhile, the carboxylic groups gain alginate a notable sensitivity to external pH stimuli, which performs a significant function in various processes such as purification, preparation, degradation, release and swelling [62]. When divalent cations (e.g. Ca^{2+}) are used as cross-linking agents, the G-blocks of alginate can undergo intermolecular cross-linking and thus form alginate hydrogels [63]. Due to the structural similarity to extracellular matrices of organisms, the alginate hydrogels can not only be used for drug delivery, but also allow wide applications in wound healing and cell transplantation [64–67].

3.2 Grafting modification of alginate

With multiple free hydroxyl and carboxyl groups on its backbone, a significant amount of researches have been carried out to modify alginate chemically in order to alter its characteristics [59, 60]. Hydrophobic modification is widely studied, which can be achieved by covalent binding of hydrophobic groups, such as long alkyl chains to the alginate backbone. And the hydrophobic modification can not only make alginate amphiphilic or hydrophobic, but also protect it from rapid dissolution and erosion in water. Table 6 summarizes the representative modification of alginates and the characteristics of alginate derivatives.

3.3 Pharmaceutical applications of grafted alginate derivatives

Contributed to the amphiphilic properties, hydrophobic modified alginate derivatives can be used as a carrier for hydrophobic drugs delivery. For example, the cholesteryl grafted alginate could load the hydrophobic pyrene by self-aggregating into nanoparticles in sodium chloride solution [73].

Chemical modification		Chemical structure	Characteristics	References
Hydroxyl groups	Oxidation	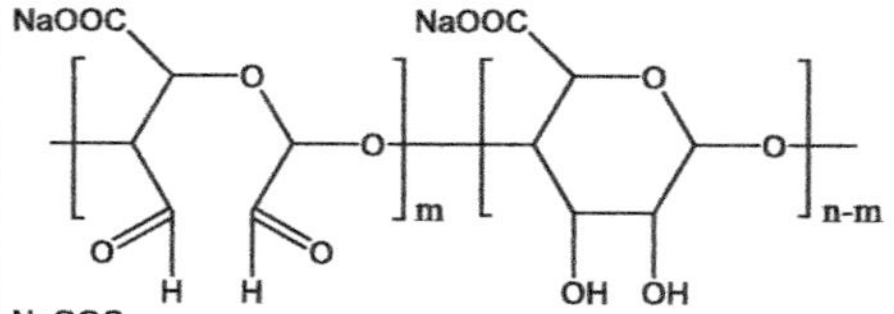	Increased reactive groups and accelerated degradation	[68]
	Reductive-amination of oxidized alginate	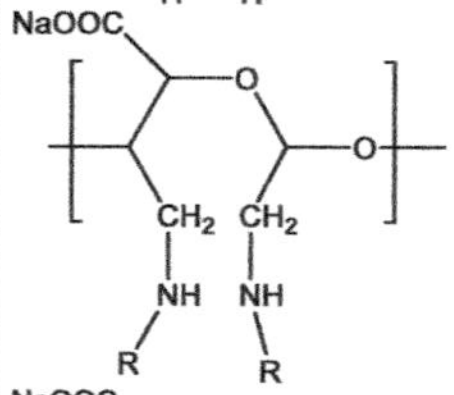	Provided convenience for the introduction of hydrophobic groups	[69]
	Sulfation	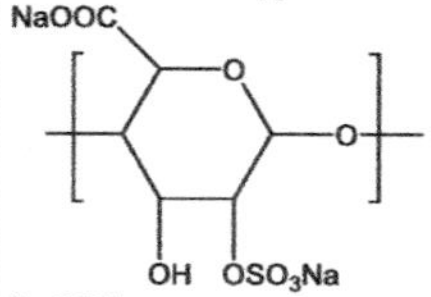	Provided blood compatibility and anticoagulant activity	[59]
	Copolymerization	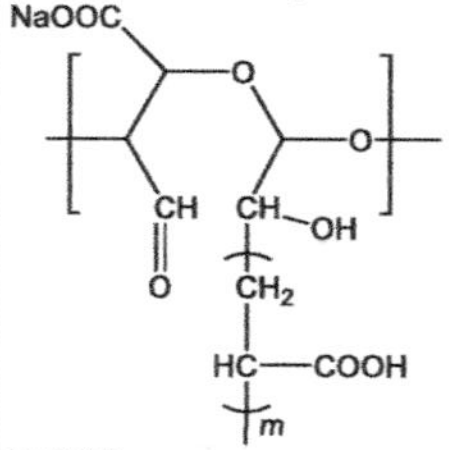	Increased ability of swelling, metal ion uptake, flocculating and degradation resistance	[70, 71]
	Cyclodextrin (CD)-linked alginate	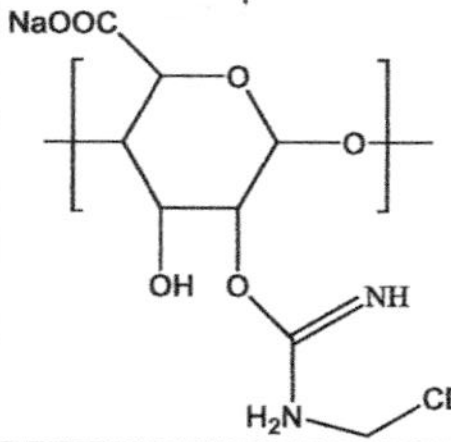	Increased ability of inclusion and gel-bead formation	[72]

Continued

Table 6 The representative chemical modification of alginates and their characteristics—cont'd

Chemical modification		Chemical structure	Characteristics	References
Carboxyl groups	Esterification	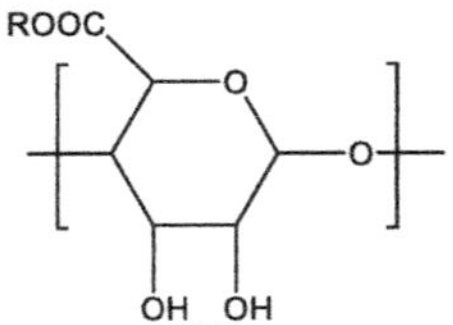	Increasing hydrophobic nature	[73, 74]
	Ugi reaction	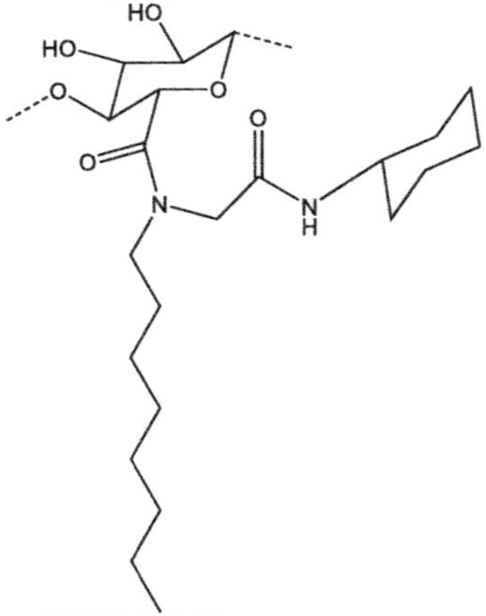	Increasing hydrophobic nature	[75, 76]
	Amidation	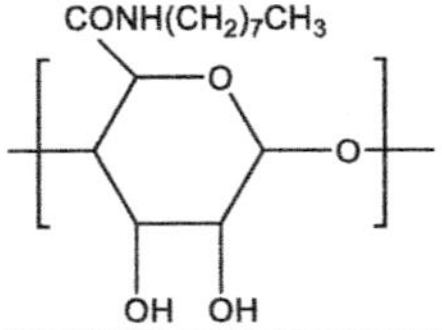	Amphiphilic properties	[77, 78]

Due to limited mechanical stiffness, the applications of alginate gels are generally restricted in pharmaceutical field. Fortunately, the grafted alginates can yield gels suitable for various applications. For example, the gels formed by poly(ε-caprolactone) (PCL)-alginate are pH sensitive, which showed little swelling at pH 1.2 (10–14%) but significant increase in volume at pH 6.8 (700–1300%), and the swelling behavior could be further adjusted by the length of PCL chains, resulting in controlled release of theophylline in gastrointestinal fluids [79]. Studies on grafted-alginate-based systems for various drug delivery are summarized in Table 7.

4 Carrageenan

4.1 Properties of carrageenan

Carrageenan (CG) is a sulphated polysaccharides originated from red seaweed. It is a natural anionic linear polysaccharide composed of galactose and anhydrogalactose units (Fig. 3). Generally, the molecular weight of CG is above 100 kDa, and it is soluble in water. Due to hydrolysis, the viscosity and gelation ability of CG can be reduced at low pH and high temperature [102, 103]. Based on the grafting degree of free hydroxyl groups, in general CG can be divided into six categories: Iota (τ)-, Lambda (λ)-, Kappa (κ)-, Theta (θ)-, Nu (ν)-, and Mu (μ)-CG [104]. And the kappa (κ), iota (τ) and lambda (λ) types of CG are commonly used in pharmaceutical industries. Due to the multiple biological activities, CG can be used as anticoagulant [105], antiviral [103], immunomodulatory, antitumor [106], and antithrombotic agents [107]. Moreover, with good biocompatibility, biodegradability, water retention capacity and gel strength, CG has been widely developed for drug delivery [6], wound healing and tissue engineering [108].

4.2 Grafting modification and pharmaceutical application of carrageenan derivatives

Due to the large number of hydroxyl/sulfate active sites, carrageenan can be modified in diverse forms. And these modifications could give carrageenan some particularly interesting characteristics. For example, compared with κ-carrageenan, the gels formed by hydroxyalkyl grafted κ-carrageenan showed decreased syneresis; alkali modification of carrageenan increased its gel strength [104, 109]; the graft copolymerization of κ-carrageenan with different reagents such as acrylic acid and poly(methacrylamide) displayed stable and super absorbent ability [110, 111]. And these modifications have greatly expanded the application of carrageenan in industry.

Table 7 Representative examples of alginate (Alg) derivatives mediated drug delivery.

Functions	Polymers	Model drugs	Dosage forms	Results	References
Hydrophobic	Oleoyl alginate ester (OAE)	Vitamin D_3	Nanoparticles	Vitamin D_3 was sustained release from nanoparticles in SIF (56.9%, 7 h)	[80]
	Poly(butyl methacrylate) modified alginate	Bovine serum albumin	Gel	Prolonged release of BSA at pH 7.2	[81]
Amphiphilic	Alkyl(C12)-alginate	Bovine serum albumin (BSA); human hemoglobin (Hb); *Helicobacter pylori* urease	Hydrogel	Protein can be released in a controlled manner by addition of surfactants or esterases; urease encapsulated in modified alginate exhibited a high level of IgG1 and IgG2a isotype response by the subcutaneous route	[82]
pH responsive	Alginate-*g*-poly (itaconic acid)	Nifedipine	Microspheres	Showed a pH-dependent release behavior	[83]
	Acetal-linked alginate	Vitamin B12, lysozyme, subtilisin, insulin, albumin, urease	Hydrogel	pH-dependent swelling behavior, the stimuli-responsive hydrogels formation and tunable pore sizes of acetal-linked alginate hydrogel served as a technology platform for oral delivery of various biological drugs	[84]

	L-Arginine grafted alginate	Bovine serum albumin	Gel beads	Showed a 3-fold higher release amount than the neat alginate beads at pH 2	[85]
pH and ionic strength sensitive	Poly((2-dimethylamino)ethyl methacrylate) modified oxidized Na-alginate	Bovine serum albumin	Gel beads	The release of BCA entrapped in the beads can be controlled by adjusting the graft ratio, pH and ionic strength	[86]
Mucoadhesive; pH sensitive	Alginate-*g*-poly (acrylic acid	7-(β-Hydroxyethyl) theophylline; ibuprofen	Microgels	The increased adhesive properties of the new polymers make them good candidates for mucoadhesive systems and the ibuprofen loaded microgels showed controlled release with pH sensitivity	[87, 88]
pH and glucose sensitive	Polyvinylpyrrolidone/ phenylboronic modified alginate	Insulin	Microcapsules	The microcapsules had no insulin release at pH 2 in the presence of glucose, while insulin was released in $<$10 min when suspended in glucose solution at pH 8	[89]
Temperature and pH responsive	Poly(*N,N*-diethyl acrylamide) grafted alginate	5-Fluorouracil (5-FU)	Microspheres	5-FU release was controlled by both the temperature (25°C $>$ 37°C) and pH of the medium (pH 7.4 and 5.5 $>$ pH 1.2)	[90]

Continued

Table 7 Representative examples of alginate (Alg) derivatives mediated drug delivery—cont'd

Functions	Polymers	Model drugs	Dosage forms	Results	References
Controlled release	Dopamine/polydopamine modified alginate	Gatifloxacin	Polydopamine particles embedded alginate hydrogel	In saline (0.9%), the release of gatifloxacin from alginate hydrogel was extremely faster (~98%, 2h) than dopamine/polydopamine particles modified alginate hydrogel (~45%, 24h)	[91]
Similar electroresponsive	Poly(acrylic acid) modified alginate	Ketoprofen	Hydrogel	The release of ketoprofen from the hydrogel could be controlled by electrical stimulus	[92, 93]
Enzyme responsive	Amylose-*g*-alginate	Fluorescein	Gel beads	The release of dye from the gel beads was increased after exposing to b-amylase	[94]
Targeting therapy	Glycyrrhetinic acid modified alginate	Doxorubicin (DOX)	Nanoparticles	Showed a 2.8- and 4.7-fold higher DOX concentration in liver relative to the non-modified nanoparticles and DOX solution, respectively	[95]

Mucoadhesive	Alginate-*g*-polyethylenglycol acrylate	Ibuprofen	Compressed tablets	Showed strong bonding to the mucus and sustained release in pH7.4 PBS	[96]
Cells transplant	Arginine-glycine-aspartic acid (RGD) grafted alginate	Growth factor; Tissue induction factors	Hydrogel	Allow for cell adhesion and migration; controlled release of the drug; promote cell movement	[97–100]
High affinity of proteins	Alginate sulfates	Fibroblast growth factors (FGFs)	Hydrogel	Improved proteins encapsulation efficiency and angiogenic activity in vivo	[101]

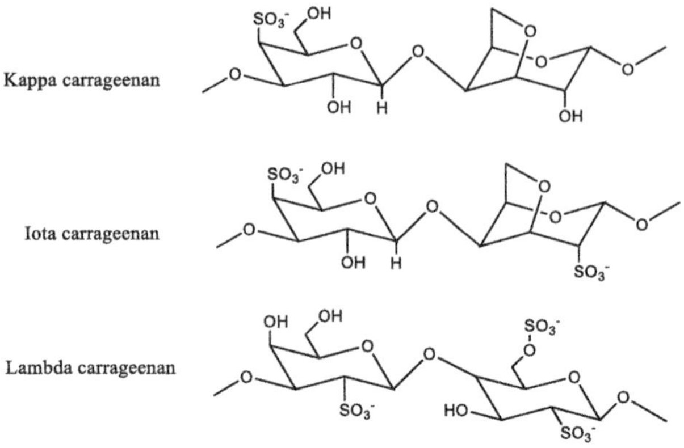

Fig. 3 Structures of kappa- (κ), iota- (τ), and lambda- (λ) carrageenan.

Another κ-carrageenan derivatives was synthesized via carboxymethyl groups modification [112], and compared to the parent polymer, the obtained carboxymethyl derivatives showed different gelling behavior with potential use for sustained drug delivery. Besides, the carboxyl groups grafted on the chain of carrageenan could endow it pH sensitivity which was beneficial for site-specific drug release [113]. In acid environment (pH 1.2–2.0), the carboxymethylated κ-carrageenan changed to a less ionized carboxylic acid, which promoted the formation of hydrogen bonds with decreased swelling capacity and thus slowed down drug release. In contrast, in neutral condition (pH 7.0–7.4), because the COOH was ionized to COO^-, the electrostatic repulsion between COO^- and SO_3^- produced a huge swelling of gel and thus accelerated drug release (Fig. 4) [108, 113]. Moreover, the specific targeting and thiolated derivatives of carrageenan can be obtained by grafting various substrates. Table 8 summarizes the applications of representative carrageenan derivatives in pharmaceutical field.

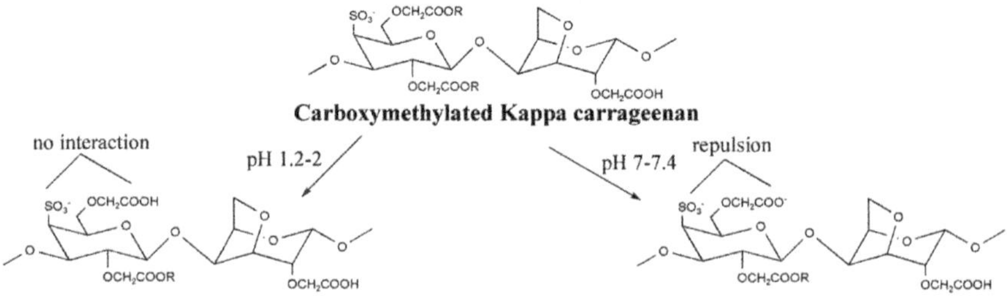

Fig. 4 Structure of carboxymethylated carrageenan (CCG) and interaction between -COOR and CCG under different pH conditions.

Table 8 Representative examples of the application of carrageenan derivatives in pharmaceutical field.

Functions	Polymers	Model drugs	Dosage forms	Results	References
pH sensitive	Carrageenan-*g*-poly (acrylic acid)	Diclofenac sodium (DS)	Hydrogel	DS was sustained release from the hydrogel, and the release rate at pH 7.4 was faster (~95%, 30 h) than that at pH 1.2 (~60%, 30 h)	[114]
	Polyacrylamide grafted-carrageenan	Ketoprofen	Hydrogel	The hydrogel beads offered pulsatile swelling and pH dependent drug release behavior: a maximum drug release of 10% at pH 1.2 and 90% at pH 7.4	[115]
	Carboxymethyleted κ-carrageenan	Dextran	Beads	FD-4 showed a low release in SGF (23%, 2 h) and high release in SIF (90%, 2 h)	[113]
pH sensitive and mucoadhesive	Lectin-functionalized carboxymethylated κ-carrageenan	Insulin	Microparticles	The release of insulin from microparticles was inhibited in SGF with controlled release in SIF; Contributed to the intestinal mucoadhesion, hypoglycemic effect of the microparticles was prolonged to 24 h in diabetic rats	[116]
Mucoadhesive	Thiolated carrageenan	Multidrug-resistance protein 2 substrate, sulforhodamine 101	Polymer solution	The κ- and τ-carrageenan-SH showed a 6.4- and 1.8-fold higher residence time on porcine intestinal mucosa, respectively; The protein 2 and sulforhodamine 101 showed a 1.38- and 1.35-fold higher cellular uptake, respectively	[117]

Continued

Table 8 Representative examples of the application of carrageenan derivatives in pharmaceutical field—cont'd

Functions	Polymers	Model drugs	Dosage forms	Results	References
Bioadhesive	Carboxymethyleted κ-carrageenan	Miconazole	Polyelectrolyte complex tablets	Exhibited excellent bioadhesive properties with the highest detachment force and prolonged drug release in pH 6.8 PBS	[118]
Targeting therapy	Folic acid (FA) modified carrageenan	Doxorubicin	Hybrid nanoporous microparticles	Increased specific surface area of the particles (9-fold) and its targeting ability to cancer cells	[119]
Sorbent materials	Isocyanate-modified carrageenan	Diclofenac (DCF) and carbamazepine (CBZ)	Polymer	The maximum adsorption capacity of CBZ and DCF was increased to 7.44–8.51 and 23.41–35.78 mg/g, respectively	[120]
Biosorbents	Glutaraldehyde-modified carrageenan	Metoprolol	Microparticles	Achieved the maximum adsorption capacity (Qm) in pH 6, and Qm could be further increased from 109 mg/g at 20°C to178 mg/g at 40°C	[121]
Improved stability	P(NIPA-*co*-AAm) or P(DMAm-*co*-AAm)-carrageenan	Lipase	Hydrogels	The storage and thermal stability of lipases was greatly improved, whose activities was still above 22% after 40th use in 5 days	[122]

5 Hyaluronic acid

5.1 Properties of hyaluronic acid

Hyaluronic acid (HA) is a natural glycosaminoglycan originated from various resources, such as bacteria, rooster combs, pig skin and human cells. It is composed of D-glucuronic acid and *N*-acetyl-glucosamine units by glycosidic bonds linking (Fig. 5) [123, 124]. HA generally exists in the form of sodium salt with negative charge and high hydrophilicity, and its molecular weight ranges from 5 to 20,000 KDa [125]. In natural, HA exists in almost all living organisms and it is the main component of connective tissue, which can stabilize cells and lubricate joints [124]. Nowadays, HA has been widely applied in medical fields, such as dermal filling [126, 127], joints lubrication [128], wound healing [129] and ophthalmic surgical assistance [130]. And in pharmaceutical field, HA has been widely investigated for various drug delivery. Besides its biocompatible, nontoxic, noninflammatory, and biodegradable properties, as a drug delivery carrier, HA can also impart solubility to hydrophobic drugs, providing protection to its "cargo" and being used as a "homing device" [131–135].

5.2 Graft modification of hyaluronic acid

In order to expand the application of HA in pharmaceutical area, its derivatives with enhanced or modulated properties have been widely explored. And the presence of multiple carboxyl, hydroxyl groups in the HA molecule and amino groups created by de-acetylation of HA makes it an ideal polymer for chemical modification (Fig. 5) [13, 124, 136]. HA is a hydrophilic polymer and the hydrophobic modifications are generally used to enhance its hydrophobicity which can be achieved by grafting hydrophobic groups, such as aliphatic/aromatic substituents [137]. The hydrophobic modification can endow HA amphiphilic properties (Fig. 5), which can self-aggregate into nanoparticles in aqueous solution with hydrophobic drugs encapsulated. Typically, the core of self-assembled aggregates are formed by the hydrophobic moiety of HA derivatives and thus the ratio of hydrophobic substituents can greatly affect their self-assembly characteristics [138]. For instance, the CMC of glycyrrhetinic acid (GA) grafted HA was affected by GA grafting degree, which was drastically decreased from 0.047 to 0.028 mg/mL when the grafting degree of GA was increased from 6% to 12% [139]. Moreover, some functional groups, which can endow polymers new property, such as stimuli responsiveness, mucoadhesion, long-acting, could also be grafted on the HA backbone for drug delivery [132, 140].

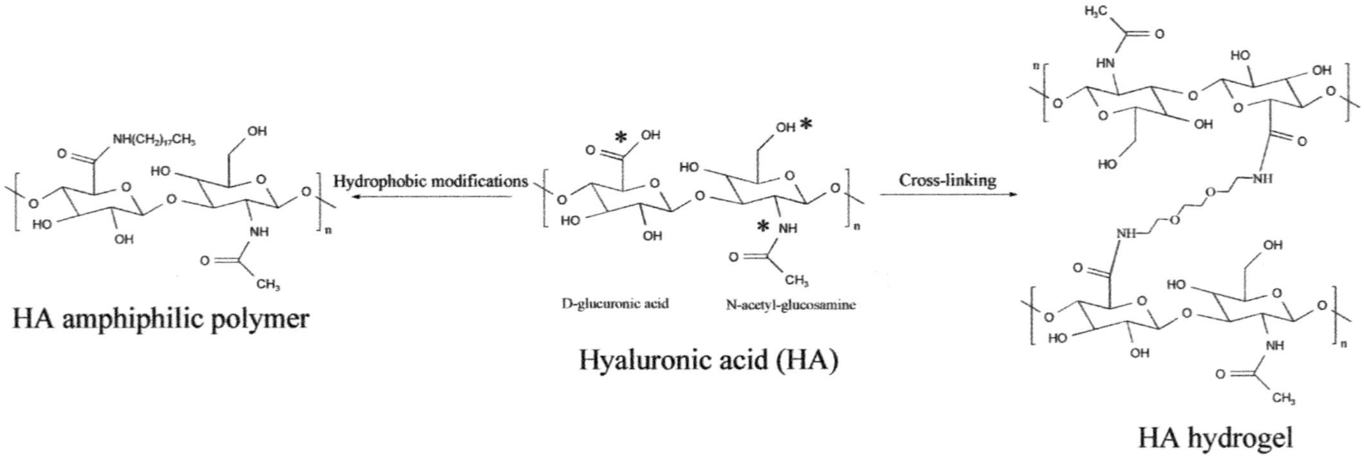

Fig. 5 Chemical structure of hyaluronic acid and its derivatives (*, modification sites).

Due to the pore size adjustability, large specific surface area and low toxicity, HA hydrogels have been widely developed for drug delivery [141]. However, compared with the synthetic polymer hydrogels, HA hydrogels show a less mechanical strength and stability [142]. Intermolecular cross-linking of carboxyl or hydroxyl groups in HA is a commonly used method to improve the mechanical strength of HA hydrogels (Fig. 5) and the commonly used cross-linking agents include glutaraldehyde, hexamethylenediamine and divinyl sulfone, carbodiimide, etc. [143, 144]. Moreover, the mechanical strength of HA hydrogel can be modulated by adjusting the weight ratio of HA and cross linker [4, 124, 136, 137]. Grafting polymers onto HA backbone can also improve the mechanical strength of HA hydrogel. For instance, with hydrogen bonding formation between nitrogen atoms (imidazole ring) and hydroxyl groups of HA, the *N*-vinyl imidazole grafted HA hydrogel showed good thermal stability and high crystallinity, with improved mechanical strength and decreased burst release of the drug [145].

5.3 The pharmaceutical applications of grafted hyaluronic acid derivatives

Particularly, HA is widely explored for cancer therapy in pharmaceutical area. And as a cancer therapy vehicle, HA shows various benefits: (1) active targeting of HA-binding receptors on cancer cells [146]; (2) facilitating rapid release of anti-cancer drugs from HA carrier due to its easy degradation in tumor cells [147]; (3) acting as a signaling molecule of inflammation irritated by cancer progression [148, 149]. And the modified HA further increased its application in cancer therapy. For example, the amphiphilic HA derivatives could self-assembly into nanoparticles with hydrophobic cancer drugs entrapped in the core. Table 9 presents representative cancer drug delivery systems based on grafted hyaluronic acid derivatives.

In addition, since the modified HA hydrogels can be used as a novel depot, they have been widely explored for protein and antibiotic delivery, especially in controlling their release. Table 10 summarizes some examples of hyaluronic acid derivatives mediated protein/antibiotic delivery systems.

6 Pectin

6.1 Properties of pectin

Pectin is a complex heteropolysaccharide composed of 1,4-linked α-D-galacturonic acid (GalA) units (Fig. 6A) and various neutral sugars, such as arabinose, galactose and rhamnose. Traditionally, pectin can be divided into two regions: "smooth regions" without side chain and "hairy regions"

Table 9 Representative cancer drug delivery systems based on grafted hyaluronic acid derivatives.

Model drugs	Polymers	Dosage forms	Results	References
Doxorubicin (DOX)	PEG-conjugated HA-ceramide (HACE)	Nanoparticles	The uptake of DOX was greatly improved in SCC7 cells, and the DOX-loaded nanoparticles showed a 73.0% reduction in in vivo clearance relative to the control group	[150]
	HA-*g*-glyceryl monostearate (HGS)	Nanoparticles	The DOX oral nanoparticles showed a 7-fold higher absorption than that of the DOX intravenous injection solution	[151]
	HA-*g*-poly (ε-caprolactone) (PCL)	Polyelectrolyte complex nanoparticles	The nanoparticles showed active targeting to the tumor tissue with efficient antitumor effect and few adverse effects	[152]
Paclitaxel (PTX)	HA-glycyrrhetinic acid conjugate	Nanoparticles	The DiR-labeled HGA nanoparticles showed a 2.88- and 1.83-fold increased accumulation in tumor and liver relative to the free DiR, respectively	[153]
	HA-deoxycholic acid conjugates (HA-ss-DOCA)	Micelles	The HA-ss-DOCA micelles showed a 2.5-fold increased accumulation in tumor than the HA-DOCA micelles	[154]
	N-Acetylcysteine modified HA	Micelles	Mucosal bioadhesion was improved (1.98-fold), and the area under the curve ($AUC_{0\text{-}24h}$) of NAC-HA-PTX micelles was 2.32- and 2.56-fold higher than the HA-PTX micelles and PTX solution (Taxol) after oral administration, respectively	[155, 156]
Epirubicin (EPI)	HA-FIT1 conjugate	Micelles	The tumor growth was drastically inhibited without side effect while the tumor volume increased 6.1-fold by EPI treating	[157]

Irinotecan (IRT)	PEGylated HA	Nanoparticles	Showed excellent antitumor activity with little systemic toxicity	[158]
α-Tocopherol	HA esterified with monostearin	Niosome	Promoted the endocytosis of nanocarrier by tumor cells and was efficient and secure for transdermal permeation	[159]
Gambogic acid (GA)	Cyclodextrin-grafted HA	Nanocomplex	Compared to free GA, GA loaded nanocomplex induced enhanced mitochondrial depolarization and apoptotic cell death in MDA-MB-435/MDR cells significantly, and effectively inhibited the MMP activities which are responsible for tumor metastasis	[160]
siRNA	HA-graft-pDMAEMA (HPD)	Nanocomplexs	After systemic administration, the nanocomplexes were selectively accumulated at the tumor site, and the quantitative analysis of the excised tumor exhibited a 6.6-fold lower signal intensity than the free siRNA (5 days post-injection)	[161]
	HA-scFv	Nanoparticles	The growth of metastasis tumor was significantly suppressed, which decreased to ~20% of the untreated control	[162]
	HA-PEI	Nanocomplexes	The nanocomplexes were mainly accumulated in the liver and tumor after tail-vein injection, and significantly suppressed the tumor growth compared to the control (5 wt% glucose solution)	[163]
DNA	HA-SS-COOH	Nanocomplexes	The nanocomplexes showed a 14-fold higher transfection efficiency than the unmodified nanocomplexes	[164]

Table 10 Representative examples of hyaluronic acid (HA) derivatives mediated protein/antibiotic delivery.

Polymers	Model drugs	Dosage forms	Results	References
HA-*g*-polyaspartylhydrazide copolymer	Bovine serum albumin (BSA)	Hydrogel	Achieved a stable and sustained release for a long time in pH 7.4 (>100 h) and the protein release rate could be controlled by adjusting the cross-linking density	[165]
HA-*g*-poly(sodium methacrylate) copolymer	α-Chymotrypsin	Freeze-dried solid containing polymer and drug	After incubated for 2 h at pH 1, the activity of the enzyme entrapped by the modified polymer was $79 \pm 4\%$, while the enzyme entrapped by the unmodified polymer was almost inactive	[166]
Thiolated HA (HA-SH)	Erythropoietin (EPO)	Microhydrogels	After subcutaneous administration, the plasma EPO concentration of HA-SH microhydrogels in SD rats could be maintained up to 7 days and the AUC was 1.68-fold higher than EPO solution	[167]
Heparin decorated HA	BMP-2	Hydrogel particles	The modified HA particles showed a 2-fold higher drug loading than the neat HA particles and avoided initial burst release with a sustained release over 13 days	[168, 169]
Sulfated methacrylated HA	TGF-β_2	Hydrogels	Compared to the wild type HA, the sulfated HA exhibited significantly slower degradation, and improved the protein sequestration with 50% lower release amount after 7 days of incubation in PBS	[170]

HA-G-poly (ε-caprolactone)	FGF1	Nanomicelles	Showed higher loading efficiency (>85%) and lower burst release (~20%, 12 h) in deionized water; better wound recovery (~82%) than native FGF1 (~76%)	[171]
HA-EDA-C18	Dexamethasone (Dex)	Microfibers	The release of Dex from microfibers was ionic sensitive which increased with the increase of salt concentration	[172]
Silane-modified HA	Vancomycin	Hydrogels	Compared with HA hydrogels, the modified HA hydrogels showed prolonged drug release (72 vs 20 h) and enhanced antibacterial efficacy	[173]
HA-*g*-pMANa	Vancomycin	Hydrogels	The drug release was pH sensitive (pH 7.0 > 5.0) with sustained release for more than 48 h	[174]
HA-*g*-PNIPAA	Cyclosporine A	Microgels	HA-PNIPAA was thermosensitive polymer, and the incorporated drug exhibited a controlled release in rabbit corneas	[175]
HA-*g*-PLA	Methylprednisolone	Microgels	The gels showed a prolonged release of methylprednisolone (96 h) in pH 7.4 PBS	[176]

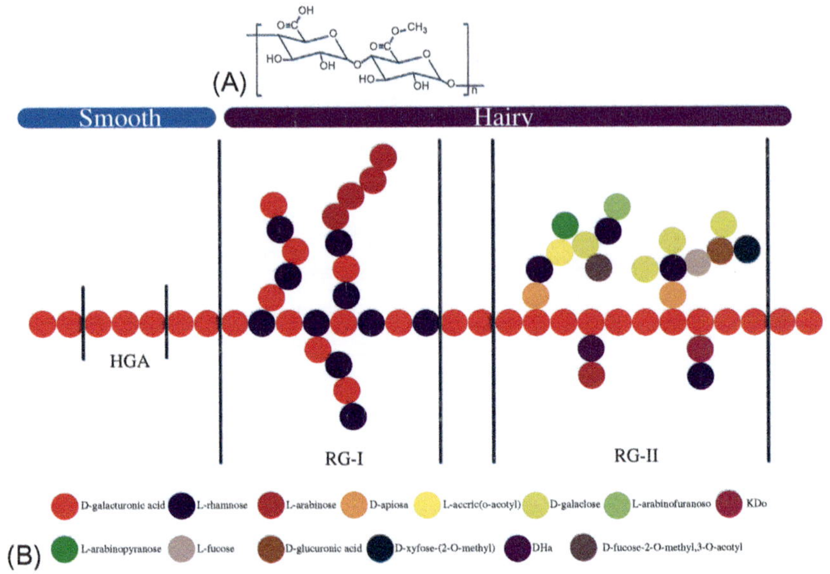

Fig. 6 (A) Basic structure of pectin. (B) The model of traditional structure of pectin, *HGA*, homogalacturonan; *RG-I*, rhamnogalacturonan-I; *RG-II*, rhamnogalacturonan-II [3].

containing non-ionic side chains (Fig. 6B) [3]. Pectins come from various sources, such as fruit pulp, plant cell wall and woody tissue fiber [177]. Generally, the univalent cationic salts of pectin are water soluble while its polyvalent cation salts present poor solubility in water. Due to the de-esterification or depolymerization, the dissolved pectin decomposes rapidly in solution and its decomposition rate is greatly affected by pH and temperature [3]. Due to its biodegradability, biocompatibility and non-toxicity, pectin has been widely used as a carrier to delivery various drugs in the form of microparticles, patches matrix and hydrogels etc. [64, 178]. Specifically, as pectin can interact with the galactose lectins on the surface of cancer cells and thus induce mitochondrial rupture [179, 180], pectins have also been developed for cancer therapy.

6.2 Graft modification of pectin

To increase pectins' application, the modification of pectin has been widely explored in recent years. Due to abundant hydroxyl and carboxyl groups as well as a certain amount of neutral sugars, various pectin derivatives have been developed. And the modification techniques of pectin can be classified as substitution (such as amidation, thiolation and sulfation) and chain elongation (such as cross-linking and grafting). With the modification, the properties of pectin, such as solubility and hydrophobicity can be modulated. Table 11 summarizes various strategies to modify pectin and the characteristics of pectin derivatives.

6.3 The pharmaceutical applications of grafted pectin derivatives

With the modified properties, pectin derivatives as advanced pharmaceutical excipients have been widely employed for drug delivery. For instance, due to the high solubility of pectin in water, drugs encapsulated in it are prone to have burst release in physiological medium. To solve this limitation, Bhatia et al. [201] decreased pectin's hydrophilicity by acetylation of its free hydroxyl groups, and the obtained ester pectin could be used to modify drug release in the gastrointestinal tract. Moreover, the hydrophobic modified pectin by aliphatic or aromatic groups could be used as stabilizing agent and emulsifying agent [202].

In the pharmaceutical area, pectin-based hydrogels have been widely explored for drug delivery. And drug release from the hydrogels can be induced by different factors, such as pH, enzymes and temperature [203, 204]. For example, poly(acrylamide) grafted pectin hydrogels showed a pH-sensitive release behavior, which was stable in gastric pH and quickly released drug in simulated intestinal pH [205]. That was because the intestinal pH led to the ionization of carboxyl groups of pectin and produced electrostatic repulsion, which expanded the hydrogel network of pectin chains, favored the liquid uptake and thus facilitated drug release. Other modified pectin, such as thiolated pectin, oxidized pectin, quaternization of pectin, have also being developed as pharmaceutical excipients and Table 12 summarizes some examples of pectin derivatives mediated drug delivery.

7 Pullulan

7.1 Properties of pullulan

Pullulan is a kind of nonionic linear polysaccharide consisting of maltotriose units by glycosidic bond linking (Fig. 7). Pullulan was firstly isolated and identified from Aureobasidium pullulans (a yeast like fungus), and could also be obtained from sweet potato, coconut by-products, de-proteinized whey, jiggery, etc. [215]. According to the strain and culture conditions, the average molecular weight of pullulan varies in the range of 5×10^3–2×10^6 Da [216]. Pullulan has excellent physicochemical properties, such as adhesiveness, safety, biocompatibility, anticoagulation, antithrombosis and anti-inflammation [217, 218]. And compared with other polysaccharides, due

Table 11 The common chemical modification methods of pectin and its characteristics.

Modified techniques and sites		Materials	Characteristics	References
Alkylation	Carboxylate groups	Alkyl/alkyl esters	Increased hydrophobicity of pectin, enhanced intrinsic viscosity	[181–183]
	Hydroxyl groups	Alkyl/fatty acid	Reduced polarity, solubility and water sorption	[184]
Amidation	Primary amino groups (-CO-NH_2)	Amino groups, amine acids	Showed good gelling properties and reduced sensitivity against Ca-ions and pH, enhanced water solubility, with thermo-reversible and desired hydrophilicity	[185–188]
Quaternization	Hydroxyl groups	3-Chloro-2-hydroxypropyltrimethylammonium chloride	Showed excellent moisture absorption and moisture retention abilities, pronounced antimicrobial effect	[189]
Thiolation	Hydroxyl groups	Thioglycolic acid	Improved mucoadhesion	[190]
	Carboxylate groups	Cysteine or 4-aminothiophenole	Improved cohesive properties, showed pH-sensitive release behaviors	[191, 192]
Sulfation	Hydroxyl groups	Sulfate groups	Improved anticoagulant, antithrombotic, contraceptive, anti-inflammatory and antitumor activity	[193, 194]
	Carboxyl groups	Sulfate groups	Increased anticoagulant and antithrombotic effects	[195]
Oxidation	Hydroxyl groups	Sodium periodate	Presented faster degradation, and increased the number of reactive groups which was beneficial for further modifications	[196]
Cross-linking	Methoxyl groups	Epichlorohydrin	Showed excellent swelling ability and rheological property	[197]
Grafting	Hydroxyl groups	Copolymers	Showed pH- and thermo-sensitive properties, controlled drug release	[198–200]

Table 12 Representative examples of pectin derivatives mediated drug delivery.

Functions	Polymers	Model drugs	Dosage forms	Results	References
pH sensitive	Methylesterase modified pectin	Indomethacin	Hydrogel	The release rate of indomethacin from the hydrogel beads was <1.0% at pH 1.2 (2 h), and increased to 12% when hydrogel beads were immersed in pH 7.4 medium for 3 h	[206]
	Amidated pectin	Progesterone	Microspheres	Microspheres showed high drug entrapment (99%), pH-dependent swelling and release behavior (2 h <5% in pH 1.2, 12 h ~70% in pH 7.4) and achieved 2.3-fold enhancement of colonic progesterone permeability	[207]
	Amidated pectins	Indomethacin, sulphamethoxazole	Hydrogel	Showed pH-dependent release behavior, which was stable in acidic solution and released in simulated colonic solution	[208]
	Polyacrylamide grafted pectin	Salicylic acid	Hydrogel	Showed limited release at pH 1.4 (~20%) and high release at pH 7.4 (~86%, 6 h)	[200]
	Poly(*N*-vinylpyrrolidone)-*g*-pectin	Theophylline	Hydrogel	Showed low release at pH 5.5 (<60%, 5 h), and high release at pH 7.4 (>90%, 3 h) which was	[198]

Continued

Table 12 Representative examples of pectin derivatives mediated drug delivery—cont'd

Functions	Polymers	Model drugs	Dosage forms	Results	References
				beneficial for colon-targeted vehicles	
Controlled release	Amidated pectin	Insulin	Hydrogel matrix patch	Sustained insulin release was maintained for up to 5-weeks, and the blood glucose level of diabetic rats was almost equivalent with normal rats after 5-weeks treatment	[209]
	Oxidized pectin	Doxorubicin	Hydrogel	The release rate of Dox from hydrogels could be adjusted by addition of adipic dihydrazide in PBS, and prevent progression of primary cancer	[210]
Mucoadhesive	L-Cysteine modified pectin	Rosuvastatin calcium	Minitablets	Disintegration time was prolonged for more than 2-fold, mucus adhesion increased for more than 5-fold with sustained release of rosuvastatin calcium over 36 h in 0.1 M HCl	[211]
	Pectin-cysteine conjugate	Insulin	Hydrogel beads	Compared with unmodified pectin, the thiolated polymer exhibited controlled swelling,	[192]

				improved adhesion (5-fold) and permeation (1.61-fold)	
	Thiolated pectin	Timolol maleate	Nanosuspension	The thiolated pectin nanoparticles showed 3.5-fold higher permeation across the excised goat cornea relative to the conventional formulation	[212]
Stomach-specific	Diethanolamine-modified pectin	Metformin HCl (MFM)	Hydrogel	Demonstrated high drug loading (68.52%), sustained drug release (Q_{8h}, 74.30%), with excellent gastro-retention ability	[213]
Gene carrier	Pectin-NH_2-Q ($Q=N^+(CH_3)_3$)	DNA	Polymer/DNA complexes	Pectin-NH_2-Q was able to interact strongly with DNA, and showed improved stability and higher transfection efficiency	[214]

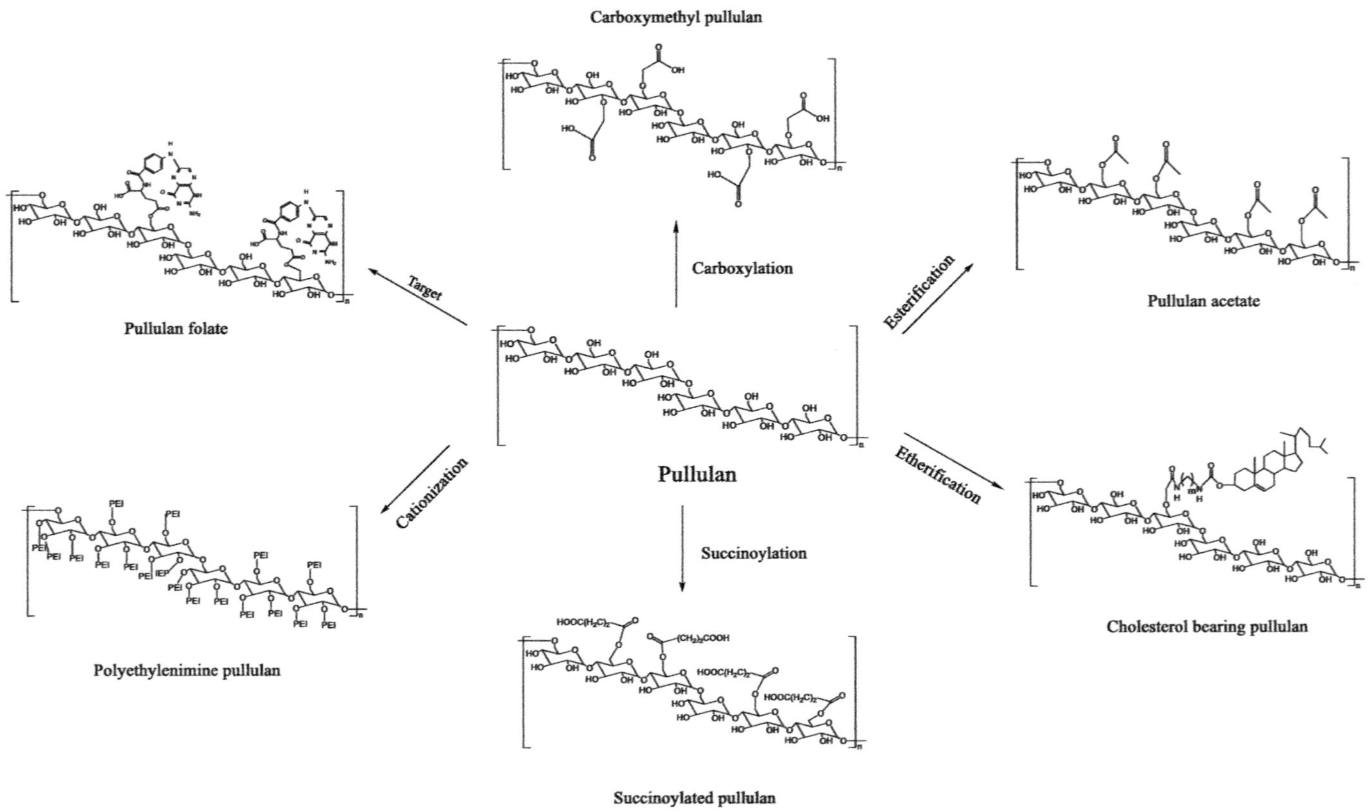

Fig. 7 Chemical structure of pullulan and its derivatives.

to its unique and regular linking pattern, pullulan possesses some distinctive properties, such as structural flexibility and high water solubility [219]. In the biomedical field, pullulan has been widely exploited for various applications, such as gene silencing, vaccination, wound healing, pharmaceutical dosages forms formation, coating and so on [219–222].

7.2 Grafting modification of pullulan

In order to expand pullulan's application, pullulan derivatives have been developed by grafting different chemical compounds on its backbone and the main modified sites are the abundant hydroxyl groups [217, 223]. Pullulan is a hydrophilic polymer and its main derivatives are hydrophobic modified. The chemical reactions involve carboxylation, esterification, etherification, succinoylation, cationization, etc. (Fig. 7). Its derivatives can improve pullulan's properties in a variable manner, with different physico-chemical properties. For example, carboxymethyl pullulan (CMP) could be obtained by cross-linking carboxylate group with the hydroxyl group of pullulan via ester linkage [224]. This modification provided pullulan negative charge, high sensitivity toward pH and ionic strength. Pullulan acetate synthesized by acetylation could self-aggregate into stable micelles in aqueous solution [225]. Another hydrophobic pullulan, cholesterol bearing pullulan (CHP), could be obtained by grafting cholesterol group to its backbone, which could form stable monodisperse nanogels (~20–30 nm) in aqueous solution [226]. The succinylated pullulan showed a pH-sensitive characteristic, and it could be further grafted with other polymers by using the new reactive site [227]. And the polyethyleneimine-pullulan derivative possessed high cationic charge density, good blood compatibility and plasma stability [228]. Besides, the hydrophobic pullulan can be further modified by other functional groups, such as biotin or folate (Fig. 6), to confer specificity and targeting to cancer cells [229–231].

7.3 The pharmaceutical application of grafted pullulan derivatives

Hydrophobic groups grafting is the main modification method of pullulan, and these derivatives can self-aggregate into nanoparticles in aqueous solution. Thus, the grafted pullulans have been widely used as nanocarriers for drug delivery. Table 13 summarizes the potential application of various pullulan derivatives mediated drug delivery.

Moreover, due to the high affinity with DNA, cationized pullulan derivatives have been widely used as carriers for genes delivery (Table 14.) [217, 259].

Table 13 Representative examples of various pullulan derivatives mediated drug delivery.

Model drugs	Polymers	Dosage forms	Results	References
Doxorubicin	Carboxymethyl pullulan	Nanoparticles	Showed a pH-dependent release behavior, with 67% drug release at pH 5.0 and 36% release at pH 7.4 after 21 h incubation	[232]
	Aminated carboxymethyl pullulan	Nanoparticles	Showed a pH-dependent release behavior, with 75% drug release at pH 5.0 (2 h) and <15% release at pH 7.4 (12 h)	[233]
	Oligo(methacryloyl sulfadimethoxine) grafted pullulan acetate	Hydrogel nanoparticles	Drug release showed pH-dependent profiles, with ~85% drug released at pH 6.2 and ~50% drug released at pH 7.4 after 12 h incubation	[225]
	Folate-decorated maleilated pullulan	Nanoparticles	Showed a slow (>100 h) and pH-dependent (pH 5.0 > 7.4) release behavior, and the folate decorated nanoparticles showed enhanced cellular uptake	[234]
	Pullulan-*g*-poly(L-lactide)	Nanogels	Drug release was thermo-responsive, the cumulative drug release was about 40% at 25°C, 65% at 37°Cand 95% at 42°Cafter 50 h incubation in pH 7.4 PBS	[235]
Epirubicin	Cholesterol modified pullulan (CHP)	Nanoparticles	Showed enhanced $t_{1/2}$ (19.33 h), C_{max} (10.45 mg/L) and AUC (48.36 mg/L)	[226]
Adriamycin	Pullulan acetate	Nanoparticles	Increased cytotoxicity to breast tumor cells at pH 6.8	[236]
	Vitamin H modified pullulan acetate	Nanoparticles	Increased the uptake of adriamycin by HepG2 cells	[237]
	O-Urocanyl pullulan	Nanoparticles	Showed accelerated drug release (72.1%) at pH 5.8 (tumor cells)	[238]

Diclofenac	Pullulan succinylate	Microspheres	The microspheres swolled up to 20-times at pH 7.4, with accelerated drug release	[239]
	Pullulan cyclodextrin	Microspheres	Polymer swolled at pH 7.4 and 82% drug was released within 1 h	[240]
Mitoxantrone	Biotin modified cholesterol-pullulan	Nanoparticles	>90% drug was released from nanoparticles at pH 3.5 (tumor cells)	[241]
Lopinavir	Pullulan acetate	Nanoparticles	The relative bioavailability of Lopinavir from nanoparticles was ~2-fold higher than the free drug and showed a higher distribution in lymphoid organs	[242]
Insulin	Concanavalin A modified carboxylated pullulan	Hydrogel	Showed a glucose-sensitive insulin release	[243]
Dextran	Poly(NIPAAm-*co*-AAm) pullulan succinylate	Microspheres	>75% drug was released from microspheres at 37°C	[227]
Lysozyme	Pullulan acetate	Microspheres	Polymer swolled up to 28-times at pH 7.4, which accelerated drug release	[244]
Methotrexate and combretastatin A4	*N*-Urocanyl pullulan	Nanoparticles	Improved drug distribution in liver and tumor, increased antitumor and antiangiogenic effects	[245]

Table 14 Representative examples of genes delivery by pullulan derivatives.

Polymers	Genes	Dosage forms	Results	References
Diethylaminoethylamine pullulan	Plasmid DNA	Hydrogels	The diethylaminoethylamine modified pullulan showed high cationic charge and strong affinity for plasmid DNA	[246]
	pSEAP	3 dimensional matrices	Showed a 150-fold increased activity in the transfection of vector expressing alkaline phosphatase	[247]
Polyethylenimine-*g*-pullulan	siRNA	Nanoparticles	The PEI-pullulan/siRNA complex showed increased level in liver and few side effects	[248]
	pGFP	Nanoparticles	Showed 46.7% gene transfection and liver binding after 3 h	[249]
	p53	Nanoplexes	Revealed excellent transfection efficiency and resulted in 90% C6 cells death	[250, 251]
Polyamidoamine (PAMAM)-pullulan conjugates	DNA	Nanoparticles	Compared to unmodified pullulan, the transfection efficiency of PAMAM modified pullulan has 2-fold enhancement in HepG2 cells	[252]
Polyethyleneimine pullulan folate	siRNA and plasmid DNA	Complex	Showed a higher gene transfection with polymer/DNA = 1/6.25 in presence of serum, while highly efficient gene silencing with polymer/DNA = 1/12.5 in presence and absence of serum	[253]
Pullulan-spermine	Plasmid EGFP-p53	Nanoparticles	Showed a 92% survival rate in U87 cells	[254]
	RFP-cDNA	Conjugate	Revealed excellent transfection efficiency and less cytotoxicity	[255]

Pullulan-*g*-poly(L-lysine)	pEGFP-N1 plasmid	Nanoparticles	High gene transfection efficiency in HEK293, HepG2 and KB cells	[256]
Pullulan-protamine	p53	Conjugate	Revealed high gene transfection and resulted in >80% death in C6 cells	[257]
Poly(β-amino) ester pullulan	Methotrexate and GFP	Conjugate	After 24 h, gene was completely distributed in tumor cells	[258]

8 Conclusions and future directions

Due to their biodegradability, renewability and nontoxicity, polysaccharides have played irreplaceable roles in the biomedical field. And the structure of natural polysaccharides can be flexibly modified to satisfy the requirement for diversified drug delivery objectives. As discussed in this chapter, the modified polysaccharides possess various novel properties, which further expand their applications in pharmaceutical field. The grafted polysaccharides provide multiple functions for nanocarriers which can be used to load different types of drugs, such as poorly water-soluble drugs, proteins/peptides, genes and vaccines. Furthermore, the polysaccharide derivatives can also control drug release in physiological fluid and improve the targeting ability and drug delivery efficacy. However, the clinical applications of polysaccharide derivatives-based drug delivery systems are still limited, which are mainly related to the safety issue and reproducibility of the grafted polymers. Therefore, in order to accelerate the clinic transfer, the systematic toxicity studies of polysaccharide derivatives under both in vitro and in vivo conditions are absolutely essential.

Acknowledgment

This project is financially supported by the National Natural Science Foundation of China (Grant No. 31870987).

References

[1] Z. Liu, Y. Jiao, Y. Wang, C. Zhou, Z. Zhang, Polysaccharides-based nanoparticles as drug delivery systems, Adv. Drug Deliv. Rev. 60 (15) (2008) 1650–1662.

[2] Y. Yu, M. Shen, Q. Song, J. Xie, Biological activities and pharmaceutical applications of polysaccharide from natural resources: a review, Carbohydr. Polym. 183 (2018) 91–101.

[3] A. Noreen, Z.-I.-H. Nazli, J. Akram, I. Rasul, A. Mansha, N. Yaqoob, R. Iqbal, S. Tabasum, M. Zuber, K.M. Zia, Pectins functionalized biomaterials; a new viable approach for biomedical applications: a review, Int. J. Biol. Macromol. 101 (2017) 254–272.

[4] C. Alvarez-Lorenzo, B. Blanco-Fernandez, A.M. Puga, A. Concheiro, Crosslinked ionic polysaccharides for stimuli-sensitive drug delivery, Adv. Drug Deliv. Rev. 65 (9) (2013) 1148–1171.

[5] J. Guan, Q. Liu, X. Zhang, Y. Zhang, R. Chokshi, H. Wu, S. Mao, Alginate as a potential diphase solid dispersion carrier with enhanced drug dissolution and improved storage stability, Eur. J. Pharm. Sci. 114 (2018) 346–355.

[6] L. Li, R. Ni, Y. Shao, S. Mao, Carrageenan and its applications in drug delivery, Carbohydr. Polym. 103 (2014) 1–11.

[7] S. Maiti, S. Jana, Polysaccharide Carriers for Drug Delivery, Woodhead Publishing, 2019.
[8] S.S. Suner, M. Sahiner, S.B. Sengel, D.J. Rees, W.F. Reed, N. Sahiner, Responsive biopolymer-based microgels/nanogels for drug delivery applications, in: A.S.H. Makhlouf, N.Y. Abu-Thabit (Eds.), Stimuli Responsive Polymeric Nanocarriers for Drug Delivery Applications, vol. 1, Woodhead Publishing, 2018 (Chapter 17).
[9] M. Swierczewska, H.S. Han, K. Kim, J.H. Park, S. Lee, Polysaccharide-based nanoparticles for theranostic nanomedicine, Adv. Drug Deliv. Rev. 99 (2016) 70–84.
[10] X. Yang, X. Shi, R. D'arcy, N. Tirelli, G. Zhai, Amphiphilic polysaccharides as building blocks for self-assembled nanosystems: molecular design and application in cancer and inflammatory diseases, J. Control. Release 272 (2018) 114–144.
[11] M. Gao, Y. Sun, Y. Kou, X. Shen, Y. Huo, C. Liu, Z. Sun, X. Zhang, S. Mao, Effect of glyceryl monocaprylate-modified chitosan on the intranasal absorption of insulin in rats, J. Pharm. Sci. 108 (11) (2019) 3623–3629.
[12] C. Liu, H. Xu, Y. Sun, X. Zhang, H. Cheng, S. Mao, Design of virus-mimicking polyelectrolyte complexes for enhanced oral insulin delivery, J. Pharm. Sci. 108 (10) (2019) 3408–3415.
[13] X. Wang, X. Gu, H. Wang, J. Yang, S. Mao, Enhanced delivery of doxorubicin to the liver through self-assembled nanoparticles formed via conjugation of glycyrrhetinic acid to the hydroxyl group of hyaluronic acid, Carbohydr. Polym. 195 (2018) 170–179.
[14] M. Amidi, W.E. Hennink, Chitosan-based formulations of drugs, imaging agents and biotherapeutics, Adv. Drug Deliv. Rev. 62 (1) (2010) 1–2.
[15] I. Aranaz, M. Mengibar, R. Harris, I. Panos, B. Miralles, N. Acosta, G. Galed, A. Heras, Functional characterization of chitin and chitosan, Curr. Chem. Biol. 3 (2) (2009) 203–230.
[16] A. Jain, A. Gulbake, S. Shilpi, A. Jain, P. Hurkat, S.K. Jain, A new horizon in modifications of chitosan: syntheses and applications, Crit. Rev. Ther. Drug Carrier Syst. 30 (2) (2013) 91–181.
[17] S. Mao, W. Sun, T. Kissel, Chitosan-based formulations for delivery of DNA and siRNA, Adv. Drug Deliv. Rev. 62 (1) (2010) 12–27.
[18] X. Zhang, H. Cheng, W. Dong, M. Zhang, Q. Liu, X. Wang, J. Guan, H. Wu, S. Mao, Design and intestinal mucus penetration mechanism of core-shell nanocomplex, J. Control. Release 272 (2018) 29–38.
[19] F. Seidi, H. Salimi, A.A. Shamsabadi, M. Shabanian, Synthesis of hybrid materials using graft copolymerization on non-cellulosic polysaccharides via homogenous ATRP, Prog. Polym. Sci. 76 (2018) 1–39.
[20] B.E. Benediktsdóttir, Ó. Baldursson, M. Másson, Challenges in evaluation of chitosan and trimethylated chitosan (TMC) as mucosal permeation enhancers: from synthesis to in vitro application, J. Control. Release 173 (2014) 18–31.
[21] V.K. Mourya, N.N. Inamdar, Trimethyl chitosan and its applications in drug delivery, J. Mater. Sci. Mater. Med. 20 (5) (2009) 1057–1079.
[22] S. Mao, X. Shuai, F. Unger, M. Wittmar, X. Xie, T. Kissel, Synthesis, characterization and cytotoxicity of poly(ethylene glycol)-graft-trimethyl chitosan block copolymers, Biomaterials 26 (32) (2005) 6343–6356.
[23] C. Liu, Y. Kou, X. Zhang, W. Dong, H. Cheng, S. Mao, Enhanced oral insulin delivery via surface hydrophilic modification of chitosan copolymer based self-assembly polyelectrolyte nanocomplex, Int. J. Pharm. 554 (2019) 36–47.
[24] M.-C. Chen, F.-L. Mi, Z.-X. Liao, C.-W. Hsiao, K. Sonaje, M.-F. Chung, L.-W. Hsu, H.-W. Sung, Recent advances in chitosan-based nanoparticles for oral delivery of macromolecules, Adv. Drug Deliv. Rev. 65 (6) (2013) 865–879.

[25] F. Föger, T. Schmitz, A. Bernkop-Schnürch, In vivo evaluation of an oral delivery system for P-gp substrates based on thiolated chitosan, Biomaterials 27 (23) (2006) 4250–4255.
[26] W. Sun, S. Mao, Y. Wang, V.B. Junyaprasert, T. Zhang, L. Na, J. Wang, Bioadhesion and oral absorption of enoxaparin nanocomplexes, Int. J. Pharm. 386 (1–2) (2010) 275–281.
[27] L. Na, J. Wang, L. Wang, S. Mao, A novel permeation enhancer: N-succinyl chitosan on the intranasal absorption of isosorbide dinitrate in rats, Eur. J. Pharm. Sci. 48 (1) (2013) 301–306.
[28] P. Mukhopadhyay, K. Sarkar, S. Bhattacharya, A. Bhattacharyya, R. Mishra, P.P. Kundu, pH sensitive N-succinyl chitosan grafted polyacrylamide hydrogel for oral insulin delivery, Carbohydr. Polym. 112 (2014) 627–637.
[29] N. Dhawan, K. Kumar, A.N. Kalia, S. Arora, N-succinyl chitosan as buccal penetration enhancer for delivery of herbal agents in treatment of oral mucositis, Curr. Drug Deliv. 11 (3) (2014) 415–425.
[30] L. Wang, L. Li, Y. Sun, J. Ding, J. Li, X. Duan, Y. Li, V.B. Junyaprasert, S. Mao, In vitro and in vivo evaluation of chitosan graft glyceryl monooleate as peroral delivery carrier of enoxaparin, Int. J. Pharm. 471 (1–2) (2014) 391–399.
[31] L. Wang, L. Li, Y. Sun, Y. Tian, Y. Li, C. Li, V.B. Junyaprasert, S. Mao, Exploration of hydrophobic modification degree of chitosan-based nanocomplexes on the oral delivery of enoxaparin, Eur. J. Pharm. Sci. 50 (3–4) (2013) 263–271.
[32] L. Wang, Y. Sun, C. Shi, L. Li, J. Guan, X. Zhang, R. Ni, X. Duan, Y. Li, S. Mao, Uptake, transport and peroral absorption of fatty glyceride grafted chitosan copolymer-enoxaparin nanocomplexes: influence of glyceride chain length, Acta Biomater. 10 (8) (2014) 3675–3685.
[33] Y. Tian, Y. Sun, X. Wang, G. Kasparis, S. Mao, Chitosan and its derivatives-based nano-formulations in drug delivery, in: A.M. Grumezescu (Ed.), Nanobiomaterials in Drug Delivery, William Andrew Publishing, 2016 (Chapter 15).
[34] R. Riva, H. Ragelle, A. Des Rieux, N.J.R. Duhem, C. Me, V. Préat, Chitosan and chitosan derivatives in drug delivery and tissue engineering, in: R. Jayakumar, M. Prabaharan, R.A.A. Muzzarelli (Eds.), Chitosan for Biomaterials II, Springer, Berlin, Heidelberg, 2011.
[35] C. Zhang, G. Qu, Y. Sun, X. Wu, Z. Yao, Q. Guo, Q. Ding, S. Yuan, Z. Shen, Q. Ping, H. Zhou, Pharmacokinetics, biodistribution, efficacy and safety of N-octyl-O-sulfate chitosan micelles loaded with paclitaxel, Biomaterials 29 (9) (2008) 1233–1241.
[36] F.Q. Hu, Y.Y. Zhang, J. You, H. Yuan, Y.Z. Du, pH triggered doxorubicin delivery of PEGylated glycolipid conjugate micelles for tumor targeting therapy, Mol. Pharm. 9 (9) (2012) 2469–2478.
[37] Z.-W. Jing, M. Luo, Y.-Y. Jia, C. Li, S.-Y. Zhou, Q.-B. Mei, B.-L. Zhang, Anti-Helicobacter pylori effectiveness and targeted delivery performance of amoxicillin-UCCs-2/TPP nanoparticles based on ureido-modified chitosan derivative, Int. J. Biol. Macromol. 115 (2018) 367–374.
[38] D. Kumar, V. Raj, A. Verma, P. Kumar, J. Pandey, Novel binary grafted chitosan nanocarrier for sustained release of curcumin, Int. J. Biol. Macromol. 131 (2019) 184–191.
[39] Y. Tian, C.J. Shi, X. Zhang, Y.J. Sun, J. Wang, Y.Y. Zhang, J.Y. Yang, L.H. Wang, L.L. Wang, S.R. Mao, Nanomicelle based peroral delivery system for enhanced absorption and sustained release of 10-hydrocamptothecin, J. Biomed. Nanotechnol. 11 (2) (2015) 262–273.
[40] F. Andrade, F. Antunes, A.V. Nascimento, S.B. Da Silva, J. Das Neves, D. Ferreira, B. Sarmento, Chitosan formulations as carriers for therapeutic proteins, Curr. Drug Discov. Technol. 8 (3) (2011) 157–172.

[41] A. Jintapattanakit, V.B. Junyaprasert, S. Mao, J. Sitterberg, U. Bakowsky, T. Kissel, Peroral delivery of insulin using chitosan derivatives: a comparative study of polyelectrolyte nanocomplexes and nanoparticles, Int. J. Pharm. 342 (1) (2007) 240–249.
[42] A. Lalatsa, N.L. Garrett, T. Ferrarelli, J. Moger, A.G. Schätzlein, I.F. Uchegbu, Delivery of peptides to the blood and brain after oral uptake of quaternary ammonium palmitoyl glycol chitosan nanoparticles, Mol. Pharm. 9 (9) (2012) 1764–1774.
[43] D. Narayanan, A. Anitha, R. Jayakumar, K.P. Chennazhi, In vitro and in vivo evaluation of osteoporosis therapeutic peptide PTH 1-34 loaded pegylated chitosan nanoparticles, Mol. Pharm. 10 (11) (2013) 4159–4167.
[44] S. Shu, X. Zhang, D. Teng, Z. Wang, C. Li, Polyelectrolyte nanoparticles based on water-soluble chitosan-poly(L-aspartic acid)-polyethylene glycol for controlled protein release, Carbohydr. Res. 344 (10) (2009) 1197–1204.
[45] Y. Liu, X.J. Cheng, Q.F. Dang, F.K. Ma, X.G. Chen, H.J. Park, B.K. Kim, Preparation and evaluation of oleoyl-carboxymethy-chitosan (OCMCS) nanoparticles as oral protein carriers, J. Mater. Sci. Mater. Med. 23 (2) (2012) 375–384.
[46] G. Maurstad, B.T. Stokke, K.M. Vårum, S.P. Strand, PEGylated chitosan complexes DNA while improving polyplex colloidal stability and gene transfection efficiency, Carbohydr. Polym. 94 (1) (2013) 436–443.
[47] O. Germershaus, S. Mao, J. Sitterberg, U. Bakowsky, T. Kissel, Gene delivery using chitosan, trimethyl chitosan or polyethylenglycol-graft-trimethyl chitosan block copolymers: establishment of structure–activity relationships in vitro, J. Control. Release 125 (2) (2008) 145–154.
[48] K. Wong, G. Sun, X. Zhang, H. Dai, Y. Liu, C. He, K.W. Leong, PEI-g-chitosan, a novel gene delivery system with transfection efficiency comparable to polyethylenimine in vitro and after liver administration in vivo, Bioconjug. Chem. 17 (1) (2006) 152–158.
[49] B. Layek, J. Singh, Amino acid grafted chitosan for high performance gene delivery: comparison of amino acid hydrophobicity on vector and polyplex characteristics, Biomacromolecules 14 (2) (2013) 485–494.
[50] J. Yan, Y.Z. Du, F.Y. Chen, J. You, H. Yuan, F.Q. Hu, Effect of proteins with different isoelectric points on the gene transfection efficiency mediated by stearic acid grafted chitosan oligosaccharide micelles, Mol. Pharm. 10 (7) (2013) 2568–2577.
[51] R. Martien, B. Loretz, A.M. Sandbichler, A.B. Schnurch, Thiolated chitosan nanoparticles: transfection study in the Caco-2 differentiated cell culture, Nanotechnology 19 (4) (2008), 045101.
[52] J.Y. Lee, S.H. Lee, M.H. Oh, J.S. Kim, T.G. Park, Y.S. Nam, Prolonged gene silencing by siRNA/chitosan-g-deoxycholic acid polyplexes loaded within biodegradable polymer nanoparticles, J. Control. Release 162 (2) (2012) 407–413.
[53] T.S.C. Li, T. Yawata, K. Honke, Efficient siRNA delivery and tumor accumulation mediated by ionically cross-linked folic acid–poly(ethylene glycol)– chitosan oligosaccharide lactate nanoparticles: for the potential targeted ovarian cancer gene therapy, Eur. J. Pharm. Sci. 52 (4) (2014) 48–61.
[54] B. Slütter, W. Jiskoot, Dual role of CpG as immune modulator and physical crosslinker in ovalbumin loaded N-trimethyl chitosan (TMC) nanoparticles for nasal vaccination, J. Control. Release 148 (1) (2010) 117–121.
[55] W. Yao, Y. Peng, M. Du, J. Luo, L. Zong, Preventative vaccine-loaded mannosylated chitosan nanoparticles intended for nasal mucosal delivery enhance immune responses and potent tumor immunity, Mol. Pharm. 10 (8) (2013) 2904–2914.
[56] J. Wei, W. Xue, X. Yu, X. Qiu, Z. Liu, pH Sensitive phosphorylated chitosan hydrogel as vaccine delivery system for intramuscular immunization, J. Biomater. Appl. 31 (10) (2017) 1358–1369.

[57] D. Pawar, K.S. Jaganathan, Mucoadhesive glycol chitosan nanoparticles for intranasal delivery of hepatitis B vaccine: enhancement of mucosal and systemic immune response, Drug Deliv. 23 (1) (2016) 185–194.
[58] T. Jiang, B. Singh, H.-S. Li, Y.-K. Kim, S.-K. Kang, J.-W. Nah, Y.-J. Choi, C.-S. Cho, Targeted oral delivery of BmpB vaccine using porous PLGA microparticles coated with M cell homing peptide-coupled chitosan, Biomaterials 35 (7) (2014) 2365–2373.
[59] S.N. Pawar, K.J. Edgar, Alginate derivatization: a review of chemistry, properties and applications, Biomaterials 33 (11) (2012) 3279–3305.
[60] Z. Tong, Y. Chen, W. Luo, Grafting derivate from alginate, in: V.K. Thakur (Ed.), Biopolymer Grafting, Elsevier, 2018 (Chapter 3).
[61] S. Mao, T. Zhang, W. Sun, X. Ren, The depolymerization of sodium alginate by oxidative degradation, Pharm. Dev. Technol. 17 (6) (2012) 763–769.
[62] L. Agüero, D. Zaldivar-Silva, L. Peña, M.L. Dias, Alginate microparticles as oral colon drug delivery device: a review, Carbohydr. Polym. 168 (2017) 32–43.
[63] M. George, T.E. Abraham, Polyionic hydrocolloids for the intestinal delivery of protein drugs: alginate and chitosan—a review, J. Control. Release 114 (1) (2006) 1–14.
[64] C.E. Beneke, A.M. Viljoen, J.H. Hamman, Polymeric plant-derived excipients in drug delivery, Molecules 14 (7) (2009) 2602.
[65] W.R. Gombotz, S.F. Wee, Protein release from alginate matrices, Adv. Drug Deliv. Rev. 64 (2012) 194–205.
[66] K.Y. Lee, D.J. Mooney, Alginate: properties and biomedical applications, Prog. Polym. Sci. 37 (1) (2012) 106–126.
[67] H.H. Tonnesen, J. Karlsen, Alginate in drug delivery systems, Drug Dev. Ind. Pharm. 28 (6) (2002) 621–630.
[68] C.G. Gomez, M. Rinaudo, M.A. Villar, Oxidation of sodium alginate and characterization of the oxidized derivatives, Carbohydr. Polym. 67 (3) (2007) 296–304.
[69] H.A. Kang, M.S. Shin, J.W. Yang, Preparation and characterization of hydrophobically modified alginate, Polym. Bull. 47 (5) (2002) 429–435.
[70] G. Sen, R.P. Singh, S. Pal, Microwave-initiated synthesis of polyacrylamide grafted sodium alginate: synthesis and characterization, J. Appl. Polym. Sci. 115 (1) (2010) 63–71.
[71] A. Sand, M. Yadav, K. Behari, Synthesis and characterization of alginate-g-vinyl sulfonic acid with a potassium peroxydiphosphate/thiourea system, J. Appl. Polym. Sci. 118 (6) (2010) 3685–3694.
[72] W. Pluemsab, N. Sakairi, T. Furuike, Synthesis and inclusion property of α-cyclodextrin-linked alginate, Polymer 46 (23) (2005) 9778–9783.
[73] L. Yang, B. Zhang, L. Wen, Q. Liang, Zhang, Amphiphilic cholesteryl grafted sodium alginate derivative: synthesis and self-assembly in aqueous solution, Carbohydr. Polym. 68 (2) (2007) 218–225.
[74] M.-C. Carr, C. Delestre, P. Hubert, E. Dellacherie, Covalent coupling of a short polyether on sodium alginate: synthesis and characterization of the resulting amphiphilic derivative, Carbohydr. Polym. 16 (4) (1991) 367–379.
[75] H. Bu, G.T.M. Nguyen, A.L. Kjøniksen, Effects of the quantity and structure of hydrophobes on the properties of hydrophobically modified alginates in aqueous solutions, Polym. Bull. 57 (4) (2006) 563–574.
[76] H. Yan, X. Chen, J. Li, Y. Feng, Z. Shi, X. Wang, Q. Lin, Synthesis of alginate derivative via the Ugi reaction and its characterization, Carbohydr. Polym. 136 (2016) 757–763.
[77] M.M. Soledad Lencina, Z. Iatridi, M.A. Villar, C. Tsitsilianis, Thermoresponsive hydrogels from alginate-based graft copolymers, Eur. Polym. J. 61 (2014) 33–44.

[78] E.F. Vall, M.C. Ller, A. Durand, S. Schimchowitsch, E. Dellacherie, C. Kelche, J.C. Cassel, M. Leonard, Synthesis and rheological properties of hydrogels based on amphiphilic alginate-amide derivatives, Carbohydr. Res. 344 (2) (2009) 223–228.
[79] I. Colinet, V. Dulong, G. Mocanu, L. Picton, D. Le Cerf, New amphiphilic and pH-sensitive hydrogel for controlled release of a model poorly water-soluble drug, Eur. J. Pharm. Biopharm. 73 (3) (2009) 345–350.
[80] L. Qian, L. Chen-Guang, H. Zhen-Hua, X. Fang-Fang, Preparation and characterization of nanoparticles based on hydrophobic alginate derivative as carriers for sustained release of vitamin D3, J. Agric. Food Chem. 59 (5) (2011) 1962–1967.
[81] B. Yao, C. Ni, C. Xiong, C. Zhu, B. Huang, Hydrophobic modification of sodium alginate and its application in drug controlled release, Bioprocess Biosyst. Eng. 33 (4) (2010) 457–463.
[82] M. Leonard, M.R. De Boisseson, P. Hubert, F. Dalen On, E. Dellacherie, Hydrophobically modified alginate hydrogels as protein carriers with specific controlled release properties, J. Control. Release 98 (3) (2004) 395–405.
[83] N. Işıklan, M. İnal, F. KurŞun, G. Ercan, pH responsive itaconic acid grafted alginate microspheres for the controlled release of nifedipine, Carbohydr. Polym. 84 (3) (2011) 933–943.
[84] A.W. Chan, R.J. Neufeld, Tuneable semi-synthetic network alginate for absorptive encapsulation and controlled release of protein therapeutics, Biomaterials 31 (34) (2010) 9040–9047.
[85] M.S. Mohy Eldin, E.A. Kamoun, M.A. Sofan, S.M. Elbayomi, l-Arginine grafted alginate hydrogel beads: a novel pH-sensitive system for specific protein delivery, Arab. J. Chem. 8 (3) (2015) 355–365.
[86] C. Gao, M. Liu, S. Chen, S. Jin, J. Chen, Preparation of oxidized sodium alginate-graft-poly((2-dimethylamino) ethyl methacrylate) gel beads and in vitro controlled release behavior of BSA, Int. J. Pharm. 371 (1) (2009) 16–24.
[87] V.R. Babu, K. Rao, M. Sairam, B.V.K. Naidu, K.M. Hosamani, T.M. Aminabhavi, pH sensitive interpenetrating network microgels of sodium alginate-acrylic acid for the controlled release of ibuprofen, J. Appl. Polym. Sci. 99 (5) (2006) 2671–2678.
[88] L. Paola, M. Mario, M. Giuditta, R. Roberto, M.I. Rotonda, Q. La Fabiana, C. Donatella, M. Luisa, Novel alginate-acrylic polymers as a platform for drug delivery, J. Biomed. Mater. Res. A 78A (3) (2006) 523–531.
[89] S. Belbekhouche, S. Charaabi, B. Carbonnier, Glucose-sensitive capsules based on hydrogen-bonded (polyvinylpyrrolidone/phenylboronic–modified alginate) system, Colloids Surf. B Biointerfaces 177 (2019) 416–424.
[90] N. Işıklan, Z. Altınışık, Development and characterization of dual sensitive poly(N,N-diethyl acrylamide) grafted alginate microparticles, Colloids Surf. A Physicochem. Eng. Asp. 575 (2019) 352–362.
[91] B. Gao, L. Chen, Y. Zhao, X. Yan, X. Wang, C. Zhou, Y. Shi, W. Xue, Methods to prepare dopamine/polydopamine modified alginate hydrogels and their special improved properties for drug delivery, Eur. Polym. J. 110 (2019) 192–201.
[92] R.V. Kulkarni, S. Biswanath, Novel pH-sensitive interpenetrating network hydrogel beads of carboxymethylcellulose-(polyacrylamide-grafted-alginate) for controlled release of ketoprofen: preparation and characterization, Curr. Drug Deliv. 5 (4) (2008) 256–264.
[93] S. Yang, G. Liu, Y. Cheng, Y. Zheng, Electroresponsive behavior of sodium alginate-g-poly (acrylic acid) hydrogel under DC electric field, J. Macromol. Sci. A Chem. 46 (11) (2009) 1078–1082.
[94] Y. Omagari, Y. Kaneko, J.I. Kadokawa, Chemoenzymatic synthesis of amylose-grafted alginate and its formation of enzymatic disintegratable beads, Carbohydr. Polym. 82 (2) (2010) 394–400.

[95] C. Zhang, W. Wang, T. Liu, Y. Wu, H. Guo, P. Wang, Q. Tian, Y. Wang, Z. Yuan, Doxorubicin-loaded glycyrrhetinic acid-modified alginate nanoparticles for liver tumor chemotherapy, Biomaterials 33 (7) (2012) 2187–2196.

[96] M. Davidovich-Pinhas, H. Bianco-Peled, Alginate–PEGAc: a new mucoadhesive polymer, Acta Biomater. 7 (2) (2011) 625–633.

[97] E. Hill, T. Boontheekul, D.J. Mooney, Designing scaffolds to enhance transplanted myoblast survival and migration, Tissue Eng. Part A 12 (5) (2006) 1295–1304.

[98] Y.M. Kolambkar, K.M. DuPont, J.D. Boerckel, N. Huebsch, D.J. Mooney, D.W. Hutmacher, R.E. Guldberg, An alginate-based hybrid system for growth factor delivery in the functional repair of large bone defects, Biomaterials 32 (1) (2011) 65–74.

[99] T. Miao, K.S. Rao, J.L. Spees, R.A. Oldinski, Osteogenic differentiation of human mesenchymal stem cells through alginate-graft-poly(ethylene glycol) microsphere-mediated intracellular growth factor delivery, J. Control. Release 192 (2014) 57–66.

[100] E.A. Silva, K. Eun-Suk, K. Hyun Joon, D.J. Mooney, Material-based deployment enhances efficacy of endothelial progenitor cells, Proc. Natl. Acad. Sci. U. S. A. 105 (38) (2008) 14347–14352.

[101] I. Freeman, A. Kedem, S. Cohen, The effect of sulfation of alginate hydrogels on the specific binding and controlled release of heparin-binding proteins, Biomaterials 29 (22) (2008) 3260–3268.

[102] J. Guan, L. Li, S. Mao, Applications of carrageenan in advanced drug delivery, in: J. Venkatesan, S. Anil, S.-K. Kim (Eds.), Seaweed Polysaccharides, Elsevier, 2017 (Chapter 15).

[103] V.D. Prajapati, P.M. Maheriya, G.K. Jani, H.K. Solanki, Carrageenan: a natural seaweed polysaccharide and its applications, Carbohydr. Polym. 105 (2014) 97–112.

[104] V.L. Campo, D.F. Kawano, D.B.D. Silva, I. Carvalho, Carrageenans: biological properties, chemical modifications and structural analysis—a review, Carbohydr. Polym. 77 (2) (2009) 167–180.

[105] G. Kindness, W.F. Long, F.B. Williamson, Anticoagulant effects of sulphated polysaccharides in normal and antithrombin III-deficient plasmas, Br. J. Pharmacol. 69 (4) (1980) 675–677.

[106] H. Yuan, J. Song, X. Li, N. Li, J. Dai, Immunomodulation and antitumor activity of κ-carrageenan oligosaccharides, Cancer Lett. 243 (2) (2006) 228–234.

[107] G. Kindness, F.B. Williamson, W.F. Long, Effect of polyanetholesulphonic acid and xylan sulphate on antithrombin III activity, Biochem. Biophys. Res. Commun. 88 (3) (1979) 1062–1068.

[108] J. Liu, X. Zhan, J. Wan, Y. Wang, C. Wang, Review for carrageenan-based pharmaceutical biomaterials: favourable physical features versus adverse biological effects, Carbohydr. Polym. 121 (2015) 27–36.

[109] A.G. Viana, M.D. Noseda, M.E.R. Duarte, A.S. Cerezo, Alkali modification of carrageenans. Part V. The iota–nu hybrid carrageenan from, Carbohydr. Polym. 58 (4) (2004) 455–460.

[110] A. Pourjavadi, A.M. Harzandi, H. Hosseinzadeh, Modified carrageenan 3. Synthesis of a novel polysaccharide-based superabsorbent hydrogel via graft copolymerization of acrylic acid onto kappa-carrageenan in air, Eur. Polym. J. 40 (7) (2004) 1363–1370.

[111] A. Pourjavadi, M. Sadeghi, H. Hosseinzadeh, Modified carrageenan. 5. Preparation, swelling behavior, salt- and pH-sensitivity of partially hydrolyzed crosslinked carrageenan-graft-polymethacrylamide superabsorbent hydrogel, Polym. Adv. Technol. 15 (11) (2004) 645–653.

[112] A. Bayquen, Synthesis and characterization of carboxymethyl derivatives of kappa-carrageenan, Carbohydr. Polym. 87 (2) (2012) 1810–1816.

[113] K.H. Leong, L.Y. Chung, M.I. Noordin, K. Mohamad, M. Nishikawa, Y. Onuki, M. Morishita, K. Takayama, Carboxymethylation of kappa-carrageenan for

intestinal-targeted delivery of bioactive macromolecules, Carbohydr. Polym. 83 (4) (2011) 1507–1515.
[114] H. Hosseinzadeh, Controlled release of diclofenac sodium from pH-responsive carrageenan-g-poly(acrylic acid) superabsorbent hydrogel, J. Chem. Sci. 122 (4) (2010) 651–659.
[115] R.V. Kulkarni, R. Boppana, G. Krishna Mohan, S. Mutalik, N.V. Kalyane, pH-responsive interpenetrating network hydrogel beads of poly(acrylamide)-g-carrageenan and sodium alginate for intestinal targeted drug delivery: synthesis, in vitro and in vivo evaluation, J. Colloid Interface Sci. 367 (1) (2012) 509–517.
[116] K.H. Leong, L.Y. Chung, M.I. Noordin, Y. Onuki, M. Morishita, K. Takayama, Lectin-functionalized carboxymethylated kappa-carrageenan microparticles for oral insulin delivery, Carbohydr. Polym. 86 (2) (2011) 555–565.
[117] W. Suchaoin, S. Bonengel, S. Hussain, C.W. Huck, B.N. Ma, A. Bernkop-Schnürch, Synthesis and in vitro evaluation of thiolated carrageenan, J. Pharm. Sci. 104 (8) (2015) 2523–2530.
[118] S. Lefnaoui, N. Moulai-Mostefa, Polyelectrolyte complex based on carboxymethyl-kappa-carrageenan and Eudragit RL 30D as prospective carriers for sustained drug delivery, Chem. Eng. Res. Des. 97 (2015) 165–174.
[119] V.E. Bosio, M.L. Cacicedo, B. Calvignac, I. León, T. Beuvier, F. Boury, G.R. Castro, Synthesis and characterization of $CaCO_3$–biopolymer hybrid nanoporous microparticles for controlled release of doxorubicin, Colloids Surf. B Biointerfaces 123 (2014) 158–169.
[120] P. Myrsini, N. Stavroula, K. George, et al., Novel isocyanate-modified carrageenan polymer materials: preparation, characterization and application adsorbent materials of pharmaceuticals, Polymers 9 (11) (2017) 595.
[121] S.G. Nanaki, G.Z. Kyzas, A. Tzereme, M. Papageorgiou, M. Kostoglou, D.N. Bikiaris, D.A. Lambropoulou, Synthesis and characterization of modified carrageenan microparticles for the removal of pharmaceuticals from aqueous solutions, Colloids Surf. B Biointerfaces 127 (2015) 256–265.
[122] H. Tümtürk, N. Karaca, G. Demirel, F. Şahin, Preparation and application of poly(N, N-dimethylacrylamide-co-acrylamide) and poly(N-isopropylacrylamide-co-acrylamide)/κ-Carrageenan hydrogels for immobilization of lipase, Int. J. Biol. Macromol. 40 (3) (2007) 281–285.
[123] R. Moy, Injections in aesthetic medicine: atlas of full-face and full-body treatment | Aesthetic Surgery Journal | Oxford Academic, Aesthet. Surg. J. 35 (4) (2014) NP98–NP99.
[124] C.E. Schant, G. Zuber, C. Herlin, T.F. Vandamme, Chemical modifications of hyaluronic acid for the synthesis of derivatives for a broad range of biomedical applications, Carbohydr. Polym. 85 (3) (2011) 469–489.
[125] M.K. Cowman, R. Mendichi, Methods for determination of hyaluronan molecular weight, in: H.G. Garg, C.A. Hales (Eds.), Chemistry and Biology of Hyaluronan, Elsevier Science Ltd., Oxford, 2004 (Chapter 3).
[126] R.S. Athre, Facial filler agents, Oper. Tech. Otolaryngol. Head Neck Surg. 18 (3) (2007) 243–247.
[127] J. Kablik, G.D. Monheit, L. Yu, G. Chang, J. Gershkovich, Comparative physical properties of hyaluronic acid dermal fillers, Dermatol. Surg. 35 (1) (2009) 302–312.
[128] L.W. Moreland, Intra-articular hyaluronan (hyaluronic acid) and hylans for the treatment of osteoarthritis: mechanisms of action, Arthritis Res. Ther. 5 (2) (2003) 54–67.
[129] Y.-K. Lin, Y. Matsumoto, Y. Kuroyanagi, S. Kagawa, A bilayer hyaluronic acid wound dressing to promote wound healing in diabetic ulcer, J. Bioact. Compat. Polym. 24 (5) (2009) 424–443.

[130] E.A. Balazs, Hyaluronan as an ophthalmic viscoelastic device, Curr. Pharm. Biotechnol. 9 (4) (2008) 236–238.
[131] J. Drobnik, Hyaluronan in drug delivery, Adv. Drug Deliv. Rev. 7 (2) (1991) 295–308.
[132] E.J. Oh, K. Park, K.S. Kim, J. Kim, J.-A. Yang, J.-H. Kong, M.Y. Lee, A.S. Hoffman, S.K. Hahn, Target specific and long-acting delivery of protein, peptide, and nucleotide therapeutics using hyaluronic acid derivatives, J. Control. Release 141 (1) (2010) 2–12.
[133] M.G. Piao, J.H. Kim, J.O. Kim, W.S. Lyoo, M.H. Lee, C.S. Yong, H.G. Choi, Enhanced oral bioavailability of piroxicam in rats by hyaluronate microspheres, Drug Dev. Ind. Pharm. 33 (4) (2007) 485–491.
[134] V.M. Platt, F.C. Szoka Jr., Anticancer therapeutics: targeting macromolecules and nanocarriers to hyaluronan or CD44, a hyaluronan receptor, Mol. Pharm. 5 (4) (2008) 474–486.
[135] L.K. Widjaja, M. Bora, P.N.P.H. Chan, V. Lipik, T.T.L. Wong, S.S. Venkatraman, Hyaluronic acid-based nanocomposite hydrogels for ocular drug delivery applications, J. Biomed. Mater. Res. A 102 (9) (2014) 3056–3065.
[136] G. Tripodo, A. Trapani, M.L. Torre, G. Giammona, G. Trapani, D. Mandracchia, Hyaluronic acid and its derivatives in drug delivery and imaging: recent advances and challenges, Eur. J. Pharm. Biopharm. 97 (2015) 400–416.
[137] S. Tiwari, P. Bahadur, Modified hyaluronic acid based materials for biomedical applications, Int. J. Biol. Macromol. 121 (2019) 556–571.
[138] T.K. Hill, A. Abdulahad, S.S. Kelkar, F.C. Marini, T.E. Long, J.M. Provenzale, A.M. Mohs, Indocyanine green-loaded nanoparticles for image-guided tumor surgery, Bioconjug. Chem. 26 (2) (2015) 294–303.
[139] X. Wang, X. Gu, H. Wang, Y. Sun, H. Wu, S. Mao, Synthesis, characterization and liver targeting evaluation of self-assembled hyaluronic acid nanoparticles functionalized with glycyrrhetinic acid, Eur. J. Pharm. Sci. 96 (2017) 255–262.
[140] K.Y. Choi, H.Y. Yoon, J.-H. Kim, S.M. Bae, R.-W. Park, Y.M. Kang, I.-S. Kim, I.C. Kwon, K. Choi, S.Y. Jeong, K. Kim, J.H. Park, Smart nanocarrier based on PEGylated hyaluronic acid for cancer therapy, ACS Nano 5 (11) (2011) 8591–8599.
[141] D.A. Ossipov, Nanostructured hyaluronic acid-based materials for active delivery to cancer, Expert Opin. Drug Deliv. 7 (6) (2010) 681–703.
[142] B.R. Brown, M.E. Hughes, C. Russo, Thermoelectricity in natural and synthetic hydrogels, Phys. Rev. E 70 (3) (2004), 031917.
[143] A. Fakhari, C. Berkland, Applications and emerging trends of hyaluronic acid in tissue engineering, as a dermal filler and in osteoarthritis treatment, Acta Biomater. 9 (7) (2013) 7081–7092.
[144] G. Huang, J. Chen, Preparation and applications of hyaluronic acid and its derivatives, Int. J. Biol. Macromol. 125 (2019) 478–484.
[145] M.H. Abu Elella, R.R. Mohamed, M.W. Sabaa, Synthesis of novel grafted hyaluronic acid with antitumor activity, Carbohydr. Polym. 189 (2018) 107–114.
[146] L. Bildstein, C. Dubernet, P. Couvreur, Prodrug-based intracellular delivery of anticancer agents, Adv. Drug Deliv. Rev. 63 (1) (2011) 3–23.
[147] G. Guclu, G. Gurdag, S. Ozgumus, Competitive removal of heavy metal ions by cellulose graft copolymers, J. Appl. Polym. Sci. 90 (8) (2003) 2034–2039.
[148] G.M. Campo, A. Avenoso, A. D'ascola, M. Scuruchi, V. Prestipino, A. Calatroni, S. Campo, Hyaluronan in part mediates IL-1beta-induced inflammation in mouse chondrocytes by up-regulating CD44 receptors, Gene 494 (1) (2012) 24–35.
[149] X. Jiang, D.J. Shapiro, The immune system and inflammation in breast cancer, Mol. Cell. Endocrinol. 382 (1) (2014) 673–682.

[150] H.-J. Cho, I.-S. Yoon, H.Y. Yoon, H. Koo, Y.-J. Jin, S.-H. Ko, J.-S. Shim, K. Kim, I.C. Kwon, D.-D. Kim, Polyethylene glycol-conjugated hyaluronic acid-ceramide self-assembled nanoparticles for targeted delivery of doxorubicin, Biomaterials 33 (4) (2012) 1190–1200.
[151] Y. Sun, H. Wu, W. Dong, J. Zhou, X. Zhang, L. Liu, X. Zhang, H. Cheng, J. Guan, R. Zhao, S. Mao, Molecular simulation approach to the rational design of self-assembled nanoparticles for enhanced peroral delivery of doxorubicin, Carbohydr. Polym. 218 (2019) 279–288.
[152] P. Huang, C. Yang, J. Liu, W. Wang, S. Guo, J. Li, Y. Sun, H. Dong, L. Deng, J. Zhang, Improving the oral delivery efficiency of anticancer drugs by chitosan coated polycaprolactone-grafted hyaluronic acid nanoparticles, J. Mater. Chem. B 2 (25) (2014) 4021–4033.
[153] J. Yao, Zhang, Qiang, Glycyrrhetinic acid-graft-hyaluronic acid conjugate as a carrier for;synergistic targeted delivery of antitumor drugs, Int. J. Pharm. 441 (1–2) (2013) 654–664.
[154] J. Li, M. Huo, J. Wang, J. Zhou, J.M. Mohammad, Y. Zhang, Q. Zhu, A.Y. Waddad, Q. Zhang, Redox-sensitive micelles self-assembled from amphiphilic hyaluronic acid-deoxycholic acid conjugates for targeted intracellular delivery of paclitaxel, Biomaterials 33 (7) (2012) 2310–2320.
[155] X. Jin, S. Asghar, M. Zhang, Z. Chen, L. Huang, Q. Ping, Y. Xiao, N-acetylcysteine modified hyaluronic acid-paclitaxel conjugate for efficient oral chemotherapy through mucosal bioadhesion ability, Colloids Surf. B Biointerfaces 172 (2018) 655–664.
[156] M. Zhang, S. Asghar, X. Jin, Z. Hu, Q. Ping, Z. Chen, F. Shao, Y. Xiao, The enhancing effect of N-acetylcysteine modified hyaluronic acid-octadecylamine micelles on the oral absorption of paclitaxel, Int. J. Biol. Macromol. 138 (2019) 636–647.
[157] K.S. Kim, S.W. Hong, H. Kim, M. Cho, S.K. Hahn, Hyaluronate–Flt1 peptide conjugate/epirubicin micelles for theranostic application to liver cancers, RSC Adv. 5 (60) (2015) 48615–48618.
[158] K.Y. Choi, E.J. Jeon, Y.Y. Hong, B.S. Lee, H.N. Jin, K.H. Min, Y.K. Sang, S.J. Myung, S. Lee, X. Chen, Theranostic nanoparticles based on PEGylated hyaluronic acid for the diagnosis, therapy and monitoring of colon cancer, Biomaterials 33 (26) (2012) 6186–6193.
[159] K. Ming, P. Hyunjin, F. Chao, H. Lin, C. Xiaojie, C. Xiguang, Construction of hyaluronic acid noisome as functional transdermal nanocarrier for tumor therapy, Carbohydr. Polym. 94 (1) (2013) 634–641.
[160] Y. Ji, S. Shan, M. He, C.C. Chu, Inclusion complex from cyclodextrin-grafted hyaluronic acid and pseudo protein as biodegradable nano-delivery vehicle for gambogic acid, Acta Biomater. 62 (2017) 234–245.
[161] Y.Y. Hong, H.R. Kim, G. Saravanakumar, R. Heo, Y.C. Su, W. Um, K. Kim, I.C. Kwon, J.Y. Lee, D.S. Lee, Bioreducible hyaluronic acid conjugates as siRNA carrier for tumor targeting, J. Control. Release 172 (3) (2013) 653–661.
[162] Y. Chen, X. Zhu, X. Zhang, B. Liu, L. Huang, Nanoparticles modified with tumor-targeting scFv deliver siRNA and miRNA for cancer therapy, Mol. Ther. 18 (9) (2010) 1650–1656.
[163] G. Jiang, K. Park, J. Kim, K.S. Kim, S.K. Hahn, Target specific intracellular delivery of siRNA/PEI-HA complex by receptor mediated endocytosis, Mol. Pharm. 6 (3) (2009) 727–737.
[164] Y. He, G. Cheng, L. Xie, Y. Nie, B. He, Z. Gu, Polyethyleneimine/DNA polyplexes with reduction-sensitive hyaluronic acid derivatives shielding for targeted gene delivery, Biomaterials 34 (4) (2013) 1235–1245.
[165] R. Zhang, Z. Huang, M. Xue, J. Yang, T. Tan, Detailed characterization of an injectable hyaluronic acid-polyaspartylhydrazide hydrogel for protein delivery, Carbohydr. Polym. 85 (4) (2011) 717–725.

[166] C. Fiorica, G. Pitarresi, F.S. Palumbo, M. Di Stefano, F. Calascibetta, G. Giammona, A new hyaluronic acid pH sensitive derivative obtained by ATRP for potential oral administration of proteins, Int. J. Pharm. 457 (1) (2013) 150–157.

[167] S.K. Hahn, J.S. Kim, T. Shimobouji, Injectable hyaluronic acid microhydrogels for controlled release formulation of erythropoietin, J. Biomed. Mater. Res. A 80A (4) (2007) 916–924.

[168] A.K. Jha, W. Yang, C.B. Kirn-Safran, M.C. Farach-Carson, X. Jia, Perlecan domain I-conjugated, hyaluronic acid-based hydrogel particles for enhanced chondrogenic differentiation via BMP-2 release, Biomaterials 30 (36) (2009) 6964–6975.

[169] X. Xu, A.K. Jha, R.L. Duncan, X. Jia, Heparin-decorated, hyaluronic acid-based hydrogel particles for the controlled release of bone morphogenetic protein 2, Acta Biomater. 7 (8) (2011) 3050–3059.

[170] Q. Feng, S. Lin, K. Zhang, C. Dong, T. Wu, H. Huang, X. Yan, L. Zhang, G. Li, L. Bian, Sulfated hyaluronic acid hydrogels with retarded degradation and enhanced growth factor retention promote hMSC chondrogenesis and articular cartilage integrity with reduced hypertrophy, Acta Biomater. 53 (2017) 329–342.

[171] T.-C. Lin, J.-H. Chen, Y.-H. Chen, T.-M. Teng, C.-H. Su, S.-H. Hsu, Biodegradable micelles from a hyaluronan-poly(epsilon-caprolactone) graft copolymer as nanocarriers for fibroblast growth factor 1, J. Mater. Chem. B 1 (43) (2013) 5977–5987.

[172] S. Agnello, L. Gasperini, R.L. Reis, J.F. Mano, G. Pitarresi, F.S. Palumbo, G. Giammona, Microfluidic production of hyaluronic acid derivative microfibers to control drug release, Mater. Lett. 182 (2016) 309–313.

[173] H.-Y. Lee, H.-E. Kim, S.-H. Jeong, One-pot synthesis of silane-modified hyaluronic acid hydrogels for effective antibacterial drug delivery via sol–gel stabilization, Colloids Surf. B Biointerfaces 174 (2019) 308–315.

[174] F.S. Palumbo, A. Bavuso Volpe, M.G. Cusimano, G. Pitarresi, G. Giammona, D. Schillaci, A polycarboxylic/amino functionalized hyaluronic acid derivative for the production of pH sensible hydrogels in the prevention of bacterial adhesion on biomedical surfaces, Int. J. Pharm. 478 (1) (2015) 70–77.

[175] Y. Wu, J. Yao, J. Zhou, F.Z. Dahmani, Enhanced and sustained topical ocular delivery of cyclosporine A in thermosensitive hyaluronic acid-based in situ forming microgels, Int. J. Nanomedicine 8 (2013) 3587–3601.

[176] G. Pitarresi, F.S. Palumbo, C. Fiorica, F. Calascibetta, M. Di Stefano, G. Giammona, Injectable in situ forming microgels of hyaluronic acid-g-polylactic acid for methylprednisolone release, Eur. Polym. J. 49 (3) (2013) 718–725.

[177] A. Rascón-Chu, G.H. Gomez-Rodriguez, E. Carvajal-Millan, A.C. Campa-Mada, Pectin in drug delivery applications, in: M.S. Hasnain, A.K. Nayak (Eds.), Natural Polysaccharides in Drug Delivery and Biomedical Applications, Academic Press, 2019 (Chapter 10).

[178] F. Munarin, M.C. Tanzi, P. Petrini, Advances in biomedical applications of pectin gels, Int. J. Biol. Macromol. 51 (4) (2012) 681–689.

[179] C.L. Jackson, T.M. Dreaden, L.K. Theobald, N.M. Tran, T.L. Beal, M. Eid, M.Y. Gao, R.B. Shirley, M.T. Stoffel, M.V. Kumar, Pectin induces apoptosis in human prostate cancer cells: correlation of apoptotic function with pectin structure, Glycobiology 17 (8) (2007) 805–819.

[180] U.V. Sathisha, S. Jayaram, M.A.H. Nayaka, S.M. Dharmesh, Inhibition of galectin-3 mediated cellular interactions by pectic polysaccharides from dietary sources, Glycoconj. J. 24 (8) (2007) 497–507.

[181] A. Fischer, M.C. Houzelle, P. Hubert, M.A.V. Axelos, C. Geoffroy-Chapotot, M.C. Carr, M.L. Viriot, E. Dellacherie, Detection of intramolecular associations in hydrophobically modified pectin derivatives using fluorescent probes, Langmuir 14 (16) (1998) 4482–4488.

[182] J.A. Klavons, R.D. Bennett, Preparation of alkyl esters of pectin and pectic acid, J. Food Sci. 60 (3) (2010) 513–515.
[183] M.C. Miralleshouzelle, P. Hubert, E. Dellacherie, Hydrophobic alkyl chains – pectin conjugates. comparative study of some physicochemical properties in relation to covalent coupling vs ionic association, Langmuir 17 (5) (2001) 1384–1391.
[184] M. Luca, B. Valeria, A. Pietro, D.L. Stefania, V. Vittoria, Physical and water sorption properties of chemically modified pectin with an environmentally friendly process, Biomacromolecules 12 (6) (2011) 2311–2318.
[185] J. Chen, W. Liu, C.M. Liu, T. Li, R.H. Liang, S.J. Luo, Pectin modifications: a review, Crit. Rev. Food Sci. Nutr. 55 (12) (2015) 1684–1698.
[186] O. Kurita, Y. Miyake, E. Yamazaki, Chemical modification of citrus pectin to improve its dissolution into water, Carbohydr. Polym. 87 (2) (2012) 1720–1727.
[187] H.J. Rha, I.Y. Bae, S. Lee, S.-H. Yoo, P.-S. Chang, H.G. Lee, Enhancement of anti-radical activity of pectin from apple pomace by hydroxamation, Food Hydrocoll. 25 (3) (2011) 545–548.
[188] A. Sinitsya, J. Čopı́kov, V. Prutyanov, S. Skoblya, V. Machovič, Amidation of highly methoxylated citrus pectin with primary amines, Carbohydr. Polym. 42 (4) (2000) 359–368.
[189] L. Fan, M. Cao, S. Gao, W. Wang, K. Peng, C. Tan, F. Wen, S. Tao, W. Xie, Preparation and characterization of a quaternary ammonium derivative of pectin, Carbohydr. Polym. 88 (2) (2012) 707–712.
[190] R. Sharma, M. Ahuja, Thiolated pectin: synthesis, characterization and evaluation as a mucoadhesive polymer, Carbohydr. Polym. 85 (3) (2011) 658–663.
[191] G. Perera, J. Barthelmes, A. Bernkop-Schnürch, Novel pectin–4-aminothiophenole conjugate microparticles for colon-specific drug delivery, J. Control. Release 145 (3) (2010) 240–246.
[192] S. Majzoob, F. Atyabi, F. Dorkoosh, K. Kafedjiiski, B. Loretz, A. Bernkop-Schnurch, Pectin-cysteine conjugate: synthesis and in-vitro evaluation of its potential for drug delivery, J. Pharm. Pharmacol. 58 (12) (2006) 1601–1610.
[193] J.C. Martinichen-Herrero, E.R. Carbonero, P.A.J. Gorin, M. Iacomini, Anticoagulant and antithrombotic activity of a sulfate obtained from a glucan component of the lichen Parmotrema mantiqueirense Hale, Carbohydr. Polym. 60 (1) (2005) 7–13.
[194] J. Wang, Q. Zhang, Z. Zhang, Z. Li, Antioxidant activity of sulfated polysaccharide fractions extracted from Laminaria japonica, Int. J. Biol. Macromol. 42 (2) (2008) 127–132.
[195] N.C. Maas, A.H.P. Gracher, G.L. Sassaki, P.A.J. Gorin, M. Iacomini, T.R. Cipriani, Sulfation pattern of citrus pectin and its carboxy-reduced derivatives: influence on anticoagulant and antithrombotic effects, Carbohydr. Polym. 89 (4) (2012) 1081–1087.
[196] M.I. Abdel-Hamid, K.S. Khairou, R.M. Hassan, Kinetics and mechanism of permanganate oxidation of pectin polysaccharide in acid perchlorate media, Eur. Polym. J. 39 (2) (2003) 381–387.
[197] R. Semde, A.J. Moes, M.J. Devleeschouwer, K. Amighi, Synthesis and enzymatic degradation of epichlorohydrin cross-linked pectins, Drug Dev. Ind. Pharm. 29 (2) (2003) 203–213.
[198] M.M. Fares, S.M. Assaf, Y.M. Abul-Haija, Pectin grafted poly(N-vinylpyrrolidone): optimization and in vitro controllable theophylline drug release, J. Appl. Polym. Sci. 117 (4) (2010) 1945–1954.
[199] A. Pourjavadi, S. Barzegar, Synthesis and evaluation of pH and thermosensitive pectin-based superabsorbent hydrogel for oral drug delivery systems, Starch-Starke 61 (3–4) (2009) 161–172.

[200] P.B. Sutar, R.K. Mishra, K. Pal, A.K. Banthia, Development of pH sensitive polyacrylamide grafted pectin hydrogel for controlled drug delivery system, J. Mater. Sci. Mater. Med. 19 (6) (2008) 2247–2253.

[201] M. Bhatia, R. Deshmukh, P. Choudhari, N. Bhatia, Chemical modification of pectins, characterization and evaluation for drug delivery, Sci. Pharm. 76 (2008) 775–784.

[202] Y. Zouambia, N. Moulai-Mostefa, M. Krea, Structural characterization and surface activity of hydrophobically functionalized extracted pectins, Carbohydr. Polym. 78 (4) (2009) 841–846.

[203] S.K. Jain, A. Jain, Y. Gupta, M. Ahirwar, Design and development of hydrogel beads for targeted drug delivery to the colon, AAPS PharmSciTech 8 (3) (2007) E34–E41.

[204] R. Mohammadinejad, H. Maleki, E. Larraneta, A.R. Fajardo, A.B. Nik, A. Shavandi, A. Sheikhi, M. Ghorbanpour, M. Farokhi, P. Govindh, E. Cabane, S. Azizi, A.R. Aref, M. Mozafari, M. Mehrali, S. Thomas, J.F. Mano, Y.K. Mishra, V.K. Thakur, Status and future scope of plant-based green hydrogels in biomedical engineering, Appl. Mater. Today 16 (2019) 213–246.

[205] T.F. Vandamme, A. Lenourry, C. Charrueau, J.C. Chaumeil, The use of polysaccharides to target drugs to the colon, Carbohydr. Polym. 48 (3) (2002) 219–231.

[206] J. Jung, R.D. Arnold, L. Wicker, Pectin and charge modified pectin hydrogel beads as a colon-targeted drug delivery carrier, Colloids Surf. B Biointerfaces 104 (3) (2013) 116–121.

[207] H.H. Gadalla, I. El-Gibaly, G.M. Soliman, F.A. Mohamed, A.M. El-Sayed, Amidated pectin/sodium carboxymethylcellulose microspheres as a new carrier for colonic drug targeting: development and optimization by factorial design, Carbohydr. Polym. 153 (2016) 526–534.

[208] O. Munjeri, J.H. Collett, J.T. Fell, Hydrogel beads based on amidated pectins for colon-specific drug delivery: the role of chitosan in modifying drug release, J. Control. Release 46 (3) (1997) 273–278.

[209] S.I. Hadebe, P.S. Ngubane, M.R. Serumula, C.T. Musabayane, Transdermal delivery of insulin by amidated pectin hydrogel matrix patch in streptozotocin-induced diabetic rats: effects on some selected metabolic parameters, PLoS One 9 (7) (2014), e101461.

[210] T. Takayuki, S. Mitsunobu, I. Hiroyuki, K. Koei, In situ gellable oxidized citrus pectin for localized delivery of anticancer drugs and prevention of homotypic cancer cell aggregation, Biomacromolecules 11 (12) (2010) 3525–3530.

[211] H. Sabine, M. Christiane, D. Sarah, L. Flavia, B.S. Andreas, Preactivated thiomers: evaluation of gastroretentive minitablets, Int. J. Pharm. 456 (2) (2013) 473–479.

[212] R. Sharma, M. Ahuja, H. Kaur, Thiolated pectin nanoparticles: preparation, characterization and ex vivo corneal permeation study, Carbohydr. Polym. 87 (2) (2012) 1606–1610.

[213] H. Bera, S. Kumar, Diethanolamine-modified pectin based core-shell composites as dual working gastroretentive drug-cargo, Int. J. Biol. Macromol. 108 (2018) 1053–1062.

[214] T. Katav, L.S. Liu, T. Traitel, R. Goldbart, M. Wolfson, J. Kost, Modified pectin-based carrier for gene delivery: cellular barriers in gene delivery course, J. Control. Release 130 (2) (2008) 183–191.

[215] S. Tabasum, A. Noreen, M.F. Maqsood, H. Umar, N. Akram, Z.I. Nazli, S.A.S. Chatha, K.M. Zia, A review on versatile applications of blends and composites of pullulan with natural and synthetic polymers, Int. J. Biol. Macromol. 120 (Pt A) (2018) 603–632.

[216] K.C. Cheng, A. Demirci, Pullulan: biosynthesis, production, and applications, Appl. Microbiol. Biotechnol. 92 (1) (2011) 29–44.

[217] R.S. Singh, N. Kaur, J.F. Kennedy, Pullulan and pullulan derivatives as promising biomolecules for drug and gene targeting, Carbohydr. Polym. 123 (2015) 190–207.

[218] R.S. Singh, G.K. Saini, J.F. Kennedy, Pullulan: microbial sources, production and applications, Carbohydr. Polym. 73 (4) (2008) 515–531.
[219] T.D. Leathers, Biotechnological production and applications of pullulan, Appl. Microbiol. Biotechnol. 62 (5–6) (2003) 468–473.
[220] V.D. Prajapati, G.K. Jani, S.M. Khanda, Pullulan: an exopolysaccharide and its various applications, Carbohydr. Polym. 95 (1) (2013) 540–549.
[221] R.S. Singh, N. Kaur, V. Rana, J.F. Kennedy, Recent insights on applications of pullulan in tissue engineering, Carbohydr. Polym. 153 (2016) 455–462.
[222] Q. Tong, X. Qian, L.T. Lim, Preparation and properties of pullulan–alginate–carboxymethylcellulose blend films, Food Res. Int. 41 (10) (2008) 1007–1014.
[223] S. Tiwari, R. Patil, S.K. Dubey, P. Bahadur, Derivatization approaches and applications of pullulan, Adv. Colloid Interface Sci. 269 (2019) 296–308.
[224] V. Dulong, G. Mocanu, D.L. Cerf, A novel amphiphilic pH-sensitive hydrogel based on pullulan, Colloid Polym. Sci. 285 (10) (2007) 1085–1091.
[225] K. Na, K.H. Lee, H.B. You, pH-sensitivity and pH-dependent interior structural change of self-assembled hydrogel nanoparticles of pullulan acetate/oligo-sulfonamide conjugate, J. Control. Release 97 (3) (2004) 513–525.
[226] S. Shen, H. Li, W. Yang, The preliminary evaluation on cholesterol-modified pullulan as a drug nanocarrier, Drug Deliv. 21 (7) (2014) 501–508.
[227] G. Fundueanu, M. Constantin, I. Oanea, V. Harabagiu, P. Ascenzi, B.C. Simionescu, Entrapment and release of drugs by a strict "on-off" mechanism in pullulan microspheres with pendant thermosensitive groups, Biomaterials 31 (36) (2010) 9544–9553.
[228] M.R. Rekha, C.P. Sharma, Hemocompatible pullulan–polyethyleneimine conjugates for liver cell gene delivery: in vitro evaluation of cellular uptake, intracellular trafficking and transfection efficiency, Acta Biomater. 7 (1) (2011) 370–379.
[229] B.C. Bae, K. Na, Self-quenching polysaccharide-based nanogels of pullulan/folate-photosensitizer conjugates for photodynamic therapy, Biomaterials 31 (24) (2010) 6325–6335.
[230] S. Kim, K.M. Park, Y.K. Jin, I.C. Kwon, H.G. Cho, D. Kang, I.T. Yu, K. Kim, K. Na, Minimalism in fabrication of self-organized nanogels holding both anti-cancer drug and targeting moiety, Colloids Surf. B Biointerfaces 63 (1) (2008) 55–63.
[231] S. Kim, Y.C. Su, K. Na, S.W. Kim, H.B. You, Insulinotropic activity of sulfonylurea/pullulan conjugate in rat islet microcapsule, Biomaterials 24 (26) (2003) 4843–4851.
[232] D.-X. Lu, X.-T. Wen, J. Liang, X.-D. Zhang, Z.-W. Gu, Y.-J. Fan, Novel pH-sensitive drug delivery system based on natural polysaccharide for doxorubicin release, Chin. J. Polym. Sci. 26 (3) (2008) 369–374.
[233] H. Li, S. Bian, Y. Huang, J. Liang, Y. Fan, X. Zhang, High drug loading pH-sensitive pullulan-DOX conjugate nanoparticles for hepatic targeting, J. Biomed. Mater. Res. A 102 (1) (2014) 150–159.
[234] F. Li, H. Zhang, C. Gu, L. Fan, Y. Qiao, Y. Tao, C. Cheng, H. Wu, J. Yi, Self-assembled nanoparticles from folate-decorated maleilated pullulan–doxorubicin conjugate for improved drug delivery to cancer cells, Polym. Int. 62 (2) (2013) 165–171.
[235] S. Seo, C. Lee, Y. Jung, N.A. Kun, Thermo-sensitivity and triggered drug release of polysaccharide nanogels derived from pullulan-g-poly(L-lactide) copolymers, Carbohydr. Polym. 87 (2) (2012) 1105–1111.
[236] K. Na, E.S. Lee, H.B. You, Adriamycin loaded pullulan acetate/sulfonamide conjugate nanoparticles responding to tumor pH: pH-dependent cell interaction, internalization and cytotoxicity in vitro, J. Control. Release 87 (1) (2015) 3–13.
[237] K. Na, T.B. Lee, K.H. Park, E.K. Shin, Y.B. Lee, H.K. Choi, Self-assembled nanoparticles of hydrophobically-modified polysaccharide bearing vitamin H as a targeted anti-cancer drug delivery system, Eur. J. Pharm. Sci. 18 (2) (2003) 165–173.

[238] H. Guo, Y. Liu, Y. Wang, J. Wu, X. Yang, R. Li, Y. Wang, N. Zhang, pH-sensitive pullulan-based nanoparticle carrier for adriamycin to overcome drug-resistance of cancer cells, Carbohydr. Polym. 111 (20) (2014) 908–917.
[239] M. Constantin, G. Fundueanu, F. Bortolotti, R. Cortesi, P. Ascenzi, E. Menegatti, A novel multicompartimental system based on aminated poly(vinyl alcohol) microspheres/succinoylated pullulan microspheres for oral delivery of anionic drugs, Int. J. Pharm. 330 (1) (2007) 129–137.
[240] G. Fundueanu, M. Constantin, D. Mihai, F. Bortolotti, R. Cortesi, P. Ascenzi, E. Menegatti, Pullulan–cyclodextrin microspheres: a chromatographic approach for the evaluation of the drug–cyclodextrin interactions and the determination of the drug release profiles, J. Chromatogr. B 791 (1) (2003) 407–419.
[241] W. Yang, M. Wang, L. Ma, H. Li, L. Huang, Synthesis and characterization of biotin modified cholesteryl pullulan as a novel anticancer drug carrier, Carbohydr. Polym. 99 (1) (2014) 720–727.
[242] P.R. Ravi, R. Vats, J. Balija, S.P.N. Adapa, N. Aditya, Modified pullulan nanoparticles for oral delivery of lopinavir: formulation and pharmacokinetic evaluation, Carbohydr. Polym. 110 (2014) 320–328.
[243] K. Lin, J. Yi, X. Mao, H. Wu, L.M. Zhang, L. Yang, Glucose-sensitive hydrogels from covalently modified carboxylated pullulan and concanavalin A for smart controlled release of insulin, React. Funct. Polym. 139 (2019) 112–119.
[244] F. Gheorghe, C. Marieta, A. Paolo, Preparation and characterization of pH- and temperature-sensitive pullulan microspheres for controlled release of drugs, Biomaterials 29 (18) (2008) 2767–2775.
[245] Y. Wang, H. Chen, Y. Liu, J. Wu, P. Zhou, Y. Wang, R. Li, X. Yang, N. Zhang, pH-sensitive pullulan-based nanoparticle carrier of methotrexate and combretastatin A4 for the combination therapy against hepatocellular carcinoma, Biomaterials 34 (29) (2013) 7181–7190.
[246] A.S. Juan, D.H. Hlawaty, Tubular cationized pullulan hydrogels as local reservoirs for plasmid DNA, J. Biomed. Mater. Res. A 83A (3) (2010) 819–827.
[247] J.A. San, H. Hlawaty, F. Chaubet, D. Letourneur, L.J. Feldman, Cationized pullulan 3D matrices as new materials for gene transfer, J. Biomed. Mater. Res. A 82A (2) (2010) 354–362.
[248] K. Jeong-Hun, T. Yoichi, K. Wakako, M. Atsushi, H.S. Mariko, Y. Tetsuji, Liver-targeted siRNA delivery by polyethylenimine (PEI)-pullulan carrier, Bioorg. Med. Chem. 18 (11) (2010) 3946–3950.
[249] M.R. Rekha, C.P. Sharma, Intracellular trafficking mechanism and cytosolic protein interactions of a non viral gene delivery vector: studies based on transferrin conjugated pullulan-PEI, Curr. Nanosci. 7 (6) (2011) 879–885.
[250] L.A. Ambattu, M.R. Rekha, Betaine conjugated cationic pullulan as effective gene carrier, Int. J. Biol. Macromol. 72 (2015) 819–826.
[251] L.A. Ambattu, M.R. Rekha, Collagen synthesis promoting pullulan–PEI–ascorbic acid conjugate as an efficient anti-cancer gene delivery vector, Carbohydr. Polym. 126 (2015) 52–61.
[252] S. Askarian, K. Abnous, S. Ayatollahi, S.A. Farzad, R.K. Oskuee, M. Ramezani, PAMAM-pullulan conjugates as targeted gene carriers for liver cell, Carbohydr. Polym. 157 (2016) 929–937.
[253] J. Wang, B. Dou, Y. Bao, Efficient targeted pDNA/siRNA delivery with folate–low-molecular-weight polyethyleneimine–modified pullulan as non-viral carrier, Mater. Sci. Eng. C 34 (2014) 98–109.
[254] T. Eslaminejad, S.N. Nematollahi-Mahani, M. Ansari, Synthesis, characterization, and cytotoxicity of the plasmid EGFP-p53 loaded on pullulan–spermine magnetic nanoparticles, J. Magn. Magn. Mater. 402 (2016) 34–43.

[255] L.B. Thomsen, J. Lichota, K.S. Kim, T. Moos, Gene delivery by pullulan derivatives in brain capillary endothelial cells for protein secretion, J. Control. Release 151 (1) (2011) 45–50.
[256] S.P. Jin, J.K. Park, J.P. Nam, W.S. Kim, C. Choi, M.Y. Kim, M.K. Jang, Preparation of pullulan—poly(L-lysine) and it's evaluation as a gene carrier, Macromol. Res. 20 (7) (2012) 667–672.
[257] S.S. Priya, M.R. Rekha, C.P. Sharma, Pullulan–protamine as efficient haemocompatible gene delivery vector: synthesis and in vitro characterization, Carbohydr. Polym. 102 (3) (2014) 207–215.
[258] Y. Liu, Y. Wang, C. Zhang, P. Zhou, Y. Liu, T. An, D. Sun, N. Zhang, Y. Wang, Core-shell nanoparticles based on pullulan and poly(beta-amino) ester for hepatoma-targeted code livery of gene and chemotherapy agent, ACS Appl. Mater. Interfaces 6 (21) (2014) 18712–18720.
[259] R.S. Singh, N. Kaur, V. Rana, J.F. Kennedy, Pullulan: a novel molecule for biomedical applications, Carbohydr. Polym. 171 (2017) 102–121.

CHAPTER 4

Pharmaceutical product development: A quality by design (QbD) approach

Md Saquib Hasnain[a], Syed Anees Ahmed[b], Aisha Khatoon[c], Mohd Afzal[d], Mohammed Tahir Ansari[e], Sidra Khatoon[f], Mohammad Tabish[g], Feras M. Al-Marshad[h], and Amit Kumar Nayak[i]

[a]Department of Pharmacy, Shri Venkateshwara University, Amroha, Uttar Pradesh, India
[b]Department of Pharmacology, Hygia Institute of Pharmaceutical Education and Research, Lucknow, Uttar Pradesh, India
[c]Molecular Cell Biology Laboratory, Integral Information and Research Centre-4 (IIRC-4), Department of Biosciences, Faculty of Science, Integral University, Lucknow, Uttar Pradesh, India
[d]Department of Chemistry, College of Science, King Saud University, Riyadh, Saudi Arabia
[e]School of Pharmacy, University of Nottingham Malaysia, Jalan Broga, Semenyih, Kajang, Selangor, Malaysia
[f]Department of Physics, University of Lucknow, Lucknow, Uttar Pradesh, India
[g]Department of Pharmacology, College of Medicine, Shaqra University, Shaqra, Saudi Arabia
[h]Department of Internal Medicine, College of Medicine, Shaqra University, Shaqra, Saudi Arabia
[i]Department of Pharmaceutics, Seemanta Institute of Pharmaceutical Sciences, Mayurbhanj, Odisha, India

1 Introduction

The pharmaceutical industry has been considered to be one of the highly controlled sectors which have consistently produced high-quality pharmaceutical products for human use to produce the desired pharmacotherapeutic effects for the treatment of various diseases. However, over the past few decades, the pharmaceutical industry has been continually faced with problems in terms of delivering quality medicinal products. According to a reported news story in The Wall Street Journal in September 2002, it has been reported that "while the pharmaceutical industry has a little secret of inventing revolutionary new medicines, it still has manufacturing standards. The fundamental problems relating to the poor quality of drug products may be due to more than one factor, such as variable starting materials, lack of manufacturing process automation and control, and poor understanding of the product and process parameters [1]. The primary sources of variability associated with the production of pharmaceuticals are primarily responsible for product retrieval, as presented in Fig. 1.

Due to the low quality of the pharmaceutical product, the first initiative was by The United States Food and Drug Administration (USFDA).Inculcation of quality paradigms in producing pharmaceuticals and Regulatory

Advances and Challenges in Pharmaceutical Technology
https://doi.org/10.1016/B978-0-12-820043-8.00017-7

Fig. 1 Sources of variability in the reliability of the drug product.

procedures. A design paper has been published in this regard in 2004, which demonstrated the Agency's vision to revolutionize quality "Pharmaceutical cGMP for the 21st Century" paradigm [2]. Following such initiative, the International Conference on Harmonization (ICH) developed various regulatory guidelines (Q8, Q9, and Q10) and set out the principle of quality by design (QbD) as a comprehensive approach that delivers high-quality, sustainable drug products.

QbD is not a new term. The idea was conceived by JM Juran, an American engineer, at the beginning of the 1970s through his famous book "Juran on Quality by Design" [3], and was later adopted by a range of technology-driven sectors such as telecommunications, automotive and aeronautical industries engaged in the production of high-quality products and services. The health care industry later adopted the idea, particularly by the manufacturers of medical devices in the 1990s. QbD entered the pharmaceutical industry very late in 2004, when the USFDA took the initiative to raise pharmaceutical development [2].

QbD is a consolidated, systematic approach to quality risk control and the use of various design methods, and including statistical analysis, to generate a quality product (Fig. 2). In 2011, the FDA associated the theoretical approach to risk management (ICH Q9) where the risk factor for product quality depends on (a) the severity of the drug on patients due to lack of effectiveness, (b) the probability of failure due to ambiguity of new processes or products, and (c) poor detectability due to improper analytical method [4, 5].

Subsequently, the FDA has provided regulatory flexibility for a few analytical procedures based on AQbD. In the pharmaceutical industry, QbD has

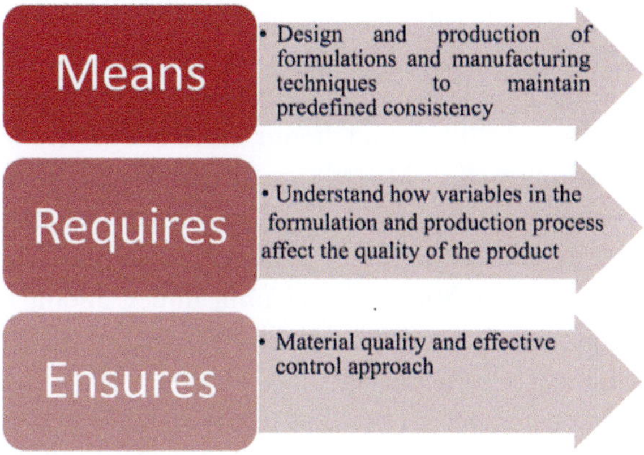

Fig. 2 What is QbD?

been mandatory for optimizing the process and product design and analytical techniques (Fig. 3) [6]. As per the International Council on Organizational Quality Guidelines, the AQbD approach to the production of analytical methods in the industry is part of the pharmaceutical output (Q8), quality risk management (Q9), and the pharmaceutical quality framework (Q10) [7–9].

The primary objective of implementing AQbD is to (a) establish a robust process, (b) reduce the probability of process failure attributed to out-of-specification (OOS), out-of-trend (OOT) performance, and out-of-control (OOC), to expedite product quality, (d) achieve regulatory flexibility in a change of method parameters within MODR (i.e., design space), and (e) to minimize the cost of analysis [10]. There have been very few reports of QbD literature on other drugs [11, 12]. However, there is a misinterpretation on AQbD, where many investigators interpret the design of experiments (DOEs) as an AQbD approach [13]. It is in contrast to the reality

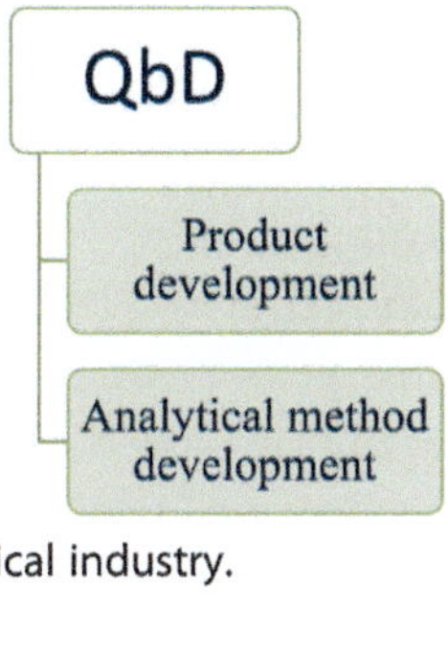

Fig. 3 QbD in the pharmaceutical industry.

that DoE is part of the QbD approach. Concerning commonly used analytical techniques in the pharmaceutical industry, chromatographic techniques such as high-performance liquid chromatography (HPLC), ultra-performance liquid chromatography (UPLC), and liquid chromatography-mass spectroscopy/mass spectrometry (LC-MS/MS) are widely used in pharmaceutical research due to their acceptable process performance (specificity, mass spectrometry) However, the development of a new analytical method using these techniques is difficult due to the lack of scientific understanding between input variables of the process (parameters, method parameters, raw materials, and reagents) and the response of the method. Due to a large number of input parameters and their impact on method response, there is often a high risk of method failure during method change [14, 15]. At this point, the AQbD process is the ideal answer for achieving a high robust analytical approach in which the effect of critical method variables (CMVs) on approach output is minimized by establishing a scientific relationship between CMVs and system responses so that AQbD can be used to design method performance as required by regulatory authorities. Therefore, AQbD may also be referred to as "Template Analytical Results."

In the pharmaceutical industry, QbD has developed the idea of creating a comprehensive production process, accelerating the product quality and production products in terms of "six sigma." The process of understanding control and capacity (PUCC) has been applied, which is a continuous improvement loop process. "Six Sigma" is a set of procedures designed to consistently develop processes that remove statistically significant defects. Because it was originally developed, six sigma has become an essential component of a variety of Total Quality Management (TQM) initiatives [16]. A large number of OOT, OOS, OOC, and Out-of-Statistics (OOSC) studies suggest that the current structure of the pharmaceutical industry is not immune to these issues. At this point, the introduction of QbD has been made mandatory in some countries, notably the European Medicines Agency (EMA) and other ICH countries [17]. Guideline for the International Conference on Hominization (ICH) Q8 (R1) defines QbD as a "Systematic approach to production that begins with desired goals and emphasizes knowledge of products and process control, based on scientific evidence and quality risk management" [18]. It suggests that scientifically developed product and process performance characteristics are required to achieve particular objectives, but not based on test performance or quality performance. QbD concepts are well established in ICH guidance Q8 (R1): pharmaceutical growth, Q9: quality risk management, and Q10: pharmaceutical quality framework (Fig. 4).

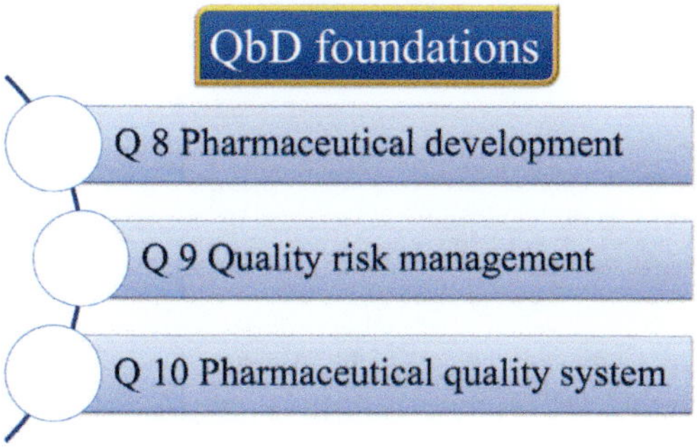

Fig. 4 QbD foundations.

2 QbD representatives

There are three main participants in QbD, including the end-user (patients), the pharmaceutical industry, and the regulatory body. Among these, the end-consumer takes the highest position in the QbD triangle. In this, patient well-being is the ultimate aim of applying QbD concepts in product development practice [19]. QbD delivers quality performance beyond customer expression, effectively reducing the number of defects at a very minimal level. Besides, there are also significant benefits for the other two stakeholders.

The pharmaceutical industry gains superior product and process comprehension, less batch failure, more reliable and successful control of the scale-up process, and a higher return on investment, leading to total cost savings. Besides, QbD-based production encourages risk-based detection and reduction, enhances yield, reduces development time, cost and efforts, reduces testing, reduces post-approval submissions, increases confidence in process validation, and batch-to-batch performance, reduces time to market along with reduced demand recalls and rejects leading to more drug availability. Besides, the QbD approach to pharmaceutical production also offers a more significant opportunity for flexible regulatory supervision and decreases post-approval submissions [20].

3 Elements of QbD

QbD plays an essential role in the design and production of the formulation process, ensuring the necessary product specifications [21, 22]. As per ICH guidelines Q8, Q9, and Q10, QbD elements are the quality target product profile (QTPP), process analytical technology (PAT), critical process

Fig. 5 Factor required during the production of pharmaceuticals.

parameters (CPP), experiment design (DoE), risk evaluation, control strategy and design space. The data and knowledge collected from the pharmaceutical production and manufacturing studies provide a scientific approach to the concept of requirements and manufacturing controls [23–25].

It addresses the different elements of design efficiency. These are the fundamental basis for the QbD approach to growth associated with the enabler [26–28]. It includes the following main factor in the production of pharmaceuticals as given below (Fig. 5).

Quality target product profile (QTPP): According to ICH Q8 [R2], QTPP is a "Prospective overview of the quality features of the drug product that would be best achieved to ensure the optimal quality, taking into account the safety and effectiveness of the drug product." It is essentially a method for the implementation of a drug strategy. Recently, QTPP is commonly used in growth planning, clinical and commercial decision-making, regulatory agency interactions, and risk management. These are the consistency characteristics that the drug product should provide to reproduce the therapeutic benefit promised on the label. QTPP guides formulation scientists to develop formulation strategies and to keep the formulation effort centered and effective.

Quality attributes: If QTPP has been identified, the next step is to identify the related critical quality attributes (CQAs). A critical quality attribute, as defined in ICH Q8 [R2], is a physical, chemical, biological or microbiological property or characteristic which should be within limits, range or distribution needed to ensure the desired quality of the product. CQAs are usually related to raw materials [drug content, excipients], intermediates [in-process materials], and the drug product. Drug product CQAs derived from QTPP are used to direct the production of the product and process.

Drug product CQAs are critical for the performance of the product, i.e., the desired quality, protection, and efficacy. This means that CQAs are subsets of QTPP that can be changed by changing the formulation or process variables. For example, QTPP may provide additional quality attributes of the drug product, such as strength and dosage type, which are not part of CQA as they would not alter during the drug development process. However, QTTP properties, such as assay, material uniformity, dissolution, and permeation flux, may also be part of CQA as they can be changed by formulation or process variables [26–28].

4 Target product quality profile

The target product profile (TPP) is widely recognized as a method for creating a conceptual basis for drug production—"planning with an end in mind." More recently, the extended use of TPPs in development planning, clinical and commercial decision-making, regulatory agency interactions, and risk management have begun to grow. The target profile is a description of the drug development program defined in the sense of the prescribing information objectives [29, 30]. TPP plays a vital role in the entire drug discovery and development process, such as [31] successful optimization of a drug candidate; [32] organizational decision-making; [33] design of clinical trial strategies; and [34] productive interaction with regulatory authorities. TPP is usually represented mainly in clinical terminology such as clinical pharmacology, indications and use, contraindications, alerts, precautions, adverse effects, substance misuse and dependency, overdose, etc. It is also arranged according to the main parts of the product mark. TPP thus relates drug development practices to particular claims intended for inclusion on the label of the drug. Target Product Quality Profile (TPQP) is a concept that is a natural extension of TPP to product quality. These are the consistency characteristics that the drug product should provide to reproduce the therapeutic benefit promised on the label. TPQP guides formulation scientists to develop formulation strategies and to keep the formulation effort centered and effective. TPQP is related to identification, assay, dosage type,

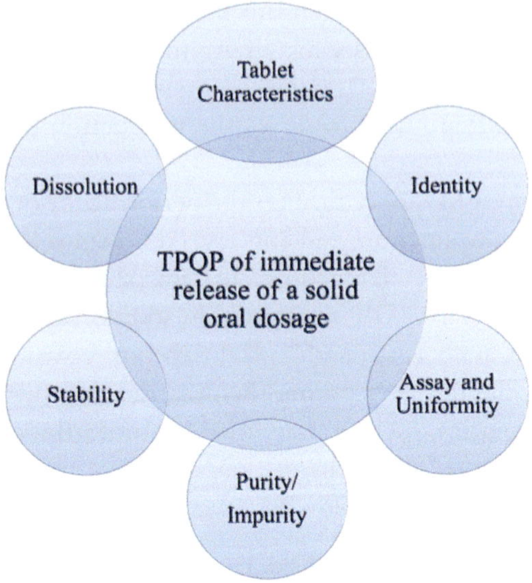

Fig. 6 Target product quality profile (TPQP) of the immediate release of a solid oral dosage.

purity, label stability. For example, a standard TPQP of immediate release of a solid oral dosage form as given below (Fig. 6) [35].

The TPQP of generic drugs can be easily determined from a reference drug. In addition to other information available from scientific literature and probably pharmacopeia, the TPQP can be used to define product specifications to some degree even before the product is produced. Predefined, high-quality product requirements make product and process design and development more analytical and persuasive.

5 Design product and manufacturing processes

5.1 Product design and development

To design and produce a durable generic product with a favorable TPQP, product development scientists need to give thoughtful deliberation to the biopharmaceutical properties of the drug material. These biopharmaceutical properties shall include physical, chemical, and biological properties [36]. Physical properties include a physical description (particle size, shape, and distribution), polymorphism, aqueous solubility, pH feature, hygroscopicity, and melting points. Pharmaceutical solid polymorphism, for example, has gained a lot of attention lately. Its effect on product quality and efficiency was addressed in recent review articles [37–39]. Chemical properties include pKa, chemical stability in solid-state and solution, and photolytic and oxidative stability, whereas biological properties have partition coefficient, membrane permeability and/or oral bioavailability. Biopharmaceutical properties should be tested for any form of interest in production and any form that could potentially be developed during processing (e.g., hydrates, anhydrates) or in vivo (e.g., less soluble salts, polymorphic forms, hydrates) [40]. The investigation of these properties is referred to as preformulation in pharmaceutical science. The preformulation studies aim to determine the appropriate salt and polymorphic form of the drug product, to evaluate and understand its essential properties, and to generate a thorough understanding of the stability of the drug under different processing and in vivo conditions, leading to an optimal drug delivery system. Pharmaceutical preformulation tests should be performed regularly to better match the dosage and processing components with the drug product and performance requirements. Biopharmaceutical evaluation offers the information required to select a solid type, to evaluate the production of a drug candidate and to decide its classification according to the Biopharmaceutical Classification System (BCS) [41–43]. BCS is a scientific method for categorizing a drug product based on its aqueous solubility, dosage and intestinal permeability [44, 45]. The BCS guidance is usually considered conservative about the class limits of solubility, permeability and dissolution requirements. Thus, enhanced consideration has been given to the possibility of modifying these boundaries and requirements [46, 47]. BCS will help deliberate efforts on production and dosage options to address the limitations of poor solubility, poor permeability, or poor stability [36]. In general, BCS Class I drugs present fewer developmental challenges. Class III BCS can be conveniently formulated. The low absorption of BCS Class II drugs can be overcome by different formulation technologies [35]. The distribution of BCS Class IV drugs is challenging.

Class V drugs, a new class developed by Amidon for developmental purposes [36], consisting of drugs with substantial pre-systemic degradation in the GI tract. Although the enteric coating technique is relatively effective in protecting the drug from stomach degradation, it may substantially affect the plasma concentration profiles. Mechanical properties, while not always studied in-depth, can have a profound effect on the production and processing of solid dosage forms [48]. A detailed understanding of the mechanical effects of nanoparticles and excipients can be helpful in (1) designing a method of processing, such as granulation or direct compression, (2) rationally selecting excipients whose properties may compensate for the properties of the drug product, and (3) continuing to help to define key material characteristics and root cause analysis during process scale-up or failure. Pharmaceutical materials may be elastic, plastic, viscoelastic, porous, soft, hard or brittle. Alternative techniques exist in the literature [42] for the testing of these mechanical properties. Awareness of the mechanical properties of drugs and excipients is expected to play a more significant role in product design and production in the future.

Drug-excipient compatibility has been described as one of the most irritating, upsetting, and perplexing formulation problems [49]. While excipients that alter the stability and bioavailability of drugs, the general principles for selecting acceptable excipients for dosage forms are not well established and excipients are sometimes selected without a systematic drug-excipient compatibility test. To prevent expensive resource waste and time delays, ICH Q8 suggests drug-excipient compatibility test achieves early drug-excipient compatibility predictions (3). Systematic drug-excipient compatibility studies provide many advantages: justifying unforeseen stability problems that typically lead to increased time and cost, optimizing formulation stability; and improving knowledge of drug-excipient interactions that can help with root cause analysis if stability problems occur.

However, despite its importance, there is no widely agreed way of conducting drug-excipient compatibility studies in this emerging field. One approach is thermal analysis [50], where the physical properties of a material (e.g., melting point) and reaction products are determined by temperature function. In contrast, the material is subject to a controlled temperature program. Another form uses isothermal stress [51, 52]. Usually, this approach involves storing drug-excipient blends or compacts with or without high-temperature moisture and evaluating the drug content or degradation product formation as a function of time. Both methods may be used together to determine the compatibility of medications with selected excipients.

5.2 Process design and development

Firmly speaking, the design and production of the process and the product cannot be isolated as the formulation cannot become a product without a process. The design of the process is the initial stage of the development of the process. The description of the commercial manufacturing processes is defined on paper, including the expected scale of production. This should include all considerations that need to be considered for the design of the process, including plant, equipment, transfer of materials and manufacturing variables [48]. The target product quality profiles are other considerations to be considered for the design of the process. Depending on the manufactured product, the type of procedure and the process expertise of the production scientists, preliminary feasibility studies may be required before completion of the design and production of the process. The choice of process form depends on the nature of the product and the properties of the materials. For example, tablet development usually involves one of two methods: direct compression or granulation. Direct compression is the clearest, easy-to-control and cheapest tablet manufacturing process. It uses two primary unit operations, mixing and compressing, to create the finished tablet.

Direct compression is used when the ingredients can be combined, put on a tablet press and turned into a high-quality tablet without any of the ingredients needing to be modified [53]. If the powder is very fine, fluffy, does not remain mixed, or does not compact, it may be granulated. Granulation is the method of gathering the particles together, forming bonds between them. Bonds are created by compression or by the use of a binding agent. Wet granulation is the method used for the manufacture of liquid to powder, and is one of the most common methods to granulate. The dry granulation process is used to create granules without the use of a liquid solution. Forming granules without moisture involves the compaction and densification of the powders. Dry granulation may be carried out on a tablet press using slow tooling or, more commonly, on a roller compactor.

Pharmaceutical research scientists have only begun to make use of computer-aided process design (CAPD) and process modeling to help the production and optimization of manufacturing processes [54]. Process simulation has been effectively used in the chemical and oil industries since, the early 1960s to facilitate the production and optimization of the design and operation of integrated processes. Similar benefits can be expected from the application of CAPD and simulation in the pharmaceutical industry. Currently, CAPD and process modeling are commonly used in the manufacture of the drug product. The use of CAPD and process simulation in the design of the drug product is minimal. This is primarily because the pharmaceutical industry has historically concentrated on new drug research and development, and the difficulty of drug product manufacturing operations is not well known. With the FDA's focus on QbD and industry and drug product cost pressures, this trend is expected to shift. The use of CAPD and process simulation can result in more efficient processes being built more efficiently and at a lower cost, resulting in better quality products.

6 Control manufacturing processes to produce consistent quality over time

Under the QbD for generic drugs, the effects of raw materials, including both the drug content and the excipients, and the process parameters on the consistency of the product, are well understood. This ensures that producers know the operational range and the established range of critical raw material attributes and process parameters. The operating range is specified as the upper and/or lower limits for the raw material features and process parameter values between which the quality and parameter are regularly monitored during production to ensure reproducibility. The proven range is defined as the upper and lower limits for the process parameter values among which the parameter is reported to produce a high-quality product that provides the therapeutic effect indicated on the label [55]. The proven range can be defined on the basis of historical and/or experimental evidence. It can also be defined on the basis of scientific and organizational judgment and expertise. In QbD, design space is characterized as a multifaceted combined application and interaction of input variables (e.g., material attributes) and process parameters which are said to deliver quality assurance. (3). Working within the approved design area of the FDA is not considered to be a move. Moving out of the design area is considered to be a change that will usually trigger a regulatory post-approval change process. The design space proposed by the applicant is subjected to regulatory evaluation and approval. Build space for generic drugs is likely to be developed in small-scale batches using experimental design (DOE) and prior experience will need to be checked on a commercial scale. The design space depends on the design theory of the equipment and the size of the batch. Control space (or standard operating ranges) is defined as the upper and/or lower limits for the critical raw material attributes and process parameters. The parameter and the material are regularly monitored during production to ensure reproducibility. The control space should be inside the design area. If the control space is much smaller than the design space, the process is considered to be robust. Otherwise, strict process control will be necessary to ensure that the process can be continuously operated within the design area.

A mixture of statistical and knowledge-based models is applied to the conventional approaches used in process monitoring. In-process research has played a significant role in the monitoring and regulation of pharmaceutical processes. If the predefined limits are not met by any in-process test outcome, the batch will be scrapped and the primary cause of the failings will be identified and corrected. If required, the process is adjusted and updated to allow the in-process or end-process test results to reach the predefined limits. The solution to QbD is more constructive. During the design phase, process steps are established, which might fail to achieve quality targets. As a first step towards QbD, process monitoring can then be developed to provide an advance indication of potential failure. Full QbD establishment requires process monitoring of critical steps to ensure that consistency is maintained [56, 57].

7 Conclusion

The present chapter provides a thorough overview of the product development using QbD approach for drug delivery of different drugs. At present, different forms of drug delivery have emerged as new drug delivery carriers and have evolved at an impressive pace and are also at a peak in the pharmacy. Current technology includes the delivery of drugs by inserting drug molecules into newer drug delivery systems, using biological vehicles and having a range of industrial as well as therapeutic advantages. The pharmaceutical QbD for generic drugs, described its basic concepts and elements, and discusses its importance in ensuring pharmaceutical quality, with a focus on dosage forms of small molecules. In pharmaceuticals, QbD is used to design and produce the consistency of a medication in order to ensure the continuous delivery of drugs and increase its performance. The drug delivery from the product is required to be administered at the site at an acceptable rate, potency and purity. There are several regulatory bodies for the production of pharmaceutical products that ensure human protection from unintended side effects, which was popular in the early 20th century.

References

[1] P. Ramesh, Transdermal delivery of drugs, Indian J. Pharm. 29 (3) (1997) 140.
[2] P.R. Rao, P.V. Diwan, Formulation and in vitro evaluation of polymeric films of diltiazem hydrochloride and indomethacin for transdermal administration, Drug Dev. Ind. Pharm. 24 (4) (1998) 327–336.
[3] P.R. Rao, P.V. Diwan, Permeability studies of cellulose acetate free films for transdermal use: influence of plasticizers, Pharm. Acta Helv. 72 (1) (1997) 47–51.
[4] S. Chatterjee, QbD considerations for analytical methods—FDA perspective, in: US IFPAC Annual Meeting, Jan 25, 2013.
[5] Y. Tang, Quality by Design Approaches to Analytical Methods-FDA Perspective, FDA/CDER, 2011.
[6] J. Ermer, Quality by design in pharmaceutical analysis, in: Conference Organized by European Compliance Academy, Vienna, Austria, 2013, pp. 1–2.

[7] Guideline IH, Pharmaceutical Development. Q8 (R2) Current Step, 2009, p. 4. 23.
[8] International Conference on Harmonization (ICH) Tripartite Guidelines, ICH Q9: Quality Risk Management, 2006.
[9] Guideline DC, Pharmaceutical Quality System. Q10, Current Step, 2008, p. 4.
[10] R. Peraman, K. Bhadraya, R.Y. Padmanabha, Analytical quality by design: a tool for regulatory flexibility and robust analytics, Int. J. Anal. Chem. 2015 (2015) 1–9.
[11] R. Peraman, B. Kalva, S. Shanka, Y.P. Reddy, Analytical quality by design (AQbD) approach to liquid chromatographic method for quantification of acyclovir and hydrocortisone in dosage forms, Anal. Chem. Lett. 4 (5–6) (2014) 329–342.
[12] R. Peraman, K. Bhadraya, Y.P. Reddy, C.S. Reddy, T. Lokesh, Analytical quality by design approach in RP-HPLC method development for the assay of etofenamate in dosage forms, Indian J. Pharm. Sci. 77 (6) (2015) 751.
[13] B.G. Sachin, R.A. Rani, G.B. Omprakash, S.T. Sanjay, Analytical method development and validation by QbD approach, Der Pharma. Sinica 6 (2015) 18–24.
[14] P. Vandana, D. Harish, Product lifecycle management in pharmaceuticals, J. Med. Mark. 12 (2012) 150–158.
[15] T. Hein, Oracle Life Sciences, Product lifecycle Management for the Pharmaceutical Industry. Improving Clinical Development and Manufacturing Processes in Pharmaceutical R and D Organization, Oracle Life Sciences, 2010.
[16] A.R. Chabukswar, S.C. Jagdale, S. Kuchekar, Six sigma: process of understanding the control and capability of ranitidine hydrochloride tablet, J. Young Pharm. 3 (2011) 15–25.
[17] L.D. Torbeck, Statistical silutions-00S, 00T, 00C, and 00SC-current GMPs demand full under-standng of out-of-control concepts, Pharm. Technol. 35 (10) (2011) 46.
[18] International Conference on Harmonization (ICH), Tripartite guidelines, ICH Q8 (R2): Pharmaceutical Development, London, 2009.
[19] J. Woodcock, The concept of pharmaceutical quality, Am. Pharm. Rev. 7 (6) (2004) 10–15.
[20] R.A. Lionberger, S.L. Lee, L. Lee, A. Raw, X.Y. Lawrence, Quality by design: concepts for ANDAs, AAPS J. 10 (2) (2008) 268–276.
[21] R.S. Kenett, D.A. Kenett, Quality by design applications in biosimilar pharmaceutical products, Accred. Qual. Assur. 13 (12) (2008) 681–690.
[22] Guideline IHT, Pharmaceutical Quality System q10. Current Step, 2008, p. 4.
[23] J.N. Sangshetti, M. Deshpande, Z. Zaheer, D.B. Shinde, R. Arote, Quality by design approach: regulatory need, Arab. J. Chem. 10 (2017) S3412–S3425.
[24] X.Y. Lawrence, G. Amidon, M.A. Khan, et al., Understanding pharmaceutical quality by design, AAPS J. 16 (4) (2014) 771–783.
[25] Guideline IHT, Pharmaceutical Development. Q8 Current Step, 2009, p. 4.
[26] US Food and Drug Administration, Guidance for Industry. Q8 (R2) Pharmaceutical Development, Food and Drug Administration, Maryland, 2009.
[27] A. Gandhi, C. Roy, Quality by Design [QbD] in pharmaceutical industry: tools, perspectives and challenges, PharmaTutor 4 (11) (2016) 12–20.
[28] V. Mogal, A review on quality by design, pharmaceutical, and biological evaluation, Pharm. Biol. Eval. 3 (2016) 313–319.
[29] J.M. Delasko, D.M. Cocchetto, L.B. Burke, Target Product Profile: Beginning Drug Development With the End in Mind. Update, 2005.
[30] US Food and Drug Administration, Guidance for Industry and Review Staff Target Product Profile—A Strategic Development Process Tool, March, FDA website, 2007.
[31] Food and Drug Administration, Final Report on Pharmaceutical cGMPs for the 21st Century—A Risk-Based Approach, 2003. http://www.fda.gov/cder/gmp/gmp2004/GMP_finalreport2004.htm. (Accessed 6 April 2005).

[32] L.X. Yu, A. Raw, R. Lionberger, R. Rajagopalan, L.M. Lee, F. Holcombe, R. Patel, F. Fang, V. Sayeed, P. Schwartz, R. Adams, US FDA question-based review for generic drugs: a new pharmaceutical quality assessment system, J. Gene Med. 4 (4) (2007) 239–246.
[33] Food and Drug Administration CDER, Guidance for Industry, Q8 Pharmaceutical Development, May 2006.
[34] M. Nasr, Risk-based CMC review paradigm, in: Advisory Committee for Pharmaceutical Science Meeting, July 20, 2004.
[35] Food and Drug Administration Office of Generic Drugs, Model Quality Overall Summary for IR Product. http://www.fda.gov/cder/ogd/OGD_Model_QOS_IR_Product.pdf. (Accessed 31 March 2006).
[36] G.E. Amidon, X. He, M.J. Hageman, in: D.J. Abraham (Ed.), Burgers Medicinal Chemistry and Drug Discovery, vol. 2.,Ch 18, Wiley-Interscience, New York, 2004.
[37] X.Y. Lawrence, M.S. Furness, A. Raw, K.P. Outlaw, N.E. Nashed, E. Ramos, S.P. Miller, R.C. Adams, F. Fang, R.M. Patel, F.O. Holcombe, Scientific considerations of pharmaceutical solid polymorphism in abbreviated new drug applications, Pharm. Res. 20 (4) (2003) 531–536.
[38] A.S. Raw, M.S. Furness, D.S. Gill, R.C. Adams, F.O. Holcombe Jr., X.Y. Lawrence, Regulatory considerations of pharmaceutical solid polymorphism in abbreviated new drug applications (ANDAs), Adv. Drug Deliv. Rev. 56 (3) (2004) 397–414.
[39] S.P. Miller, A.S. Raw, L.X. Yu, Scientific considerations of pharmaceutical solid polymorphism in regulatory applications, in: Polymorphism, Wiley-VCH Verlag GmbH & Co. KGaA, 2006, pp. 385–403.
[40] G.E. Amidon, Intrinsic properties of drug substance, excipients, and process selection, in: FDA OGD Regulatory Science Training Series. Manufacturing Science and Scale-Up Challenges, 2005.
[41] D. Sun, L.X. Yu, M.A. Hussain, D.A. Wall, R.L. Smith, G.L. Amidon, In vitro testing of drug absorption for drug'developability'assessment: forming an interface between in vitro preclinical data and clinical outcome, Curr. Opin. Drug Discov. Devel. 7 (1) (2004) 75.
[42] L.X. Yu, An integrated absorption model for determining dissolution, permeability, and solubility limited absorption, Pharm. Res. 16 (1999) 1884–1888.
[43] X.Y. Lawrence, C.D. Ellison, A.S. Hussain, Predicting human oral bioavailability using in silico models, in: R. Krishna (Ed.), Applications of Pharmacokinetic Principles in Drug Development, Kluwer, New York, 2004.
[44] G.L. Amidon, H. Lennernäs, V.P. Shah, J.R. Crison, A theoretical basis for a biopharmaceutic drug classification: the correlation of in vitro drug product dissolution and in vivo bioavailability, Pharm. Res. 12 (3) (1995) 413–420.
[45] Food and Drug Administration CDER, Guidance for industry, Waiver of In Vivo Bioavailability and Bioequivalence Studies for Immediate-Release Solid Oral Dosage Forms Based on a Biopharmaceutics Classification System, CDER/FDA, August 2000.
[46] L.X. Yu, G.L. Amidon, J.E. Polli, H. Zhao, M.U. Mehta, D.P. Conner, V.P. Shah, L.J. Lesko, M.L. Chen, V.H. Lee, A.S. Hussain, Biopharmaceutics classification system: the scientific basis for biowaiver extensions, Pharm. Res. 19 (7) (2002) 921–925.
[47] J.E. Polli, L.X. Yu, J.A. Cook, G.L. Amidon, R.T. Borchardt, B.A. Burnside, P.S. Burton, M.L. Chen, D.P. Conner, P.J. Faustino, A.A. Hawi, Summary workshop report: biopharmaceutics classification system—implementation challenges and extension opportunities, J. Pharm. Sci. 93 (6) (2004) 1375.
[48] G.E. Amidon, Physical and mechanical property characterization of powders, in: H.G. Brittain (Ed.), Physical Characterization of Pharmaceutical Solid, Marcel Dekker, New York, 1995, pp. 281–320.

[49] S. Jain, Mechanical properties of powders for compaction and tableting: an overview, Pharm. Sci. Technol. Today 2 (1) (1999) 20–31.
[50] H. Dubin, Formulation frustrations, Drug Deliv. Technol. 5 (8) (2005) 108–114.
[51] R.K. Verma, S. Garg, Selection of excipients for extended-release formulations of glipizide through drug–excipient compatibility testing, J. Pharm. Biomed. Anal. 38 (4) (2005) 633–644.
[52] K.C. Waterman, R.C. Adami, Accelerated aging: prediction of chemical stability of pharmaceuticals, Int. J. Pharm. 293 (1–2) (2005) 101–125.
[53] A.T. Serajuddin, A.B. Thakur, R.N. Ghoshal, M.G. Fakes, S.A. Ranadive, K.R. Morris, S.A. Varia, Selection of solid dosage form composition through drug–excipient compatibility testing, J. Pharm. Sci. 88 (7) (1999) 696–704.
[54] M. Gibson, Product optimization, in: M. Gibson (Ed.), Pharmaceutical Preformulation and Formulation, Taylor & Francis, New York, 2001, pp. 295–330.
[55] M.D. Tousey, The granulation process 101: basic technologies for tablet making, Pharm. Technol. 26 (2002) 8–13.
[56] D.P. Petrides, A. Koulouris, P.T. Lagonikos, The role of process simulation in pharmaceutical process development and product commercialization, Pharm. Eng. 22 (1) (2002) 56–65.
[57] F.A. Menard, Quality by design in generic drug development, in: Presentation to FDA Office of Generic Drugs, 2006, p. 25.

CHAPTER 5

Kokum butter and rice bran oil-based oleogels as novel ocular drug delivery systems

Somali Dhal[a], Dilshad Qureshi[a], Biswaranjan Mohanty[b], Samarendra Maji[c], Arfat Anis[d], Doman Kim[e], Preetam Sarkar[f], and Kunal Pal[a]

[a]Department of Biotechnology and Medical Engineering, National Institute of Technology Rourkela, Rourkela, Odisha, India
[b]Department of Pharmaceutics, Institute of Pharmacy and Technology, Salipur, Cuttack, Odisha, India
[c]Department of Chemistry, Faculty of Engineering and Technology, SRM Institute of Science and Technology, SRM Nagar, Kattankulathur, Tamil Nadu, India
[d]SABIC Polymer Research Center, Department of Chemical Engineering, King Saud University, Riyadh, Saudi Arabia
[e]Department of International Agricultural Technology & Institute of Green BioScience and Technology, Seoul National University, Seoul, Republic of Korea
[f]Department of Food Process Engineering, National Institute of Technology, Rourkela, India

1 Introduction

Garcinia indica, commonly known as kokum, is an ancient fruit that is widely found in the western coasts of India. The fruits are used for the preparation of cooled drinks (sharbat). The various parts of the kokum tree are used for culinary, pharmaceutical, nutraceuticals, confectionery, and industrial applications. The different parts of the tree have also found applications in ayurvedic medicine [1]. The seeds of the plant consist of vegetable fat, which is regarded as Kokum Butter (KB). The fat is composed of ~33–44% of di-saturated and mono-saturated glycerides [2]. KB remains as solid at room temperature. It has been used for the production of stearic acid [3]. The presence of the saturated glycerides imparts gelation capability to KB. The fat has been used for converting apolar liquids into solids. Thus, it has been explored for inducing gelation of the liquid oils [4]. It appears as light gray to yellow and appears as greasy solid textured mass. It has a bland taste and mild odor [5]. Drug delivery vehicles with different properties have been developed using KB. Solid lipid nanoparticles of KB has been used for the targeted delivery of nevirapine to the brain tissues [6]. KB has been successfully used to synthesize delivery vehicles for pregabalin to treat anal pruritis and other perianal disorders [7]. Kumar et al. [8] have stated that KB can be used as a natural excipient for designing controlled drug delivery systems [8].

Advances and Challenges in Pharmaceutical Technology
https://doi.org/10.1016/B978-0-12-820043-8.00006-2

The presence of linoleic and linolenic acids in KB helps to improve the transdermal permeation of the drug molecules [9].

Rice bran oil (RBO) is obtained from the rice bran. RBO contains high levels of phytosterols, gamma-oryzanol, and sterolins, which leads to immune-stimulation effects. Due to the aforesaid reason, RBO has been reported to modulate the immune system [10]. RBO imparts antioxidant properties against free radicals due to its specific micronutrient composition [11, 12]. RBO-based bigels have been explored as matrices for controlled delivery applications [13]. Recently, Eleftheriadis et al. [14] had reported the use of RBO for the synthesis of self-nanoemulsifying drug delivery systems for the delivery of poorly soluble drugs [14].

Gel-based products have been categorized into hydrogels, emulgels, bigels, and oleogels (also known as organogels), depending on the nature of the fluid component [15]. The term oleogel is used to refer to any organogels containing palatable oil rather than an organic solvent [16]. The oleogelation of plant oils could be achieved using a multitude of molecules such as fatty alcohols, fatty acids, and wax esters [17, 18]. Oleogels have a non-polar dispersion medium such as mineral oil, organic solvents, fixed oil, *etc.* that is gelled with an agent referred to as oleogelators [19]. The facile preparation methods and inherent long-term stability of oleogel make it suitable for use as the drug delivery system [20]. The use of oleogels has been successfully explored in developing topical and transdermal drug delivery systems [21]. In an experiment by Tessari et al. [22], the authors have developed an oleogel-based transdermal formulation of the drug aceclofenac. The formulation was found to deliver the said drug across the skin into the systemic circulation. The researchers concluded that the prepared formulation could help to avoid gastric irritation related side effects that are caused due to the oral administration of aceclofenac [22].

In the current study, we propose to formulate the oleogels comprising of KB and RBO. Though the KB and RBO based oleogel have recently been proposed for food applications [23], there is no literature on its probable pharmaceutical applications. Hence, a series of oleogels were prepared by varying the concentration of KB and RBO. The oleogels were thoroughly characterized to understand the physicochemical, mechanical, and thermal properties. Further, the ability of the oleogels as drug delivery matrices were evaluated. For this purpose, CPH was loaded within the oleogels. The *in vitro* drug release, *ex vivo* corneal permeation, and *in vivo* corneal tolerance studies were carried out to understand the suitability of the oleogels as ocular drug delivery systems. Lastly, the antimicrobial activity of the CPH-loaded oleogels against the model microbe was also investigated.

2 Materials and methods

2.1 Materials

RBO was procured from Patanjali Ayurved Ltd., Haridwar, India. KB (*Garcinia indica* seed butter) was procured by Keynote International, Mumbai, India. CPH was received as a gift from Alkem Laboratories Ltd. (India). *Escherichia coli* NCIM 2563 was obtained from NCIM (Pune, India). Deionized water was used throughout the study. The freshly excised goat cornea was obtained from a local butcher's shop.

2.2 Methods

2.2.1 Determination of critical gelation concentration (CGC) and preparation of the oleogels

Different proportions of KB were added to the specified amounts of RBO to determine the critical gelation concentration (CGC) of KB. The mixtures were heated together at 60°C and subsequently cooled down (5°C; 15min). The formation of the oleogel was confirmed by inverting the sample container after the incubation at 5°C. The CGC of KB for inducing the gelation of RBO was 19.0% (*w*/w). After determining the CGC, five oleogel samples were prepared by varying the concentration of the KB in the range of 20.0% (*w*/w) and 30.0% (w/w) (Table 1, Fig. 1). The method of preparation of the oleogels was the same as that of the determination of

Table 1 Composition of oleogel samples for 20g.

	Weight (g)		
Sample name	**KB**	**RBO**	**CPH**
KR1	4.00	16.00	0.00
KR2	4.50	15.50	0.00
KR3	5.00	15.00	0.00
KR4	5.50	14.50	0.00
KR5	6.00	14.00	0.00
KR1D	4.00	16.00	0.10
KR2D	4.50	15.50	0.10
KR3D	5.00	15.00	0.10
KR4D	5.50	14.50	0.10
KR5D	6.00	14.00	0.10

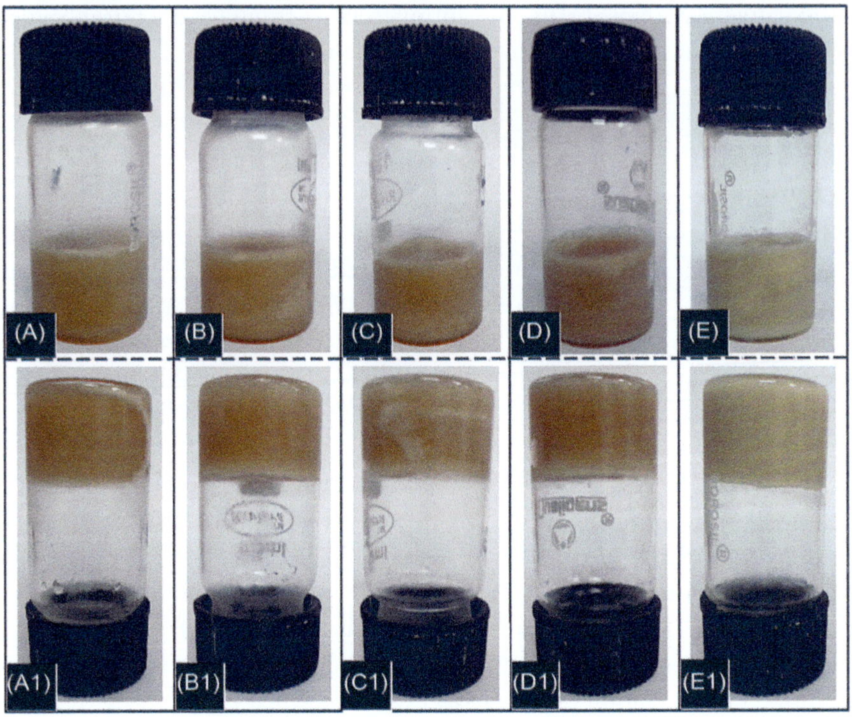

Fig. 1 Pictographs of the oleogels: (A) KR1, (B) KR2, (C) KR3, (D) KR4, and (E) KR5; Invert tube test of the oleogels: (A1) KR1, (B1) KR2, (C1) KR3, (D1) KR4, and (E1) KR5.

the CGC. In gist, the mixture of KB and RBO was kept in the water bath at 60 °C until the dissolution of the KB. Then, the hot solutions were kept in the thermal cabinet (5 °C) for 15 min to induce the oleogelation. The prepared oleogels were observed visually, and the texture of the oleogels was sensed by pressing the oleogels between the index finger and the thumb [24]. 0.5% (*w*/w) of the model drug (CPH) was incorporated within the oleogels to formulate the drug-loaded oleogels. The preparation method followed for CPH-loaded oleogels was akin to that used for pristine oleogels.

2.2.2 Colorimetric analysis

Reflective colorimeter (Flex EZ, Hunter Lab, United States) was used to quantify the color parameters of the prepared oleogels. A blank reading using an empty glass tube was recorded to remove the effects of the sample holder. The sample holder was then filled with the samples up to 1 cm. The CIELAB color parameters (L^*, a^*, and b^*) of the oleogels were recorded using the device. The readings were taken in triplicates [25]. The whiteness index (WI) was calculated using the formula given in Eq. (1) [26]. The color parameters of the oleogels have been reported as a mean $\pm$ standard deviation. ANOVA analysis and Tukey's HSD multiple comparison test was performed to check for significant differences at $p < 0.05$ level.

$$WI = 100 - \sqrt{(100 - L^*)^2 + a^{*2} + b^{*2}} \qquad (1)$$

where L^*, a^*, and b^* are the color parameter values of the oleogel [27].

2.2.3 Microscopic analysis

The distribution of the liquid and the solid phases within the oleogels were analyzed using a confocal laser scanning microscope (CLSM) (TCS SP8, Leica, Wetzlar, Germany). The samples for CLSM were prepared using fluorol yellow 088 dye. 0.01% (*w*/w) of the dye was added to the RBO and homogenized for 1 min to form a solution. The dye-containing RBO was used for the preparation of the oleogels. The oleogels were transferred into in-lab developed specially-designed wells for CLSM microscopy. The prepared oleogel samples were excited at a wavelength of 488 nm, and the resulted images were recorded at 515 nm [13].

The microstructural features of the oleogels were examined under a bright-field microscope (BFM) (Leica Microsystems GmbH, Wetzlar, Germany) and phase-contrast microscope (PCM) (ZEISS Primovert equipped with ZEISS Axiocam ERc 5s camera). The sample preparation for both microscopic analyses was done in the same manner. In short, a droplet of each of the molten oleogels was put on a glass slide and covered with a clean coverslip [24]. This helped to make thin smears of the oleogels over the slides. The microstructure of the smears of the oleogels was then visualized under the microscope [28].

2.2.4 X-ray diffraction (XRD) study

The oleogels were scanned in the 2θ range of 5° 2θ and 50° 2θ using an X-ray diffractometer (Bruker D8 Advance X-Ray Diffraction System, United States) to study the crystal nature of the oleogels [29]. The molten oleogels were spread over glass coverslips and then transferred to the thermal cabinet (5 °C). The crystallized samples were then scanned using X-rays generated from a cobalt source (1.789 Å). The test was carried out at a scan rate of 5° 2θ/min.

2.2.5 Fourier transform infrared (FTIR) spectroscopy analysis

The presence of specific chemical groups in the oleogels was characterized using an FTIR spectrophotometer (AlphaE ATR-FTIR, Bruker, Germany). The samples were analyzed in the transmittance mode, and the spectra were obtained in the wavenumber range of 4500 and 250 cm^{-1} [30]. The FTIR analysis is mainly used to study the chemical interactions between the functional groups present in the oleogels [31].

2.2.6 Thermal analysis

Gelation kinetics

The gelation kinetics of the oleogels was studied using a previously reported in-lab developed device [32]. The device records the change in the temperature as a function of time. In short, 10 g of each of the oleogels was transferred into glass culture bottles (15 mL). The oleogels were heated to 60 °C and then inserted into the sample holding slot of the testing device. The device was then incubated in a thermal cabinet (5 °C) for 1.5 h. The output of the thermal probe was acquired in a laptop using a LabVIEW program.

Differential scanning calorimetry (DSC) analysis

The thermal characteristics of the oleogels were measured using a differential scanning calorimeter (DSC 200 F3, Maia, Netzsch, Germany) [32]. The oleogels (15 mg) were sealed within the aluminum pans. The oleogels were heated in the temperature range of 0 °C and 100 °C at a scanning rate of 5 °C min^{-1} [13]. The analysis was performed in the inert atmosphere of the nitrogen (N_2) gas.

2.2.7 *In vitro* drug release study

The *in vitro* drug release study of the oleogels was performed using the diffusion cell of Franz's type. This experiment is useful in predicting the human tissue (e.g., dermal, cornea) penetration of the drug molecules that are loaded in the prepared oleogels [33]. The dialysis membrane (MWCO: 10 kDa) was used as the semi-permeable for the experiment. CPH-loaded oleogels (0.50 g) were dispensed into the donor chamber. On the other hand, the receptor chamber housed 12.0 mL of simulated tear fluid (STF; pH 7.4). The temperature of the diffusion cells was maintained at 37 °C using a water-jacket around the receptor chamber. The receptor fluid was continuously stirred throughout the study at 100 rpm to uphold the sink conditions. The diffusion of CPH from the oleogel across the dialysis membrane was quantified by analyzing the sequentially collected samples of the receptor fluid from the sampling port. The receptor fluid (1 mL) was sampled from the receptor chamber and replaced with the STF of equal volume [34]. UV-visible spectrophotometer (Model: UV-1700, Shimadzu Corporation, Japan) was used for analyzing the drug content in the receptor fluid by recording the absorbance at 277 nm. The experiments were conducted in triplicates. The drug release profiles were obtained by plotting cumulative percent drug release (CPDR) as a function of time. The total time of the experiment was 3 h.

2.2.8 Biological studies

Ex vivo corneal permeation study

The corneal permeation study was conducted similarly as that of the *in vitro* drug release studies. The only exception was the use of a freshly excised goat cornea in place of the dialysis membrane. The rest of the procedure was the same. The experiment was conducted for 3 h.

In vivo corneal tolerance study

The *in vivo* corneal tolerance test was conducted in triplicates using albino rabbits. For the purpose, rabbits of 1.5–2.0 kg were used. 0.5 g of the samples were inserted into the right ocular sacs. Saline (100 μL) was introduced in the left eye of the rabbit, which served as the control. Thereafter, at regular intervals (1, 6, 24, and 72 h), the eyes of the rabbit were monitored. The test was carried out with the agreement of the Institute Animal Ethical Committee vide permission number 49/IAEC/IPT/19.

Antimicrobial study

The antimicrobial activity of the CPH-loaded oleogels was studied and compared with that of blank oleogels (without CPH). *Escherichia coli* (*E. coli*) (Gram-negative bacteria) was used as the model microbe for the antimicrobial studies. 100 μL of the bacterial suspension (1×10^8 CFU/mL) was spread on the nutrient agar plates using a spreader. Thereafter, wells (5 mm) were made in the center of the agar plates, which were then loaded with the drug-loaded oleogels (0.2 g). The inoculated agar plates were subsequently incubated at 37 °C for 24 h. After the incubation period, the zone of inhibition was measured. The test was performed to verify whether the drug retains its antimicrobial property after being incorporated within the oleogels.

3 Results and discussion

3.1 Preparation of the oleogels

KB was dissolved in RBO (60 °C) and subsequently cooled to 5 °C. The proportion of KB was varied in the range of 5.0% (*w*/w) and 20.0% (w/w) to determine its minimum amount needed (CGC) to induce gelation of the RBO [35]. At concentrations below 19.0% (w/w), KB was not able to induce the gelation of the RBO. Since the hydrophobic interactions between the KB (gelator) and RBO (liquid phase) provide stability to the oleogels, it is expected that the interactions are not strong enough to form a stable oleogel below the CGC. However, the addition of KB at a concentration of 19.0% (*w*/w) formed stable oleogels. Hence, the CGC of KB for the structuring of RBO can be regarded as 19.0% (w/w). The molten mixture of KB and RBO appeared to be brown. The color of the formulation can be attributed to the RBO. The initiation of the oleogelation of the RBO by KB led to the initial formation of a cloudy phase, and subsequently, an opaque phase during the cooling process. After cooling, the samples were transformed into semisolid consistency, which were regarded as oleogels. It was observed that upon inverting the glass culture bottles, the oleogels did not flow under the effect of gravity (Fig. 1A1–E1) [29]. The oleogels were pale yellow (Fig. 1) and had a characteristic odor. They also exhibited a smooth and oily texture. The hardness of the oleogels increased with an increase in the KB concentration. This can be explained by an increase in the solid content within the oleogels when the KB concentration was increased [36].

3.2 Colorimetric analysis

The CIELAB color parameters L^* (lightness-darkness), a^* (redness-greenness), and b^* (blueness-yellowness) of the oleogels have been shown in Fig. 2. A significant increment in the L^* component values was observed with a corresponding rise in KB concentration in the oleogels ($p < 0.05$). No negative values were recorded for a^* coordinate, which suggests that the oleogels did not exhibit a green tone. The b^* component values were similar in KR1 and KR2 ($p > 0.05$). However, as the concentration of KB was further increased in the oleogels, a monotonous increase in the b^* component values was observed in KR3, KR4, and KR5 ($p < 0.05$). This can explain the rise in the yellow hue of the oleogels as the KB concentration was increased. All three parameters were taken into account to calculate the WI of the oleogels (Eq. 1). The variations in the WI values of the prepared oleogels

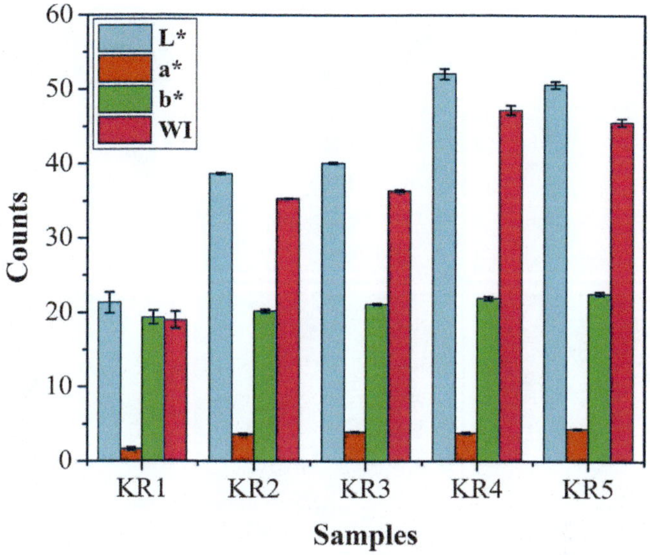

Fig. 2 Column plot showing variations in the CIE Hunter Lab coordinates L^*, a^*, and b^* of the oleogels.

(KR1-KR5) were statistically significant ($p<0.05$). The WI values were highest in KR4 (47.20 ± 0.64) and KR5 (45.58 ± 0.52), which suggested a positive correlation with KB concentration.

3.3 Microscopic analysis

Confocal micrographs, acquired using fluorol yellow 088 (chromophore dye), of the oleogels were used to confirm the organization of the hydrophobic phases. The oleogel showed a continuous fluorescent phase and dispersed black globular structures. It can be concluded that the solid fat molecules of KB did not accommodate the dye molecules within its dense structure [13]. In sample KR1 (Fig. 3A–E), the amount of the KB was low. Hence, the dispersed-phase particles appeared as scattered. However, in the case of KR3 and KR4, aggregation of the individual globular structures could be observed, which indicates the formation of the oleogels by forming a network-like structure [13,29].

The formation of a network through the globular structures in the oleogels was further confirmed by phase-contrast microscopy (PCM) (Fig. 3F–J). The micrographs showed the presence of two distinct phases in the prepared oleogels. There was a dispersed globular phase, which was consistently present in a continuous phase [13]. The dispersed phase represents the particles

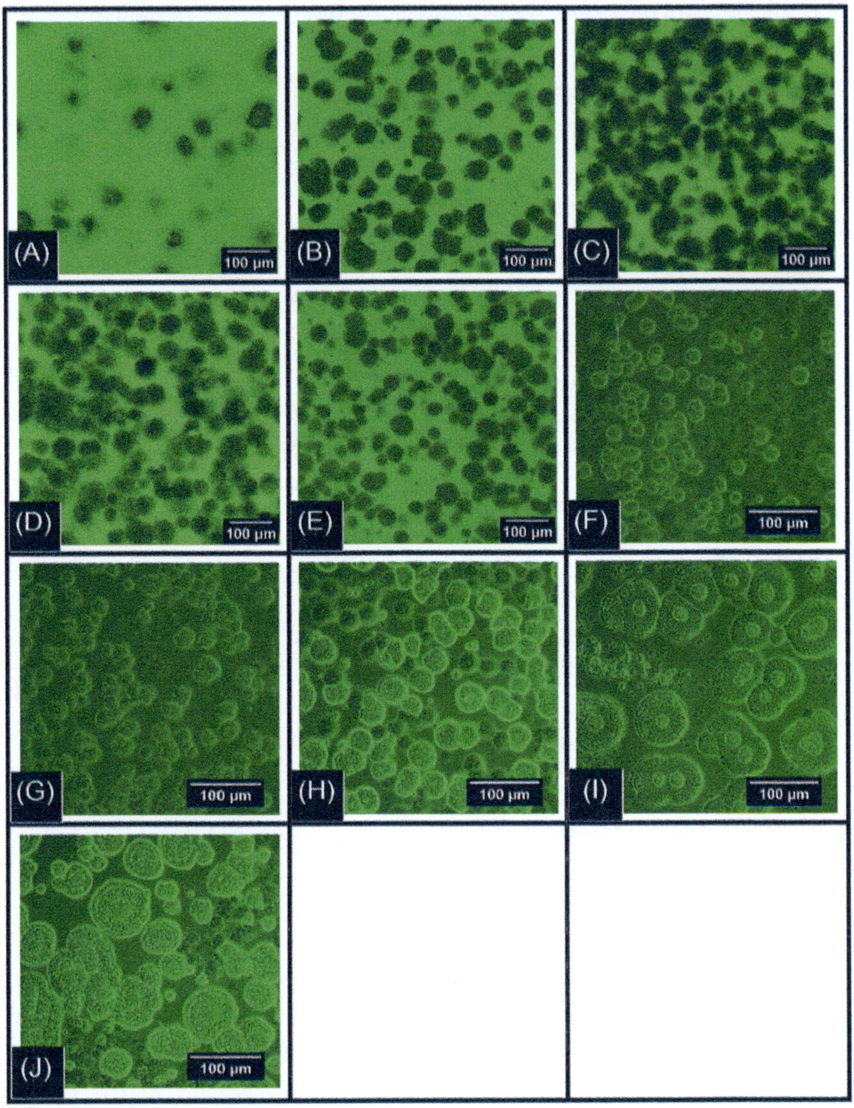

Fig. 3 Confocal micrographs of the oleogels: (A) KR1, (B) KR2, (C) KR3, (D) KR4, and (E) KR5. Phase-contrast micrographs of the oleogels: (F) KR1, (G) KR2, (H) KR3, (I) KR4, and (J) KR5.

(solid fat) of KB. The RBO acted as the non-polar dispersion medium. With an increase in the KB concentration, the size of the crystal globules increased in the order of KR1 < KR2 < KR3 < KR4 < KR5. The network formation by the globules was low in KR1, but with an increase in the gelator concentration, the aggregation of the globules was observed. However,

KR4 showed the formation of a unique structure. Within the gelator globule, there was a formation of another globular structure. Thus, it can be said that there was a possibility of a two-stage crystallization process in KR4. Similar observations were also acquired from the bright-field microscopy (results are not shown).

3.4 XRD study

The XRD profiles of the oleogels are shown in Fig. 4. The KB/RBO oleogels had a single and broad peak, which indicated that the oleogels were amorphous. The amorphous nature of oleogels is due to the presence of triglycerides that were present in the KB and the RBO (2014) ([37]). Nagavekar et al. [23] reported similar results while working with olive oil, RBO, and flaxseed oil [23]. The diffraction peak was observed ~22.58° 2θ in KR1, KR2, KR3, and KR5, while the same shifted to a higher angle in KR4 (22.64° 2θ). The intensity of this peak increased consistently from KR1 to KR4 and reduced abruptly in KR5. The peak-fitting tool of the Origin Pro 2016 software was used to determine the underlying peaks of the XRD profiles [38]. The Gauss function was used for the fitting of the peaks. The peak fitting analysis revealed the presence of five underlying peaks (Fig. 5). Table 2 summarizes the position of each of the peaks obtained. Bragg's law was used to calculate the d-spacing for the peaks (Eq. 2) [39]. The average d-spacing was similar in KR1 (4.03 Å) and KR4 (4.01 Å) (Table 2). The samples KR2, KR3, and KR5 had average d-spacing values of 3.99 Å > 3.95 Å > 3.72 Å, respectively. The results indicated that the KR5 oleogel, which had the highest content of KB, exhibited compact crystal structures. The peak positions and d-spacings obtained after the deconvolution of the XRD profiles suggested the presence of stearic acid in KB [40]. The peaks further revealed that the stearic acid is existing in C-form (monoclinic structure). This showed that during the oleogelation process, the native structure of the KB was conserved [37].

The crystallite size was determined using Scherrer's equation (Eq. 3) [39]. The average crystallite size was nearly equal in the samples KR1 (6.32 nm), KR3 (6.24 nm), and KR4 (6.50 nm). The samples KR2 and KR5 showed significantly high average crystallite sizes of 10.60 and 10.25 nm, respectively. The results suggested that the growth of the solid fat crystals in KR2 and KR5 was more appropriate. The average lattice strain values were observed to be low in KR4 and KR5, which suggested the presence of little crystal defects in these samples (Eq. 4; Table 2) [39].

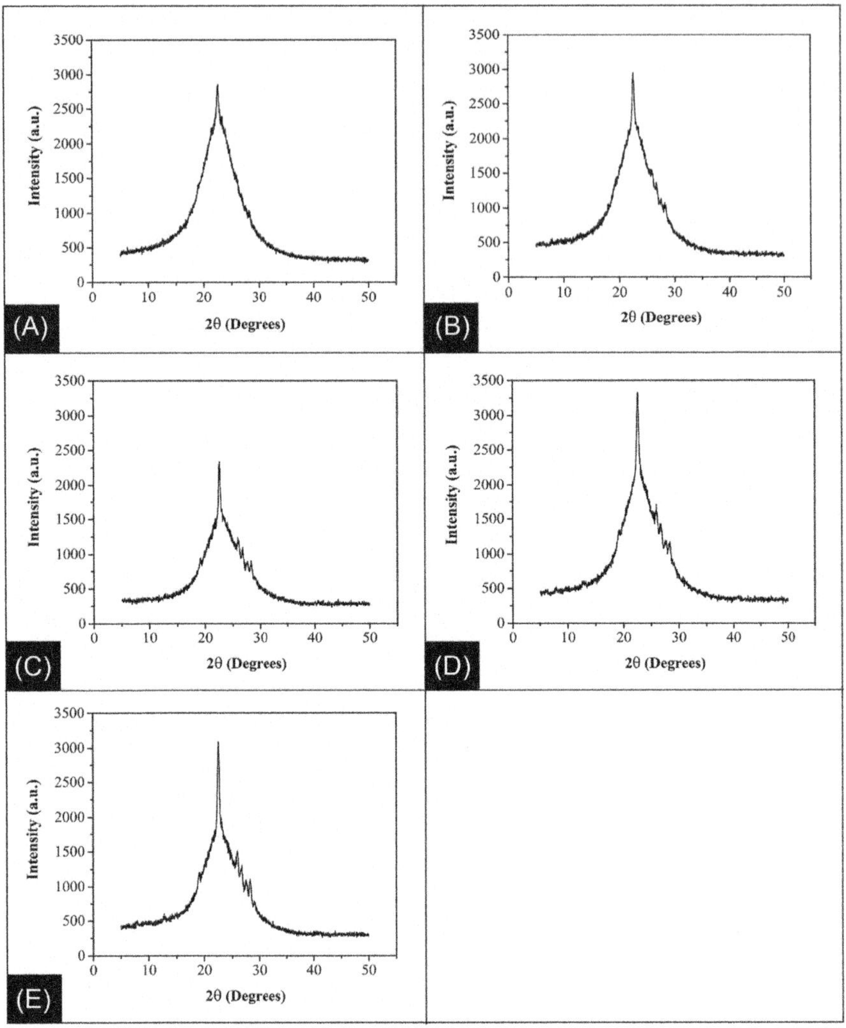

Fig. 4 XRD diffractograms of the oleogels: (A) KR1, (B) KR2, (C) KR3, (D) KR4, and (E) KR5.

Considering all the crystal aspects, it seems that the sample KR5 had a stable network as compared to the others.

$$n\lambda = 2d\sin\theta \tag{2}$$

where n = order, λ = wavelength of radiation (1.789 Å), d = d-spacing, and θ = peak position.

$$\text{Crystallite size (D)} = \frac{0.94^{*}\lambda}{\beta\cos\theta} \tag{3}$$

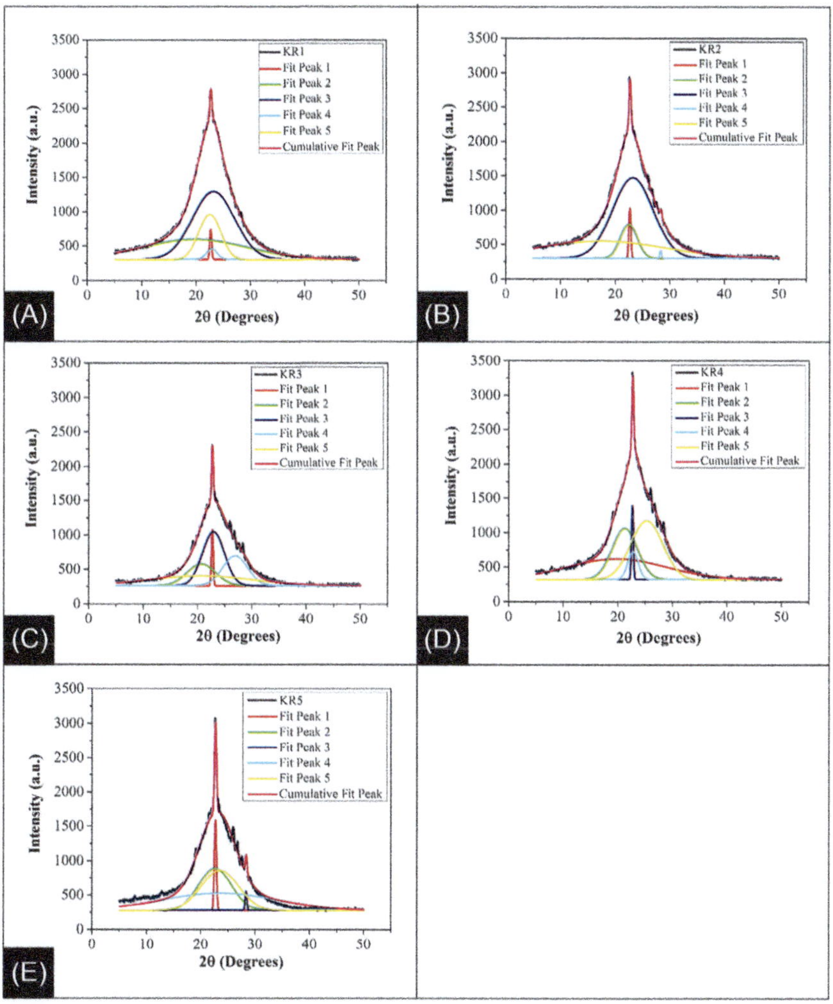

Fig. 5 Deconvoluted diffraction profiles of the oleogels: (A) KR1, (B) KR2, (C) KR3, (D) KR4, and (E) KR5.

where $\lambda = 1.789$ Å (wavelength for Cobalt), $\beta =$ Full width at half maximum (FWHM), and $\theta =$ peak position.

$$\varepsilon = \frac{\beta}{4 \tan \theta} \tag{4}$$

where ε represents is the lattice strain of the material, β is the FWHM, and θ represents Bragg's diffraction angle.

Table 2 XRD peak parameters obtained after deconvolution.

Samples	Peaks	Peak position (x_c)	FWHM	Height (Intensity, a.u.)	*d*-spacing (Å)	Crystallite size (nm)	Lattice strain
KR1	Peak1	22.559	0.416	431.922	3.938	23.620	0.009
	Peak2	19.656	22.681	289.331	4.513	0.430	0.571
	Peak3	23.112	9.068	988.245	3.845	1.080	0.194
	Peak4	22.850	2.236	150.427	3.889	4.400	0.048
	Peak5	22.376	4.735	660.030	3.970	2.080	0.105
Average			**7.827**	**503.991**	**4.030**	**6.322**	**0.185**
KR2	Peak1	22.591	0.436	736.500	3.933	22.550	0.010
	Peak2	22.428	3.607	493.868	3.970	2.720	0.079
	Peak3	23.162	8.711	1172.017	3.837	1.130	0.186
	Peak4	28.311	0.379	109.998	3.150	26.260	0.007
	Peak5	17.432	27.547	249.656	5.083	0.350	0.784
Average			**8.136**	**552.408**	**3.994**	**10.602**	**0.213**
KR3	Peak1	22.594	0.389	810.963	3.932	25.240	0.009
	Peak2	20.626	6.139	311.782	4.303	1.600	0.147
	Peak3	22.835	4.739	783.278	3.891	2.080	0.102
	Peak4	26.875	5.397	427.045	3.315	1.840	0.099
	Peak5	20.626	22.559	144.207	4.303	0.430	0.541
Average			**7.845**	**495.455**	**3.948**	**6.238**	**0.180**
KR4	Peak1	19.765	21.066	295.310	4.488	0.460	0.528
	Peak2	21.214	5.448	746.221	4.185	1.800	0.127
	Peak3	22.631	0.403	1074.893	3.926	24.370	0.009
	Peak4	22.783	2.225	405.353	3.900	4.420	0.048
	Peak5	25.142	6.902	859.144	3.539	1.430	0.135
Average			**7.208**	**676.184**	**4.007**	**6.496**	**0.169**

KR5	Peak1	22.574	0.457	1311.107	3.936	21.490	0.010
	Peak2	22.574	7.010	612.404	3.936	1.400	0.153
	Peak3	28.289	0.372	275.301	3.152	26.740	0.006
	Peak4	23.414	24.666	245.107	3.796	0.400	0.519
	Peak5	23.414	8.115	587.162	3.796	1.210	0.171
Average			**8.124**	**606.216**	**3.723**	**10.248**	**0.171**

3.5 FTIR analysis

The chemical interactions between the components, i.e., RBO and KB, of the oleogels were analyzed using FTIR spectroscopy [41]. Fig. 6 represents the FTIR spectra of the oleogels. The FTIR spectrum of the control (KR1) showed a feeble vibrational signal at ~3740 cm^{-1}, which signifies the presence of O—H stretching in free fatty acid [42]. A small absorption band obtained at 3003 cm^{-1} can be accredited to the stretching vibration of

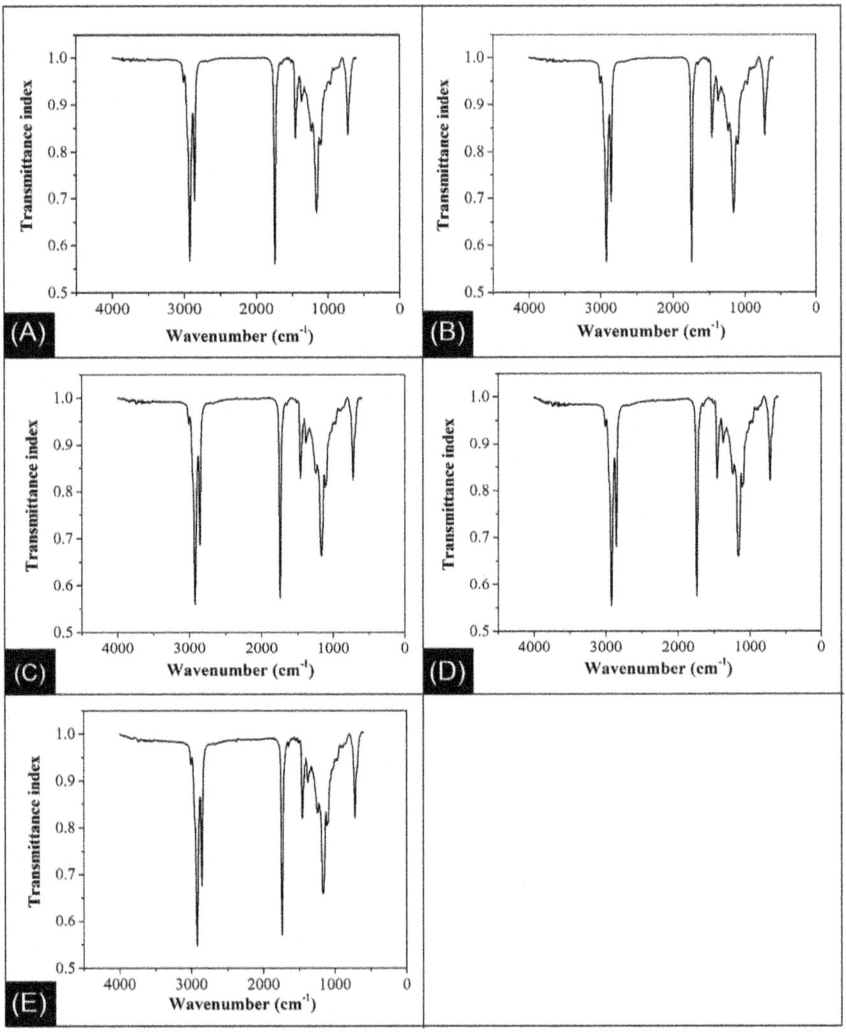

Fig. 6 FTIR spectra of the oleogels: (A) KR1, (B) KR2, (C) KR3, (D) KR4, and (E) KR5.

cis-olefin (=C—H) double bonds [43]. The distinct vibrational signals corresponding to the asymmetric and symmetric C—H stretching of methylene groups were recorded at 2922 and 2852 cm^{-1}, respectively [44]. The absorption bands observed at 1742 cm^{-1} can be assigned to the C=O stretching in triglyceride ester [41]. The sharp bands corresponding to the scissoring (bending) vibration of methylene (1461 cm^{-1}) and methyl (1375 cm^{-1}) groups were also noticed [45]. Furthermore, the vibrational signals that were observed at 1236 and 1159 cm^{-1} can be assigned to the C—O stretching vibrations of esters [46]. The strong peak corresponding to the methylene (—CH_2) rocking was observed at 720 cm^{-1} [47].

The FTIR spectra of the remaining oleogels (KR2 to KR5) also demonstrated the presence of the above-mentioned vibrational signals. Surprisingly, no significant shifting of the absorption peaks (except the signal recorded at 1236 and 1159 cm^{-1}) was observed with an increase in the KB content in the oleogels. Although the strong peak corresponding to the ester C—O stretching vibration was observed at 1236 cm^{-1} in KR2, the same signal shifted to a higher wavenumber position by 4 cm^{-1} in KR3 (1240 cm^{-1}). As the concentration of KB was increased further in KR4 and KR5, the peak displaced by 10 cm^{-1} in these oleogels (1246 cm^{-1}) in comparison to KR1. In the case of the vibrational signal recorded at 1159 cm^{-1}, no significant shifting was observed in its position in KR2 and KR3. However, as compared to KR1, the signal shifted by 4 cm^{-1} to a higher wavenumber (1163 cm^{-1}) in KR4 and KR5.

3.6 Thermal studies

3.6.1 Gelation kinetics

The crystallization kinetics is characterized by two processes, namely, nucleation kinetics and crystal growth kinetics that occur during the gelation process. Nucleation kinetics refers to the rate of creation of a stable nucleus, whereas the growth kinetics refers to the rate of formation of a macroscopic gel from a stable nucleus. The time required for the induction of nucleation of the gelators was observed to be dependent on the composition of the oleogels. There was an increase in the cloudiness of the hot homogenous transparent solution after the initiation of the nucleation process. Finally, as the temperature decreased, the solution became opaque that resulted in the achievement of the equilibrium phase where the gelation process was completed. The kinetics data (at the initial stages) were modeled using exponential decay function (Eq. 5) [39]. The correlation coefficient (R^2) of the modeled data was >0.99, which suggested that there was a good fit between

Table 3 Exponential decay modeling parameters.

Parameters	Samples				
	KR1	KR2	KR3	KR4	KR5
a (°C),	52.004	52.010	52.102	52.102	52.004
k (°C/ms)	0.709	0.698	0.765	0.791	0.905
R^2	1.000	1.000	1.000	1.000	1.000

the model data and the experimental values (Table 3). The crystallization rate (*k*) value was observed to be similar in KR1 (0.709°C/ms) and KR2 (0.698°C/ms). However, a monotonous increment in the crystallization rate of KB was observed from KR3-KR5. It is also clear that as the concentration of the KB was increased, the time for crystallization was delayed (Fig. 7). A careful observation of the kinetics profiles of the oleogels revealed that in sample KR1, KR3, and KR4, there was a slight increase in the temperature before achieving the equilibrium stage [28]. This increment in the temperature towards the end of the oleogelation process can be associated with the secondary crystallization of the KB molecules. As compared to KR1 and KR3, the exothermic event towards the end of the crystallization profile appeared to be the most pronounced in KR4. This observation further strengthened our confidence in the hypothesis that we made from the PCM analysis regarding secondary crystallization of KB molecules during the oleogelation process.

$$y(t) = a^* e^{-kt} \tag{5}$$

where $y(t)$ is the temperature at time t, $a =$ is the initial temperature, i.e., the temperature at time $t = 0$, and $k =$ decay constant.

3.6.2 DSC analysis

The DSC thermograms of the oleogels are shown in Fig. 8. The thermograms of the oleogels showed a melting endotherm (heating cycle) and crystallization exotherm (cooling cycle). The DSC thermograms showed endothermic peaks during the heating cycle. The endothermic peaks are suggestive of the process of the alteration in the polymorphic states. The thermal properties of the oleogels might get affected due to the level of unsaturation present in the formulations [48]. DSC has been used to characterize the polymorphism of fat molecules present in vegetable fats. As KB has a high content of stearic acid (~45%), the endothermic peaks obtained in the DSC profiles of the oleogels can be related to changes in the

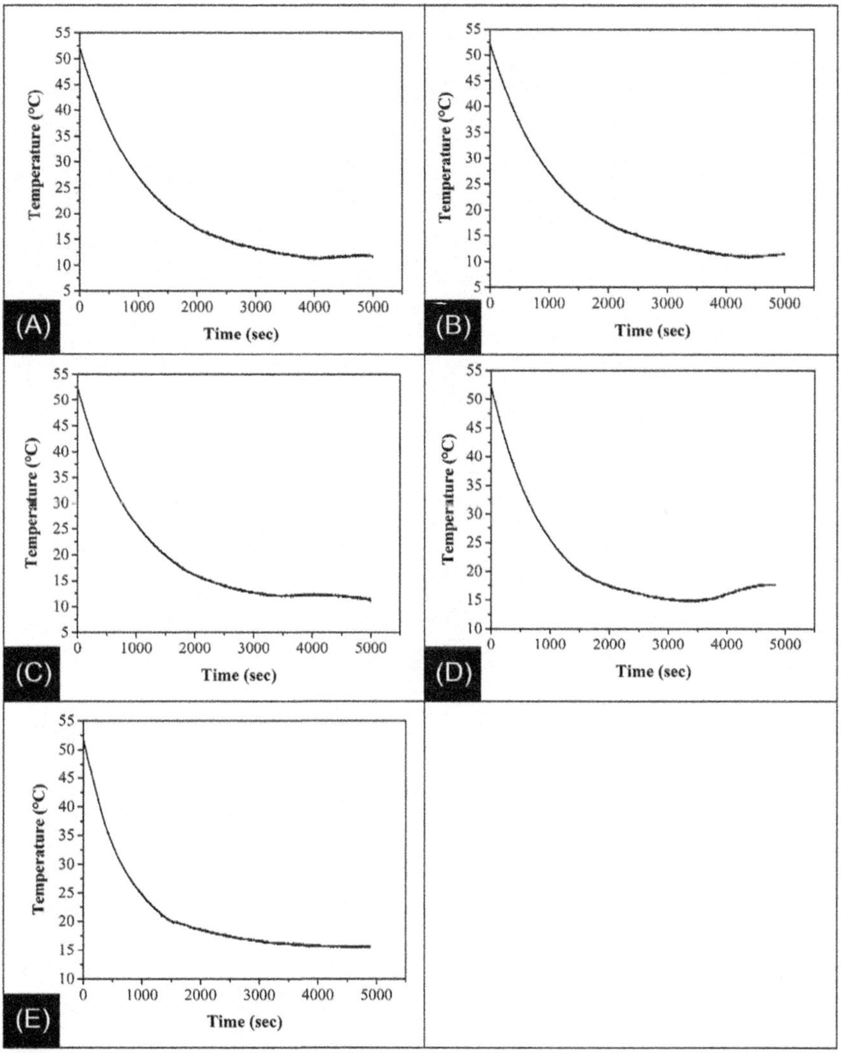

Fig. 7 Crystallization profiles of the oleogels at 5 °C: (A) KR1, (B) KR2, (C) KR3, (D) KR4, and (E) KR5.

polymorphic states of the stearic acid. Among the four polymorphs (A, B, C, and E) of stearic acid, A, B, and C forms are the common polymorphs. The transition between the B and C form usually takes place at ~32 °C [49]. The DSC profile of KR2 showed an endothermic peak at ~48 °C, which suggests the transition between the A and C polymorphs of the stearic acid crystal (Table 4). The peaks obtained at ~33 °C and ~34 °C in the thermograms

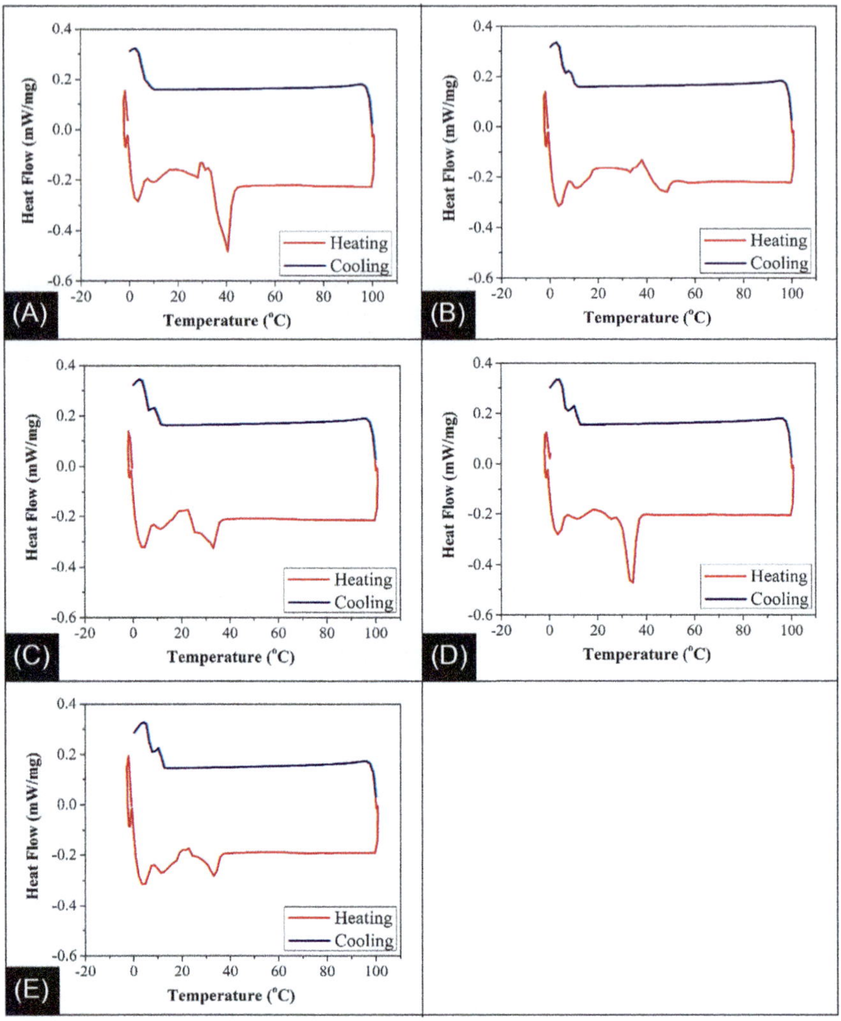

Fig. 8 DSC thermograms of the oleogels: (A) KR1, (B) KR2, (C) KR3, (D) KR4, and (E) KR5.

of KR3, KR4, and KR5 can be attributed to the melting of the B polymorphic form of the stearic acid [50]. The small peaks obtained in the temperature range of 25 °C and 30 °C indicated the contraction lattice structure of the C-form of stearic acid [51]. The cooling curves appeared to be much simpler than that of the heating curves. Supercooling effect is referred to as the difference between the melting and the crystallization temperatures, which occurs due to the thermal hysteresis. The supercooling

Table 4 Major peaks obtained in the heating and cooling cycle of DSC thermograms.

Samples	Peak temperature (°C)							
	Heating cycle						Cooling cycle	
	Peak 1	Peak 2	Peak 3	Peak 4	Peak 5	Peak 6	Peak 7	Peak 8
KR1	3.70	10.29	28.39	31.46	37.38	40.72	6.54	2.80
KR2	3.66	11.55	33.36	35.81	45.38	48.55	7.78	2.70
KR3	4.96	11.53	25.65	28.27	32.09	33.35	9.04	2.79
KR4	2.21	3.57	11.5	25.62	33.2	34.61	10.30	4.05
KR5	3.61	4.95	11.51	24.39	32.05	33.322	10.31	4.06

effect plays a vital role in the process of crystallization [48]. The cooling cycle had only one exothermic peak in KR1 (Fig. 8), which was suggestive of the recrystallization of the C polymorphic form of stearic acid crystal. No additional peaks were recorded at higher temperatures.

3.7 *In vitro* drug release study

The *in vitro* release profiles of CPH-loaded oleogels are represented in Fig. 9A. This study was done to check the diffusion of the model drug across the dialysis membrane into a fluid reservoir. The release of the drug was studied for a period of 3 h. The investigation of the release profiles suggested that an increase in the KB content resulted in the slower diffusion of the drug molecules across the semi-permeable dialysis membrane. This can be related to the entrapment of the drug molecules within the solid fat structure. The drug release data (60 min) was modeled using the Korsmeyer-Peppas (KP) model (Fig. 9B; Table 5) to critically analyze the release mechanism. The value of "k" provides information on the diffusion of the drug molecules within the gel matrix, whereas the value of "n" indicates the nature of diffusion. The KP model (Eq. 6) is a simple relationship that describes the release of the drug from a gel formulation [52]. The correlation coefficient of the model fitting was >0.99, which suggested that the model data and the experimental data are highly correlated. Similar "k" values were observed for KR1D (0.081), KR2D (0.088), KR3D (0.083), and KR4D (0.086). Among all the oleogels, KR5D exhibited the lowest "k" value (0.056), which may be related to the entrapment of the drug molecules within the crystalline fat matrix of KB (as observed in XRD studies). The diffusion exponent (n) values were in the range of 0.45 and 0.89 [48]. This indicated that the release of the drug from the oleogels was non-Fickian diffusion-mediated (anomalous diffusion) [53], which may be attributed to the combination of Fickian diffusion and gel network relaxation process [54].

Peppas-Sahlin (PS) model was further used to model the release data (Fig. 9C). This model is represented by a bi-exponential equation (Eq. 7). The term ($k_d \cdot t^m$) on the left side of the equation represents the Fickian diffusional contribution (F), whereas the term ($k_r \cdot t^{2m}$) on the right side represents the case II-relaxational contribution (R) to the drug release. The correlation of the fitting was >0.99, which suggested a good fit between the model data and the experimental data. The Fickian diffusion parameter (k_d) value of KR1D (0.103) was observed to be moderate. However, the k_d parameter value increased abruptly in KR2D (0.111) followed by a

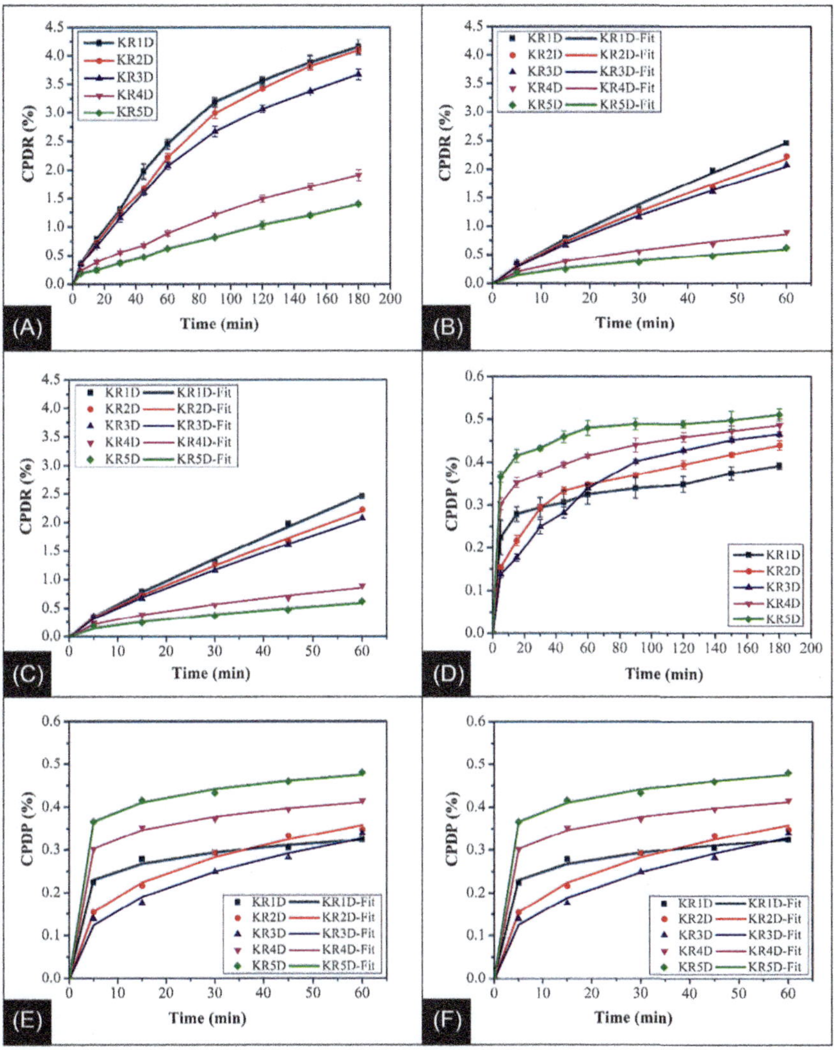

Fig. 9 *In vitro* drug release study of the CPH-loaded oleogels: (A) CPDR profiles, (B) Korsmeyer-Peppas model fitting of CPDR profiles, and (C) Peppas-Sahlin model fitting of CPDR profiles; *Ex vivo* corneal permeation study of the CPH-loaded oleogels: (D) CPDP profiles, (E) Korsmeyer-Peppas model fitting of CPDP profiles, and (F) Peppas-Sahlin model fitting of CPDP profiles.

monotonous decrease in KR3D (0.104), KR4D (0.088), and KR5D (0.057), respectively. The relaxation mediated diffusion parameter (k_r) varied slightly with the composition of the oleogels (Table 5). It can be observed that the k_d values for all oleogels were higher than k_r values. This indicated that the release of CPH from the oleogel matrices was controlled

Table 5 CPDR and CPDP model parameters.

Study	Model	Parameters	Samples				
			KR1D	KR2D	KR3D	KR4D	KR5D
In vitro drug release	Korsmeyer-Peppas	k	0.081	0.088	0.083	0.086	0.056
		n	0.833	0.786	0.783	0.561	0.577
		R^2	0.999	0.999	0.999	0.995	0.990
	Peppas-Sahlin	k_d	0.103	0.111	0.104	0.088	0.057
		k_r	0.014	0.008	0.009	0.012	0.008
		m	0.565	0.586	0.569	0.421	0.431
		R^2	0.999	0.999	1.000	0.997	0.992
Ex vivo corneal permeation	Korsmeyer-Peppas	K	0.185	0.091	0.067	0.250	0.310
		N	0.137	0.335	0.389	0.122	0.105
		R^2	0.999	0.998	0.995	1.000	1.000
	Peppas-Sahlin	K_d	0.000	0.000	0.059	0.191	0.225
		K_r	0.185	0.091	0.013	0.061	0.086
		m	0.068	0.168	0.291	0.094	0.079
		R^2	0.999	0.998	0.996	1.000	1.000

significantly by diffusion process as compared to the relaxation process of the oleogel network structure [55].

$$C = k \cdot t^n \tag{6}$$

where C is the cumulative percent drug release (CPDR), k represents diffusion constant, t is time, and n denotes diffusion exponent.

$$C = k_d \cdot t^m + k_r \cdot t^{2m} \tag{7}$$

where C = cumulative percent drug release, m = diffusion exponent, k_d = Fickian diffusion parameter, k_r = relaxation mediated diffusion parameter, and t = time.

3.8 Biological studies

3.8.1 Ex vivo *corneal permeation study*

The cornea consists of three major barriers: epithelium, stroma, and endothelium [56,57]. The typical route for ocular drug delivery is the administration of the drug formulations in the cul-de-sac of the eye, which presents the primary way of ocular drug delivery [58]. The corneal permeation study of the CPH-loaded oleogels was performed using freshly excised goat cornea. The corneal permeation profiles of the drug molecules have been represented in Fig. 9D. It was observed that the cumulative percentage of drug permeation increased with an increase in the KB content in the oleogels. However, the natural barriers provided by the cornea resulted in lower drug permeation as compared to *in vitro* drug diffusion studies. The permeation profiles up to 60 min were modeled using KP (Eq. 6) (Fig. 9E) and PS (Eq. 7) (Fig. 9F) models. The diffusion constant (K) parameter of the KP model showed a monotonous decrease in its values from KR1D to KR3D, which was followed by an increment in KR4D and KR5D. The diffusion exponent (N) values of all the oleogels were observed to lower than 0.45, which suggested the Fickiandiffusion-mediated permeation of the CPH molecules (Table 5). However, the PS model parameters indicated that the relaxation process also played a role in the diffusion of the CPH molecules across the corneal tissue (Table 5). Similarly to the *in vitro* diffusion study, the values of diffusion constant (K_d) were comparatively higher than that of polymer relaxation constant (K_r). Hence, it can be concluded that Fickian diffusion significantly contributed to the release of CPH molecules through the goat corneal tissue.

3.8.2 *In vivo* corneal tolerance study

Draize's eye irritation test is usually employed to test the ocular safety and toxicity of the ophthalmic formulations. Hence, in our study, the oleogels were exposed to the ocular tissue of the rabbit. As the rabbit's eyes are more susceptible to the irritating substances than that of humans, they have been used as the model for conducting the irritation study [59]. The right eye was treated with test material (here, oleogels) within the cul-de-sac, and the left eye was treated with normal saline. The observations were recorded at different times for a period of 72h. Fig. 10 shows the observations recorded for the representative oleogel (KR5). At the end of 72h, no chemosis of the cornea and other ophthalmic tissues was observed. This indicated the suitability of the oleogels for ocular drug delivery applications [60].

3.8.3 *Antimicrobial study*

CPH is a broad-spectrum antibiotic [61], which can act on both Gram-negative and Gram-positive bacteria [62]. The blank oleogels (without CPH) and the CPH-loaded oleogels were tested for their antimicrobial

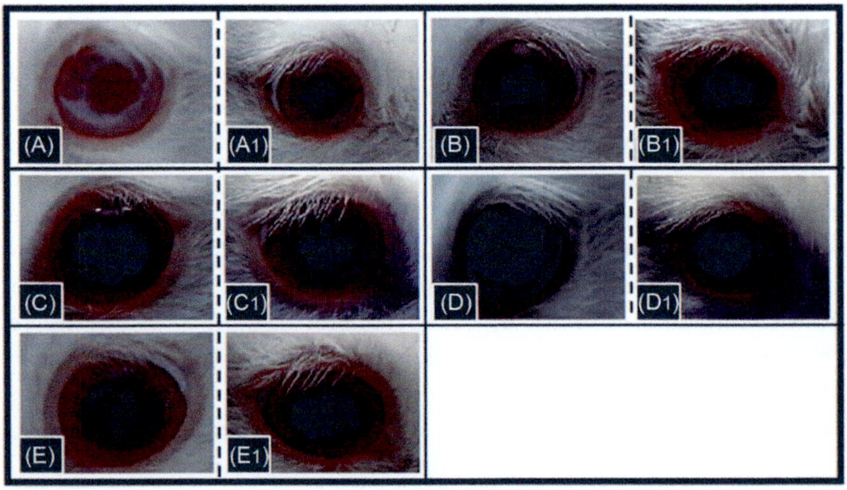

Fig. 10 Draize's eye irritation test conducted in the rabbit model as per the test protocol. Demonstrative pictures of the eye administered with KR5: (A) Left eye administered with normal saline at 0h, (A1) Right eye administered with KR5 at 0h, (B) Left eye administered with normal saline at 1h, (B1) Right eye administered with KR5 at 1h, (C) Left eye administered with normal saline at 24h, (C1) Right eye administered with KR5 at 24h, (D) Left eye administered with normal saline at 48h, (D1) Right eye administered with KR5 at 48h, (E) Left eye administered with normal saline at 72h, and (E1) Right eye administered with KR5 at 72h.

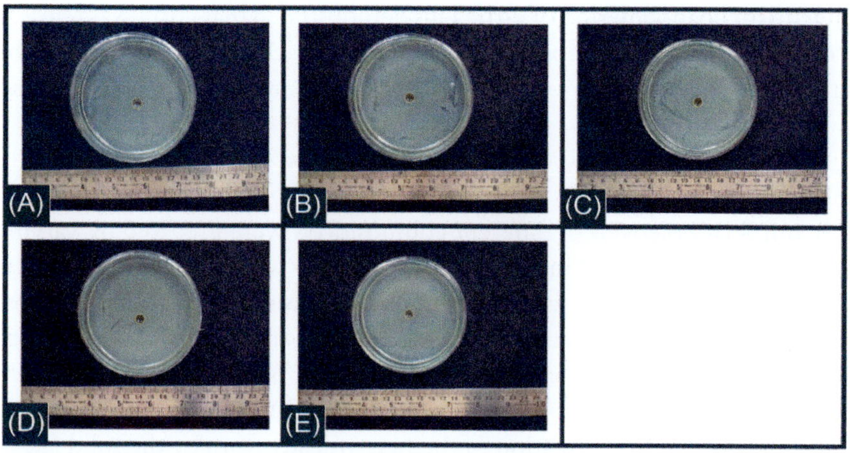

Fig. 11 Antimicrobial activity of pristine oleogels against *E. coli*: (A) KR1, (B) KR2, (C) KR3, (D) KR4, and (E) KR5.

activities against *E. coli*. The blank oleogels did not result in the formation of a zone of inhibition (Fig. 11). This suggested that the oleogels did not have any inherent antimicrobial activity. However, the CPH-loaded oleogels showed a clear zone of inhibition around the wells (Fig. 12), when observed

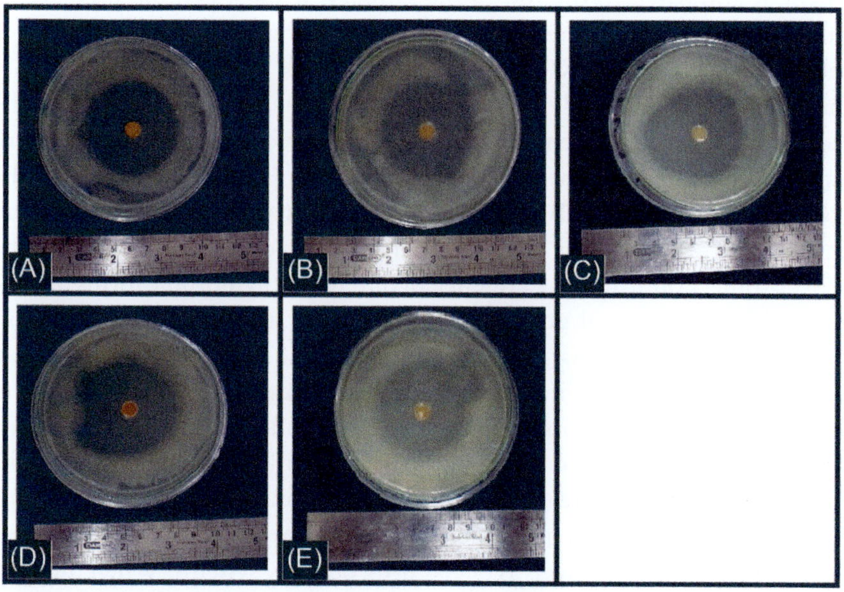

Fig. 12 Antimicrobial activity of drug-loaded oleogels against *E. coli*: (A) KR1D, (B) KR2D, (C) KR3D, (D) KR4D, and (E) KR5D.

after an incubation period of 24 h. This demonstrated that CPH was released from the drug-loaded oleogels in its active form such that it could inhibit the growth of *E. coli*.

4 Conclusion

KB and RBO oleogels of different compositions were prepared by varying the concentration of KB in the range of 20.0% (*w*/w) and 30.0% (w/w). The prepared oleogels were investigated in-depth using colorimetry, microscopy, XRD, FTIR spectroscopy, and thermal studies. The colorimetric study showed that the oleogels containing high KB content (KR2-KR5) exhibited more whiteness as compared to control (KR1). The microscopic analyses revealed the formation of the network structure by the solid fats of the KB, which crystallized in the form of globular architectures. The density of the globular structures increased in the oleogels with an increase in the concentration of the gelator (herein, KB). The formulation KR4 showed the formation of a unique microstructure that indicated the occurrence of a two-step crystallization process. The hypothesis of two-step crystallization in KR4 was confirmed by the crystallization kinetics study. The XRD analysis showed that the average lattice strain values of KR4 and KR5 were lower as compared to the rest of the samples. It suggested the presence of little crystal defects in these samples. Considering the different aspects, it seemed that the sample KR5 had a stable network as compared to the other samples. The FTIR analysis of the oleogels indicated the presence of the stearic acid crystals, which were involved in the formation of the network structure for the entrapment of the oil molecules [63]. The presence of the stearic acid crystals in KB and its polymorphic states in the oleogels were further confirmed using thermal analysis. A closer observation of the DSC profiles confirmed the presence of different polymorphic forms of stearic acid in their natural polymorphic form of B and C [64]. This was also suggestive of the fact that the components of the formulated oleogels had thermal compatibility [65]. The *in vitro* drug release and *ex vivo* corneal permeation profiles were modeled using the KP and PS models. The *in vitro* release of the CPH molecules from the oleogels followed an anomalous diffusion mechanism, whereas the corneal permeation of the drug was predominantly Fickian-diffusion mediated. The corneal permeation study indicated the suitability of the oleogels as ocular drug delivery systems. When the pristine oleogels were tested for the antimicrobial activities, they did not show any zone of inhibitions. However, the CPH-loaded oleogels were found to be active against *E. coli*. This concluded that the oleogels could be used for delivering drugs into the eye in their biologically active form.

Acknowledgments

The authors acknowledge the infrastructural support provided by the National Institute of Technology, Rourkela, India.

Animal ethical permission

The animal study was carried out with the permission of the Institute Animal Ethical Committee vide permission number 49/IAEC/IPT/19.

Disclosure of interest

The authors declare that this article content has no conflict of interest.

Funding

The present study has been conducted under Indo-Korea joint research program of Department of Science and Technology, Government of India (Sanction order # INT/Korea/P-37, June 15, 2017) and under the framework of International Cooperation Program managed by the NRF (2016K1A3A1A19945059), Republic of Korea.

References

[1] F. Mohammed, S.V. Joshi, S.T. Bairy, S.K.K. Narayana, Analytical standards of Vrukshamla (Garcinia indica Choisy) Beeja Taila-Kokum Butter, J. Ayu. Med. Sci. 2 (3) (2017), https://doi.org/10.5530/jams.2017.2.21.
[2] M. Sowmya, A. Kuna, M.R. Sahoo, P. Devi, D.M. Mayengbam, M. Sreedhar, Formulation and sensory evaluation of value added products developed with underutilized Garcinia indica fruit, J. Pharmacogn. Phytochem. 8 (1) (2019) 435–439.
[3] S. Krist, Kokum butter, in: Vegetable Fats and Oils, Springer, 2020, pp. 399–403.
[4] S.B. Swami, N. Thakor, S. Patil, Kokum (Garcinia indica) and its many functional components as related to the human health: a review, J. Food Sci. Technol. 2 (4) (2014) 130–142, https://doi.org/10.18805/ajdfr.DR-1493.
[5] S. Suna, Ö.U. Çopur, A new approach: replacement and alternative foods for food industry, in: Alternative and Replacement Foods, Elsevier, 2018, pp. 1–30.
[6] S. Lahkar, M. Kumar Das, Surface modified kokum butter lipid nanoparticles for the brain targeted delivery of nevirapine, J. Microencapsul. 35 (7–8) (2018) 680–694, https://doi.org/10.1080/02652048.2019.1573857.
[7] E.D. Ehrenpreis, Method for Treating Anal Pruritis and Other Perianal Disorders, 2014. Google Patents.
[8] T. Kumar, S.K. Gupta, M.K. Prajapati, D. Tripathi, Natural excipients: a review, Asian J. Pharm. Life Sci. 2 (1) (2012) 4423.

[9] T.-I.C. Wang, B.V. Biundo, Testosterone Booster Transdermal Compositions, 2015. Google Patents.
[10] S. Sierra, F. Lara-Villoslada, M. Olivares, J. Jiménez, J. Boza, J. Xaus, Increased immune response in mice consuming rice bran oil, Eur. J. Nutr. 44 (8) (2005) 509–516, https://doi.org/10.1007/s00394-005-0554-y.
[11] S. Kochhar, Stable and healthful frying oil for the 21st century, Inform 11 (2000) 642–647.
[12] T. Wang, K.B. Hicks, R. Moreau, Antioxidant activity of phytosterols, oryzanol, and other phytosterol conjugates, J. Am. Oil Chem. Soc. 79 (12) (2002) 1201–1206, https://doi.org/10.1007/s11746-002-0628-x.
[13] S.R. Paul, D. Qureshi, Y. Yogalakshmi, S.K. Nayak, V.K. Singh, I. Syed, K. Pal, Development of bigels based on stearic acid–rice bran oil oleogels and tamarind gum hydrogels for controlled delivery applications, J. Surfactant Deterg. 21 (1) (2018) 17–29, https://doi.org/10.1002/jsde.12022.
[14] G.K. Eleftheriadis, P. Mantelou, C. Karavasili, P. Chatzopoulou, D. Katsantonis, M. Irakli, D.G. Fatouros, Development and characterization of a self-nanoemulsifying drug delivery system comprised of rice bran oil for poorly soluble drugs, AAPS Pharm. Sci. Tech. 20 (2) (2019) 78.
[15] A. Shakeel, F.R. Lupi, D. Gabriele, N. Baldino, B. De Cindio, Bigels: a unique class of materials for drug delivery applications, Soft Mater 16 (2) (2018) 77–93, https://doi.org/10.1080/1539445X.2018.1424638.
[16] A.G. Marangoni, N. Garti, Edible Oleogels: Structure and Health Implications, Elsevier, 2018.
[17] M. Davidovich-Pinhas, Oleogels, in: Polymeric Gels, Elsevier, 2018, pp. 231–249.
[18] S.S. Sagiri, K. Rao, Natural and bioderived molecular gelator–based oleogels and their applications, in: Biopolymer-Based Formulations, Elsevier, 2020, pp. 513–559.
[19] R. Balasubramanian, A.A. Sughir, G. Damodar, Oleogel: a promising base for transdermal formulations, Asian J. Pharm. 6 (1) (2014) 1–9, https://doi.org/10.22377/ajp.v6i1.67.
[20] S. Murdan, Organogels in drug delivery, Expert Opin. Drug Deliv. 2 (3) (2005) 489–505, https://doi.org/10.1517/17425247.2.3.489.
[21] R. Hamed, A.A. AbuRezeq, O. Tarawneh, Development of hydrogels, oleogels, and bigels as local drug delivery systems for periodontitis, Drug Dev. Ind. Pharm. 44 (9) (2018) 1488–1497, https://doi.org/10.1080/03639045.2018.1464021.
[22] L. Tessari, L. Ceciliani, A. Belluati, G. Letizia, U. Martorana, L. Pagliara, G. Torri, Aceclofenac cream versus piroxicam cream in the treatment of patients with minor traumas and phlogistic affections of soft tissues: a double-blind study, Curr. Ther. Res. Clin. Exp. 56 (7) (1995) 702–712, https://doi.org/10.1016/0011-393X(95)85140-2.
[23] N. Nagavekar, A. Kumar, K. Dubey, R.S. Singhal, Supercritical carbon dioxide extraction of kokum fat from Garcinia indica kernels and its application as a gelator in oleogels with oils, Ind. Crop. Prod. 138 (2019), https://doi.org/10.1016/j.indcrop.2019.06.022, 111459.
[24] S. Satapathy, V.K. Singh, S.S. Sagiri, T. Agarwal, I. Banerjee, M.K. Bhattacharya, K. Pal, Development and characterization of gelatin-based hydrogels, emulsion hydrogels, and bigels: a comparative study, J. Appl. Polym. Sci. 132 (8) (2015), https://doi.org/10.1002/app.41502.
[25] B.Z. Hmar, S. Mishra, R.C. Pradhan, Physico-chemical, mechanical and antioxidant properties of kendu (*Diospyros melanoxylon* Roxb.), Curr. Res. Nutr. Food Sci. 5 (3) (2017) 214–222.
[26] P.B. Pathare, U.L. Opara, F.A.-J. Al-Said, Colour measurement and analysis in fresh and processed foods: a review, Food Bioprocess Technol. 6 (1) (2013) 36–60.

[27] B. Saberi, R. Thakur, Q.V. Vuong, S. Chockchaisawasdee, J.B. Golding, C.J. Scarlett, C.E. Stathopoulos, Optimization of physical and optical properties of biodegradable edible films based on pea starch and guar gum, Ind. Crops Prod. 86 (2016) 342–352.
[28] K. Uvanesh, S. Sagiri, K. Senthilguru, K. Pramanik, I. Banerjee, A. Anis, K. Pal, Effect of span 60 on the microstructure, crystallization kinetics, and mechanical properties of stearic acid oleogels: an in-depth analysis, J. Food Sci. 81 (2) (2016) E380–E387, https://doi.org/10.1111/1750-3841.13170.
[29] S.S. Sagiri, U. Kasiviswanathan, G.S. Shaw, M. Singh, A. Anis, K. Pal, Effect of sorbitan monostearate concentration on the thermal, mechanical and drug release properties of oleogels, Korean J. Chem. Eng. 33 (5) (2016) 1720–1727, https://doi.org/10.1007/s11814-015-0295-4.
[30] H.S. Mansur, C.M. Sadahira, A.N. Souza, A.A. Mansur, FTIR spectroscopy characterization of poly (vinyl alcohol) hydrogel with different hydrolysis degree and chemically crosslinked with glutaraldehyde, Mater. Sci. Eng. C 28 (4) (2008) 539–548, https://doi.org/10.1016/j.msec.2007.10.088.
[31] P. Goyal, R. Dhar, S.S. Sagiri, K. Uvanesh, K. Senthilguru, G. Shankar, S.S. Ray, Synthesis and characterization of novel dual environment-responsive hydrogels of hydroxyethyl methacrylate and methyl cellulose, Des. Monomers Polym. 18 (4) (2015) 367–377, https://doi.org/10.1080/15685551.2015.1012626.
[32] K. Uvanesh, S.S. Sagiri, I. Banerjee, H. Shaikh, K. Pramanik, A. Anis, K. Pal, Effect of tween 20 on the properties of stearate oleogels: an in-depth analysis, J. Am. Oil Chem. Soc. 93 (5) (2016) 711–719, https://doi.org/10.1007/s11746-016-2810-0.
[33] L. Bartosova, J. Bajgar, Transdermal drug delivery in vitro using diffusion cells, Curr. Med. Chem. 19 (27) (2012) 4671–4677, https://doi.org/10.2174/092986712803306358.
[34] G.S. Shaw, K. Uvanesh, S. Gautham, V. Singh, K. Pramanik, I. Banerjee, K. Pal, Development and characterization of gelatin-tamarind gum/carboxymethyl tamarind gum based phase-separated hydrogels: a comparative study, Des. Monomers Polym. 18 (5) (2015) 434–450, https://doi.org/10.1080/15685551.2015.1041075.
[35] R.G. Weiss, P. Terech, Molecular gels, Materials With Self-Assembled Fibrillar Networks, first, Springer, The Netherlands, 2006, pp. XV–978.
[36] F.G. Gandolfo, A. Bot, E. Flöter, Structuring of edible oils by long-chain FA, fatty alcohols, and their mixtures, J. Am. Oil. Chem. Soc. 81 (1) (2004) 1–6.
[37] S. Sagiri, V.K. Singh, K. Pal, I. Banerjee, P. Basak, Stearic acid based oleogels: a study on the molecular, thermal and mechanical properties, Mater. Sci. Eng. C 48 (2015) 688–699, https://doi.org/10.1016/j.msec.2014.12.018.
[38] X.D. Hou, T.J. Smith, N. Li, M.H. Zong, Novel renewable ionic liquids as highly effective solvents for pretreatment of rice straw biomass by selective removal of lignin, Biotechnol. Bioeng. 109 (10) (2012) 2484–2493, https://doi.org/10.1002/bit.24522.
[39] D. Qureshi, H. Behera, A. Anis, D. Kim, K. Pal, Effect of polyglycerol polyricinoleate on the polymorphic transitions and physicochemical properties of mango butter, Food Chem. 323 (2020), https://doi.org/10.1016/j.foodchem.2020.126834, 126834.
[40] Y. Guo, K. Fukushi, S. Hirayama, H. Machida, Q. Meng, S. Akasaka, A. Fujimori, Thermal stability of ordered multi-particle layers of long-chain phosphonate-modified nanodiamond with superior heat-resistance, Colloids Surf. A Physicochem. Eng. Asp. 556 (2018) 227–238, https://doi.org/10.1016/j.colsurfa.2018.08.021.
[41] V.K. Singh, A. Anis, S. Al-Zahrani, D.K. Pradhan, K. Pal, FTIR, electrochemical impedance and iontophoretic delivery analysis of guar gum and sesame oil based bigels, Int. J. Electrochem. Sci. 9 (2014) 5640–5650.
[42] X. Jiang, S. Li, G. Xiang, Q. Li, L. Fan, L. He, K. Gu, Determination of the acid values of edible oils via FTIR spectroscopy based on the OH stretching band, Food Chem. 212 (2016) 585–589, https://doi.org/10.1016/j.foodchem.2016.06.035.

[43] J. Zúñiga-Diaz, E. Reyes-Dorantes, A. Quinto-Hernandez, J. Porcayo-Calderon, J. - Gonzalez-Rodriguez, L. Martinez-Gomez, Oil extraction from "Morelos Rice" bran: kinetics and raw oil stability, J. Chem. 2017 (2017) 1–9.

[44] J.Y. Chen, H. Zhang, J. Ma, T. Tuchiya, Y. Miao, Determination of the degree of degradation of frying rapeseed oil using Fourier-transform infrared spectroscopy combined with partial least-squares regression, Int. J. Anal. Chem. 2015 (2015), https://doi.org/10.1155/2015/185367.

[45] I. Irnawati, S. Riyanto, S. Martono, A. Rohman, Determination of sesame oil, rice bran oil and pumpkin seed oil in ternary mixtures using FTIR spectroscopy and multivariate calibrations, Food Res. 4 (2019) 135–142, https://doi.org/10.26656/fr.2017.4(1).260.

[46] M. Moharam, L. Abbas, A study on the effect of microwave heating on the properties of edible oils using FTIR spectroscopy, Afr. J. Microbiol. Res. 4 (19) (2010) 1921–1927.

[47] D. Üçüncüoğlu, V.A. Küçük, A multivariate data approach for FTIR-ATR monitoring of virgin olive oils: lipid structure during autoxidation, OCL 26 (2019) 42.

[48] S.P. Kodela, P.M. Pandey, S.K. Nayak, K. Uvanesh, A. Anis, K. Pal, Novel agar–stearyl alcohol oleogel-based bigels as structured delivery vehicles, Int. J. Polym. Mater. 66 (13) (2017) 669–678, https://doi.org/10.1080/00914037.2016.1252362.

[49] M. Mirmehrabi, S. Rohani, Measurement and prediction of the solubility of stearic acid polymorphs by the UNIQUAC equation, Can. J. Chem. Eng. 82 (2) (2004) 335–342, https://doi.org/10.1002/cjce.5450820214.

[50] K. Sato, Physical and molecular properties of lipid polymorphs—a review, Food Struct. 6 (2) (1987) 7.

[51] L.F.L. Silva, W. Paschoal, G.S. Pinheiro, J.G. da Silva Filho, P.T. Freire, F.F. de Sousa, S.G. Moreira, Understanding the effect of solvent polarity on the polymorphism of octadecanoic acid through spectroscopic techniques and DFT calculations, CrystEngComm 21 (2) (2019) 297–309, https://doi.org/10.1039/C8CE01402G.

[52] H.K. Shaikh, R. Kshirsagar, S. Patil, Mathematical models for drug release characterization: a review, World J. Pharm. Pharm. Sci. 4 (4) (2015) 324–338.

[53] P. Sriamornsak, R.A. Kennedy, Effect of sodium fluorescein on release characteristics of a macromolecule from calcium alginate gel beads, Carbohydr. Polym. 84 (3) (2011) 1208–1212, https://doi.org/10.1016/j.carbpol.2010.12.065.

[54] S. Dash, P.N. Murthy, L. Nath, P. Chowdhury, Kinetic modeling on drug release from controlled drug delivery systems, Acta Pol. Pharm. 67 (3) (2010) 217–223.

[55] R.B. Baggi, N.B. Kilaru, Calculation of predominant drug release mechanism using Peppas-Sahlin model, part-I (substitution method): a linear regression approach, Asian J. Pharm. Technol. 6 (4) (2016) 223–230, https://doi.org/10.5958/2231-5713.2016.00033.7.

[56] S. Li, E. Kim, J.A. Bonanno, Fluid transport by the cornea endothelium is dependent on buffering lactic acid efflux, Am. J. Phys. Cell Physiol. 311 (1) (2016) C116–C126,-https://doi.org/10.1152/ajpcell.00095.2016.

[57] S. Wang, C.E. Ghezzi, R. Gomes, R.E. Pollard, J.L. Funderburgh, D.L. Kaplan, In vitro 3D corneal tissue model with epithelium, stroma, and innervation, Biomaterials 112 (2017) 1–9, https://doi.org/10.1016/j.biomaterials.2016.09.030.

[58] O.P. Sharma, V. Patel, T. Mehta, Nanocrystal for ocular drug delivery: hope or hype, Drug Deliv. Transl. Res. 6 (4) (2016) 399–413, https://doi.org/10.1007/s13346-016-0292-0.

[59] K.R. Wilhelmus, The Draize eye test, Surv. Ophthalmol. 45 (6) (2001) 493–515.

[60] D. Quereshi, S. Dhal, D. Das, B. Mohanty, A. Anis, H. Shaikh, T.T. Hanh Nguyen, D. Kim, P. Sarkar, K. Pal, Neem seed oil and gum arabic-based oil-in-water emulsions as potential ocular drug delivery system, J. Disper. Sci. Technol. (2019) 1–14.

[61] S.A.R. Naqvi, S. Roohi, A. Iqbal, T.A. Sherazi, A.F. Zahoor, M. Imran, Ciprofloxacin: from infection therapy to molecular imaging, Mol. Biol. Rep. 45 (5) (2018) 1457–1468, https://doi.org/10.1007/s11033-018-4220-x.

[62] S.M. Sreedharan, R. Singh, Ciprofloxacin functionalized biogenic gold nanoflowers as nanoantibiotics against pathogenic bacterial strains, Int. J. Nanomedicine 14 (2019) 9905.
[63] K. Rehman, M.H. Zulfakar, Recent advances in gel technologies for topical and transdermal drug delivery, Drug Dev. Ind. Pharm. 40 (4) (2014) 433–440, https://doi.org/10.3109/03639045.2013.828219.
[64] F. Araiedh, F. Ducos, A. Houas, N. Chaoui, Kinetic study of the photocatalytic degradation of the C-polymorph of a stearic acid microcrystal grown on an amorphous titania surface scattered with anatase microdomains, J. Photochem. Photobiol. A Chem. 353 (2018) 458–463, https://doi.org/10.1016/j.jphotochem.2017.12.006.
[65] S. Doktorovova, R. Shegokar, E.B. Souto, Role of excipients in formulation development and biocompatibility of lipid nanoparticles (SLNs/NLCs), in: Nanostructures for Novel Therapy, Elsevier, 2017, pp. 811–843.

CHAPTER 6

Effective formulation strategies for poorly water soluble drugs

Kamla Pathak
Faculty of Pharmacy, Uttar Pradesh University of Medical Sciences, Etawah, Uttar Pradesh, India

1 Introduction

Oral administration is the most convenient and popular route of drug delivery due to flexibility in the design of dosage form, ease of administration, least sterility constraints, high patient compliance and cost effectiveness [1]. However, the key issue remains poor oral bioavailability owing to various factors including poor aqueous solubility, poor dissolution rate, extensive first-pass metabolism, low drug permeability and susceptibility to efflux mechanisms. The primary cause of poor oral bioavailability is ascribed to poor solubility and low permeability. Aqueous solubility has been designated as a critical parameter primarily responsible for the poor performance of oral drug delivery. Poor solubility can lead to unsatisfactory pharmacokinetic (ADME) properties leading to low oral bioavailability, low half-life, fast pre-systemic metabolism and of course difficulty in formulation [2].

As repeatedly documented, more than 40% of the new chemical entities developed via pharmaceutical research are practically insoluble in water. The molecules exhibit poor drug absorption and hence poor bioavailability. Any drug to be absorbed must be present in the form of an aqueous solution at the site of absorption. For oral liquid pharmaceutical formulations water is the solvent of choice and most of the drugs are either weakly acidic or weakly basic in nature, hence poor aqueous solubility renders a critical issue. This has garnered considerable interest among pharmaceutical researchers and numerous approaches are available and reported in literature to enhance the solubility of poorly water-soluble drugs. The technique is chosen based on the physicochemical properties of drug, nature of excipients, and the intended dosage form. Low aqueous solubility and consequently low dissolution rate is a matter of concern especially for BCS class II (low solubility and high permeability) drugs. The bioavailability of such drugs may be

Advances and Challenges in Pharmaceutical Technology
https://doi.org/10.1016/B978-0-12-820043-8.00004-9

improved by enhancing the solubility and hence the dissolution rate of the drug(s) in the gastrointestinal fluids [3].

Continual discovery and development of new drug molecules alone are not sufficient to achieve therapeutic excellence. Therefore, modified (improved) formulations of existing drugs are gaining significant importance. Alternative dosage form, making ester/salt of drugs, designing prodrug/active metabolite of drug, different route(s) of administration are few changes that the pharmaceutical industries are exploring for 505(b)(2) fillings [4]. Sizable number of poorly soluble drugs in the global market provides profitable strategies for pharmaceutical industries to file NDA of improved formulations amenable to faster dissolution and enhanced bioavailability.

2 BCS to DCS

The US FDA has documented a Biopharmaceutics Classification System of drugs that serves as a guide for predicting the drug absorption based on the solubility and intestinal permeability of the drugs. Solubility of a drug is based on its highest-dose strength in an immediate release product. Accordingly, a drug is considered highly soluble when its highest dose strength is soluble in 250mL or less of aqueous media over the pH range of 1–7.5. The volume of 250mL is generated from the bioequivalence study procedures that describe administration of a drug product to fasting human volunteers with a glass of water [1]. On the other hand, the intestinal permeability classification is based on intravenous administration. The existing oral drugs have been classified into four classes: class I – highly soluble and highly permeable, class II – poorly soluble and highly permeable, class III – poorly soluble and highly permeable, and class IV – poorly soluble and poorly permeable.

Later, using the biopharmaceutics classification system (BCS) as the reference point, a revised classification system for oral drugs was developed. The new system is focused on drug developability. The parameters considered in the revised system include intestinal solubility, the compensatory nature of permeability and solubility in the small intestine, and an estimate of the particle size needed to overcome dissolution rate limited absorption. The system was validated by comparison with literature the *in vivo* performance of a number of drugs. The observations on the drugs were consistent with the revised classification, termed as Developability Classification System (DCS), showing it to be of greater value in predicting the factors critical to *in vivo* performance of the drugs, than BCS system [5]. The DCS was designed to omit the gap between the BCS and the need for early evaluation of drug candidates with respect to their amenability for oral delivery [6]. The DCS applied solubility of drug in fasted state simulated intestinal fluid to estimate intestinal solubility, evaluated the compensatory nature of permeability and solubility during oral absorption and provided a way of estimating the critical particle size at which dissolution becomes rate-limiting to absorption. Building on the framework, a refined developability classification system (rDCS) is now proposed. The rDCS stratified into standard investigations is applied to all candidates, and customized investigations. Standard investigation of solubility and permeability can be performed according to in-house methods, and the results compared with standard sets of fasted state human intestinal fluid solubility and human effective jejunal permeability, which have been generated specifically for rDCS. Customized investigations are used when there is possibility of supersaturation/precipitation (weak bases; salts of weak acids) and to assess dissolution versus permeation limited absorption. Additionally, the rDCS offers simplistic visualization of the results, enabling practical comparison of drug candidates and formulation approaches. The concept has been effectively applied to six marketed compounds (acyclovir, danazole, albendazole, dipyridamole, dantrolene and piroxicam) for which there is rich database of information. Furthermore, the rDCS has been applied to 20 pipeline compounds from past and current research projects at Bayer AG. The rDCS was able to forecast the results in humans correctly in 80% cases [7].

3 Pharmaceutical strategies for effective formulation development

Plethora of formulation strategies could be used to enhance the bioavailability of BCS class II drugs, either by enhancing the dissolution rate or by presenting the drug in solution and maintaining it in solution in the gastrointestinal lumen. Formulation strategies are not very successful in improving the absorption of BCS class III and IV drugs that are limited by poor membrane permeability. The logical approach for improvement in bioavailability of class IV drugs is to revert back to the lead optimization phase of drug discovery and redesign the drug candidate with optimized physiochemical properties. The molecular modifications could be salt formation, polymorphs generation, new congener/analogs and prodrug design [8]. Each drug molecule is characterized by physicochemical and biological properties, some of which are desirable while others are undesirable. Molecular modification affects the physicochemical properties of the drug that in turn influences the localization of the drug.

3.1 Salt formation and pH modification

Salt formation of the ionizable drug molecule is a common strategy to correct the suboptimal drug properties so as to improve physical properties, aqueous solubility, dissolution, stability, alter gastrointestinal absorption, reduce toxicity or enhance organoleptic properties [9, 10]. Identifying a suitable salt form for an ionizable small molecule is often the cost-efficient optimization approach. Consequently, about 50% of the US FDA approvals constitute active pharmaceutical ingredients in salt forms [11]. The aqueous solubility of a drug is particularly important for the development of parenteral formulations. Appropriate counterions can be screened on the basis of solubility studies, in order to find out the best candidate. In many cases, the drug solubility can be modulated by altering the pH of a solution. A well-known example is phenytoin whose solubility is significantly increased by the addition of sodium hydroxide, in order to permit parenteral administration at desired concentration [12].

3.2 Prodrug development

When an adequate pharmaceutical formulation is unable to overcome pharmaceutical drawbacks, a chemical modification of the active molecule is necessary to correct its pharmacokinetic insufficiencies. The aim is to convert the active molecule into a clinically acceptable drug, and the prodrug design offers an alternative to improve physicochemical, biopharmaceutical and pharmacokinetic attributes. Initially, the term prodrug was used to define any compound that undergoes biotransformation before manifesting its pharmacological effects. Later the term drug latentiation was introduced which essentially is chemical modification of a biologically active compound to form a new compound that, upon *in vivo* enzymatic attack, will liberate the parent compound. The advanced drug latent systems like CDS, ADEPT GDEPT/VDEPT, PDEPT, PELT, ODDS and LEAPT are some examples [13]. The classification of prodrugs is presented in Fig. 1.

Over the past decade, a sizeable number of approved prodrugs have been marketed (Table 1), substantiating the importance of this tool on drug design. Prodrug approach can be used to resolve many undesirable properties, including poor aqueous solubility, insufficient oral absorption, chemical

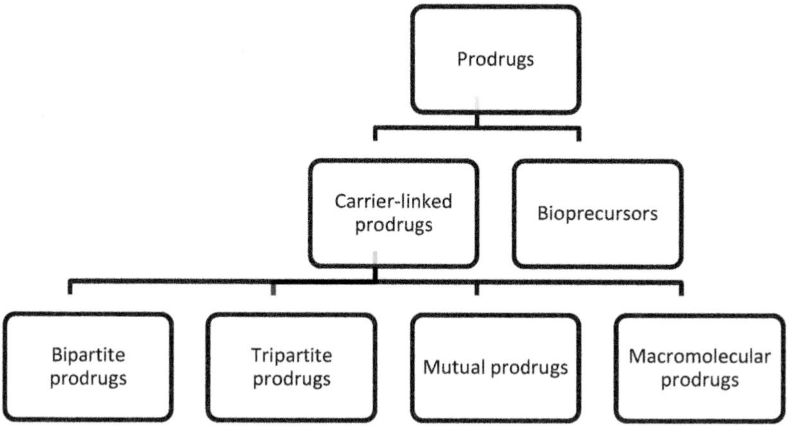

Fig. 1 Classification of prodrugs.

Table 1 A list of approved prodrugs that have been marketed.

Prodrug moiety	Brand name	Manufacturer (year)
Baloxavir marboxil	Xofluza	Roche (2018)
Krintafel	Tafenoquine	GlaxoSmithKline (2018)
Akynzeo	Fosnetaupitant	Helsinn (2018)
Valbenazine tosylate	Ingrezza	Neurocrine Biosciences (2017)
Deutetrabenazine	Austedo	Teva Pharmaceutical Industries (2017)
Deflazacort	Emflaza	Marathon Pharmaceuticals (2016)
Telotristat ethyl	Xermelo	Lexicon Pharmaceuticals (2016)
Isavuconazolium sulfate	(Cresemba®)	Astellas (2015)
Tedizolid phosphate	(Sivextro®)	Cubist (2014)
Sofusbuvir	(Sovaldi®)	Gilead (2013)
Dimethyl fumarate	(Tecfidera®)	Biogen IDEC (2013)
Azilsartan medoxomil	(Edarbi®)	Takeda (2011)
Gabapentin enacarbil	(Horizant®)	Xenoport (2011)
Dabigatran etexilate	(Pradaxa®)	Boehringer Ingelheim (2010)
Prasugrel	(Effient®)	Eli Lilly (2009)
Fesoterodine fumarate	(Toviaz®)	Pfizer (2008)
Lisdexamfetamine dimesylate	(Vynase®)	New River (2007)

instability, presystemic metabolism, local absorption/irritation, low half-life, toxicity. Additionally, problems related to drug formulation and optimum delivery can also be overcome by prodrug design [14, 15]. The prodrug approach is used to mask the charge or increase the lipophilicity of the molecule, leading to favorable passive membrane permeation, or to implicate chemical modifications of the molecule that would facilitate uptake by intestinal transporters. For example covalently-linked lipophilic carboxylate and/or phosphonate ester functional groups are used to increase passive diffusion. Once absorbed, the prodrug gets rapidly cleaved by biological enzyme to release the parent drug. The prodrug approach primarily focuses on BCS III drug molecules, as exemplified by enalaprilat, adefovir and gabapentin. In some cases, the prodrug may result in a change of BCS classification, as is the case for gabapentin (BCS III) where esterification of the molecule to gabapentin enacarbil resulted in a BCS II prodrug.

While the prodrug approach has been successfully employed to overcome many drug delivery issues, it presents various formulation challenges that arise due to modulation of physicochemical properties in comparison to the parent drug. These include (i) chemical instability of the drug-promoiety linker, (ii) generation of degradative by-products that may be reactive intermediates leading to secondary degradation pathways, (iii) insufficient aqueous solubility of prodrug and degradation products, (iv) polymorphism, and (v) disruption of solid state crystallinity [16]. Table 2 summarizes the research reports on prodrugs primarily focusing on drug delivery aspect.

3.3 Cosolvency and use of surfactants

Cosolvency is a popular approach for improving the solubility of poorly aqueous soluble drugs in pharmaceutical liquid dosage forms. The solubility of weak electrolytes and non-polar molecules in water can be increased by the addition of water miscible solvent in which the drug has good solubility. The cosolvent system works by reducing the interfacial tension between the aqueous solution and hydrophobic solute [22]. By disrupting the self-association of water molecules, cosolvents reduce the ability of water to squeeze out non-polar, hydrophobic compounds, thus increasing solubility [23]. The most frequently used GRAS listed cosolvents for oral/parenteral use include propylene glycol, glycerin, ethanol, low molecular weight polyethylene glycols, dimethylsulfoxide and dimethylacetamide [24–26]. Cosolvent solubilization approach has been used to enhance the solubility of seven antidiabetic drugs: glyburide, gliclazide, glipizide, repaglinide, glimepiride, roziglitazone and pioglitazone. The aqueous solubility of the drugs was recorded as less than 0.04 mg/mL. The solubility of glipizide, gliclazide and repaglinide was increased by 3–6 times in phosphate buffer, pH 7.4 by use of cosolvents. The combined effect of cosolvent and buffer was synergistic and the solubility of sulfonylureas and repaglinide increased

Table 2 Highlights of drug delivery via prodrugs.

Objective and approach	Result highlights	Ref
The oleate prodrug of docetaxel (DTX-OA) was synthesized and efficiently incorporated into SNEDDS using core-matching technology with a high drug-loading rate to enable effective oral delivery of DTX.	DTX-S-OA SNEDDS offered superior characteristics over DTX SNEDDS • Uniform droplet size of about 30 nm • High drug loading capability • Slower drug release • Higher (>2-fold) drug solubilization in the aqueous phase after 60 min lipolysis • Increased membrane permeability and intestinal bioadhesion • Enhanced intestinal lymphatic • 2.0-fold higher bioavailability	[17]
A range of carboxylic ester prodrugs of bexarotene (BEX) were designed and synthesized. The activated ester prodrug approach was also applied to retinoic acid (RA)	All the esters showed • Improved association with chylomicrons substantiating potential for delivery to the intestinal lymphatic system • 4-(hydroxymethyl)-1,3-dioxol-2-one ester prodrug of BEX was able to increase the exposure of the mesenteric lymph nodes to BEX 17-fold compared to when BEX itself was administered. • In case of RA the exposure of the mesenteric lymph nodes was increased 2.4-fold with cyclic activated prodrug.	[18]
A unique prodrug/nanoparticle formation strategy was designed to facilitate semi-solid prodrug nanoparticles of the highly water-soluble nucleoside reverse transcriptase inhibitor: emtricitabine	*In vitro* to *in vivo* extrapolation modeling predicted sustained prodrug release, with activation in relevant biological environments, representing a first step toward complete injectable long acting regimens containing nucleoside reverse transcriptase inhibitors.	[19]

Continued

Table 2 Highlights of drug delivery via prodrugs—cont'd

Objective and approach	Result highlights	Ref
Prodrugs of atenolol were made that can be (a) formulated in aqueous solutions and maintain stability over a long period of time and (b) bitterless and capable of undergoing intra-conversion in physiological environment to provide the parent active drug, atenolol, in a programmable manner.	The QM calculations at different levels demonstrated that the efficiency of atenolol ProD 1-ProD 2 is largely sensitive to the pattern of substitution on the carbon-carbon double bond and nature of the alkyl group on amide nitrogen. • Prodrugs were bitterless, stable in neutral aqueous solutions, released the parent drug in a sustained manner. • $t_{1/2}$ of prodrug was largely affected by the pH of the medium. The $t_{1/2}$values in 1N HCl, buffer pH 2, and were 2.53, 3.82, and 133h respectively.	[20]
A lipidic formulation of a glycerolipidic prodrug of didanosine was aimed to improve its bioavailability. Three types of lipidic formulations (emulsions, mixed micelles and liposomes) were used to encapsulate the prodrug.	The anhydrous prodrug showed polymorphic behavior. • Solubilities in water even in the presence of taurocholate micelles was very low (max 244 μg/mL) • Was amphiphilic (log $P=2$). • Was perfectly incorporated in DPPC multilamellar liposomes up to a ratio of 1:5 (mol:mol) prodrug:DPPC • The freeze-dried liposomal preparation could be formulated as gastro-resistant capsules to prevent didanosine from acidic degradation.	[21]

enormously. However, cosolvent alone caused significant enhancement in case of glitazones and the presence of buffer negatively affected the solubilization potential of the cosolvents. Up to 763, 316, 153, 524, 297, 792 and 513 times increase in solubility was reported for gliclazide, glyburide, glimepiride, glipizide, repaglinide, pioglitazone and rosiglitazone, respectively [27]. Some other examples of use of cosolvents in formulation development are listed in Table 3.

Table 3 A cross section of reports on use on cosolvency as an approach for formulation development.

Drug + cosolvent(s) screened	Result	Ref
Etoricoxib + PEG 400, PG, glycerin	Among various solvent-cosolvent blends investigated, water-PEG 400 showed highest solubilization potential (1.4358) in comparison with PG (1.3080) and glycerin (1.0154). PEG may be used to develop liquid dosage forms.	[28]
19-nor-1α,25dihydroxy-vitamin D_2 + water, ethanol, and PG	An injectable solution was formulated using the ternary cosolvent system.	[29]
Tenoxicam + DMSO/polyethoxylated castor oil/ethanol system	DMSO/polyethoxylated castor oil/ethanol system in volume ratio of 5:4:1 was used to develop intramuscular injection of the drug. The drug showed high solubility (20.73 mg/mL) and the system had low viscosity (10.0 Cp). Pharmacokinetic study revealed a bioequivalence (F_{rel}) of 0.89.	[30]
Diclofenac sodium + benzyl alcohol	Low concentration benzyl alcohol (7.5% v/v) was used as co-solvent for diclofenac sodium in aqueous medium and due to this, injection is less painful then propylene glycol based injection.	[31]
Valdicoxeb + PEG400 and PEG600	Both PEG-400 and PEG-600 showed exponential increase in solubility with increase in cosolvent fraction. The increment in solubility was found to be 80-fold and 59-fold for PEG 400 and PEG 600 at 25% cosolvent level.	[32]
Ketoprofen	The solubility increased up to 85.66 times (maximum) in case of ethanol while it was 33 times (minimum) in case of glycerol. Using selected cosolvents and hydrotropes, aqueous injections of the drug was formulated.	[33]
Tipredane	A topical solution of the drug (0.1% w/w) was formulated using polyethylene glycol 400 (PEG 400), propylene glycol, and water. The solvent mixture was non-irritating to the skin	[34]

The use of surfactants to improve the dissolution performance of poorly soluble drug products is probably the oldest method. Surfactants act by reducing the surface tension thereby improving the dissolution of poorly soluble drugs in water. When the concentration of surfactant exceeds its critical micelle concentration which is in the range of 0.05–0.10% for most surfactants; micelle formation occurs. Micellization results in enhanced solubility of poorly soluble drugs. Surfactants also improve wetting of solids and increase the disintegration of solids into finer particles [35]. Commonly used non-ionic surfactants include polysorbates, polyoxyethylated glycerides, polyoxyethylated castor oil, mono- and di-fatty acid esters of low molecular weight polyethylene glycols and lauroyl macroglycerides. Surfactants also find application in stabilizing microemulsions and suspensions of drugs [36]. Examples of drugs that use micellar solubilization as a successful strategy to improve solubility are antidiabetic drugs namely glyburide, gliclazide, glimepiride, repaglinide, glipizide, rosiglitazone and pioglitazone [37]. In a comparative study effect of various solubilization techniques, namely micellar solubilization, cyclodextrin complexation and cosolvency, was estimated on the solubility of valdecoxib. Highest solubility (70 fold) was observed with use of Cremophor EL followed by Tween 80 and sulfobutyl ether-7-β-cyclodextrin. It was concluded that surfactants with higher HLB values were better solubilizers. Solubilization capacity increased with increase in hydrocarbon chain of surfactant, suggesting hydrocarbon core of micelles as locus of solubilization [32]. In a study by Maheshwari et al. the dissolution rate of naproxen was enhanced using micellar solubilization technology. Cloud point temperature of the non-ionic surfactants was used as the basis for the solubility study because the entrapment of the drug into the surfactant micelle is best at this temperature. Naproxen was solubilized in polysorbate 80 micelles at cloud point temperature; lactose was dissolved in micellar dispersion and the dispersion was directly spray-dried to obtain microparticles, which was subsequently converted into tablets using suitable excipients. Spray-dried naproxen powder and tablets made out of it by direct compression method, exhibited superior dissolution rate over controls at all pH conditions. Hence better dissolution characteristics were conferred by the micellar solubilization of naproxen and rapid wetting of microparticles [38].

3.4 Micronization

Micronization is a size reduction technique where the resulting particle size distribution is less than 10 μm. Milling, spray drying and supercritical fluid

(SCF) technologies are mostly employed micronization techniques which can enhance the solubility and dissolution rates. Although these techniques produce particles of micron size, the particle size distribution is non-homogenous distribution due to the agglomeration of the particles which decreases the surface available for dissolution [39, 40]. Furthermore, formation of amorphous regions by use of these techniques is leads to poor stability. *In-situ* micronization is a novel technique where microcrystallization and surface modification can be affected using hydrophilic polymers which act as stabilizing agents to enhance wetting properties, stabilize the microcrystals and prevent their crystal growth [41]. The microcrystals produced by this method can be incorporated in solid, liquid and semisolid dosage forms. Specifically, these can be formulated as dry powder inhalers or aerosols because of improved aerodynamic behavior [42]. Varied research on successful application of micronization techniques has generated extensive literature and the techniques have been successfully exploited commercially by pharmaceutical industry. One can also find numerous patents on this technique and select patents embracing formulation and drug delivery strategies have been compiled in Table 4.

3.5 Solid dispersions

A solid dispersion is essentially a drug–polymer two-component system in which the mechanism of drug dispersion is the key to understanding its behavior. Primarily an amorphous product is produced. With continual increase in the number of FDA approved products in recent years, solid dispersion is now firmly established as a platform technology for the formulation of poorly-soluble drugs. Solid dispersion technology has been successfully applied to develop formulations with a high drug load (e.g. 375 mg per tablet in Incivek) and/or containing drugs with a high tendency to crystallize (as indicated by the high melting point of 291 °C of ivacaftor in Kalydeco) [43]. Extensive literature on solid dispersions of drugs using variety of carriers is documented and is not being included in the chapter. At least three methods of preparing solid dispersions have been successfully used in commercial production (Table 5).

There are instances where neither hot melt extrusion technique nor spray drying technique is applicable for preparation of solid dispersions, specifically for the APIs with high melting point (>200 °C) and limited solubility in volatile organic solvents. For such drugs either an alternative method for generation of amorphous solid dispersion is required; or a

Table 4 Select patents embracing formulation and drug delivery strategies based on micronization.

Patent number (year) *Title*	Invention
US10258573 (2019) *Micronized insulin and micronized insulin analogs prepared under acidic conditions, and methods of manufacturing the same under acidic conditions*	A method of preparing inhalable insulin suitable for pulmonary delivery includes: dissolving an insulin raw material in an acidic solution to form a dissolved insulin solution; titrating the dissolved insulin solution with a buffer solution to form a suspension comprising micronized insulin particles; and stabilizing the micronized insulin particles.
US10105398 (2018) *Micronized placental tissue compositions and methods of making and using the same*	Described herein are compositions composed of micronized placental components and pharmaceutical compositions thereof. The compositions have numerous medical applications.
US9744178 (2017) *Respirable agglomerates of porous carrier particles and micronized drug*	Embodiments of the present invention provide a dry powder composition comprising porous carrier particles associated with micronized drugs wherein an ordered mixture between the micronized drug(s) and the carrier particle results, such that the micronized drug(s) adhere strongly to the carrier particles forming a stable ordered mixture of respirable agglomerates. Embodiments of the present invention further comprise a spray-drying process to create the respirable agglomerates.
US 9314447 (2016) *Reduced dose pharmaceutical compositions of fenofibrate*	The present invention provides a reduced dose oral pharmaceutical composition comprising a mixture of nano-particulate fenofibrate and micronized fenofibrate and one or more pharmaceutically acceptable excipients.
	Novel compositions, formulations and dosage forms comprising stabilized micronized opioid particles are disclosed. Exemplary

US 9339463 (2016) *Micronized opioid compositions having a specific particle size distribution*	opioids include oxycodone, oxymorphone, hydrocodone, and hydromorphone, including as free bases or as salts. Stabilized micronized opioid particles having an average dia of less than 90 µm are disclosed. Methods for micronizing an opioid to provide stabilized micronized opioid particles are also disclosed.
US 9040030 (2015) *Mascara compositions containing an oil-dispersible micronized wax and a water-dispersible wax*	The invention relates to a composition comprising at least one water-dispersible wax and at least one oil-dispersible micronized wax which, if desired, can be prepared at room temperature.
US8846099 (2014) *Parenteral composition comprising microspheres with a diameter between 10 and 20 µm*	The invention relates to pharmaceutical formulations suitable for targeting particular tissue and/or organ(s) with a formulated active ingredient and a biodegradable excipient, wherein 90% particles have a diameter between 10 and 20 µm and the formulation is substantially free of particles with a diameter >50 µm and <5 µm, such that where the formulation is administered upstream of the target tissue the ability of the active to pass through the target tissue and pass into systemic circulation is restricted.
WO2014174233A1 (2014) *Preparation of drug particles by micronization*	The invention relates to a process for preparing crystalline micronized particles of tiotropium bromide. The process involves suspending the drug in a water immiscible anti-solvent in which the drug has little or no solubility and micronizing the suspension. The resulting drug particles were physically stable with regard to agglomeration and/or aggregation on storage.

Continued

Table 4 Select patents embracing formulation and drug delivery strategies based on micronization—cont'd

Patent number (year) *Title*	Invention
US8512753 (2013) *Micronized particles of low-dosage strength active agents for powder formulations for inhalation*	The invention discloses micronized particles of a low-dosage strength active ingredient, to be used in dry powder formulations for inhalation, with particular properties can easily and homogenously disperse in a dry powder formulation to be administered by means of a dry powder inhaler device.
US8356606 (2013) *Production of micronized encapsulated tobacco particles for tobacco flavor delivery from an oral pouch*	A method of providing a comestible plant material taken orally by coating a micronized plant material with a first layer of polysaccharides or proteins, then a second layer of poly-saccharides or proteins is disclosed. A bi-encapsulated plant material produced by this process is also disclosed.
US 8435946 (2013) *Orally dosed pharmaceutical compositions comprising a delivery agent in micronized form*	Solid pharmaceutical compositions suitable for the oral delivery of pharmacologically active agents, e.g. peptides, along with crospovidone/povidone; and a delivery agent for the active agent are disclosed. The compositions utilized micronized forms of the delivery agent to enhance bioavailability of pharmacologically active agent calcitonin.
US 8513240 (2013) *Micronized tanaproget and compositions containing same*	The present invention provides compositions containing micronized tanaproget and various excipients. The compositions are useful in contraception and hormone replacement therapy and other hormone-dependent tumors.
KR20120104235A (2012) *Antibiotic microparticles for inhalation*	The invention relates to inhalation powder formulations containing fine particles consisting of antibiotics and magnesium stearate for the treatment of bacterial infections associated with the manufacturing process and pulmonary disease of the fine particles.

US7704525 (2010) *Lipid microparticles by cryogenic micronization*	This invention relates to lipid microparticles enriched in amphiphilic components, which promote the incorporation of peptides and/or protein. A cryogenic micronization manufacturing process for their preparation is also disclosed.
US 7829115 (2010) *Pharmaceutical composition based on micronized progesterone, preparation method and uses thereof*	The present invention relates to a pharmaceutical composition comprising micronized progesterone, soya bean lecithin, and oil, to the method for the preparation thereof and to the uses thereof for treating insufficiency of progesterone secretion.
ES2296831T3 (2008) *Method of preparing microparticles for use in pharmaceutical compositions for inhalation*	The patent discloses a method of preparing microparticles exhibiting delayed dissolution for use in a pharmaceutical composition for pulmonary administration. It describes milling particles of the active substance in the presence of particles of the material hydrophobic so that particles of hydrophobic material fuse to the surfaces of the particles of active substance.
US7101563 (2006) *Micronized vitamin C formulation*	A stable topical application, comprising micronized L-ascorbic acid in a non-aqueous carrier having a pH between 5.5 and 7.5 is disclosed. The composition may comprise papain as an enzymatic exfoliant in addition to humectants, emollients, viscosity increasing agents, surfactants, and preservatives. The composition is useful as UV protectant, promoters of collagen synthesis and in the removal and/or treatment of wrinkles.
WO2005025536A2 (2003) *Dry powder composition comprising co-jet milled particles for pulmonary inhalation*	The present invention relates to methods of making composite active particles comprising a pharmaceutically active material for pulmonary inhalation, the method comprising a jet milling process.

Continued

Table 4 Select patents embracing formulation and drug delivery strategies based on micronization—cont'd

Patent number (year) *Title*	Invention
US6555135 (2003) *Pharmaceutical compositions comprising co-micronized fenofibrate*	The co-micronization of the fenofibrate and excipient was carried out by mixing them followed by micronization of the mixture in an air-jet mill. The product obtained had a mean particle size of less than 15 μm. The co-micronized powder may be processed into solid dosage forms (tablets or capsules) for oral administration.
WO1993013752A1 (1993) *Improved process for preparing micronized polypeptide drugs*	The invention discloses solid particle aerosol formulations of polypeptide drugs made by lyophilizing drug solution(s) containing milling stabilizers. The lyophilized drug was milled in fluid energy mills Pure nitrogen was introduced to eliminate particles of greater than 0.1 μm to transport the drug. Milling stabilizers in the solution and abrasion-resistant fluid energy mills were meant to reduce insoluble contaminants and inactive fractions in the milled product.

Table 5 Commercial methods of solid dispersion preparation.

Technology	Applicable to	Examples
Melt extrusion	Drugs with not-very-high melting points	Lopinavir, ritonavir, itraconazole
Spray drying	Drugs soluble in at least one volatile solvent	Tacrolimos, Telaprevir, Ivacaftor
Coprecipitation	Drugs with high melting point and limited solubility in common organic solvents	Vemurafenib, Praziquantel

non-solid dispersion approach (nano-milling or cyclodextrin complexation) must be adopted. Other techniques classified as solvent based/free methods can be used for generation of amorphous solids. Solvent-free methods involve supercooling of a material in the liquid state, vapor condensation and disruption of a crystal lattice in the solid state via grinding. On the other hand, solvent-based methods include freeze drying, precipitation of the API from solution and solvent evaporation by various techniques. Till date, these methods are limited to small scale in laboratories.

Coprecipitation of an API and a polymer is an attractive option for problematic APIs that is a scalable technique as well. For example solid dispersion of vemurafenib (Zelboraf®, Roche) is produced at large scale via solvent-controlled precipitation process [44]. However, improvement of the coprecipitation process is required to make the technology more amenable for commercial manufacturing.

A robust and reliable, third-platform technology for solid dispersion generation is based on microfluidization [45]. The solvent-controlled precipitation process, involves dissolving the API and polymer in a common solvent followed by gradual addition of a second, antisolvent in which the ingredients are insoluble. The interaction under carefully controlled conditions in a microreactor forms API-polymer coprecipitates constituting agglomerates of nanoparticles. The suspensions containing the coprecipitates may be isolated and spray dried to yield nanocomposite microparticles. The technology uses polar, high-boiling solvents (dimethylformamide and dimethylacetamide) and in comparison to hot melt extrusion, it is a low-temperature process that can be utilized for thermosensitive APIs. The formulation and processing variables have a direct impact on the attributes of the solid dispersion generated. The method can be regarded as a highly attractive option for enhancing the bioavailability of BCS Class II drugs with poor aqueous solubility. The methodology has been patented as a United States patent (by the same research team) describing the method for producing amorphous solid dispersions in a nanoparticulate form, through solvent controlled co-precipitation, using microfluidization/microreaction technology [46].

The selection of amorphous solid dispersion manufacturing method is often based on the properties of the drug candidate [47]. For example, lyophilization, is often used to process molecules with high aqueous solubility but is not suitable for insoluble small molecules. Recently, a novel solvent controlled precipitation technology known as microprecipitated bulk powder, has been successfully developed for compounds that cannot be processed by spray drying or melt extrusion [48]. Based on an efficient high-throughput screening, the method assists selection of polymer type, drug loading and solvent to antisolvent ratio required for microprecipitated bulk powder formulation development. The miniaturized coprecipitation screening (MiCoS) involves mixing of drug and polymer in dimethylacetamide a 96-well platform, followed by controlled precipitation to generate microprecipitated bulk powder, filtration/washing, drying and high throughput characterization. MiCoS finds application both in early and late stage formulation development. In early stage research programs, it allows quick screening of microprecipitated bulk powder formulation with minimal amount of API to enable efficacy and toxicology studies. In late stage drug development programs, microprecipitated bulk powder formulation optimization could be realized through wide experimental design space [49].

For accelerating formulation development of amorphous solid dispersions, Hovione has developed proprietary software, Ternarius®. Based on incorporation of diffusion, solvent evaporation kinetics and thermodynamics into a mathematical model, it allows the formulators to identify promising formulations without the need to consume the API [50]. Thus, this tool reduces the number of physical experiments, thereby saving both time and the expensive API.

Not only APIs, the solid dispersion approach has also been used to improve the dissolution rate of herbal extracts (*K. parviflora* extracted with dichloromethane). The homogeneous powdered solid dispersions were prepared by solvent evaporation using hydroxypropyl methylcellulose and polyvinyl alcohol-polyethylene glycol grafted copolymer (PVA-co-PEG). The solubility enhancement was at least 5 fold [51]. The study proves versatility of the solid dispersion technique both for solubility enhancement and formulation development.

3.6 Complexation with beta cyclodextrins

The cyclodextrin molecules are characterized by a hydrophilic surface and a hydrophobic cavity in which a guest molecule can be entrapped. The cyclodextrin molecules form inclusion complex with a wide variety of solids, liquids and gaseous compounds [52]. Generally speaking, the hydrophobic molecules have higher affinity for the hydrophobic cyclodextrin cavity in aqueous medium. The entrapment primarily modulates the solubility and stability of the guest molecules. The entrapment takes place predominantly by hydrophobic interactions between the walls of the hydrophobic cyclodextrin cavity and the guest molecules [53]. However, there exists a possibility of other forces, such as vander Waals and dipole–dipole interactions. Despite various factors and different forces involved in the complexation of the drug with cyclodextrins, the preparation of complexes is a rather simple process. Several methods based on the properties of the guest (drug molecule) and the nature of the selected cyclodextrin are available that can be used to obtain cyclodextrin–guest complexes.

A sizable number of cyclodextrin (CD) derivatives with extended physiochemical characteristics and inclusion capacity are now available. These can be classified as hydrophilic, hydrophobic and ionic derivatives of CD. The hydrophilic CDs, namely methyl-β-cyclodextrin, 2,6-di-*O*-methyl-β-CD and 2,3,6-per-*O*-methyl-β-CD, enhance the release of poorly water-soluble drugs thereby increasing the drug absorption across the biological membranes. Abundant literature on solubility enhancement of poorly water soluble drugs using CDs and hydrophilic CD derivates is available and has not been cited in this chapter. The amorphous hydrophilic CDs, such as hydroxyalkylated-β-CD or hydroxypropyl-β-CD, are useful for reducing polymorphic transitions and crystallization rates of poorly water-soluble drugs on storage, which can consequently affect high dissolution profile and oral bioavailability of the drugs [54]. A hydroxyalkylated β-CD derivative has relatively high aqueous solubility (above 60g β-CD in 100mL water), satisfactory inclusion ability and proven safety profile [55]. The hydrophobic CDs are used to modulate drug release and are employed as sustained-release carriers for water-soluble drugs.

CDs have been successfully employed in aqueous dermal formulations, nasal drug delivery systems [56] and eye drop preparations [57]. They can also be used to reduce the bitterness, irritancy, and foul smell of drugs [58, 59]. The poorly water soluble anticancer drug, Pt (IV)-bis (benzoate), was entrapped in β-CD to enhance its solubility [60]. The multicomponent inclusion complex of chloramphenicol with β-CD and amino acid (glycine/cysteine) improved its aqueous solubility, retained its antimicrobial activity and reduced its reactive oxygen species generation capacity in leucocytes [61]. The encapsulation of drug-CD inclusion complexes in liposomes have recently been reviewed by Gharib et al. [62]. CD was used to increase the entrapment of hydrophobic drug in the aqueous compartment of liposomes and the liposomal vesicle offered protection to the CD/drug inclusion complex until drug release. Anethole, a primary constituent of fennel and anise essential oils, was used as a model of a volatile hydrophobic drug to be entrapped in liposome [63].

In the pharmaceutical industry CDs have been employed as complexing agents to enhance aqueous solubility of poorly soluble drugs, increase their bioavailability and stability. Studies in both animals and humans have affirmed that CDs can be used to improve drug delivery from almost any kind of formulation. However, one cannot derive acceptable outcome simply by addition of CDs to existing formulations, without optimization. Currently, there are around 30 different pharmaceutical drug products across the globe containing drug-CD complexes on the market [64]. Increasing knowledge on the reactive ability of the CDs to form supramolecular structures with a variety of architectures (drug-conjugates, self-assembled systems, cross-linked networks, poly (pseudo) rotaxanes) makes it possible to use CDs not only as drug solubilizers but also as true delivery systems [65]. Table 6 below briefly describes the supramolecular structures of CDs.

3.7 Liquisolid systems

Liquisolid technique relates to the conversion of liquid medications into apparently dry, non-adherent, free flowing and compressible powder mixtures by blending the liquid medication (s) with suitable excipients, namely carriers and coating materials [71]. The liquid medication is first absorbed into the interior structure of the carrier. Once the interior framework gets saturated, the surface of carrier particles gets covered with the liquid layer, which is instantly adsorbed by the fine coating material. Thus, an apparently dry and free flowing and compressible powder mixture is formed. Usually water-miscible organic solvents with high boiling point, such as propylene glycol and polyethylene glycol (PEG) 400, are used as the liquid vehicles for the drug solution/dispersion. Porous materials with large specific surface area and high liquid absorption capacity (lactose, starch and various grades of cellulose) are used as carriers [72]. In contrast excipients with very fine

Table 6 Supramolecular structures of CDs.

Supramolecular structure of CD	Description	Ref
Polymeric chains and CDs in water produce supramolecular host – guest assemblies called polypseudorotaxanes.	The guest polymers thread the host CD molecules such as beads on a string. Once formed, these molecular necklaces aggregate into larger assemblies that reveal themselves in a significant increase of turbidity. The phenomenon is enthalpy driven and its kinetics depends strongly on choice of reactants, concentration, solvent nature, addition of cosolutes, and temperature.	[66]
β-cyclodextrin (β-CD) self-aggregates	At critical aggregation concentration (cac) between 2 and 3 mM of CD in water, aggregates were formed as revealed by dynamic and static light scattering. Cryo-TEM revealed polymorphism depending on the β-CD concentration: polydisperse nearly spherical objects with diameters of about 100 nm are present at lower concentrations, whereas micrometer planar aggregates were predominant at higher concentrations.	[67]
Chemically grafted or cross-linked CD (*CD-functionalized materials*)	Combination of CDs and hydrogels in a single material leads to synergic properties: the **hydrophilic network** enhances biocompatibility and prevents dilution in the physiological medium increasing the stability of the inclusion complexes, while CDs finely tune the mechanical features and the stimuli-responsiveness and provide affinity-based regulation of drug loading and release. These can be used as DDS for hydrophobic and hydrophilic drugs to be administered via almost any route. Medical devices (catheters, prosthesis, vascular grafts, bone implants) can also benefit from surface grafting or thermofixation of CDs. The synthetic approaches include (*i*) direct cross-linking through condensation with di- or multifunctional reagents, (*ii*) copolymerization of CD derivatives with acrylic/vinyl monomers, and (*iii*) grafting of CDs to preformed medical devices.	[68]

Continued

Table 6 Supramolecular structures of CDs—cont'd

Supramolecular structure of CD	Description	Ref
CD-drug conjugate	Drug is covalently bound to CD unlike inclusion complexes. One of the advantages of CD–drug conjugates is that they can survive passage through the stomach and small intestine, however, drug release will be triggered by enzymatic degradation of CDs in the colon.	[69]
Cyclodextrin-based nanosponges	Features include: • Nanostructured three-dimensional network. • Highly porous spherical nanoparticles either crystalline or amorphous; swelling properties. • The polarity, dimension of the polymer mesh and release of entrapped molecule can be easily tuned by varying the type of cross-linker and degree of cross-linking. • Site-specific targeting can be achieved by conjugating various ligands on the surface of nanosponge. • Offer controlled release, are biologically safe and biodegradable. • Can form complexes with lipophilic or hydrophilic molecules. • Improve the aqueous solubility of poorly water-soluble molecules, protect degradable substances and as innovative carrier in pharmaceuticals, cosmetics and protein/peptide delivery.	[70]

particle size and highly adsorptive property are used as coating materials [73]. The technique has been used to enhance the dissolution rate of low dose, poorly soluble drugs such as famotidine, clonazepam etc. The feasibility of the liquisolid technique for high dose, water insoluble drugs has been attempted either by use additives (PVP, HPMC and PEG 35000) that enhance absorptive capacity of carriers [74] or by use of novel carries that have high specific surface are and absorptive capacity like Neusilin® [75]. In a research reported by Sharma and Pathak, liquisolid formulation of paclitaxel was developed using novel highly porous liquisolid carriers (modified polysaccharides) to enhance the bioavailability of orally administered paclitaxel. Modified polysaccharides namely co-grinded treated guar gum, co-grinded treated tamarind kernel powder and co-grinded treated locust bean gum were developed by sequentially subjecting the corresponding polysaccharides to wetting, drying and co-grinding with mannitol (1:1). The liquisolid systems were directly compressed to produce liquisolid tablets that enhanced the dissolution and hence the oral bioavailability of the drug [76]. Table 7 compiles a cross section of the liquisolid systems developed for a variety of drugs.

Table 7 A cross section of liquisolid systems of drugs.

Drug **& formulation approach**	**Outcome**	**Ref**
Olmesartan medoximil Liquisolid compacts were prepared using Acrysol El 135 as a solvent, Avicel PH 102, Fujicalin and Neusilin as carrier and Aerosil as coating material in different ratios.	Liquisolid compacts with 80% w/w of Acrysol EL 135 to the drug showed significant higher drug release rates than conventional tablets. Among carriers used Fujicalin and Neusilin were found to be more effective carrier materials for liquid adsorption.	[77]
Telmisartan Liquisolid compacts were prepared using Transcutol HP as vehicle, Avicel PH102 as carrier, and Aerosil 200 as a coating material.	The optimized formulation showed pH independent release profile with significant improvement ($P < 0.005$) in dissolution compared to plain drug and conventional marketed formulation.	[78]

Continued

Table 7 A cross section of liquisolid systems of drugs—cont'd

Drug & formulation approach	Outcome	Ref
Olanzapine Non-volatile solvent used were Tween 80 and propylene glycol, Microcrystalline cellulose and silica were used as carrier and coating materials, respectively.	Liquisolid formulations showed higher drug dissolution rate than directly compressible tablets and marketed tablets	[79]
Curcumin Drug-loaded liquisolid systems were prepared using vehicles (PEG200, PEG400 and Tween80). The carrier was microcrystalline cellulose PH102 and coating material was Aerosil.	Higher cumulative drug release was observed from liquisolid systems than directly compressed tablets.	[80]
Ritonavir Liquisolid pellets (LP) were formulated using microcrystalline cellulose (carrier), Kollidon_ CL-SF (coating and disintegrating material) and high load (30%, w/w) of Kolliphor_ EL or PEG 400 (non-volatile solvent) by extrusion-spheronization	Partial conversion of RTV polymorph I to the less soluble form II occurred during the preparation LP containing Kolliphor_ EL achieved higher drug dissolution min compared to LP made with PEG 400 and Norvir tablet.	[81]
Felodipine The effects of Kollidon®CL-SF as a coating and disintegrating material and the type of non-volatile solvent, PEG 400 or Cremophor®EL, was analyzed on dissolution behavior.	Cremophor®EL was found more effective in comparison to PEG 400 due to the formation of softer and more porous structures. The amount of Kollidon®CL-SF also showed remarkable positive effect on the drug dissolution characteristics.	[82]
Spironolactone Three non-volatile liquids Capryol™ 90, Synperonic® PE/L61 in combination with Solutol® HS-15 at a ratio of 1:1, and Kollicoat® SR 30 D were used.	Liquisolid powder formulations formulated with a combination of Synperonic® PE/L61–Solutol®HS-15 showed highest dissolution characteristics.	[83]

Liquisolid technique was initially designed to enhance the dissolution of poorly water-soluble drugs. However, of lately, studies have demonstrated the utilization of liquisolid technique for preparation of sustained release formulations of drugs. The primary advantage of using liquisolid technique in extending drug release is the potential to achieve zero order release kinetics [84]. However, its main limitation would be high tablet weight, attributable to the high dose of drug used in the sustained release formulations (usually higher than that in conventional tablets) [85]. The concept underlying sustainment of drug release via liquisolid technique is based on the use of either hydrophobic carriers (Eudragit® RL and RS) instead of hydrophilic carriers, or use of release retarding agents (HPMC) in the liquisolid formulations [86]. Besides, as the specific surface area value of the commonly used hydrophobic carriers are usually lesser than that of the hydrophilic carriers, the amount of coating material required to convert wet carrier particles to apparently dry and free flowing powders will be generally higher.

Another category of formulative approach is development of orodispersible liquisolid system which is the combination of liquisolid technique and orodispersible system. Orodispersible liquisolid compacts of aceclofenac were made using different dissolution enhancers (propylene glycol, polyethylene glycol 400 and Tween 80 in 1:1 ratio with drug), followed by addition of diluents, superdisintegrants (crosscarmellose sodium, crosspovidone and sodium starch glycolate) in various ways and in combinations; finally addition of glidants and lubricants. All liquisolid compacts disintegrated within 3 min with enhanced dissolution properties over the conventional tablet of aceclofenac. Liquisolid compacts with sodium starch glycolate added intragranularly and cross povidone extra granularly showed highest dissolution rate [87].

3.8 Nanosizing strategies

Nanosizing refers to the reduction in particle size of the active pharmaceutical ingredient (API) to the sub-micron range, with the final particle size typically being 100–200 nm. The particle size reduction can significantly increase the dissolution rate of the API, which in turn can lead to substantial increase in bioavailability [88]. The use of media milling technology to formulate poorly-water-soluble drugs as nanocrystalline particles offers the opportunity to address the performance characteristics of BCS class II drugs. NanoCrystal® Technology is an attrition process wherein large micron size drug crystals are media milled in a water-based stabilizer solution to yield nanometer-sized drug crystals. Nanoparticle dispersions are stable and typically have particles of mean diameter of less than 200 nm with 90% of the particles being less than 400 nm. The formulation consists of water, drug and one or more GRAS excipients (stabilizers). These liquid nanodispersions exhibit acceptable shelf-life [89].

Nanocrystalline particles are suitable delivery system for all commonly used routes of administration, namely oral, injectable (IV, IM and SC) and topical applications. Additionally, the aqueous dispersions of nanoparticles can be processed into tablets, capsules, fast-melts and lyophilized products for sterile product applications [90]. Nanoparticle based drug products have demonstrated improved bioavailability and enhanced drug exposure for oral and parenteral dosage forms. Pharmacokinetic profiles for injectables vary from being rapidly soluble in the blood, to gradual dissolving, after which macrophage uptake and subsequent release greatly prolong drug delivery, while minimizing peak height. For several drug classes, this leads to improved safety, this permits higher dosing and improved efficacy.

Methods of nanocrystals manufacture involve crystallization, building nanocrystals up from the supersaturated solution state and downsizing larger particles by milling or homogenization. First generation standard top down technologies produce nanocrystals in the size range of 200–1000 nm. These processes become tedious when a particle size of less than 100 nm is desirable. However, ultrafine drug nanocrystals with a size below 100 nm are clearly superior to the larger nanocrystals of the first generation and offer many potential advantages. In this context, second generation combinative technologies such as H42, H69 and H96 production, processes have been developed to generate ultrafine nanocrystals on the lab scale. However, these processes still need to be established on a larger scale [91]. Nanonization using supercritical fluid technology is discussed in next section. Table 8 compiles some significant studies reporting nanonization as an advantageous formulative approach.

Nanosuspensions are fast and easy to produce, enhance the bioavailability of poorly soluble drugs and feature many beneficial characteristics. However, nanosuspensions are physically metastable and to overcome long term stability issues nanocrystals can be converted into solid dosage forms by various methods. Transformation of nanosuspensions into solid dosage forms opens up manifold options for the dosage form development with tailor-made drug release profiles.

Drug	Nanonization approach	Outcome	Ref
Bicalutamide	Combination of anti-solvent precipitation and spray drying method	The lactose loaded particles had a mean size of 330 nm within a narrow distribution. Dissolution rate of bicalutamide nanodispersion was significantly faster than that of commercial products. It increased to 94% in 10 min while both commercial formulas Casodex and bicalutamide tablets dissolved 60% and 38% respectively at the same period.	[92]
Miconazole	Nanoground with hydroxypropylcellulose in combination with sodium dodecyl sulfate	Drug particle mean size of 140–170 nm was achieved. Nanosuspension was stable at 5 °C, whereas storage at 25 °C resulted in particle agglomeration and single crystal growth.	[93]
Isradipine	Sonoprecipitation by varying the amplitude and the time length of ultrasonication.	Sonoprecipitation with HPMC resulted in nanocrystals with enhanced the drug release rate. Stability of nanosuspension was depended on amplitude and time of ultrasonication since these factors affected on the size of nanoparticles.	[94]
Rutin	Lyophilization and incorporated into tablets	exhibited a higher particle dissolution rate than microcrystal-loaded tablets and tablets already on the market	[95]
Baicalein	Combination of antisolvent crystallization and high-pressure homogenization.	exhibited significantly enhanced dissolution, confirmed *in vivo*: pulmonary baicalein was rapidly and extensively absorbed through the pulmonary mucosa, showing pharmacokinetics parameters that were identical to those obtained after the intravenous injection of this drug	[96]
Apigenin	Combination of bead milling and subsequent high-pressure homogenization were formulated into a hydrogel for topical application.	The antioxidant capacity of apigenin was demonstrated by evaluating its antiradical scavenging activity. The fast release profile from nanometric range particles was also proven	[97]

3.9 Melt sonocrystallization

Melt sonocrystallization is a particle engineering technique that uses ultrasonic energy to the melt during crystallization. This technique affects both the surface and structure of crystalline product and can be used to confer desirable characteristics to the drug crystals. Melt sonocrystallization produces smaller crystal size in comparison to conventional crystallization [98]. The technique has been utilized for a variety of drugs wherein the drug melt dispersed in water was insonated at select frequency for improvement in solubility, dissolution rate and compression properties; for e.g. ibuprofen [99] and celecoxib [100]. Sonocrystallization utilizes ultrasound waves in the frequency of 20–100 kHz for induction of crystallization. The technique not only improves the nucleation rate but is also an efficient method for size reduction and controlling the size distribution of APIs [101]. There are several reports on use of ultrasonic energy during crystallization to achieve nucleation at moderate supersaturation during the crystallization process and to obtain desired crystal habit [102]. Melt sonocrystallized piroxicam (MSCP) was prepared by melting the drug at 200 °C and pouring it in deionized water maintained at 75–80 °C followed by sonication for 4 min (frequency of 33 ± 3 kHz at 80% amplitude). The MSCP had superior micromeritic and rheological properties than native piroxicam. The solubility and intrinsic dissolution rate were enhanced by 76.45% and 33.33%, respectively. The MSCP was formulated as immediate release tablets and pharmacokinetic evaluation in rats reported 2.08-fold increase in bioavailability [103]. Reports on melt sonocrystallization have been tabulated in Table 9.

3.10 Supercritical fluid technology

Supercritical Fluid Technology (SCF) technology has emerged as an important solubilization technology whose application in particle size reduction has increased considerably. Supercritical fluids are recognized as the fluids whose temperature and pressure are greater than its critical temperature (Tc) and critical pressure (Tp), allowing it to assume the properties of both a liquid and a gas. At near-critical temperature, SCFs are highly compressible allowing substantial changes in pressure to alter the density and mass transport characteristics of the fluid that largely determine its solvent power. Once the drug particles are solubilized within the SCF, they may be recrystallized to significantly small particle sizes. The flexibility and precision offered by SCF processes allows micronization of drug particles within

Table 9 Melt sonocrystallization reports on drugs.

Drug	Crystallization conditions	Outcome	Ref
Rosiglitazone	Melt was poured in deionized water and simultaneously subjected to ultrasonic energy for 20 min at amplitude 80%.	After 30 min the drug release for rosiglitazone pure was 40.36 ± 0.40 and 97.48 ± 0.29 for MSC rosiglitazone.	[104]
Paracetamol Indomethacin Mefanamic acid	Molten mass was poured into double distilled water at 50 °C and sonicated for 4 min using an ultrasonic bath at an amplitude of 80%, with the power rating of 60 W.	Equilibrium solubility of MSC form of drugs (Fig. 2) was higher than the original forms. The intrinsic dissolution rate was approximately 1.5 times higher in comparison to original form of drugs.	[105]
Valdicoxeb	Molten mass was poured in a vessel containing 50 mL of deionized water maintained at 60 °C using thermostatic water bath and sonicated for 15 min using probe ultrasonicator (Ikasonic U 200 s control) at amplitude of 80% and cycle of 0.8 per second.	MSC valdecoxib agglomerates are irregular in shape with pores. Solubility of MSC Valdecoxib was 6.45 times higher than the original powder	[106]
Naproxen	Molten mass was poured in aqueous phase and system was sonicated for three different time level at three different levels of amplitude using probe ultra-sonicator at a speed of 600 rpm.	The product obtained after MSC showed higher rate of dissolution and better compressibility than crystalline drug. Percent yield and solubility was found to decrease with increase in the amplitude.	[107]

Continued

Table 9 Melt sonocrystallization reports on drugs—cont'd

Drug	Crystallization conditions	Outcome	Ref
Carbamazepine	Carbamazepine melt was poured in deionized water maintained at 60 °C and simultaneously subjected to ultrasonic energy.	Crystals with different crystal habits as needles, plates, and some hollow tubes with porous surface and cracks were generated. Higher specific surface area and thereby increase in saturated solubility.	[108]
Flurbiprofen (FBP)	FBP melt was poured in deionized water at 25 °C and sonicated for 4 min at an amplitude of 60% and cycle of 40 s on and 10 s off	The solubility was increased by 35%. The intrinsic dissolution rate was increased by 2-fold. Dissolution rate studies revealed that 90% of the drug was released within 20 min for MSC FBP compared with 60% release for FBP The relative bioavailability of MSCFBP was increased by 2-fold.	[109]
Simvastatin	The drug was melted in a vessel on a paraffin oil bath maintained at 190 °C. Molten mass was poured in deionized water maintained at 80 °C using thermostatic water bath and sonicated for 7 min using probe ultrasonicator at amplitude of 80% and cycle of 0.8 per second. The dried product was compressed into tablets.	Chemical integrity of drug was retained and showed a sound flow property. The controlled release formulation displayed a zero order release of drug (96.45%) over 12 h.	[110]

Fig. 2 Scanning electron micrographs of original forms (OF) of drugs and their melt sonocrystallized forms (MSC). *PCM,* paracetamol; *IMC,* indomethacin; *MA,* mefenamic acid.

narrow particle size distribution, often to submicron levels. The existing SCF processes have the ability to create nanosuspensions of particles 5–2000 nm in diameter.

In order to improve the dissolution properties of poorly water-soluble drugs, the drugs are subjected either to micronization or prepared as composite particles using SCF technology with carbon dioxide. Solubility in carbon dioxide is the key when using this method as it affects the supersaturation of the materials in the solvent as well as the mass transfer of that solvent, both of which are both critical to the micronization of the materials and the formation of the composite particles [111]. Since the first report of the SCF technology in 1879, various variants of the technology have been developed. Different processes of particle formation are categorized based on the behavior of SCF as *solute* (particle formation from gas-saturated solutions, PGSS); *solvent* (rapid expansion of supercritical solutions, RESS); *anti-solvent* (supercritical anti-solvent, SAS), gaseous anti-solvent (GAS), precipitation with compressed anti-solvent (PCA), aerosol solvent extraction system (ASES), solution enhanced dispersion by supercritical fluids (SEDS), suspension-enhanced dispersion by supercritical fluids (SpEDS); supercritical anti-solvent with enhanced mass transfer (SAS-EM) and *others* such as depressurization of an expanded liquid organic solution (DELOS) and the supercritical-assisted atomization (SAA) process, etc. All these have been exhaustively discussed in various reviews. The SCF acts as a re-precipitation aid for rapid and uniform nucleation of solute in all the above listed methods of particle formation. The performance efficiency of this technology is based on proper solvent selection and adjustment of critical process parameters during operations. RESS, SAS and, to a lesser extent, PGSS are the most studied processes. It is known that RESS can be used with CO_2 soluble molecules, while SAS can process non-soluble molecules. However, the selection is not so simple and a good knowledge of operating conditions and phase equilibrium thermodynamics is required.

Indeed, this technology has found application in almost all drug categories administered by oral, intravenous, ophthalmic, pulmonary, transdermal, and as polymeric implants. Revercheon et al. described the production of various nanoconstructs such as nanoparticles, nanowires, nanofibers, nanotubes using SCF based techniques [112]. Byrappa et al. explored the adaptive properties of SCF for the synthesis of advanced nanomaterials including carbon nanotubes, quantum dots, fullerenes, magnetic particles, nanocomposites (e.g., peptide/hydroxyapatite), phosphors and gold nanoshells for drug delivery and other biomedical applications namely imaging, cancer theranostics and sensing [113].

Standard processes such as crushing/milling and crystallization/precipitation are widely used for particle generation in pharmaceutical industry. However, SCF technology presents an interesting route for particle formation, which avoids most of the drawbacks of the traditional methods. Supercritical processes produce micro- or nanoparticles with narrow size distribution, and can also be used to achieve microencapsulation, surface coating of an active substance particle with a polymer or co-crystallization with excipients or host molecules like cyclodextrins.

The bioavailability of drugs depends on their absorption by the gastrointestinal tract, which is governed by their dissolution and membrane permeation rates. Micronization/nanonization based on SCF techniques leads to an increase in the specific surface area (typically several tens of m^2/g) that can significantly contribute to improvement in bioavailability [114]. This is particularly advantageous for development of dry powder inhaler formulations wherein the drug is directly delivered to the lungs and a mean particle size of 2–5 μm is desirable. Furthermore, SCF technology can be explored for development of sustained/controlled release systems via microencapsulation [115, 116] coating and formation of composite particles [117]. Numerous studies with respect to particle generation and delivery systems using the SCF technology are documented in literature. Few reports on the micronization of APIs alone (Table 10) and others on drug impregnated in various water-soluble polymeric matrices are listed in Table 11.

Occasionally, in SCF technology, certain problems are posed due to the drug's characteristics. Those can be overcome by combining SC-CO_2 with other technologies, such as the formation of coacervates or emulsions, and use of other equipment such as milling or ultrasound energy. The characteristic advantage of SCF technology is that it is considered to be green chemistry. SC-CO_2 can improve the solubility of poorly water-soluble drug substances using few or no organic solvents and with little or no heating.

3.11 Lipid based formulations

The utility of lipid based formulations (LBFs) to enhance the absorption of poorly water-soluble, lipophilic drugs has been recognized for many years. Co-administration with lipids has proven effective in significantly enhancing bioavailability of some compounds with low aqueous solubility [136]. Yet, lipid-based delivery technologies have not been widely employed in commercial oral products. Lipids can impact drug transport and fate in the gastrointestinal tract through multiple mechanisms including enhancement of solubility and dissolution kinetics, enhancement of permeation through the intestinal mucosa, and triggering drug precipitation upon lipid emulsion depletion (e.g., by digestion). The effect of lipids on drug absorption is currently not quantitatively predictable, in part due to the multiple complex dynamic processes that can be impacted by lipids [137].

Table 10 Micronization and nanonization of some drugs using SCF technology.

API	Highlight	Ref
Artemisinin	SCF technology resulted in a significant decrease in particle size as compared to untreated artemisinin. The dissolution rate of the micronized form improved in comparison to the untreated form, but showed no difference in comparison to mechanically ground artemisinin.	[118]
Cefuroxime axetil	Nanoparticles obtained were between 158 and 513 nm. More than 90% of the nano-particles dissolved in 3 min and complete dissolution occurred within 20 min, while the commercial drug achieved only about 50% during 60 min.	[119]
Diclofenac	Micronization resulted in the average particle size between 10.92 and 1.33 μm. The morphology of the processed particles changed to quasi-spherical while the virgin particles of diclofenac were irregular.	[120]
Digitoxin	The particle size of digitoxin was decreased from 0.2–8 μm to 68–458 nm by RESS technique.	[121]
Curcumin	A swirl mixer was employed to produce the micronized curcumin with polyvinyl pyrrolidone (PVP) by the supercritical anti-solvent process to improve the bioavailability of curcumin. The dissolution of curcumin/PVP particles is faster than that of raw curcumin.	[122]
Ibuprofen Naproxen	The RESOLV process produced particles less than 100 nm. The aqueous suspension of drug nanoparticles was protected against particle agglomeration and precipitation by oligomeric stabilizers.	[123]
Lidocaine, Griseofulvin, Benzoic acid	The solubility of the drug substance in supercritical CO_2 had a major effect on the average diameter of the particle prepared by RESS process, and the particle diameter decreased with the solubility for all the drugs and operating conditions.	[124]
Raloxifene	The particle size was reduced from 45 μm to 19 nm by RESS. The dissolution rate increased 7-fold after the process	[125]

Table 11 A cross section of research reports on composite particles using SCF technology.

API	Polymer and solvent	Result	Ref
5-Aminosalicylic acid	Eudragit®S100 Acetone, DMSO	5-ASA loaded colon-specific nanoparticles. 5-ASA was embedded into EU S100 in an amorphous state after SEDS processing and the SEDS process did not induce degradation of 5-ASA.	[126]
Silymarin	PVPK17 Ethanol, DCM	The $AUC_{0\text{-}\alpha}$ increased by 2.4-fold after oral administration of silymarin solid dispersion prepared by SEDS as compared to silymarin suspension.	[127]
5-Flurouracil	PEG, PLA–PEG Acetone, DCM, water	5-FU-loaded-poly (L-lactic)-PEG/PEG nanoparticles reverse emulsion-SEDS had potential anti-tumor applications as a controlled drug release dosage form without harmful drug toxicity.	[128]
Ibuprofen	Kollidon CL, Carbon dioxide	The dissolution of the SCF-prepared IBU dispersions improved significantly. The product produced remarkably high blood drug concentrations and a rapid absorption rate. Negative binding (molecular modeling) energy between IBU and polymer suggested a relatively stable system.	[129]
Camptothecin	Dextran, DMSO	Tumor targeted delivery	[130]
Azacytidine	PLA DCM, DMSO	Drug systems were prepared by precipitation from solutions, using supercritical CO_2 antisolvent technique operating in a semi-continuous mode. Sustained delivery of the drug over a period of various hours was obtained.	[131]

Continued

Table 11 A cross section of research reports on composite particles using SCF technology—cont'd

API	Polymer and solvent	Result	Ref
Carbamazepine	PEG Acetone	Solvent-free solid dispersions were developed by gas antisolvent crystallization technique using supercritical carbon dioxide as processing medium. Better crystal shape, together with remarkable enhancement of the *in vitro* drug dissolution rate.	[132]
Chlorhexidine diacetate	PEMA, THFMA	*In vitro* dissolution study revealed that the drug release rate was almost seven times faster from the SCF processed (self-curing polymeric system based on poly (ethyl methacrylate) and tetrahydrofurfuryl methacrylate) system than conventional cured samples.	[133]
Indomethacin	Eudragit RS 100, Magnetite, PLGA, PMMA, DCM	SAS technique was used to precipitate drug loaded magnetically responsive polymer particles. The SAS-FM technique was used to produce magnetite encapsulated polymeric nanoparticles of controllable size. Magnetically responsive drug release was observed.	[134]
Insulin	PLA PEG 6000	The insulin loaded nanoparticles PEG/PLA were spherical (400–600 nm), smooth surfaced particles with compact internal structure. Protein bioactivity demonstrated maintenance of >80% of insulin hypoglycemic activity.	[135]

Lipids provide the pharmaceutical formulator with options for developing systems for improving and controlling the absorption of poorly water-soluble drugs. The formulation options include lipidic solutions, emulsions, suspensions, microemulsions, mixed micelles, SEDDS, SMEDDS, liposomes, thermosoftening matrices and thixotropic vehicles. The excipients used for LBFs are such that the LBFs can be used for both lipophilic and hydrophilic drugs. For successful development of LBFs the foremost challenge is to choose appropriate formulation excipients that not only have adequate solvent capacity to solubilize the entire dose, but which also ensure that the formulation maintains its solvent properties in the intestine after dilution and digestion [138]. There is a considerable risk of precipitation of drug during the various intermediate stages of drug transfer, for example, from the solution state to the micellar state. The extent of this precipitation is dependent upon the formulation—it is, therefore, crucial to conduct *in vitro* studies to challenge the formulation and help predict the likelihood of precipitation and/or guide the appropriate excipient selection. Some of the important parameters to consider in LBFs include: screening for solubility in excipients, biorelevant media and lipid-digestion products, excipient compatibility, and finally, the risk of precipitation upon dispersion and digestion.

Another key consideration in oral formulation design is the safety and regulatory status of proposed lipid excipients. Not all lipid excipients are generally recognized as safe (GRAS), hence, special attention to their maximum daily intake should be considered as soon as possible in the development of a LBF. Table 12 lists the categories of excipients used for LBFs.

This is a critical factor for new molecular entities because high quantities of excipients could be required during dose escalation studies. Apart from these categories of excipients, liposoluble antioxidants such as tocopherols and butylated hydroxytoluene or hydroxyanisole are also needed as additives in LBFs. Because lipids are prone to lipid peroxidation, which generates free radicals that can adversely affect API stability, antioxidant is sometimes incorporated in the formulation.

LBFs are primarily applied for BCS Class II and Class IV drug substances, which are characterized by solubility limited absorption [139]. The lipid formulation classification system classifies LBFs into four main types, depending on LBF composition (Table 13). Each type is characterized by a set of advantages and drawbacks. For example, the glyceride-rich types 1 and 2, usually have poor drug solvent capacity but are unlikely to lose that solvency upon dispersion in the intestinal fluids, whereas the solvent- and surfactant-rich LBF types III and IV respectively, can dissolve

Table 12 Excipients used in LBFs.

Category	Properties and applications	Examples
Triglycerides (foundational excipients for LBFs)	Are vegetable oils composed of triglyceride esters of fatty acids. Have limited solvent power, but after digestion, the fatty acids released form mixed micelles with bile salts can dissolve a portion of the API and can act as carriers for the suspended API.	corn oil and sesame oil
Mixed glycerides and polar oils	Are partially hydrolyzed triglycerides that are generally better solvents than triglycerides. These excipients help to form self-emulsifying systems but can be sensitive to digestion. Other esters such as propylene glycol or sorbitan esters of fatty acid may be valuable additives in cases of chemical incompatibility.	glycerol monocaprylocaprate
Water-insoluble surfactants include non-ionic polyethoxylated or polyglycerylated esters of fatty acid	Are not hydrophilic enough to be soluble in water but form a good oil/water interface. Are dispersible in water and are used to formulate self-emulsifying systems.	linoleoyl polyoxylglycerides
Water-soluble surfactants	Most commonly used excipients for formulation of self-emulsifying drug delivery systems or self-micro-emulsifying drug delivery systems. Above their critical micelle concentration, these excipients spontaneously form micellar solutions that help solubilization of the API.	polysorbate 20 and 80
Cosolvents	Increase the solvent capacity of the formulation for drugs and aid the dispersion of systems containing a high proportion of water-soluble surfactants. As they lose their solvent power during dilution in gastrointestinal fluids, their use is limited.	Ethanol, propylene glycol, and polyethylene glycol

Table 13 The lipid formulation classification according to Pouton [140].

Excipients in formulation	Content of formulation (%, w/w) Type I	Type II	Type IIIA	Type IIIB	Type IV
Oils: triglycerides or mixed mono- and diglycerides	100	40–80	40–80	<20	–
Water-insoluble surfactants (HLB < 12)	–	20–60	–	–	0–20
Water-soluble surfactants (HLB > 12)	–	–	20–40	20–50	30–80
Hydrophilic cosolvents (e.g. PEG, propylene glycol, transcutol)	–	–	–	20–50	0–50

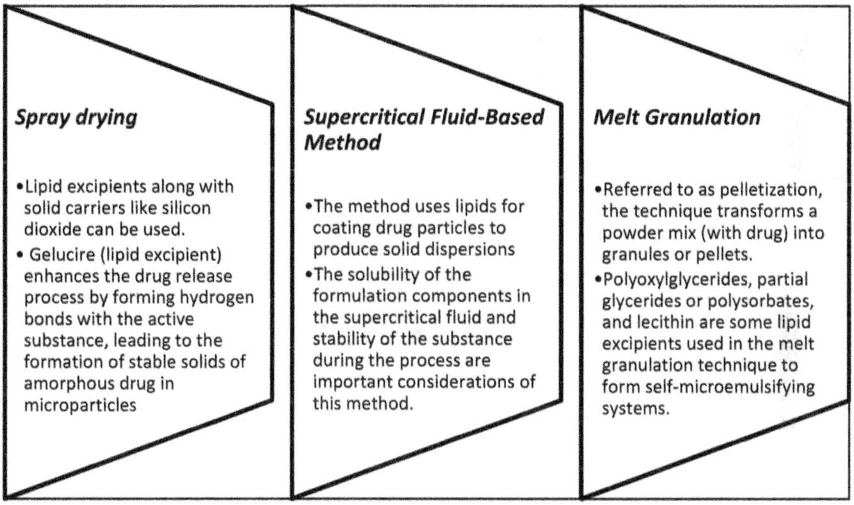

Fig. 3 Highlights of methods used for lipid based formulation preparation.

higher drug concentrations but suffer from significant phase changes and potential drug precipitation upon dispersion.

One aspect of LBF is that a significant fraction of the surfactants used may be digestible, which should be considered during formulation development [140]. LBFs can be used to deliver various types of drugs from ranging from new chemical entities, proteins and peptides and nucleic acids (DNA, siRNA) [141–143].

The formulation of LBFs primarily utilizes three methods namely spray drying [144], Supercritical fluid-based method [145] and melt granulation [146]. The highlights of each process are depicted in Fig. 3.

The physicochemical stability of LBFs is just as crucial as with any other formulation. While chemical stability is handled the same as it would be with other dosage forms, guided by ICH guidelines, the testing for physical stability is different. It is necessary to conduct stress studies to challenge the robustness of the formulation. Cycling studies are commonly used with cycles of high and low temperature to stress the formulation. Additional tests to challenge the LBF's robustness to ensure the API does not precipitate out in *in-vivo* or *in-vitro* conditions can also be performed.

4 Conclusion

The solubility improvement of drugs is a crucial aspect for their development and performance. A plethora of solubility enhancement approaches are available and judicious selection of the strategy is significantly important for formulation development. Selection of method depends upon various drug characteristics like physicochemical nature, solubility, absorption site, pharmacokinetic behavior and the dosage form requirement. The regulatory requirements like safety profile and maximum permissible limit of excipients are also critical aspects. Access to the best solution for solubility enhancement would ensure optimal bioavailability of the drug from its formulation. Furthermore, the technology used must be scalable to ensure commercial viability.

References

[1] S.R.K. Yellela, Pharmaceutical technologies for enhancing oral bioavailability of poorly soluble drugs, J. Bioequivalence Bioavailability 2 (2010) 28–36.
[2] D. Jain, D. Pathak, K. Pathak, Pharmaceutical technologies based on biopharmaceutical classification systems for product development, Pharmazie 64 (8) (2009) 483–490.
[3] G.L. Amidon, H. Lennernäs, V.P. Shah, J.R. Crison, A theoretical basis for a biopharmaceutic drug classification: the correlation of in vitro drug product dissolution and in vivo bioavailability, Pharm. Res. 12 (3) (1995) 413–420.
[4] J.M. Butler, J.B. Dressman, The developability classification system: application of biopharmaceutics concepts to formulation development, J. Pharm. Sci. 99 (2010) 4940–4954.
[5] US Department of Health and Human Services, Food and Drug Administration, Center for Drug Evaluation and Research, Guidance for Industry, Applications Covered by Section 505(b)(2), Rockville, MD 20857, 1999.
[6] J. Rosenberger, J.M. Butler, J.B. Dressman, A refined developability classification system, J. Pharm. Sci. 107 (2018) 2020–2032.
[7] J. Rosenberger, J.M. Butler, U. Muenster, J.B. Dressman, Application of a refined developability classification system, J. Pharm. Sci. 108 (2019) 1090–1110.

[8] J.H. Michael, R.T. Dhiren, L.S. James, T. Ronald, E.H.K. Borchardt, Optimizing the "drug-like" properties of leads in drug discovery, in: Biotechnology: Pharmaceutical Aspects, vol. 4, Springer-Verlag, New York, 2007, pp. 99–130.
[9] A.T. Serajuddin, Salt formation to improve drug solubility, Adv. Drug Deliv. Rev. 59 (7) (2007) 603–616.
[10] D.P. Elder, R. Holm, H.L. Diego, Use of pharmaceutical salts and cocrystals to address the issue of poor solubility, Int. J. Pharm. 453 (2013) 88–100.
[11] G.S. Paulekuhn, J.B. Dressman, C. Saal, Trends in active pharmaceutical ingredient salt selection based on analysis of the Orange Book database, J. Med. Chem. 50 (26) (2007) 6665–6672.
[12] R.R. Carmichael, C.D. Mahoney, L.P. Jeffrey, Solubility and stability of phenytoin sodium when mixed with intravenous solutions, Am. J. Hosp. Pharm. 37 (1980) 95–98.
[13] P. Ettmayer, G.L. Amidon, B. Clement, B. Testa, Lessons learned from marketed and investigational prodrugs, J. Med. Chem. 47 (2004) 2393–2404.
[14] H.K. Han, G.L. Amidon, Targeted prodrug design to optimize drug delivery, AAPS PharmSciTech 2 (2000) E6.
[15] A.T.D.A. Silva, M.C. Chung, L.F. Castro, R.V.C. Güido, E.I. Ferreira, Advances in prodrug design, Mini Rev. Med. Chem. 5 (2005) 893–914.
[16] R.G. Strickley, R. Oliyai, Formulation challenges of prodrugs, in: V.J. Stella, R.T. Borchardt, M.J. Hageman, R. Oliyai, H. Maag, J.W. Tilley (Eds.), Prodrugs. Biotechnology: Pharmaceutical Aspects, vol. V, Springer, New York, 2007.
[17] W. Cui, S. Zhang, H. Zhao, C. Luo, B. Sun, Z. Li, M. Sun, Q. Ye, J. Sin, Z. He, Formulating a single thioether-bridged oleate prodrug into a self-nanoemulsifying drug delivery system to facilitate oral absorption of docetaxel, Biomater. Sci. 7 (2019) 1117–1131.
[18] J.B. Lee, A. Zgair, J. Malc, T.H. Kim, M.G. Kim, J. Ali, C. Qin, W. Feng, M. Chiang, X. Gao, G. Voronin, A.E. Garces, C.L. Lau, T.-H. Chan, A. Hume, T.M. McIntosh, F. Soukarieh, M. Al-Hayali, E. Cipolla, H.M. Collins, D.M. Heery, B.S. Shin, S.D. Yoo, L. Kagan, M.J. Stocks, T.D. Bradshaw, P.M. Fischer, P. Gershkovich, Lipophilic activated ester prodrug approach for drug delivery to the intestinal lymphatic system, J. Control. Release 286 (2018) 10–19.
[19] J.J. Hobson, A. Al-khouja, P. Curley, D. Meyers, C. Flexner, M. Siccardi, A. Owen, C.F. Meyers, S.P. Rannard, Semi-solid prodrug nanoparticles for long-acting delivery of water-soluble antiretroviral drugs within combination HIV therapies, Nat. Commun. 10 (2019), 1413.
[20] R. Karaman, A. Qtait, K.K. Dajani, S.A. Lafi, Design, synthesis, and in vitro kinetics study of atenolol prodrugs for the use in aqueous formulations, Sci. World J. 2014 (2014), 248651. 13 pages https://doi.org/10.1155/2014/248651.
[21] M. Lalanne, K. Andrieux, A. Paci, M. Benard, M. Re, C. Bourgaux, M. Ollivon, D. Desmaele, P. Couvreur, Liposomal formulation of a glycerolipidic prodrug for lymphatic delivery of didanosine via oral route, Int J. Pharm. 344 (2007) 62–70.
[22] S.H. Yalkowsky, T.J. Roseman, Solubilization of drugs by cosolvents, in: S.H. Yalkowsky (Ed.), Techniques of Solubilization of Drugs, Marcel Dekker, New York, 1981, pp. 91–102.
[23] J.W. Millard, F.A. Alvarez-Núñez, S.H. Yalkowsky, Solubilization by cosolvents: establishing useful constants for the log-linear model, Int. J. Pharm. 245 (2002) 153–166.
[24] P.R. Babu, C.V. Subrahmanyam, J. Thimmasetty, R. Manavalan, K. Valliappan, S.A. Kedarnath, Solubility enhancement of cox-II inhibitors by cosolvency approach, Dhaka Univ. J. Pharm. Sci. 7 (2008) 119–126.
[25] S. Nema, R.J. Washkuhn, R.J. Brendel, Excipients and their use in injectable products, PDA J. Pharm. Sci. Technol. 51 (4) (1997) 166–171.

[26] J.T. Rubino, S.H. Yalkowsky, Co-solvency and deviations from log-linear solubilization, Pharm. Res. 4 (1987) 231–236.
[27] N. Seedhar, M. Kanojia, Co-solvent solubilization of some poorly-soluble antidiabetic drugs, Pharm. Dev. Technol. 14 (2009) 185–192.
[28] A.K. Nayak, P.P. Panigrahi, Solubility enhancement of etoricoxib by cosolvency approach, ISRN Phys. Chem. (2012), 820653. 5 pages.
[29] D. Stephens, L. Chui, L.E. Pec, D. Robinson, Statistical Experimental approach to cosolvent formulation of a water-insoluble drug, Drug Dev. Ind. Pharm. 25 (1999) 961–965.
[30] M.-K. Yeh, L.-C. Chang, A.H.-J. Chiou, Improving tenoxicam solubility and bioavailability by cosolvent system, AAPS PharmSciTech 10 (2009) 166–171.
[31] Y. Mehmood, Formulation development and evaluation of diclofenac sodium injection using benzyl alcohol (co-solvent), mixed solvency concept, Edorium J. Drug Res. 1 (2015) 1–8.
[32] A. Modi, P. Tayade, A comparative solubility enhancement profile of valdecoxib with different solubilization approaches, Indian J. Pharm. Sci. 69 (2007) 274–278.
[33] A.K. Singhai, S. Jain, N.K. Jain, Cosolvent solubilization and formulation of an aqueous injection of ketoprofen, Pharmazie 51 (1996) 737–740.
[34] S.A. Varia, M.M. Faustino, A.B. Thakur, C.S. Clow, A.T.M. Serajuddin, Optimization of cosolvent concentration and excipient composition in a topical corticosteroid solution, J. Pharm. Sci. 80 (1991) 872–875.
[35] V.R. Vemula, V. Lagishetty, S. Lingala, Solubility enhancement techniques, Int. J. Pharm. Sci. Rev. Res. 5 (2010) 41–51.
[36] C.D. Rangel-Yagui, A. Pessoa, L.C. Tavares, Micellar solubilization of drugs, J. Pharm. Pharm. Sci. 8 (2) (2005) 147–163.
[37] C.H. Hsu, Z. Cui, R.J. Mumper, M. Jay, Micellar solubilization of some poorly soluble antidiabetic drugs, AAPS PharmSciTech 9 (2) (2008) 939–943.
[38] P.D. Maheswari, D. Rambhau, M.L. Narasu, Micellar solubilization in the formulation development of poorly soluble naproxen, Pharm. Reg. Affairs 2 (2013) 108–112.
[39] K. Yohei, W. Koichi, N. Manabu, Y. Shizuo, O. Satomi, Formulation design for poorly water-soluble drugs based on biopharmaceutics classification system: basic approaches and practical applications, Int. J. Pharm. 420 (2011) 1–10.
[40] A.S. Sunday, C. Simon, Particle engineering techniques for inhaled biopharmaceuticals, Adv. Drug Del. Rev. 58 (2006) 1009–1029.
[41] N. Rasenack, B.W. Muller, Dissolution rate enhancement by in-situ micronization of poorly water-soluble drugs, Pharm. Res. 19 (2002) 1894–1900.
[42] H. Steckel, N. Rasenack, B.W. Muller, In-situ micronization of disodium cromoglycate for pulmonary delivery, Eur. J. Pharm. Biopharm. 35 (2003) 173–180.
[43] S. Baghel, C. Helen, O.R. Niall, Polymeric amorphous solid dispersions: a review of amorphization, crystallization, stabilization, solid-state characterization, and aqueous solubilization of biopharmaceutical classification system class II drugs, J. Pharm. Sci. 105 (2016) 2527–2544.
[44] N. Shah, R. Iyer, H.-J. Mair, D.S. Choi, H. Tian, Improved human bioavailability of vemurafenib, a practically insoluble drug, using an amorphous polymer-stabilized solid dispersion prepared by a solvent- controlled coprecipitation process, J. Pharm. Sci. 102 (2013) 967–981.
[45] D. Íris, M.L. Corvo, P. Serôdio, J. Vicente, J.F. Pinto, Production of nano-solid dispersions using a novel solvent-controlled precipitation process—benchmarking their in vivo performance with an amorphous micro-sized solid dispersion produced by spray drying, J. Pharm. Sci. 105 (2016) 203–214.
[46] A Method of Preparing Amorphous Solid Dispersion in Submicron Range by Co-Precipitation, United States Patent (2017) 20170209372.

[47] N. Shah, H. Sandhu, D.S. Choi, O. Kalb, S. Page, N. Wyttenbach, Structured development approach for amorphous systems, in: R.O. Williams III, A.B. Watts, D.A. Miller (Eds.), Formulating Poorly Water Soluble Drugs, Springer, New York, 2012, pp. 267–310.
[48] N. Shah, H. Sandhu, W. Phuapradit, R. Pinal, R. Iyer, A. Albano, A. Chatterji, S. Anand, D.S. Choi, K. Tang, H. Tian, H. Chokshi, D. Singhal, W. Malick, Development of novel microprecipitated bulk powder (MBP) technology for and comparison with crystallization tendency from undercooled melts, J. Pharm. Sci. 99 (2012) 3826–3838.
[49] Q. Hu, D.S. Choi, H. Chokshi, N. Shah, H. Sandhu, Highly efficient miniaturized coprecipitation screening (MiCoS) for amorphous solid dispersion formulation development, Int. J. Pharm. 450 (2013) 53–62.
[50] D. Íris, J.L. Santos, J. Pinto, M. Temtem, Screening methodologies for the development of spray-dried amorphous solid dispersions, Pharm. Res. 32 (2015) 222–237.
[51] Y. Weerapol, S. Tubtimsri, C. Jansakul, P. Sriamornsak, Improved dissolution of *Kaempferia parviflora* extract for oral administration by preparing solid dispersion via solvent evaporation, Asian J. Pharm. Sci. 12 (2017) 124–133.
[52] H.M.C. Marques, A review on cyclodextrin encapsulation of essential oils and volatiles, Flavour Fragr. J. 25 (2010) 313–326.
[53] A.R. Hedges, W.J. Shieh, C.T. Sikorski, Use of cyclodextrins for encapsulation in the use and treatment of food products, in: S.J. Risch, G.A. Reineccius (Eds.), Encapsulation and Controlled Release of Food Ingredients, American Chemical Society, Washington, DC, 1995, pp. 60–73.
[54] K. Uekama, F. Hirayama, H. Arima, in: H. Dodziuk (Ed.), Pharmaceutical applications of cyclodextrins and their derivatives. Cyclodextrins and their complexes: chemistry, analytical methods, applications, Wiley-VCH, Weinheim, 2006, pp. 381–389.
[55] C. Garnero, A. Zoppi, D. Genovese, M. Longhi, Studies on trimethoprim: hydroxypropyl-Î²-cyclodextrin: aggregate and complex formation, Carbohydr. Res. 345 (2010) 2550–2556.
[56] H. Kublik, T.K. Bock, H. Schreier, B.W. Müller, Nasal absorption of 17b-estradiol from different cyclodextrin inclusion formulations in sheep, Eur. J. Pharm. Biopharm. 42 (1996) 320–324.
[57] T. Loftsson, E. Stefánsson, Effect of cyclodextrins on topical drug delivery to the eye, Drug Dev. Ind. Pharm. 23 (1997) 473–481.
[58] Q. Zhao, J. Temsamani, S. Agrawal, Use of cyclodextrin and its derivatives as carriers for oligonucleotide delivery, Antisense Res. Dev. 5 (1995) 185–192.
[59] T. Irie, K. Uekama, Cyclodextrins in peptide and protein delivery, Adv. Drug Deliv. Rev. 36 (1999) 101–123.
[60] M. Ravera, E. Gabano, S. Bianco, G. Ermondi, G. Caron, et al., Host–guest inclusion systems of Pt(IV)-bis(benzoato) anticancer drug candidates and cyclodextrins, Inorg. Chim. Acta 432 (2015) 115–127.
[61] V. Aiassa, A. Zoppi, M.C. Becerra, I. Albesa, M.R. Longhi, Enhanced inhibition of bacterial biofilm formation and reduced leukocyte toxicity by chloramphenicol: ß-cyclodextrin: N-acetylcysteine complex, Carbohydr. Polym. 152 (2016) 672–678.
[62] R. Gharib, H. Greige-Gerges, A. Jraij, L. Auezova, C. Charcosset, Preparation of drug-in-cyclodextrin-in-liposomes at a large scale using a membrane contactor: application to trans-anethole, Carbohydr. Polym. 154 (2016) 276–286.
[63] R. Gharib, L. Auezova, C. Charcosset, H. Greige-Gerges, Drug-in-cyclodextrin-in-liposomes as a carrier system for volatile essential oil components: application to anethole, Food Chem. 218 (2017) 365–371.
[64] T. Loftsson, P. Jarho, M. Másson, T. Järvinen, Cyclodextrins in drug delivery, Expert Opin. Drug Deliv. 2 (2005) 335–351.

[65] F.J. Otero-Espinar, J.J. Torres-Labandeira, C. Alvarez-Lorenzo, J. Blanco-Méndez, Cyclodextrins in drug delivery systems, J. Drug Delivery Sci. Technol. 20 (2010) 289–301.
[66] A. Becheri, P.L. Nostro, B.W. Ninham, P. Baglion, The curious world of polypseudorotaxanes: cyclodextrins as probes of water structure, J. Phys. Chem. B 107 (2003) 3979–3987.
[67] M. Bonini, S. Rossi, G. Karlsson, M. Almgren, P.L. Nostro, P. Baglioni, Self-assembly of β-Cyclodextrin in water. Part 1: Cryo-TEM and dynamic and static light scattering, Langmuir 224 (2006) 1478–1484.
[68] A. Concheiro, C. Alvarez-Lorenzo, Crosslinked ionic polysaccharides for stimuli-sensitive drug delivery, Adv. Drug Deliv. Rev. 65 (2003) 1188–1203.
[69] F. Hirayama, K. Uekama, Cyclodextrin-based controlled drug release system, Adv. Drug Deliv. Rev. 36 (1999) 125–141.
[70] A.P. Sherje, B.R. Dravyakar, D. Kadam, M. Jadhav, Cyclodextrin-based nanosponges: a critical review, Carbohydr. Polym. 173 (2017) 37–49.
[71] S. Spireas, S.M. Bolton, Liquisolid systems and methods of preparing same, US5968550, 1999.
[72] S. Spireas, Liquisolid system and method of preparing same. US Patent 6423339B1, 2002.
[73] S.S. Spireas, C.I. Jarowski, B.D. Rohera, Powdered solution technology: principles and mechanism, Pharm. Res. 9 (1992) 1351–1358.
[74] Y. Javadzadeh, B. Jafari-Navimipour, A. Nokhodchi, et al., Liquisolid technique for dissolution rate enhancement of a high dose water-insoluble drug (carbamazepine), Int. J. Pharm. 341 (2007) 26–34.
[75] C.M. Hentzschel, A. Sakmann, C.S. Leopold, Suitability of various excipients as carrier and coating materials for liquisolid compacts, Drug Dev. Ind. Pharm. 37 (2011) 1200–1207.
[76] V. Sharma, K. Pathak, Liquisolid system of paclitaxel using modified polysaccharides: in vitro cytotoxicity, apoptosis study, cell cycle analysis, in vitro mitochondrial membrane potential assessment, and pharmacokinetics, Int. J. Biol. Macromol. 137 (2019) 20–31.
[77] S.T. Prajapati, H.H. Bulchandani, D.M. Patel, S.K. Dumaniya, C.N. Patel, Formulation and evaluation of liquisolid compacts for olmesartan medoxomil, J. Drug Deliv. 2013 (2013), 870579. 9 pages https://doi.org/10.1155/2013/870579.
[78] N. Chella, N. Narra, T.R. Rao, Preparation and characterization of liquisolid compacts for improved dissolution of telmisartan, J. Drug Deliv. 2014 (2014), 692793. 10 pages https://doi.org/10.1155/2014/692793.
[79] R. Korni, S. Voodikala, C.S.R. Gonugunta, V. Jayanti, Liquisolid technique: an approach to enhance the dissolution rate of olanzapine, Indian J. Pharm. Sci. 80 (2018) 1003–1010.
[80] V. Sharma, K. Pathak, Effect of hydrogen bond formation/replacement on solubility characteristics, gastric permeation and pharmacokinetics of curcumin by application of powder solution technology, Acta Pharm. Sin. B 6 (2016) 600–613.
[81] B.D. Espíndola, A.O.R. Beringhs, D. Sonaglio, H.K. Stulzer, M.A.S. Silva, H.G. Ferraz, B. Ramos, Liquisolid pellets: a pharmaceutical technology strategy to improve the dissolution rate of ritonavir, Saudi Pharm. J. 27 (2019) 702–712.
[82] B.R. Pezzini, A.O. Beringh, H.G. Ferraz, M.A. Silva, H.K. Stulzer, D. Sonaglio, Liquisolid technology applied to pellets: evaluation of the feasibility and dissolution performance using felodipine as a model drug, Chem. Eng. Res. Des. (2016) 1–8.
[83] A.A. Elkordy, X.N. Tan, E.A. Essa, Spironolactone release from liquisolid formulations prepared with Capryol™ 90, Solutol HS-15 and Kollicoat SR 30 D as nonvolatile liquid vehicles, Eur. J. Pharm. Biopharm. 83 (2013) 203–223.

[84] Y. Javadzadeh, L. Musaalrezaei, A. Nokhodchi, Liquisolid technique as a new approach to sustain propranolol hydrochloride release from tablet matrices, Int. J. Pharm. 362 (2008) 102–108.
[85] A. Nokhodchi, C.M. Hentzschel, C.S. Leopold, Drug release from liquisolid systems: speed it up, slow it down, Expert Opin. Drug Deliv. 8 (2011) 191–205.
[86] I.D. Gonjari, A.B. Karmarkar, A.H. Hosmani, Evaluation of in vitro dissolution profile comparison methods of sustained release tramadol hydrochloride liquisolid compact formulations with marketed sustained release tablets, Dig. J. Nanomater. Biostruct. 4 (2009) 651–661.
[87] A.V. Yadav, A.S. Shete, A.P. Dabke, Formulation and evaluation of orodispersible liquisolid compacts of aceclofenac, Indian J. Pharm. Edu. Res. 44 (2010) 227–235.
[88] F. Kesisoglou, S. Panmai, Y. Wu, Nanosizing-oral formulation development and biopharmaceutical evaluation, Adv. Drug Deliv. Rev. 59 (2007) 631–644.
[89] E. Merisko-Liversidge, G.G. Liversidge, E.R. Cooper, Nanosizing: a formulation approach for poorly-water-soluble compounds, Eur. J. Pharm. Sci. 18 (2003) 113–120.
[90] E. Merisko-Liversidge, G.G. Liversidge, Nanosizing for oral and parenteral drug delivery: a perspective on formulating poorly-water soluble compounds using wet media milling technology, Adv. Drug Deliv. Rev. 63 (2011) 427–440.
[91] S. Dahiya, Drug nanonization: an overview of industrially feasible top-down technologies for nanocrystal production, Bull. Pharm. Res. Inst. 7 (2017) 144–151.
[92] C. Li, C. Li, J.-F. Chen, Formation of bicalutamide nanodispersion for dissolution rate enhancement, Int. J. Pharm. 404 (2011) 257–263.
[93] A.M. Cerdeira, M. Mazzotti, B. Gander, Miconazole nanosuspensions: influence of formulation variables on particle size reduction and physical stability, Int. J. Pharm. 396 (2010) 210–218.
[94] T.T.-D. Tran, P.H.-L. Tran, M.N.U. Nguyen, K.T.M. Tran, M.N. Pham, P.C. Tran, T.V. Vo, Amorphous isradipine nanosuspension by the sonoprecipitation method, Int. J. Pharm. 474 (2014) 146–150.
[95] R. Mauludin, R.H. Müller, C.M. Keck, Development of an oral rutin nanocrystal formulation, Int. J. Pharm. 370 (2009) 202–209.
[96] R. Zhao, C.P. Hollis, H. Zhang, L. Sun, R.A. Gemeinhart, T. Li, Hybrid nanocrystals: achieving concurrent therapeutic and bioimaging functionalities toward solid tumors, Mol. Pharm. 8 (2011) 1985–1991.
[97] L. Al-Shaal, R. Shegokar, R.H. Müller, Production and characterization of antioxidant apigenin nanocrystals as a novel UV skin protective formulation, Int. J. Pharm. 420 (2011) 133–140.
[98] A.H. El-Kamel, Improvement of physiochemical and biopharmaceutical properties of flurbiprofen using melt sonocrystallization technique, Drug Dev. Res. 69 (2008) 34–41.
[99] M. Maheshwari, H. Jahagirdar, A. Paradkar, Melt sonocrystallisation of ibuprofen: effect on crystal properties, Eur J. Pharm. Sci. 25 (2005) 41–48.
[100] A. Paradkar, M. Maheshwari, R. Kamble, I. Grimsey, P. York, Design and evaluation of celecoxib porous particles using melt sonocrystallization, Pharm. Res. 23 (2006) 1395–1400.
[101] Crystallization process using ultrasound, US patent, 20020031577.
[102] A. Fini, M.J. Fernàndez-Hervàs, M.A. Holgado, L. Rodriguez, C. Cavallari, N. Passerini, O. Caputo, Fractal analysis of beta-cyclodextrin-indomethacin particles compacted by ultrasound, J. Pharm. Sci. 86 (11) (1997) 1303–1309.
[103] P.S. Gupta, V. Sharma, K. Pathak, Melt sonocrystallized piroxicam for oral delivery: particle characterization, solid state analysis, and pharmacokinetics, Expert Opin. Drug Deliv. 10 (1) (2013) 17–32.
[104] V.A. Jagtap, G. Vidyasagar, S.C. Dvivedi, Solubility enhancement of rosiglitazone by using melt sonocrystallization technique, J. Ultrasound 17 (1) (2014) 27–32.

[105] B. Kumar, V. Sharma, K. Pathak, Effect of melt sonocrystallization on pharmacotechnical properties of paracetamol, indomethacin and mefenamic acid characterized by dynamic laser scattering and its impact on solubility, Drug Dev. Ind. Pharm. 39 (5) (2013) 687–695.
[106] P.D. Chaudhari, P.S. Uttekar, Melt-sonocrystallization: a novel particle engineering technique for solubility enhancement, Int. J. PharmTech. Res. 1 (2009) 111–120.
[107] P. Kamble, L. Nangare, Garge, Effect of variables on naproxen agglomerates engineered by melt sonocrystallization, J. Adv. Pharm. Edu. Res. 5 (2015) 21–32.
[108] T. Deshmukh, M. Deshmukh, P. Deshmukh, Jadhav, Design and development of melt sonocrystallization technique for carbamazepine, Ind. J. Pharm. Edu. Res. 47 (2013) 199–205.
[109] A.H. El-Kamel, Improvement of physicochemical and biopharmaceutical properties of flurbiprofen using melt sonocrystallization technique, Drug Dev. Res. 69 (2008) 34–41.
[110] S. Tripathy, B.K. Singh, D.K. Patel, N.N. Palei, Hydrophilic matrix based oral controlled release tablets of melt sonocrystallized simvastatin, J. Appl. Pharm. Sci. 6 (2016) 41–47.
[111] T. Yasuji, H. Takeuchi, Y. Kawashima, Particle design of poorly water-soluble drug substances using supercritical fluid technologies, Adv. Drug Deliv. Rev. 60 (3) (2008) 388–398.
[112] E. Reverchon, R. Adami, Nanomaterials and supercritical fluids, J. Supercrit. Fluids 37 (2006) 1–22.
[113] K. Byrappa, S. Ohara, T. Adschiri, Nanoparticles synthesis using supercritical fluid technology-towards biomedical applications, Adv. Drug Deliv. Rev. 60 (2008) 299–327.
[114] A. Baldelli, M.A. Boraey, D.S. Nobes, R. Vehring, Analysis of the particle formation process of structured microparticles, Mol. Pharm. 12 (2015) 2562–2573.
[115] D.L. Tomasko, H. Li, D. Liu, X. Han, M.J. Wingert, L.J. Lee, K.W. Koelling, A review of CO_2 applications in the processing of polymers, Ind. Eng. Chem. Res. 42 (2003) 6431–6456.
[116] K. Mishima, Biodegradable particle formation for drug and gene delivery using supercritical fluid and dense gas, Adv. Drug Deliv. Rev. 60 (2008) 411–432.
[117] Y.-C. Tien, C.-S. Su, L.-H. Lien, Y.-P. Chen, Recrystallization of erlotinib hydrochloride and fulvestrant using supercritical antisolvent process, J. Supercrit. Fluids 55 (2010) 292–299.
[118] T.V. Nijlen, K. Brennan, G.V. Mooter, N. Blaton, R. Kinget, P. Augustijns, Improvement of the dissolution rate of artemisinin by means of supercritical fluid technology and solid dispersions, Int. J. Pharm. 254 (2003) 173–181.
[119] J. Varshosaz, F. Hassanzadeh, M. Mahmoudzadeh, A. Sadeghi, Preparation of cefuroxime axetil nanoparticles by rapid expansion of supercritical fluid technology, Powder Technol. 189 (2009) 97–102.
[120] A.Z. Hezave, F. Esmaeilzadeh, The effects of RESS parameters on the diclofenac particle size, Adv. Powder Technol. 22 (2011) 587–595.
[121] C. Atila, N. Yildiz, A. Caliml, Particle size design of digitoxin in supercritical fluids, J. Supercrit. Fluids 51 (2010) 404–411.
[122] K. Chhouk, W.H. Kanda, S.-I. Kawasaki, M. Goto, Micronization of curcumin with biodegradable polymer by supercritical anti-solvent using micro swirl mixer, Front. Chem. Sci. Eng. 12 (2018) 184–193.
[123] P. Pathak, M.J. Meziani, T. Desai, Y.-P. Sun, Nanosizing drug particles in supercritical fluid processing, Am. Chem. Soc. 126 (2004) 10842–10843.
[124] J.-T. Kim, H.-L. Kim, C.-S. Ju, Micronization and characterization of drug substances by RESS with supercritical CO_2, Korean J. Chem. Eng. 27 (2010) 1139–1144.

[125] A. Keshavarz, J. Karimi-Sabet, A. Fattahi, A.A. Golzary, M. Rafiee-Tehrani, F.A. Dorkoosh, Preparation and characterization of raloxifene nanoparticles using rapid expansion of supercritical solution (RESS), J. Supercrit. Fluids 63 (2012) 169–179.
[126] D. Hu, L. Liu, W. Chen, S. Li, Y. ZhaoY, A novel preparation method for 5-aminosalicylic acid loaded Eudragit S100 nanoparticles, Int. J. Mol. Sci. 13 (2012) 6454–6468.
[127] J. Yang, Y. Zhao, N. Feng, Y. Zhang, Y. Lui, B. Dang, Improved dissolution and bioavailability of silymarin delivered by a solid dispersion prepared using supercritical fluids, Asian J. Pharm. Sci. 10 (3) (2015) 194–202.
[128] C. Zhang, G. Li, Y. Wang, F. Cui, J. Zhang, Q. Huang, Preparation and characterization of 5-fluorouracil-loaded PLA-PEG/PEG nanoparticles by a novel supercritical CO_2 technique, Int. J. Pharm. 436 (2012) 272–281.
[129] F. Han, W. Zhang, Y. Wang, Z. Xi, L. Chen, S. Li, L. Xu, Applying supercritical fluid technology to prepare ibuprofen solid dispersions with improved oral bioavailability, Pharmaceutics 11 (2019). pii 67.
[130] X. Zhao, D. Wang, Y. Zu, R. Jiang, D. Zhao, Y. Li, et al., Preparation and characterization of camptothecin-loaded folate-conjugated dextran nano-particles for tumor-targeted drug delivery using supercritical antisolvent method, J. Control. Release 152 (Suppl. 1) (2011) e90–e92.
[131] A. Argemi, A. Vega, P. Subra-Paternault, J. Saurina, Characterization of azacytidine/poly(L-lactic) acid particles prepared by supercritical antisolvent precipitation, J. Pharm. Biomed. Anal. 50 (2009) 847–852.
[132] M. Moneghini, I. Kikic, D. Voinovich, B. Perissutti, J. Filipović-Grčić, Processing of carbamazepine-PEG 4000 solid dispersions with supercritical carbon dioxide: preparation, characterisation, and in vitro dissolution, Int. J. Pharm. 222 (2001) 129–132.
[133] K. Gong, M. Braden, M.P. Patel, I.U. Rehman, Z. Zhang, J.A. Darr, Controlled release of chlorhexidine diacetate from a porous methacrylate system: supercritical fluid assisted foaming and impregnation, J. Pharm. Sci. 96 (2007) 2048–2056.
[134] P. Chattopadhyay, R.B. Gupta, Supercritical CO_2 based production of magnetically responsive micro- and nanoparticles for drug targeting, Ind. Eng. Chem. Res. 41 (2002) 6049–6058.
[135] N. Elvassore, A. Bertucco, P. Caliceti, Production of insulin-loaded poly (ethylene glycol)/poly (l-lactide) (PEG/PLA) nanoparticles by gas antisolvent techniques, J. Pharm. Sci. 90 (2001) 1628–1636.
[136] C.J. Porter, N.L. Trevaskis, W.N. Charman, Lipids and lipid-based formulations: optimizing the oral delivery of lipophilic drugs, Nat. Rev. Drug Discov. 6 (3) (2007) 231–248.
[137] O. Rezhdo, L. Speciner, R.L. Carrier, Lipid-associated oral delivery: mechanisms and analysis of oral absorption enhancement, J. Control. Release 240 (2016) 544–560.
[138] A. Siew, Lipid-based formulations offer a means of addressing the physicochemical and biological challenges of poorly soluble APIs, Pharm. Technol. 42 (2018) 26–29.
[139] B.J. Boyd, C.A.S. Bergstrom, Z. Vinarov, M. Kuentz, J. Brouwers, P. Augustijns, et al., Successful oral delivery of poorly water-soluble drugs both depends on the intraluminal behavior of drugs and of appropriate advanced drug delivery systems, Eur. J. Pharm. Sci. 137 (2019) 104–167.
[140] C.W. Pouton, Formulation of poorly water-soluble drugs for oral administration: physicochemical and physiological issues and the lipid formulation classification system, Eur. J. Pharm. Sci. 29 (2006) 278–287.
[141] M. Rawat, D. Singh, S. Saraf, S. Saraf, Lipid carriers: a versatile delivery vehicle for proteins and peptides, Yakugaku Zasshi 128 (2008) 269–280.
[142] A.J. Almeida, E. Souto, Solid lipid nanoparticles as a drug delivery system for peptides and proteins, Adv. Drug Deliv. Rev. 59 (2007) 478–490.

[143] W. Choi, J. Kim, S. Choi, J. Park, W.S. Ahn, C. Kim, Low toxicity of cationic lipid-based emulsion for gene transfer, Biomaterials 25 (2004) 5893–5903.

[144] B. Chauhan, S. Shimpi, A. Paradkar, Preparation and evaluation of glibenclamide-polyglycolized glycerides solid dispersions with silicon dioxide by spray drying technique, Eur. J. Pharm. Sci. 26 (2005) 219–230.

[145] C. Thies, R. Dos Santos, V.V. Richard, H. Vandevelde, J.-P. Rolland, A. Benoit, Supercritical fluid-based coating technology 1: process considerations, J. Microencapsul. 20 (2003) 87–96.

[146] A. Royce, J. Suryawanshi, J. Shah, K. Vishnupad, Alternative granulation technique: melt granulation, Drug Dev. Ind. Pharm. 22 (1996) 917–924.

CHAPTER 7

Formulation strategies to improve the bioavailability of poorly absorbed drugs

Ram Kumar Sahu[a] and Jiyauddin Khan[b]
[a]Department of Pharmaceutical Science, Assam University (A Central University), Silchar, Assam, India
[b]School of Pharmacy, Management and Science University, Selangor Darul Ehsan, Malaysia

1 Introduction

The oral administration is most chosen route for intake of drug. It has several benefits like cost effective, easily availability and ease of compliance over other dosage form. Even though several advantages of oral administration, its poor bioavailability is still challenging for researchers. The deprived bioavailability affects the drug potency which changes the therapeutic potency of drug in patients. There are numerous factors such as mostly polar solubility, drug penetrability, presystemic metabolism, rate of dissolution, metabolism by first pass and vulnerability to efflux mechanisms affects the oral bioavailability of drugs. The low permeability and poor solubility of dug caused major factor for low oral bioavailability. It is essential to recognize the root sources of low oral absorption, and there is great demand to improve the bioavailability of such drug molecule by making modification in the formulation process [1–3].

The movement of drug from the site of administration into the systemic blood circulation represent the rate of absorption. The systemic therapeutic efficacy of drug is depending upon the amount of drug crossing the biological membranes present between administration site and blood vessels. The existence of phospholipid bilayer over biological cell membrane, suggesting the lipophilic nature of cell membrane. The drug molecule should have appropriate hydrophilic and lipophilic properties to produce maximum therapeutic efficacy. The hydrophilic properties make drug to dissolve in the aqueous environments surrounding the biological membranes, while lipophilic properties to partition into the biological membranes in order to accomplish passive absorption by the transcellular pathway. Neighboring epithelial/endothelial cells are associated by close intersections, which are

Advances and Challenges in Pharmaceutical Technology
https://doi.org/10.1016/B978-0-12-820043-8.00009-8

navigated by aqueous channels/fenestrae through which just little water-soluble molecules (<600 Da) can go to get ingested by means of the paracellular pathway [4, 5].

To overcome the poor absorption of poorly soluble drug molecule by considering formulation into three various forms namely amorphous formulations, lipid formulations and crystalline solid formulations.

1.1 Crystalline solid formulations

The micronization and salt formation of the crystalline compound of molecule enhance the surface area. This overcome the challenges of the molecule solubilization in particular solvent. This can enhance the rate of dissolution of the drug. The micronization technique is utilized to make the particle size of crystalline drug about 2–5 μm. Moreover, the novel nanocrystal technology utilizes to decrease the crystalline particle size upto 100–250 nm by employing dense gas technologies, ball-milling and many more technologies has been developed for size reduction of particles of drug molecule. But these technologies have certain limitation for size reduction of drug particles. The neutral compounds encounter problem to form salt, and it persist the issues of dissolution of drug. While in some cases the drugs having poor wettability, and in such condition particle size reduction of drug are not desirable.

1.2 Amorphous formulations

The melt extrusion and spray drying are used to prepared amorphous formulation of drug substances. The polymer and surfactant are incorporated into the formulation, which afford surface activity during dispersion.

1.3 Lipid formulations

Presently, the researchers have more attentions on lipid-based carrier systems for the delivery of the poor absorptive drug molecule. In this system, the poorly absorptive drug into incorporated into inert lipid vehicles namely surfactant dispersions, oils, solid lipid nanoparticles, emulsions, solid dispersions, micro/nanoemulsifying formulations, microemulsions, self-emulsifying formulations, nanoemulsions and liposomes. Table 1 demonstrated the various methods used for formulation strategies of poorly absorbed drug along with their merits and demerits [6–11].

Table 1 Strategies for the formulation of poorly absorbed drugs [6].

Method used	Merits	Demerits
Micronization of drug by conventional	**(a)** Reliable method, simple and easy to operate **(b)** Solid dosage form are manufacture humble	**(a)** Non uniform distribution of particle size **(b)** Dissolution rate not linear
Nanocrystals obtained by ball-milling	**(a)** This method has developed and product available in the market **(b)** old technology and easy to prepare solid dosage form	**(a)** Incorporation of additional method to evade aggregation of nanocrystals
Dense gas method used for production of nanocrystals	**(a)** Substitute the other nanocrystal processing scheme	**(a)** Technology require modification for better performance **(b)** Incorporation of additional method to evade aggregation of nanocrystals
Drug molecule immobilized in polymer solution—"Solid solutions"	**(a)** Easy to function **(b)** Novel extrusion methods proposed solvent free continuous process **(c)** Rapid and nonstop methods, minimum cost	**(a)** Formulation are physically unstable **(b)** Crystallization process of drug molecule or polymer are not stable
Self-dispersing process of solid solutions with surfactants	**(a)** Accumulation of drug molecules are eliminated	**(a)** Formulation are unstable, and easily crystallize
Formulation prepared with solid lipid nanoparticles or Nanoparticles	**(a)** Sustained release dosage form **(b)** Minimum variation	**(a)** Drug loading in formulation are lower **(b)** Drug extract out after polymorphic conversion **(c)** Water content are high

Continued

Table 1 Strategies for the formulation of poorly absorbed drugs [6]—cont'd

Method used	Merits	Demerits
Lipid solutions methods (LFCS Type I lipid systems)	**(a)** It is humble and toxic free drug delivery system **(b)** Mostly effective in the lipophilic activities **(c)** Controlled release drug system	**(a)** Solvent capability is poor **(b)** This method is limited to highly lipid drug **(c)** It require encapsulation
Self-emulsifying drug delivery systems (SEDDS) and SMEDDS (LCFS Type II or Type III lipid systems)	**(a)** Excellent novel drug delivery scheme **(b)** Having rapid absorption and minimize the variability **(c)** Drug absorption is not affected by digestion	**(a)** Due to its surfactant not suitable for chronic use, and sealed in soft gel or hard gel capsule **(b)** Maximum loss of solvent on dispersion (Type III)
Solid or semisolid SEDDS	**(a)** Free flowing powder used for the preparation **(b)** Easy to filed in capsules and also compressed into tablet form **(c)** Leakage associated problem can be overcome	**(a)** Due to its surfactant not suitable for chronic use **(b)** The products are unstable **(c)** Sometimes drug may be crystallized
Prepared by surfactant cosolvent systems (LFCS Type IV lipid systems)	**(a)** Due to surfactant properties have comparatively high solvent capacity for different drugs **(b)** Uniform dissolution to micellar solution, minimized irritancy of drug properties	**(a)** Due to its surfactant not suitable for chronic use **(b)** Higher loss solvent capacity **(c)** Possibility of drug precipitation on dilution

2 Lipid formulation approaches

The drug deliver by lipid-based system has been recognized for the improved delivery of the products having low bioavailability. The products of this delivery system can be formulated in better way by choosing proper excipients on the basis of their physicochemical properties like melting point, HLB value, fatty acid composition, disposability and digestibility.

The solubility, dissolution characters, stability and compatibility of various excipient are required to investigate before formulation. The suitable in vivo models are used to determine the efficacy of formulation. Finally, the drug loading and dissolution profile are employed for the optimization of the concluding formulation [12]. The various lipid based oral formulations are available in the market and their list is displayed in Table 2. According to properties of drug, the lipid formulation methods are selected and details are mention on Table 3.

2.1 Nanosuspensions

The Nanosuspension products are considered most appropriate for the drug molecules have higher log *P* value, does and melting point. Additionally, the drugs should be insoluble in water and organic solution are most suitable for the nanosuspensions formulation. The drug Atovaquone having poor bioavailability (10–15%) due to its rate limited absorption. Therefore, malaria, HIV and leishmanial infections of patient are treated with the Atovaquone at higher doses i.e., 750 mg twice a day. While the nanosuspensions of Atovaquone increase the oral bioavailability by 2.5-fold when equated to the marketed product of Wellvone and it is available as micronized drug [14].

2.2 Microemulsion and SMEDDS

The SMEDDS and microemulsions and can be prepared and manufactured in simple way and by using physically stable isotropic blends of surfactants, oil, co-surfactants and solubilized drug substance. These formulations are delivered in capsules or soft gelatin [15]. Microemulsions and SMEDDS are best approach to recover the oral bioavailability of drug and enhance therapeutic efficacy of drug molecules. The activity of the Cytochrome P450 metabolizing enzymes get suppressed by the surfactant present in the emulsion. The lymphatic absorption of the drug molecule is enhanced from the gastrointestinal tract due to the presence of long-chain triglycerides and glyceryl monooleate and also avoid the first-pass metabolism of the drugs.

The SMEDDS facilitate to improve the bioavailability of carvedilol upto 413% higher compared to marketed tablets by preventing it from first pass metabolism. Further it has been observed in increased in the lymphatic transport of active constituent having lipophilic properties when coadministered with long chain fatty acids. After absorption drug into the enterocytes, the drug easily entered into the chylomicrons due to re-esterification of fatty

Table 2 Lipid-based oral administration formulation available in the market [12, 13].

Brand name and dosage form	Active constituents	Medicinal uses	Manufacturer
Agenerase® (Oral suspension)	Amprenavir molecule	Antiviral HIV activity	Glaxo SmithKline Pvt. Ltd
Neoral (Capsules)	Cyclosporin A/I molecule	Anticancer activity	Novartis Pvt. Ltd
Kaletra (Tablets and oral suspension)	Lopinavir and Ritonavir molecule	Antiretroviral activity	Abott Pvt. Ltd
Rocaltrol® (Oral suspension and capsule)	Calcitriol molecule	Regulation of calcium	Roche Company
Lamprene (Tablets)	Clofazamine molecule	Antileprosy drugs	Alliance laboratories
Accutane (Tablets)	Isotretinoin molecule	Anticomedogenic activity	Roche Company
Gengraf (Capsules)	Cyclosporin A/III molecule	Anticancer activity	Abott Pvt. Ltd
Restandol (Capsules)	Testosterone undecanoate molecule	Testosterone replacement therapy	Merck Sharp & Dohme Ltd
Cipro® (Tablets and oral suspension)	Ciprofloxacin molecule	Antibiotic activity	Bayer Pvt. Ltd
Norvir (Tablets, oral powder, capsules and oral suspension)	Ritonavir molecule	Antiretroviral activity	Abott Pvt. Ltd
Lipirex (Tablets)	Atorvastatin molecule	Antihypercholesterolemia and antihyperlipidaemia	Highnoon Laboratories Limited
Sustiva (Tablets and capsules)	Efavirenz molecule	Antiretroviral activity	Bristol-Meyers Pvt. Ltd
Fenogal (Capsules)	Finofibrate molecule	Antihypercholesterolemia and antihyperlipidaemia	Genus Pharmaceuticals

Juvela (Capsules)	Tocopherolnicotinate molecule	Antihypertension, antihyperlipidemic	Eisai Company
Fortovase (Soft gelatin capsules)	Saquinavir molecule	Antiretroviral activity	Hoffmann-la Roche Inc.
Sandimmune (Capsules, oral solution and injection)	Cyclosporin A/I molecule	Immune suppressant	Novartis Pvt. Ltd
Convulex (Capsules)	Valproic acid molecule	Antiepileptic	Pharmacia Pvt. Ltd
Targretin (Capsules)	Bexarotene molecule	Antineoplastic agent	Ligand Pvt. Ltd

Table 3 Classification system of the lipid formulation [12].

Type I dosage form	Type II dosage form	Type IIIA (fine emulsion formulation)	Type IIIB (microemulsion formulation)	Type IV dosage form
Surfactant are not incorporating in oils	Incorporating oils and water with the addition of insoluble surfactants	Containing surfactants, Oils, cosolvents (excipient of water-soluble and water-insoluble)	Containing surfactants, Oils, cosolvents (excipient of water-soluble and water-insoluble)	Formulation incorporating water-soluble surfactants and cosolvents without oils
Non-dispersing systems	Emulsion are formulated by SEDDS	Formulation of SEDDS/SMEDDS prepared with water-soluble constituents	Formulation of SEDDS/SMEDDS prepared with water-soluble constituents and little oil content	Formulation disperses typically to form a micellar solution
Digestion is necessary	Formulation can be digested	Formulation not required digestion	Formulation not required digestion	Partial digestion

Table 4 Formulation prepared by micron emulsion technology to improved bioavailability drugs [18].

Drug	Therapeutic properties	Method
Paclitaxel	Anticancer	SMEDDS
Fenofibrate, Fluvastatin	Antihyperlipidemic	SMEDDS
Rapamycin, Cyclosporin	Immunosuppresive	SMEDDS
Nifedipine	Antihypertensive	SMEDDS
Ibuprofen, Naproxen	Analgesic	SMEDDS
Tipranavir	Anti-HIV	SMEDDS
Progesterone	Hormones	SMEDDS
Vitamins (A, D, E, K)	Nutritional supplement	SMEDDS
Melatonin	Immunomodulatory activity	SMEDDS
Acyclovir	Antiretroviral activity	SMEDDS

acids present in the formulations. Earlier it has been noted that the active constituents having log P above 5, and above 50 mg/g triglyceride solubility demonstrated prominent lymphatic transport. The drug which transported through the lymph can avoid hepatic first pass metabolism and enhance the bioavailability of drug [14–17]. The research work conducted on microemulsions and self-emulsifying systems for increment of bioavailability of drug are illustrated in Table 4.

2.3 Solid lipid nanoparticles (SLNs)

Mostly the SLN dosage from are formulated to increase the oral bioavailability of drug along with sustained release. The SLN formulation are best model to improve the bioavailability of drug by avoiding the degradation of drug into hepatic first-pass metabolism. Additionally, the intestinal lymphatic transport of drugs can be improved by the SLN methods. The drug molecule is transported to the systemic circulation by passing through the intestinal lymphatics. This transportation of drug molecule bypasses the presystemic hepatic metabolism and thus improved the bioavailability of drug by SLN method. It has been observed that the drug having high lipophilic properties like long chain triglycerides arrive systemic circulation through the lymphatics. The higher log P value (4.3) and good solubility in oils of Lovastatin make it prominent substrate for transportation to intestinal lymphatic. The delivery of lipophilic drugs namely halofantrine and ontazolast through lipid system improved their bioavailability by circulating these through lymphatic. Further oral administration of SLN of vinpocetine showed 4.17 times higher $AUC_{0\rightarrow\infty}$ compared to vinpocetine solution [14].

2.4 Polymeric nano- and microparticles

The polymeric nano and microparticles are prepared in the form of spheres and capsules. The drug molecules having hydrophilic and lipophilic properties can be consolidated or ensnared into these polymeric carriers. While in this carrier system accomplishment efficacy of drug is one the most vital concern. The drug formulated into these polymeric carriers system are more stable compared to liposomes in respect to in vivo and during storage. The major drawback of the preparation of these polymeric carries need organic solvents, and therefore production in large scale are moderately difficult. Besides, it is now and again critical to see suitable polymers that are affirmed as biodegradable, nontoxic and nonimmunogenic. Some synthetic polymer matrices have likewise been suspected to apply inconvenient impact on peptides and proteins fused during the manufacturing process [15].

2.5 Delivery of drug through vesicular systems namely liposomes, niosomes etc.

The liposomes are spherical shapes and having compact phospholipid bilayers inside the spherical particles. The presence of phospholipid bilayers easily allows to enter the lipophilic drug into lipid bilayers compartment while hydrophilic drug enter into the aqueous compartment. The release of drug from the liposomes, pharmacokinetic parameters and stability depend on the structure, surface charge, size and drug solubility. The liposomes are stable and safe compared to other novel drug delivery system. The various studies suggested that the liposomes products showed improved therapeutic efficacy compared to non-liposomal formulations. The biocompatible and biodegradable properties of liposomes make it more reliable formulation. But the liposomes formulation has associated with some disadvantages like low encapsulation efficiency, low stability and difficult scale-up [15].

Liposomes formulation can retain maximum time into the biological fluids due to lipoidal nature and improved vesicle uptake of drug by the target cells. The oral administration of liposomes formulation diminishes the toxicity of targeted cells, and increased the bioavailability of drugs. There is various mechanism has been developed for the liposome's formulation for its therapeutic efficacy. The cefotaxime (acid liable) drug incorporated inside the liposomes which provide protection layer over the drug and protect drug from degradation in acidic medium of the stomach. Moreover, liposomes increase the intestinal permeability of drug due to coating by the lipid component. The oral bioavailability of cefotaxime liposomes are 2.7 time higher than the bioavailability of its aqueous solution, while 2.3 times higher compared to its physical mixture. Some researchers noted that liposomes formulation gathered at brush-border membrane of enterocytes. This assist to enhanced the drug concentration gradient across the intestinal epithelium and facilitate to prominent absorption of insulin into the systemic circulation [14, 18].

The Niosomes belongs to nonionic surfactant vesicles and considered better drug delivery system for the hydrophilic drug. The Niosomes formulation have better retention at gastrointestinal tract for specific drug when administered through orally. The lipophilic properties of these formulation simplify the entry of drug into systemic circulation neutralizing the barrier capacity of mucosa, and improved the improved bioavailability of drug at absorption site. The 2 times higher oral bioavailability of acyclovir niosomal formulation in rabbits was observed compared to its free solution [14]. Table 5 summarizes some work done on marketed nanotechnology formulation.

2.6 Phytosomes

The formation of complex between botanical substances and natural phospholipids are termed as Phytosomes. In liposomes the phosphatidylcholine molecules are surrounded by the water-soluble compound without any chemical bond. While in Phytosomes, molecular complex between phosphatidylcholine and the botanical extract by 1:1 or a 2:1 conditional on the drug material complexed, involving chemical bonds (hydrogen bonds). Hence the complexes formed in Phytosomes showed better absorption and bioavailability of drug compared to the liposome's formulation. The Phytosomes of plant extract or isolated polar chemical constituent such terpenoids, flavonoids, tannins etc. demonstrated improved absorption when administered orally. Further the lipid components of Phytosomes enhance the drug permeability through biological membrane and lead to increase the bioavailability of the drug. The oral bioavailability of Phytosomes of standardized *Silybum marianum* extract or silybin showed significantly increased bioavailability equated to their pure standardized extract. The increment in the bioavailability of silybin Phytosomes due to increase in lipophilic properties of silybin-phospholipid complex and this complex deliver higher potency of silybin [14].

Table 5 Marketed Nanotechnology formulation having enhanced oral bioavailability [13].

Manufacturer	Methods	Description
Baxter company	Nanoedge technique	In this drug particle size are reduced to nanorange by microprecipitation, homogenization and emulsions methods
BioSante company	Calcium phosphate-based nanoparticles (CAP) technique	The nano particles formed by CAP methods to enhanced the oral bioavailability of hormones/proteins
Elan company	NanoCrystal technique	This method produced drug particle size less than 1000 nm by using wet-milling, and fine particles are stable without any agglomeration
Eurand company	Biorise technique	In this method the nanocrystals or amorphous drug formed by breaking of the crystal lattice. The drug particles are stabilized with biocompatible carriers
SkyePharma company	Insoluble drug delivery (IDD) technique	The micro nano meter or droplet water-insoluble drug core might be more stabilized by phospholipids
PharmaSol company	Nanostructured lipid carriers (NLC) technique	The high-pressure homogenization method employed for the preparation of NLC solid particles, while lipid-drug conjugate nanoparticles deliver high-loading capacity for hydrophilic drugs for oral delivery

3 Conclusion

The poorly absorbed drugs have low therapeutic efficacy, but it can be improved by enhancing gastrointestinal solubilization which alter pharmacokinetic profiles of drug. Consequently, it facilitates to enhance the drug molecule into systemic circulation and give rapid therapeutic action. The formulation of these drugs can be prepared into different formulation namely nanosuspensions, solid lipid nanoparticles, polymeric nano- and microparticles, liposomes, niosomes, phytosomes, microemulsion and SMEDDS. The bioavailability of hydrophobic drugs can be improved by delivering drug into lipid-based drug delivery systems. The bioavailability of poorly water soluble drug can be increased by formulating into lipid nanoparticle formulations and these formulation deliver better therapeutic applications, biodegradability and bio-acceptability.

References

[1] I.G. Orellana, Strategies to improve oral drug bioavailability, Expert Opin. Drug Deliv. 2 (3) (2005) 419–433.
[2] P.D. Leeson, A.M. Davis, Time-related differences in the physical property profiles of oral drugs, J. Med. Chem. 47 (25) (2004) 6338–6348.
[3] C.R. Gardner, C.T. Walsh, O. Almarsson, Drugs as materials: valuing physical form in drug discovery, Nat. Rev. Drug Discov. 3 (11) (2004) 926–934.
[4] B. Peterson, M. Weyera, J.H. Steenekamp, J.D. Steyn, C. Gouws, J.H. Hamman, Drug bioavailability enhancing agents of natural origin (bioenhancers) that modulate drug membrane permeation and pre-systemic metabolism, Pharmaceutics 11 (33) (2019) 1–46.
[5] S. Chillistone, J. Hardman, Factors affecting drug absorption and distribution, Anaesth. Intensive Care Med. 9 (2008) 167–171.
[6] S. Gupta, R. Kesarla, A. Omri, Formulation strategies to improve the bioavailability of poorly absorbed drugs with special emphasis on self-emulsifying systems, ISRN Pharm. 2013 (2013) 848043.
[7] E. Merisko-Liversidge, G.G. Liversidge, E.R. Cooper, Nanosizing: a formulation approach for poorly-water-soluble compounds, Eur. J. Pharm. Sci. 18 (2) (2003) 113–120.
[8] A.M. Kaushal, P. Gupta, A.K. Bansal, Amorphous drug delivery systems: molecular aspects, design, and performance, Crit. Rev. Ther. Drug Carrier Syst. 21 (3) (2004) 133–193.
[9] M. Shulman, M. Cohen, A. Soto-Gutierrez, et al., Enhancement of naringenin bioavailability by complexation with hydroxypropoyl-β-cyclodextrin, PLoS One 6 (4) (2011) e18033.
[10] M. Anwar, M.H. Warsi, N. Mallick, et al., Enhanced bioavailability of nano-sized chitosan-atorvastatin conjugate after oral administration to rats, Eur. J. Pharm. Sci. 44 (3) (2011) 241–249 (PubMed] [Google Scholar).
[11] B.P. Divyakant, R.M. Valay, N.T. Alok, A.P. Arpita, P.T. Hetal, Development and characterization of solid lipid nanoparticles for enhancement of oral bioavailability of Raloxifene, J. Pharm. Bioallied Sci. 4 (supplement) (2012) 14–16.
[12] S. Kalepu, M. Manthina, V. Padavala, Oral lipid-based drug delivery systems—an overview, Acta Pharm. Sin. B 3 (6) (2013) 361–372.
[13] S.B. Gavhane, S. Mantry, S.A. Joshi, G.Y. Dama, S. Mohanto, Enhancement of poor oral absorption drug via lipid formulation: self-emulsifying drug delivery system, Int. J. Pharm. Sci. Res. 11 (3) (2020) 1042–1056.

[14] H. Thankkar, B. Patel, T. Sneha, A review on techniques for oral bioavailability enhancement of drugs, Int. J. Pharm. Sci. Rev. Res. 4 (3) (2010) 203–233.
[15] B.K. Nanjwade, D.J. Patel, R.A. Udhani, F.V. Manvi, Functions of lipids for enhancement of oral bioavailability of poorly water-soluble drugs, Sci. Pharm. 79 (4) (2011) 705–727.
[16] V.B. Patravale, A.A. Date, R.M. Kulkarni, Nanosuspensions: a promising drug delivery strategy, J. Pharm. Pharmacol. 56 (2004) 827–840.
[17] A. Mullertz, Lipid-based drug delivery systems: choosing the right in vitro tools, Am. Pharm. Rev. 10 (4) (2007) 102–110.
[18] A. Samad, Y. Sultana, M. Aqil, Liposomal drug delivery systems: an update review, Curr. Drug Deliv. 4 (4) (2007) 297–305.

CHAPTER 8

Process analytical technology (PAT) tools: Uses in pharmaceutical manufacturing

Sabya Sachi Das[a], Saad Alkahtani[b], Amit Kumar Nayak[c], and Md Saquib Hasnain[d]
[a]Department of Pharmaceutical Sciences and Technology, Birla Institute of Technology, Mesra, Ranchi, Jharkhand, India
[b]Department of Zoology, College of Science, King Saud University, Riyadh, Saudi Arabia
[c]Department of Pharmaceutics, Seemanta Institute of Pharmaceutical Sciences, Mayurbhanj, Odisha, India
[d]Department of Pharmacy, Shri Venkateshwara University, Gajraula, Uttar Pradesh, India

1 Introduction

Process analytical technologies (PAT) is defined as a system or tool to design, analyze, and control various processes (in- or out-line) involved in manufacturing stages, by aptly measuring different critical quality attributes of raw materials, with the aim of achieving best quality final products [1]. The prime principle of PAT signifies as, that product quality cannot be tested, rather it should be developed by design and this could be efficiently achieved by an increase in knowledge of processes and control over manufacturing techniques [2]. Traditionally, the pharmaceutical manufacturing process included the production of finished products, trailed by laboratory studies to validate the product quality. However, frequent procedure optimization, repeated manufacturing complications, and the chances of failed lots, were some of the major disadvantages of this method. These issues were discussed and to overcome these issues PAT was established as a prime tool for pharmaceutical current good manufacturing practices (CGMPs), by the Food and Drug Administration (FDA) [3]. Initially, the United States FDA (US-FDA), and Centre for Drug Evaluation and Research (CDER) proposed the PAT initiative, with the prime aim of attaining better health and cost profits by applications of modern procedures in manufacturing industries of pharmaceutical sectors [4].Quality-by-Design (QbD) is a well-recognized tactics applied in developing and manufacturing of active pharmaceutical ingredients (API)or drug products and constantly demonstrates regular performance for producing quality products. The understanding found during

Advances and Challenges in Pharmaceutical Technology
https://doi.org/10.1016/B978-0-12-820043-8.00007-4

the manufacturing have helped for justifying the formation of a significant process or stage, regulate approaches and set procedures within the design space [5]. The materials developed inside the design space would provide adequate products. All of the principles and perceptions associated with QbD have been applied for the development of various analytical approaches, and have been designated as Analytical QbD (A-QbD).

1.1 Significance of PAT

PAT has significantly acted for controlling the production cost, resulted in developing more effective production methods, and also minimized the need of finishing discards. In an era of globe in which the financial matters have arrived as a triage of verdicts, the cost control has turn out to be compactly tangled with patient care and treatment, PAT have demonstrated other significant benefits. Incessant monitoring permits extra organized processes and a better control of intervening production stages. For instance, during production of vaccines and protein separation approaches, the constant application of PAT could significantly increase worth of final product [6]. As a direct concern through initiatives of the "cGMPs for 21st Century", the pharmaceutical industries have been pressurized by the authorities to look after the issues associated with the inadequate processes understanding, disorganizations in interim processes and constant progress in processes by adopting PAT. From the day, FDA issued its PAT initiative, some of the pharmaceutical enterprises have showed their eagerness in discussing about their efforts for implementing PAT [7]. For encouraging industries to implement PAT, the FDA familiarized the concept of "safe harbor" or "research exemption", aimed to inspire the industries to explore the tools that would offer improved informations of various processes deprived of the distress of having an adverse impact over the capability to produce final products that meets all facets of the quality control strategies of the companies [8]. Technology dealers have urged that, numerous of their pharma clients have been in the planning stage of understanding and approving PAT, but some of them have really executed new manufacturing technologies in their respective industries [9]. Novartis had applied sensor technologies parallel with the old processes and also had performed lots of ground-breaking and critical works. The understanding of any significant process or technology depends mostly upon the establishment and crucial fundamentals of that technology implemented by the company and afterwards it could be forwarded to others [10]. A group of researchers from Eli Lilly, demonstrated few concepts and perceptions regarding various analytical techniques which was further improved for instantaneous analysis of different stages during manufacturing processes in pharmaceutical industries. Analytical techniques like Fourier transforms infrared (FTIR) is used for analyzing reactions, Near-IR spectroscopy (NIRS)is used for examining the dryness and consistency of products, whereas the nuclear magnetic resonance (NMR) spectroscopy, mass spectroscopy (MS) and chromatography techniques such as high pressure/performance liquid chromatography (HPLC), gas chromatography (GC) are used for analysis and identification of products [11]. PAT has been effectively executed by TechniKrom in the control/mixing stages of critical liquids and has assisted the industrial clients to observe processes and validate that the alterations in materials would affect the production outputs. Moreover, the applications of NIRS has significantly increased in PAT-associated analysis and has been mostly used for analyzing the on-line processes. Interestingly, NIRS have been able to measure more than one product simultaneously and allows a person to have a follow-up on different ongoing processes during chemical reactions [12]. Similarly, strong alterations in the NIR spectrums of different polymorphs of a single API allows researchers to compute the formation of polymorph during the phases of pre-formulation and formulation. This skill is vital for an establishment of drug as final product, as modifications in system could cause significant changes in behavior of the drug *in-vivo* [13]. A group of researchers from Sigma-Aldrich Biotechnology implemented a DOE (design of experiment) method and developed a media of cell culture fora wide-range of CHO (Chinese hamster ovary) cell lines, that were used by the biopharmaceutical organizations for producing protein-mediated conjugates or delivery systems. They utilized statistical softwares to recognize the optimized culture media in their resources and developed systems to further enhance the cellular growth. It was an extremely significant technique which allowed the investigators for easily and precisely determining the API content in a particular formulation although overlooked the physical factors of the samples or sampling circumstances [14]. PAT has been widely used in chemical, petrochemical and pharmaceutical or biopharmaceutical sectors, as it has offered a platform for evaluating the effects of various parameters used for simulation analysis and in-line processes. With the growth in business, the manufacturers have focused particularly into the process and equipments types, for expanding their market. These systems have been conventionally related to the petrochemical industries, which have severely expanded their equipment lines to involve features and new measuring methods [15]. Additionally, the pharmaceutical industries realized that the chemical/petrochemical industries offered a splendid source for process analyzers. Moreover, the pharmaceutical industries should have a watch on the vendors. The vendors such as Thermo technologies, Aspen Technology and Pavilion Technologies have exhibited crucial role over the pharma sectors with analysis equipments [16] (Table 1).

Table 1 Significant applications of various PAT tools for industrial aspects.

Analyzing technique	Applications	Outcomes
NMR spectroscopy	Waste water analysis for the presence of any organic matter	↑ Productivity ↓ Time for analysis
NIRS	Detection of raw material and simultaneous quality control	↑ Recognition time Cost effective
Raman spectroscopy, attenuated total reflectance (ATR) and FTIR	Concurrent determination of concentration of solute and crystal polymorph form	Quantitative information about the effects of addition of reactant over the crystal polymorphic states
In-line NIRS	Catalysis reactions including transformation of acetone to methyl isobutyl ketone	Affects the productivity, selectivity, and yield of methyl isobutyl ketone
PSD analyzer or SFT	Distribution of particle size	Quantitative and qualitative information about particle size and other related features
DIA	Reduction behavior of particles	Applicable in milling process
XRD, FTIR	API, drug or excipient identification	Qualitative information of any polymorph formed during intermediate reactions
HPLC, RP-HPLC	Identification of bioactives in developed products	↑ Chances of raw material or API quantification in a dosage form
MRI, CLSM	Surface and near-surface topographies	Improves the clarity of surface alterations in complexes or blend dosage forms

2 Applications of PAT in pharmaceutical industries

Approval and implementations of PAT in the pharmaceutical industries is completely based upon the novelties in the analytical processes, abilities of the working persons to adopt and examine huge amount of data's [17]. The foremost purpose of PAT is that the quality of product is made at earlier stages, rather being examined before the product release [5]. Furthermore, outline of PAT generally comprises of risk management, analyzing by sensors to monitor, control, design and performance of the processes [18].In pharmaceutical industries, various types of analytical methods and equipments such as FTIR, UV-spectroscopy, GC, HPLC, X-ray diffraction (XRD)spectroscopy, and NIRS are used for effective monitoring. It has been observed that the PAT methods are applied indifferent manufacturing unit processes, including dispensing, milling, coating, blending, compression and packaging. Applications of PAT in major industrial processes or sectors have been summarized in Fig. 1.

Fig. 1 Schematic illustration representing various industrial applications of PAT.

2.1 Application of PAT tools in coating process

In pharmaceutical industries, film coating is a crucial and well-recognized process. Commonly, the process of polymeric film coating is performed in perforated coating pans, while sugar coating is usually directed in non-perforated coating pans. Porter and colleagues reported a good outline of various types of pan coaters [19]. According to principles of QbD, timely measurements plays an important role for ensuring product quality and is essentially required during or at the end of coating procedure to assist determination of critical processing parameters (CPPs), critical material attributes (CMAs), and critical quality attributes (CQAs). The prime objective of these measurements is to regulate or observe the coating process, evaluate quality of coating and assess the coating phases [20]. PAT has been implemented majorly to monitor and regulate the process, and for analyzing technologies. Moreover, various commonly used approaches and modeling technologies, for analyzing and characterizing the coating process was reported by Sahni and Chaudhuri [21]. Recently, novel PAT tools used to characterize and regulate the coating processes have been developed. PAT can be effectively used to measure and analyze four types of processes, including in-line (without removing the sample/s), on-line (sample is distracted from the central process, investigated, and might be reverted back), at-line (removing the sample and analyzing the process closely) and off-line (removal of sample and analyzing apart from the process) [1]. The in-line or on-line measuring approaches provides fastest result, while at-line measurements are found to be long and consumes more time. To observe the different features and quality of film coating, various techniques including light and electron microscopy (L/EM) [22], Magnetic resonance imaging (MRI) [23], confocal laser scanning microscopy (CLSM) [24], NMR [25], positron emission tomography (PET) [26], NIRS, Raman spectroscopy (RS) and laser induced break-down spectroscopy [27–29], have been used. Each technique has been found significant and provides vital informations regarding the processes involved during coating method, but each technique has its specific limits. Many of these processes or methods are found to be destructive and have been found to be incapable to solve multiple layer coating with single measurement, thus an apt multivariate analysis model might be required for data analysis [30].

2.2 PAT tools in biopharmaceuticals processing

Some other approaches are also involved in PAT tools, including statistical processes, analytical chemistry, mathematical tools and control schemes, that are allows understanding of QbD to improve the product quality. The CQAs such as product yield, needs to be controlled possibly by effectively controlling the possible critical process parameters (CPPs), including time, temperature and pH, thus improves the quality of product. Moreover, the PAT tools are implemented in the biopharmaceutical processes in four significant phases [31]. In the initial phase, the quality-associated information is congregated by three different methods, including on-line detecting, omics-centered analysis, and CQAs assessment. In the second phase, the informations acquired in the first phase are analyzed either using multivariate analysis or any other specific robust tools. In the third phase, the procedure is regulated with the help of programmed control devices. In the last and fourth phase, the human decisions related to the discharging of the batch is taken (whether tom discharge or not) based upon the data produced to establish precised set of product standards [32].

2.3 Application of PAT tools in injectable processing

PAT tools have also represented as a significant and potential tool in manufacturing of injectables by the manufacturers as it helps to regulate the manufacturing process. The method of acoustic resonance spectroscopy was applied for determining the viscosity and density of solvents used in preparation of injectables [33]. The calorimetry technique was used for detecting the moisture content and dissolution profile drugs or therapeutic compounds [34]. The dielectric spectroscopy was used for determining the concentration of salt or salt conjugates present in the solvent medium used in injectable manufacturing [35]. A NIRS method was well developed to measure the concentration oflysine clonixinate present in the injectable solution (100 mg/2 mL) [18]. Techniques such as FT-NIRS and partial least square (PLS) regression was significantly used to quantify the concentrations of caffeic acid, baicalin, chlorogenic acid, chenodeoxycholic acid, ursodesoxycholic acid, and luteoloside present in a traditional Chinese medicine, termed as *Tanreqing* injections [36].In another studies, the concentration of 5-fluoruracil, (loaded in infusion pump)and adriamycin and epirubucin (loaded in polyethylene syringes) were analyzed using HPLC technique [37, 38].The antiviral drug, ganciclovir, was effectively quantified by means of PLS algorithm, including three calibrating sets of the drug solutions having a concentration range of 0.8–10 mg/mL [39]. Similar methods were used for quantification of two drugs, Irinotecan and topotecan and simultaneously used RS for more clear identifications [40]. As compared to other delivery systems, the injectable-based dosage forms or delivery systems are found to be based on formulation process, however the safety profile still remain a concern for manufacturers and regulatory authorities. So, to overcome these issues, PAT and QbD significantly improved the safety profile of pharmaceutical products and thus improved their stability level as per required by the regulatory bodies.

2.4 Application of PAT in hot melt extraction process

As concerned for PAT tools, the optical spectroscopy has found significant potentials as an in-line process analysis tool for hot melt extraction (HME) process. This approach improved the safety concerns of the products, met the prerequisite conditions with minor deviations, reduced the batch loss. In 2003, Apruzzese et al. firstly demonstrated the significance applications of in- and on-line processing of NIRS for HME. In their study, the residence time distribution (RTD) was identified with the help of VIS-NIRS probe attached with an extruder, that helped to detect the color dye [41]. In an another study, in-line NIRS (transmittance mode) was used to measure the drug (carbamazepine) content and the plasticizer (polyethylene glycol) concentrations simultaneously in a blend of polymer melt of poly (vinylpyrrolidone-*co*-vinyl acetate) co-polymer, having diverse opacity [42]. Saerens et al., in their study used NIRS tool to measure in-line process parameters like of concentration of APIs, behavior of polymer-drug conjugates, and molecular interfaces during the HME process. They also used RS to indorse their outcomes [43]. Researchers from this group, further published their study related with NIRS as tool to monitor the HME processes and showed the abilities of NIRS as an important PAT tool to observe the critical HME factors and the whole process [44]. Coates et al. established the usage of in-line RS for HME process. This study involved numerous probes, involving Raman probe, that were attached with extruder die panel of a particular screw extruder to evaluate the varying configuration of polymer-polymer composites. A comparative analysis between RS and other methods was also performed by the same researchers, and emphasized the issues associated with each of the method [45]. Moreover, the in-line NIRS was also used to observe the formation of co-crystals formed in the presence of two types of APIs during a HME process [46]. One of the study showed that the middle-infrared spectroscopy (MIRS) was used during the HME process. The study was performed to demonstrate the significant potentials of MIRS a vital tool, used for *in-situ* measurements during polymerization phases of HME processes. It was observed that MIRS was appropriate for the identification of the composition of the polymer (polystyrene/polyethylene) blends ranging from 0% to 100% and also determined the end-point of polymer alterations. In this study, the MIRS was found as an apt technique for monitoring the in-line processing and product quantification to a certain level [47].In view of the numerous advantages and limitations of each of the reported approaches or techniques, it is very hard to just indorse one technique that could be appropriate to all situations. It is quite impossible to constrict it down to a particular or a single approach which could state all the potential concerns. Therefore, a particular technique should be chosen by the user as per the situation and suitability to attain the purpose.

2.5 Application of PAT in milling process

The prime objective of PAT is to improve our essential facts associated with the manufacturing procedures. In last decades, researchers have reported with few non-intrusive PAT tools to measure the particle size distribution (PSD), including dynamic image analysis (DIA), focused beam reflectance measurement (FBRM), and spatial filtering technique (SFT).The major advantages of these non-intrusive techniques was that they did not interfere with the ongoing process and allowed to monitor and regulate the real-time processes. Furthermore, these techniques also eliminated the additional steps of sample preparation essential for conventional off-line techniques. In these techniques, a sidewise stream of particles was often required to get isolated, as the conventional stream processes were vastly concentrated with flow rates which led to entangling of the detectors [48]. Harvill et al., showed the capability of instantaneous measurements with the help of laser diffraction to perform proper analysis and modification of inconsistent feeding rate of a classifier mill [49]. Likewise, Chan et al. established an at-line and in-line laser diffraction arrangements and with its help measured the PSD of the microspheres. As per the results, they recommended at-line process to be more apt and dependable for measuring the particle size, as there was a considerable amount of breakdown of agglomerates was observed while measuring through in-line laser diffraction techniques [50]. In an another study, the researchers performed a comparative study between various modes of DIA, including on-line, at-line and in-line, to observe the behavior of reduction in particle size in a conical mill. Interestingly, results showed that the PSD measurements were found to be identical with the use of all the three modes for different materials [51]. Furthermore, for better understanding and control of dry milling procedure in pharmaceutical industries, feedback controlling systems could be attached to the conical mill parallelly with DIA sensors [52]. In a study it was observed that during a unremitting granulation-drying-milling procedure, the FBRM was used as an in-line concurrent PAT tool for monitoring PSD of granules. Apart from above arrangements, a self-scrubbing rotating disc was assembled, which helped to reduce the adhering of smaller particles with the probe window. The results of this study showed an important correlation ($P < 0.0001$) between D20, D50, and D80 particle size. Thus, the authors proposed and established a significant correlation within the sieve data and real-time feedback controlling systems [53]. Additionally, the SFT has also been simultaneously suggested for identifying the particle size in processes like coating, mixing, fluidized beds, high-shear wet granulation, dry granulation, and spray drying [48, 54].

2.6 PAT tools for monitoring crystallization

The formation of crystals (crystallization) during manufacturing process is considered to as unworthy phase and thus designing, optimization, and controlling of the various phases is very important to regulate the crystallization phase. Various PAT tools and methods have been reported which could be used to observe and regulate the crystallization procedure in the pharmaceutical industries, to illustrate the crystalline constituents, and to increase the understanding associated with different procedures, so as to confirm safety and better quality products [55–57]. The various novel PAT techniques that could be used to meet the requirements of the regulatory guidelines should effectively monitor and control the CPPs such as polymorphic transformations [58] and phase conversions [59] during the pharmaceutical productions, monitor and control the CQAs such as particle size [60], identify the APIs final products by using spectroscopic method or XRD method, quantify the formed polymorphs [61] and measure the crystallinity [62] and monitor the stability of formulations [63].The governing authorities have signified the vital features, have well-justified and provided their adverse influences over the current good manufacturing practice (CGMP) agreements concerning strength, identity, stability and purity of the APIs or final drug products, and their contrary effects over the drug safety [64, 65]. It is very essential to identify and establish the concept that the design and application of a PAT tools or systems for pharmaceutical applications have developed unique opportunities for innovations in scientific field. Apart from the industrial aspects, QbD in formulation perspectives [66], has been efficiently used for various product development including nanoemulsifying systems [67], analyzing drug regimens [68–71], identifying products [72–74], process optimization [75], nano-pharmaceutical product development [76], establishment of multi-particulate drug delivery systems [77], biopharmaceutical development [78] and other biomedical applications. So, based upon the definite usage and objective, a diverse set of QbD and PAT components might be used to attain the required processes and understanding of product, which would definitely support for implementing the regulatory decisions.

3 Commercial and regulatory aspects of PAT

PAT is a new technology to pharma sector, but it is an old and long-standing technique to other industrial sectors, including food and beverages, chemicals and petrochemicals *etc.* The drug companies established in United States have had the capability to practice the PAT related process control systems, but they were unable to do so as it was very costly in terms of the equipment cost, cost to establish quality and validation systems. Till date, there are very few industrial leaders who have demonstrated the significance of PAT, therefore the pharmaceutical companies are impending PAT with ambiguity [7]. Even though PAT had increased the production efficacy and production rate, this might not be observed from a business viewpoint as a robust incentive for modification due to the awareness that the execution costs might compensate the profit for the investments in several cases, particularly for small scale pharmaceutical enterprises which manufactures the drug products which are already competing with constricted or fictional margins [9]. As per the advantages and limitations of these technologies, the industries and companies should decide when the investment would let them to gain profits from the process development and when it doesn't [10]. Nowadays, the business models are varying and the significance of various process and tool involved in the product manufacturing plays crucial role in regulating the financial assets of the pharmaceutical industries and thus a lower risk of product quality would compensate the cost of PAT implementation [11,12].

Moreover, the regulatory agencies have overcome as a major contributor towards the reduction in the implementations of PAT, as it has become a major concern to the industrial experts that how the regulatory organizations would respond during the analysis of a particular equipment facility. There is a major apprehension inside the industries that there is deficiency of global harmonization of regulatory prospects related to PAT tools.

However, the FDA has contributed in international symposiums for developing harmonization over the PAT approaches [13]. FDA have never intended to examine through the research data's collected for a prevailing product for the perseverance of assessing the appropriateness of any PAT tool [14]. Also, the PAT tools are mostly utilized by the companies to just illustrate that their products are in good conditions after the implementation of the prevailing processes, rather than reshaping and adjusting their processes [79]. Although PAT is not a regulatory law, yet the FDA significantly contends the industrial authorities to adopt it for better performance and quality product.

4 Conclusion and future perspectives

The usage and application of process analytical techniques (PAT) could provide enormous benefits and assistances to the pharmaceutical industries, as it could help to increase the quality, safety and stability of the products, while conveying superior advantages related to utilizations and economic values. PAT also have provided better understanding of raw materials by characterizing it both physically and chemically understanding of raw materials as well as various parameters essential for manufacturing processes. Overally, these tolls helps to produce more robust method, better quality products, better control over processes and saves time which eventually results in development of a novel brand appearance for the company. The goal of PAT approaches should be to execute robust methods which are flexible enough to establish a definite level of inconsistency in CQAs by adjusting the conditions of processes. These tools could be efficiently applicable to various unit operations on not only pharmaceutical organizations but also other sectors. Although, there are some limitations and challenges in their implementations, but it could be sorted with the help of guidelines ascribed by the regulatory agencies like FDA. The suitability and applicability of PAT would definitely help the researchers and manufacturers to reduce the unwanted issues and to gain superior product quality in near future.

References

[1] FDA, Guidance for Industry: PAT-A Framework for Innovative Pharmaceutical Development, Manufacturing, and Quality Assurance, 2004, Accessed at: http://www/fdagov/cder/guidance/6419fnlpdf. (Accessed 19 May 2020).
[2] FDA, Guidance for Industry PAT-A Framework for Innovative Pharmaceutical Development, Manufacturing, and Quality Assurance, U.S. Food and Drug

Administration—Center for Drug Evaluation and Research, Center for Veterinary Medicine and Office of Regulatory Affairs, 2004. Available from: http://www.fda.gov/downloads/Drugs/GuidanceComplianceRegulatoryInformation/Guidances/ucm070305.pdf. (Accessed 19 May 2020).
[3] U.S. Department of Health and Human Services Food and Drug Administration, 2020. www.fda.gov/Drugs/DevelopmentApprovalProcess/Manufacturing/QuestionsandAnswersonCurrentGoodManufacturingPracticescGMPforDrugs/UCM071836. (Accessed 19 May 2020).
[4] U.S. Department of Health and Human Services, Food and Drug Administration, A Presentation at the Advisory Committee for Pharmaceutical Science by Dr. Ajaz S. Hussain, 2020. http://www.fda.gov/ohrms/dockets/ac/cder01.htm#Pharmaceutical%20Science. (Accessed 19 May 2020).
[5] U.S. Department of Health and Human Services, Food and Drug Administration Guidance for industry, PAT—A Framework for Innovative Pharmaceutical Development, Manufacturing and Quality Assurance, 2020. http://www.fda.gov/downloads/Drugs/GuidanceComplianceRegulatoryInformation/Guidances/ucm070305.pdf. (Accessed 19 May 2020).
[6] S. Banerjee, N. Bhatwadekar, P. Mahajan, P. Karode, A. Mishra, Process Analytical Technology (PAT): Boon to Pharmaceutical Industry, 2020. Available at: http://www.pharmainfo.net/reviews/processanalytical-technology-pat-boon-pharmaceutical-industry. (Accessed 19 May 2020).
[7] N.G. Anderson, In-process assays, in-process controls, and specifications, Practical Process Research and Development. A Guide for Organic Chemists, second ed., Elsevier Inc., USA, 2012, pp. 189–207.
[8] J. Ganguly, G. Vogel, Process analytical technology, J. Int. Soc. Pharm. Eng. 26 (2006) 1–9.
[9] B.H. Bunker, H.Y. Wang, Bioprocess monitoring and computer control: key roots of the current PAT initiative, Biotechnol. Bioeng. 95 (2) (2006) 226–261.
[10] J. Methfessel, J.R. Roy, A practical approach to PAT implementation, Pharm. Technol. Eur. 18 (3) (2006).
[11] R. Kamble, S. Sharma, V. Varghese, K.R. Mahadik, Process analytical technology (PAT) in pharmaceutical development and its application, Int. J. Pharm. Sci. Rev. Res. 23 (2) (2013) 212–223.
[12] Slow adoption of PAT for bio processing—challenges emanate for technology development and design of processes, Genet. Eng. Biotechnol. News 26 (16) (2006). Available at: www.genengnews.com/gen-articles/slow-adoption-ofpat-for-bioprocessing/1891/. (Accessed 19 May 2020).
[13] P. Scott, Process analytical technology in the pharmaceutical industry: a toolkit for continuous improvement, PDA J. Pharm. Sci. Technol. 60 (1) (2006) 17–53.
[14] D.G. Phadtare, A.R. Pawar, R.B. Saudagar, A brief review on quality by design and process analytical technology, Asian J. Pharm. Anal. 6 (2) (2016) 122–130.
[15] C.W. Randall, PAT pending technology tools for developing improved pharmaceutical manufacturing processes, in: Presented at the Advancing Manufacturing Summit, West Lafayette, vol. 7, Issue 12, Dec, 2004.
[16] K. Muteki, K. Yamamoto, G.L. Reid, M. Krishnan, De-risking scale-up of a high shear wet granulation process using latent variable modeling and near infrared spectroscopy, J. Pharm. Innov. 6 (3) (2011) 142–156.
[17] J.T. Park, R. Shah, K.A. Brorson, A.S. Rathore, Process analytical technology scope, Biotechnol. Bioeng. 105 (2) (2009) 285–295.
[18] R. López-Arellano, E.A. Santander-García, J.M. Andrade-Garda, G. Alvarez-Avila, J.A. Garduño-Rosas, E.A. Morales-Hipólito, Quantification of lysine clonixinate in intravenous injections by NIR spectroscopy, Vib. Spectrosc. 51 (2009) 255–262.

[19] S.C. Porter, A.L. Vea, Coating of pharmaceutical dosage forms, in: Remington the Science and Practice of Pharmacy, twenty-second ed., Pharmaceutical Press, London, 2012, pp. 977–987.
[20] S. Porter, G. Sackett, L. Liu, Development, optimization, and scale-up of process parameters: pan coating, in: Y. Qiu, Y. Chen, G.G.Z. Zhang, L. Liu, W.R. Porter (Eds.), Developing Solid Oral Dosage Forms, Academic Press, San Diego, CA, 2009, pp. 761–805.
[21] E. Sahni, B. Chaudhuri, Experimental and modeling approaches in characterizing coating uniformity in a pan coater: a literature review, Pharm. Dev. Technol. 17 (2) (2012) 134–147.
[22] P. Seitavuopio, J. Heinamaki, J. Rantanen, J. Yliruusi, Monitoring tablet surface roughness during the film coating process, AAPS PharmSciTech 7 (2) (2006), E31.
[23] J.C. Richardson, R.W. Bowtell, K. Mäder, C.D. Melia, Pharmaceutical applications of magnetic resonance imaging (MRI), Adv. Drug Deliv. Rev. 57 (8) (2005) 1191–1209.
[24] M.J. Lee, D.Y. Seo, H.E. Lee, I.C. Wang, W.S. Kim, M.Y. Jeong, G.J. Choi, In line NIR quantification of film thickness on pharmaceutical pellets during a fluid bed coating process, Int. J. Pharm. 403 (1–2) (2011) 66–72.
[25] E. Fukushima, Nuclear magnetic resonance as a tool to study flow, Annu. Rev. Fluid Mech. 31 (1) (1999) 95–123.
[26] D.J. Parker, A.E. Dijkstra, T.W. Martin, J.P.K. Seville, Positron emission particle tracking studies of spherical particle motion in rotating drums, Chem. Eng. Sci. 52 (13) (1997) 2011–2022.
[27] M. Wirges, A. Funke, P. Serno, K. Knop, P. Kleinebudde, Development and in-line validation of a process analytical technology to facilitate the scale up of coating processes, J. Pharm. Biomed. Anal. 78–79 (0) (2013) 57–64.
[28] C.V. Moltgen, T. Herdling, G. Reich, A novel multivariate approach using science-based calibration for direct coating thickness determination in real-time NIR process monitoring, Eur. J. Pharm. Biopharm. 85 (3) (2013) 1056–1063.
[29] J. Muller, K. Knop, J. Thies, C. Uerpmann, P. Kleinebudde, Feasibility of Raman spectroscopy as PAT tool in active coating, Drug Dev. Ind. Pharm. 36 (2) (2010) 234–243.
[30] K. Knop, P. Kleinebudde, PAT-tools for process control in pharmaceutical film coating applications, Int. J. Pharm. 457 (2) (2013) 527–536.
[31] J. Glassey, K. Gernaey, C. Clemens, T. Schulz, R. Oliveira, G. Striedner, C. Mandenius, Process analytical technology (PAT) for biopharmaceuticals, Biotechnol. J. 6 (2011) 369–377.
[32] S. Challa, R. Potumarthi, Chemometrics-based process analytical technology (PAT) tools: applications and adaptation in pharmaceutical and biopharmaceutical industries, Appl. Biochem. Biotechnol. 169 (2013) 66–76.
[33] G. Lévêque, J.Y. Ferrandis, J.V. Est, B. Cros, An acoustic sensor for simultaneous density and viscosity measurements in liquids, Rev. Sci. Instrum. 71 (2000) 1433–1440.
[34] A. Marabi, G. Mayor, A. Raemy, I. Bauwens, J. Claude, A.S. Burbidge, R. Wallach, I.S. Saguy, Solution calorimetry: a novel perspective into the dissolution process of food powders, Food Res. Int. 40 (2007) 1286–1298.
[35] C.C. Fagan, C. Everard, C.P. O'Donnell, G. Downey, D.J. O'Callaghan, Prediction of inorganic salt and moisture content of process cheese using dielectric spectroscopy, Int. J. Food Prop. 8 (2005) 543–557.
[36] W. Li, L. Xing, L. Fang, J. Wang, H. Qu, Application of near infrared spectroscopy for rapid analysis of intermediates of tanreqing injection, J. Pharm. Biomed. Anal. 53 (2010) 350–358.
[37] P. Bourget, A. Amin, A. Moriceau, B. Cassard, F. Vidal, R. Clement, La Spectroscopie Raman (SR): un nouveloutiladapté au contrôle de qualitéanalytique des préparations injectables en milieu de soins. Comparaison de la SR aux techniques CLHP et

UV/visible-IRTF appliquée à la classe des anthracyclinesencancérologie, Pathol. Biol. 60 (2012) 369–379.

[38] P. Bourget, A. Amin, F. Vidal, C. Merlette, F. Lagarce, Comparison of Raman spectroscopy vs. high performance liquid chromatography for quality control of complex therapeutic objects: model of elastomeric portable pumps filled with a fluorouracil solution, J. Pharm. Biomed. Anal. 91 (2014) 176–184.

[39] A. Amin, P. Bourget, F. Vidal, F. Ader, Routine application of Raman spectroscopy in the quality control of hospital compounded ganciclovir, Int. J. Pharm. 474 (2014) 193–201.

[40] A. Amin, P. Bourget, F. Ader, F. Vidal, C. Neuzillet, A. Baillet- Guffroy, Contribution and limits of a non- intrusive Raman spectroscopic method compared with HPLC for routine application to pre- delivery analytical control of two major camptothecin analogs: irinotecan and topotecan, J. Raman Spectrosc. 46 (2015) 1283–1290.

[41] F. Apruzzese, J. Pato, S.T. Balke, L.L. Diosady, In-line measurement of residence time distribution in a co-rotating twin-screw extruder, Food Res. Int. 36 (5) (2003) 461–467.

[42] A.L. Kelly, S.A. Halsey, R.A. Bottom, S. Korde, T. Gough, A. Paradkar, A novel transflectance near infrared spectroscopy technique for monitoring hot melt extrusion, Int. J. Pharm. 496 (1) (2015) 117–123.

[43] L. Saerens, L. Dierickx, T. Quinten, P. Adriaensens, R. Carleer, C. Vervaet, et al., In-line NIR spectroscopy for the understanding of polymer-drug interaction during pharmaceutical hot-melt extrusion, Eur. J. Pharm. Biopharm. 81 (1) (2012) 230–237.

[44] T. De Beer, A. Burggraeve, M. Fonteyne, L. Saerens, J.P. Remon, C. Vervaet, Near infrared and Raman spectroscopy for the inprocess monitoring of pharmaceutical production processes, Int. J. Pharm. 417 (1–2) (2011) 32–47.

[45] P.D. Coates, S.E. Barnes, M.G. Sibley, E.C. Brown, H.G.M. Edwards, I.J. Scowen, In-process vibrational spectroscopy and ultrasound measurements in polymer melt extrusion, Polymer 44 (19) (2003) 5937–5949.

[46] A.L. Kelly, T. Gough, R.S. Dhumal, S.A. Halsey, A. Paradkar, Monitoring ibuprofen-nicotinamide cocrystal formation during solvent free continuous cocrystallization (SFCC) using near infrared spectroscopy as a PAT tool, Int. J. Pharm. 426 (1–2) (2012) 15–20.

[47] D. Fischer, K. Sahre, M. Abdelrhim, B. Voit, V.B. Sadhu, J. Pionteck, et al., Process monitoring of polymers by in-line ATR-IR, NIR and Raman spectroscopy and ultrasonic measurements, C. R. Chim. 9 (11 – 12) (2006) 1419–1424.

[48] F.T. Silva Ana, A. Burggraeve, Q. Denon, P.V. Meeren, N. Sandler, T.V.D. Kerkhof, M. Hellings, C. Vervaet, J.P. Remon, J.A. Lopes, T. De Beer, Particle sizing measurements in pharmaceutical applications: comparison of in-process methods versus off-line methods, Eur. J. Pharm. Biopharm. 85 (2013) 1006–1018.

[49] T.L. Harvill, J.H. Hoog, D.J. Holve, In-process particle size distribution measurements and control, Part. Part. Syst. Charact. 12 (1995) 309–313.

[50] L.W. Chan, L.H. Tan, P.W.S. Heng, Process analytical technology: application to particle sizing in spray drying, AAPS Pharm. Sci. Tech. 9 (2008) 259–266.

[51] R.N. Venkateshwar, P. Schirg, X. Gao, A. Virdis, G. Imanidis, M. Kuentza, Different modes of dynamic image analysis in monitoring of pharmaceutical dry milling process, Int. J. Pharm. 391 (2010) 107–114.

[52] Y. Weili, B.C. Hancock, Evaluation of dynamic image analysis for characterizing pharmaceutical excipient particles, Int. J. Pharm. 361 (2008) 150–157.

[53] V. Kumar, M.K. Taylor, A. Mehrotra, W.C. Stagner, Real-time particle size analysis using focused beam reflectance measurement as a process analytical technology tool for a continuous granulation–drying–milling process, AAPS Pharm. Sci. Tech. 14 (2013) 523–530.

[54] D. Petrak, S. Dietrich, G. Eckardt, M. Köhler, In-line particle sizing for real-time process control by fibre-optical spatial filtering technique (SFT), Adv. Powder Technol. 22 (2011) 203–208.
[55] H. Wu, A.S. Hussain, In use of PAT for active pharmaceutical ingredient crystallization process control, in: P. Horacek, M. Simandl, P. Zitek (Eds.), Conference Proceedings of the 16th International Federation of Automatic Control (IFAC) World Congress, Prague, Czech Republic, 2005.
[56] H. Wu, M. White, M.A. Khan, Quality-by-design (QbD): an integrated process analytical technology (PAT) approach for a dynamic pharmaceutical co-precipitation process characterization and process design space development, Int. J. Pharm. 405 (2011) 63.
[57] H. Wu, M. White, R. Berendt, R.D. Foringer, M. Khan, Integrated process analytical technology approach for nucleation induction time measurement and nucleation mechanism assessment for a dynamic multicomponent pharmaceutical antisolvent crystallization system, Ind. Eng. Chem. Res. 53 (2014) 1688.
[58] K. Iwata, T. Kojima, Y. Ikeda, Solid form selection of highly solvating TAK-441 exhibiting solvate-trapping polymorphism, Cryst. Growth Des. 14 (2014) 3335.
[59] G.G. Zhang, D. Law, E.A. Schmitt, Y. Qiu, Phase transformation considerations during process development and manufacture of solid oral dosage forms, Adv. Drug Deliv. Rev. 56 (2004) 371.
[60] F. Neville, A. Seyfaee, Real-time monitoring of in situ polyethyleneimine-silica particle formation, Langmuir 29 (2013) 14681.
[61] M. Tiwari, G. Chawla, A.K. Bansal, Quantification of olanzapine polymorphs using powder X-ray diffraction technique, J. Pharm. Biomed. Anal. 43 (2007) 865.
[62] A. Siddiqui, Z. Rahman, S. Bykadi, M.A. Khan, Chemometric methods for the quantification of crystalline tacrolimus in solid dispersion by powder X-ray diffractrometry, J. Pharm. Sci. 103 (2014) 2819–2829.
[63] P.S. Rao, U.K. Ray, S. Hiriyanna, S.V. Rao, H.K. Sharma, V.K. Handa, K. Mukkanti, Identification of oxidative degradation impurities of olanzapine drug substance as well as drug product, J. Pharm. Biomed. Anal. 56 (2011) 413.
[64] A.Y. Lee, D. Erdemir, A.S. Myerson, Crystal polymorphism in chemical process development, Annu. Rev. Chem. Biomol. Eng. 2 (2011) 259.
[65] Q. Zang, D.A. Keire, L.F. Buhse, R.D. Wood, D.P. Mital, S. Haque, S. Srinivasan, C.M. Moore, M. Nasr, A. Al-Hakim, Identification of heparin samples that contain impurities or contaminants by chemometric pattern recognition analysis of proton NMR spectral data, Anal. Bioanal. Chem. 401 (2011) 939.
[66] S. Beg, M.S. Hasnain, M. Rahman, S. Swain, Introduction to quality by design (QbD): fundamentals, principles, and applications, in: Pharmaceutical Quality by Design, Academic Press, 2019, pp. 1–17.
[67] S. Das, A. Singh, S. Kar, R. Ghosh, M. Pal, M. Fatima, N. Singh, S. Singh, Application of QbD framework for development of self-emulsifying drug delivery systems, in: Pharmaceutical Quality by Design, Academic Press, 2019, pp. 297–350.
[68] M.S. Hasnain, S. Rao, M.K. Singh, N. Vig, A. Gupta, A. Ansari, P. Sen, P. Joshi, S.A. Ansari, Development and validation of LC-MS/MS method for the quantitation of lenalidomide in human plasma using Box-Behnken experimental design, Analyst 138 (5) (2013) 1581–1588.
[69] M.S. Hasnain, M.N. Javed, M.S. Alam, P. Rishishwar, S. Rishishwar, S. Ali, A.K. Nayak, S. Beg, Purple heart plant leaves extract-mediated silver nanoparticle synthesis: optimization by Box-Behnken design, Mater. Sci. Eng. C 99 (2019) 1105–1114.
[70] T. Waghule, V.K. Rapalli, G. Singhvi, P. Manchanda, N. Hans, S.K. Dubey, M.S. Hasnain, A.K. Nayak, Voriconazole loaded nanostructured lipid carriers based topical delivery system: QbD based designing, characterization, in-vitro and ex-vivo evaluation, J. Drug Deliv. Sci. Technol. 52 (2019) 303–315.

[71] P.N. Bhatt, M.S. Hasnain, A.K. Nayak, B.F. Hassan, S. Beg, Development and validation of QbD-driven bioanalytical LC-MS/MS method for the quantification of paracetamol and diclofenac in human plasma, Anal. Chem. Lett. 8 (2018) 677–691.
[72] A.K. Nayak, D. Pal, J. Pradhan, M.S. Hasnain, Fenugreek seed mucilage-alginate mucoadhesive beads of metformin HCl: design, optimization and evaluation, Int. J. Biol. Macromol. 54 (2013) 144–154.
[73] S. Beg, K. Kohli, S. Swain, M.S. Hasnain, Development and validation of RP-HPLC method for quantitation of amoxicillin trihydrate in bulk and pharmaceutical formulations using box-behnken experimental design, J. Liq. Chromatogr. Relat. Technol. 35 (3) (2012) 393–406.
[74] A.K. Nayak, S. Khatua, M.S. Hasnain, K.K. Sen, Development of diclofenac sodium-loaded alginate-PVP K 30 microbeads using central composite design, Daru 19 (5) (2011) 356–366.
[75] S. Beg, S. Swain, M. Rahman, M.S. Hasnain, S.S. Imam, Application of design of experiments (DoE) in pharmaceutical product and process optimization, in: Pharmaceutical Quality by Design, Academic Press, 2019, pp. 43–64.
[76] M.N. Javed, S.U. Alam, A. Waziri, F.H. Pottoo, A.K. Yadav, M.S. Hasnain, F.A. Almalki, QbD applications for the development of nanopharmaceutical products, in: Pharmaceutical Quality by Design, Academic Press, 2019, pp. 229–253.
[77] M.S. Hasnain, S.A. Ahmed, S. Beg, M.T. Ansari, A.K. Nayak, "Quality by design" approach for development of multiparticulate drug delivery systems, in: Pharmaceutical Quality by Design, Academic Press, 2019, pp. 351–365.
[78] A.K. Nayak, S.A. Ahmed, S. Beg, M. Tabish, M.S. Hasnain, Application of quality by design for the development of biopharmaceuticals, in: Pharmaceutical Quality by Design, Academic Press, 2019, pp. 399–411.
[79] S. Beg, S. Swain, M. Rahman, M.S. Hasnain, S.S. Imam, Application of design of experiments (DoE), Pharmaceutical Product and Process Optimization, Pharmaceutical Quality by Design, Academic Press, USA, 2019, pp. 43–64.

CHAPTER 9

3D cell culture for pharmaceutical application

Shalini Dasgupta and Ananya Barui
Centre for Healthcare Science and Technology, Indian Institute of Engineering Science and Technology, Shibpur, Howrah, India

1 Introduction

The state of well-being for any human suffering from any disease or deficiency depends on the pharmaceutical industry. Whether a drug is useful to cure the disease conditions, trials of drugs on *in vitro* models is expected to be the norm. The drug development process is extremely lengthy and failure of drugs at preclinical process involves huge cost. The primary reason behind drug failure predominantly depends on inappropriate *in vitro* experimental design which fails to provide adequate information regarding the drug safety [1]. In this regard, development of suitable *in vitro* model is utmost important in the pharmaceutical industry for understanding the drug efficacy, toxicity or tolerance. Cell culture basically provides a similar representation of the drug effect within the human body in a very simplistic way. It investigates the physiological and pathological conditions outside the human body. Pharmaceutical technology has improved by leaps and bounds ever since the first antibiotic was discovered. However, its progress is still dependent upon *in vitro* drug trials on different kinds of cell cultures. Unless the *in vitro* assays are effective, permission will not be obtained for further stages of human trials. While 2D cell cultures are still widely used all over the world, 3D culture is gaining fast popularity [2].

The traditional 2D culture system consists of a single layer of cells growing on the flat surface. Till date the 2D culture is important for the first line of drug screening process. However, such systems are unable to mimic the multi-cellular complexity of human body. The over simplified model of 2D culture platform is not suitable for developing personalized medicine by considering the uniqueness and variation of the population [3]. To overcome such limitation, researchers have improved the culture technique and developed 3D culture system which can mimic the complex physiology of human body. Moreover, considering the rising costs and ethical concerns of

Advances and Challenges in Pharmaceutical Technology
https://doi.org/10.1016/B978-0-12-820043-8.00003-7

in vivo trial methods, *in vitro* simulations are still popular platforms for drug trials. It not only reduces the cost of healthcare industry but also diminishes the chances of fatalities and errors before the trials can reach a critical stage. While the entire replacement of human trials may still be a long way to go, however, it can be expedited to the nearest point with the help of three dimensional (3D) cell culture systems. Not only will they open up new vistas for organ regeneration and tissue healing, but they will also have a lucrative foot hold in understanding and scanning for better anti-cancerous drugs and other such diseases.

The 3D cell culture systems would also improve our understanding of drug delivery and efficiency and thus help to overcome a major hurdle that has often been called the death valley of any pharmaceutical trial. Usually when a potential lead is tested and singled out for further causes; it is done so because of its overall characteristics and its efficacy in eliminating the symptoms of the disease in 2D culture systems. However, it has been a paradox to simply ponder that a single sheet of one kind of cells is considered enough to check how a certain bioactive compound will act within the premises of a multi-organ system with its dimensions and features changed.

Thus, preparing a better assay system for *in-vitro* studies becomes an urgent need. With rising environmental concerns and animal and human safety standards becoming stricter by the day, a simple monolayer culture can no longer be considered as the standard for exploring a drug's effectiveness in comparison to a real human body. Only a 3D culture platform, which can to some extend replicate the delicate and quite complex systems to the smallest molecular level, without causing harm to any person, can fulfill the preliminary aim of pharmaceutical applications [4].

2 Comparison of 2D and 3D cultures

The previous decades have seen critical endeavors toward the advancement and ascent of complex cell cultures as frameworks that better simulate the functioning of the human organ types and tissue systems [5]. Now, 3D cell culture networks are developing, not just as another instrument in early medication disclosure, yet in addition as potential therapeutics fit for treating any ailment [6]. The basic model of 2D and 3D culture systems is illustrated in Fig. 1A.

Various *in vitro* test techniques have been created dependent on human cell and tissue cultures. Most of the assays depend with respect to the development of the tissue as standardized 2D single cell-type culture, which are

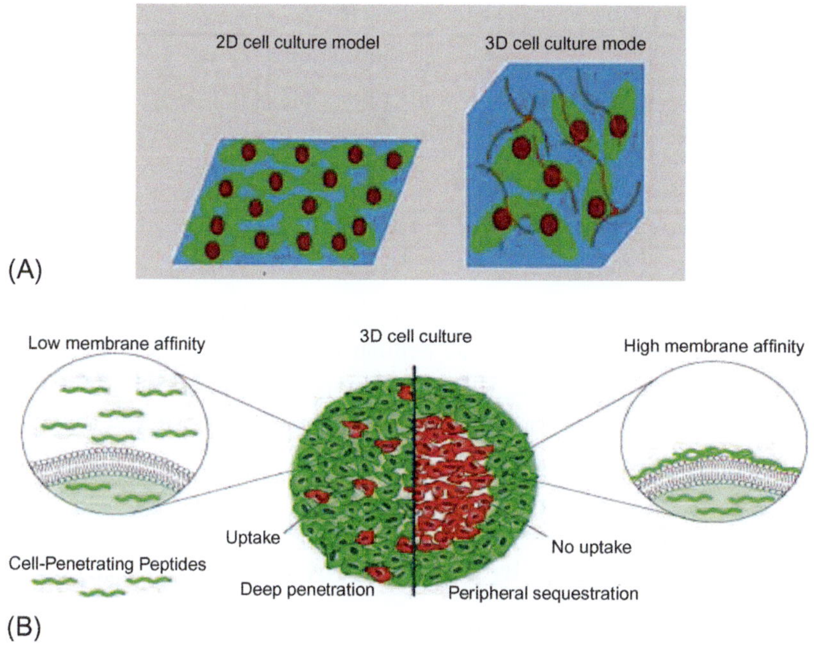

Fig. 1 (A) Comparison of 2D *vs* 3D cell culture model [7] (B) Structure of a spheroid with uniform cell growth around a focal point [8].

quite easy to use and take into account basic and productive refined molecular processes [9]. 2D cultures fill in as the principle workhorse for fundamental cell study and are being utilized for diagnosis of drug action (pharcodynamics and pharmacokinetics) and toxicity *in vitro* [10]. For tumor screenings, cytotoxicity assays are commonly conducted on built up cell cultures developed in monolayer cultures since they grow and develop in an exponential rate [11].

However, they may always not be an accurate reflection of efficacy when the drug proceeds for human trials. 2D cultures have only the particular cells of the target arranged in monolayer form and thus are unable to take into account various external factors like the sudden change in the interaction of molecules in cellular and extracellular environments, different morphologies, differences in polarity, and randomized cell division [12].

The major limitations of 2D culture systems are:

(a) Such systems are unable to represent the real environment of cells present in human body.

(b) 2D platforms cannot predict the effects the success of drug trial and cost of the processes.

Table 1 Comparison between 2D and 3D cell culture.

2D cell culture	3D cell culture
Cheaper and easier to handle	More detailed and qualitative
Easier to obtain	Unique and mimics conditions of the body
It has lot of comprehensive work	Unique field
Easier to replicate	Requires specialized training
Easier to measure cell statistics	More range of tests can be conducted

(c) Accumulation of cellular waste products in culture media may induce toxicity and affect the overall cellular health.

Due to such drawbacks, the 3D culture system is gaining popularity especially in the pharmaceutical industries. The comparison between 2D and 3D culture system is represented in Table 1.

3 3D culture system in pharmaceutical applications

In last few decades the pharmaceutical industry has witnessed tremendous progress [13]. According to recent market research, it has been assessed that it might take almost 3 billion dollars to get another medication from the point of conception to the market [14]. Such a high cost is certainly not feasible for many aspiring pharmaceutical companies, especially when there are high rates of failure involved as well. Only 1 in 5000 drugs actually makes it to the market. And it takes 12 years of extensive research to transform an idea into a feasible medicine for treatment to the market [15]. This is because results obtained in 2D culture do not show proper efficacy in human trials. It is possible to avoid a high failure rate by altering and improving test plan and practices, and upgrading from the simplistic 2D culture to 3D culture [16]. The 2D culture assays are not enough to test the complications of drug delivery, nutrient movement and also overall microenvironment and adhesion factors which are starkly different in a 3D setting [17]. This occurs mostly due to an enormous distance between the theories prepared for the lab as compared to actual presence of the drugs in the organs [18]. While we simple see a cause and effect reaction to any changes we make in a cell culture experiment, the human body is much more complicated than that. Basic 2D cell cultures have not had the option to give all the precise conditions and factors that impact the viability of a specific medication in complex tissue [19]. In such a circumstance, a proficient 3D culture mimics a tissue framework with smaller scale vessels and cell development and in this way can give a closer and progressively better comparative aftereffect of the impact of the medication contrasted with that of the genuine human body [20].

To streamline this circumstance, specialists depend on a wide scope of methods and model frameworks that copy significant highlights of human physiology. Every one of these methodologies can be used to find some kind of harmony between ease of use, cost, and similarity to *in vivo* conditions. Moreover, there is an enormous interest for quicker improvement in existing medication because of dire situations worldwide, for example, developing pandemic infections and the expansion in immunology [21]. Malignant growth is another significant medical problem, especially in the United States and other Western nations [22]. As malignant growth turns out to be increasingly normal and widespread, the interest for new anticancer medications increases [23]. In the course of recent decades, huge advancements in oncology and immunology have been made. This is mostly because of the way that preclinical examinations need proficiency with regards to distinguishing molecules released in the pathways that can change the result of malignant growth advancement and movement [24].

Recent developments in biological sciences, scaffold fabrication systems, and vascularization have empowered the improvement of a large scope of 3D tissue innovations [25]. There are many techniques involved in the preparation of 3D scaffolds, depending on the type of tissues and organs required. These incorporate the following techniques:

1. Spheroids [26]—When a group of cells aggregate together to grow all around each other forming a spherical core with proper nutrient absorption and waste removal system.
2. Organoids [27]—When different kinds of cells aggregate together like spheroids, but instead form a miniature organ unit, they are called organoids.
3. Platforms [2]—When cells are grown on biocompatible firms or 3-dimensional structure which are incorporated in the body over a particular area, they are known as 3D platforms
4. Hydrogels [28,29]—A composite material consisting of usage of a set of polymer chains which can absorb water and take up ions as a scaffold.
5. 3D bioprinting [30]—The process of printing different kinds of structures or scaffolds for usage in the implementation of tissue engineering, pharmaceutical technology and disease research.

These 3D culture techniques, albeit distinctive on a fundamental level and conventions, are utilized to re-establish the morphological, practical, and micro-anatomy of various tissues and parts of the body [7].

3.1 Spheroids

Multi-cellular spheroid cultures were at first mentioned by Sutherland and his collaborators in 1970 to re-create the environmental reactions of the tissues and organs to radiotherapy [31]. From that point forward, spheroid cultures have been used to prepare numerous different kinds of tissues, including undifferentiated organs [32]. Besides, carcinogenic spheroid cultures with different kinds of cells adjusted to the study of various diseases and as of late to check the progression of cancer cells. The spheroid model makes up for a large number of the inadequacies observed in single layer culture [33]. For example, spheroids can create slopes of oxygen demand, supplements, metabolites, and solvent movement, along these lines making mixed co-cultures. Depending upon the type of nutrient penetration, the structure of the spheroid can vary. The structure of simple spheroid model is depicted in Fig. 1B.

Moreover, spheroids have a definite shape and structure with specific adhesion toward other cells. There are some difficulties related with spheroid culture, including the improvement and preparation of similar spheroids, the development of the dimensional scaffold a little seed number of cells, the exact control of explicit proportions of various cells used in spheroid bead culture, and the absence of solid, straightforward, institutionalized, and good output perfect tests for utilizing such techniques [8,34]. The main technique can be to just seed the microcarrier spheroids onto a well and coat them with the cells [35]. Be that as it may, legitimate uniform covering may not happen.

The subsequent methodology is to utilize a technique known as hanging drop plates (HDPs) to advance the development of multi-cellular spheroids [36]. At the point when the seeded cells are apportioned into the highest point of a HDP well, cells are isolated on the surface of the bead framed beneath the gap of the HDP well base opening, in the long run, helping in formation of the shape of spheroids. As the plates have low intensity, the HDP can likewise be utilized for spheroid co-culture, wherein various cell types are included either at the hour of beginning administering or successively. Nonetheless, an unmistakable admonition of this methodology is that spheroids are required to move from the item to a secondary area for tests.

Another common way is to utilize a known bioreactor which can be used to contain cells to arrange it into such bulbous cultures [37]. This methodology allows enormous scale creation of spheroids. Bioreactor has impediments related with fluidic stream actuated shear stress, just as non-consistency in size of spheroids delivered. Here the various processes of spheroid culture in hanging drop method.

One more technique is to utilize smaller scale surfaces as the platforms to handle the cell adhesion and movement, along these lines empowering spheroid societies. This methodology offers a wide scope of nano-scale platforms engraved onto a level substrate for the determination of suitable examples and cement properties for an assortment of cell types. Like low-attachment plates, these micro-designed plates have minimal well design and plate design variety, which makes them agreeable with HTS [38]. One important provision of air pockets that may effectively frame and affect during its growth is a concern that needs to be monitored. The pipetting in this case without care can destroy the delicate pattern of the surface of these cultures.

3.2 Organoids

The definition of an organoid is "an assortment of organ-explicit cell types that creates from immature microorganisms or organ forebears and self-composes through cell arranging and spatially confined ancestry responsibility in a way like *in vivo* [39]." The organoids are ordered into tissue and undifferentiated organs, contingent upon how the organ buds are shaped. Tissue structure organoids allude to stromal cell-free culture and generally apply to epithelial cells in view of their natural capacity to self-compose into tissue-like structures. These organoids can be used to check for drug delivery and deep penetration issues that cannot be checked in 2D culture. Undifferentiated cell organoids are produced from either undeveloped immature microorganisms or actuated pluripotent foundational microorganisms or essential immature microorganisms, for example, neonatal tissue undifferentiated organisms or tissue-occupant grown-up undifferentiated organisms [40]. For now, a few lab prepared organoid cultures have been set up to take after different tissues, including practical organ-like structures which can be used for vascularized cardiovascular fix, cerebral cortex, thymus, kidney, lung, and retina [41]. The image of Fig. 2 represents the different types of organs that can be regenerated using spheroids.

Various methodologies have been utilized to get organoids. The primary methodology is to straightforwardly prepare the cells as a single layer on another layer of stem cells or an ECM-covered surface scaffold, so the organoids are shaped after the cells separately [42]. The subsequent methodology is to utilize a precisely bolstered culture to permit the further separation of essential tissues. For instance, the epidermal cells can additionally separate and self-collect into a stratified version of the issue when the prepared

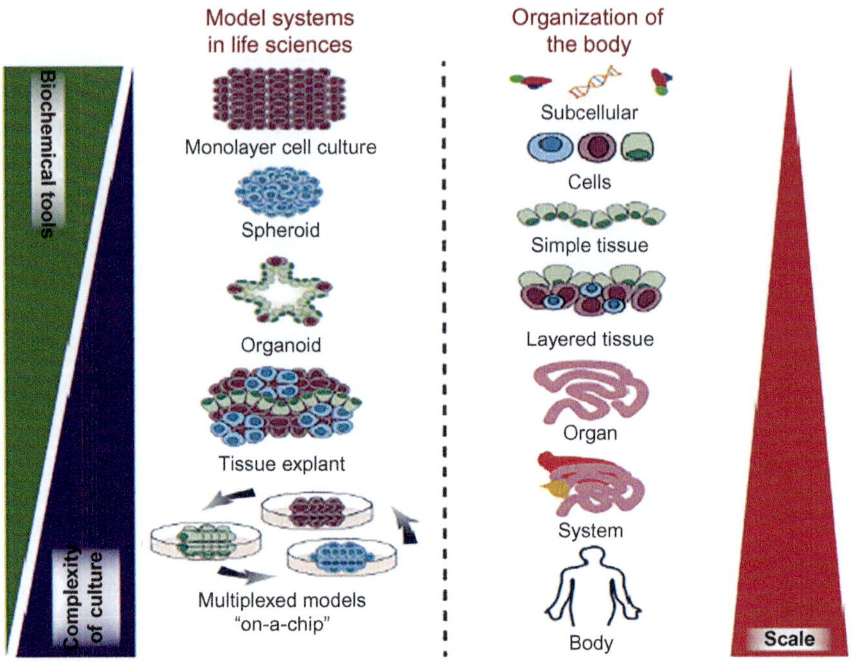

Fig. 2 Possibility of different organs regenerating *via* organoids [42].

culture has been in kept with the air-fluid interface over a certain stretch of time (35 weeks) [43]. The third approach is to create spheroid using embrogenic cells on some of the low-bond plates or through hanging drop culture, like spheroid societies.

4 ECM like scaffold and hydrogels

Frameworks allude to manufactured multi-dimensional scaffolds made of a huge assortment of biocompatible items with various pore sizes, penetrability, upper layer chemistry, and mechanical attributes intended to impersonate the microanatomy of outer tissues [44]. They might be made of different natural examples like hyaluronic acid [45], chitosan [46], collagen [47], polycaprolactone [48] or even out of composite material reasonable for an organic domain, for example, poly-urethane [49]. These scaffolds and hydrogels have ca lot of use in pharmaceutical research, as they can be used to store or have a coating of different kinds of drugs, which can be released upon application within the human body. The presence of scaffolds can also be used to test structural viability of the proteins that are given as supplements. Hydrogels can be made of natural and polymeric frameworks combined with cells and can be injected directly into the affected organ. Organic platforms generally utilize normally determined ECM, for example, usage of collagen to advance the cell connection and rearrangement into multi-dimensional scaffolds [50].

Contrasted with engineered platforms, Matrigel can give an all the more physiologically pertinent microenvironment of solvent development variables, hormones, and different particles with which cells associate in an *in vivo* condition [51]. This enables to develop an environment, fundamentally similar to the human body. However, matrigels have some fundamental disadvantages as well. The impediments related with Matrigels usually occur due to fluctuation during assembling of the matrigel and the comprehensive intricacy in structure [52]. Often every structure has some minor differences which are regularly not well characterized, making it hard to decide precisely which components are advancing cell work. Other common gels, for example, fibrin, hyaluronate, chitosan, alginate [53], or collagen fibrils have likewise been utilized for 3D cell culture; nonetheless, these regular gels have less flexibility to advance the preparation of multi-dimensional scaffolds rather than matrigel. [54].

Polymeric platforms utilize engineered scaffolds or hydrogels or similar compounds to produce the backbone structure for 3D scaffolds [55]. The gel-like scaffolds utilized for 3Dculture incorporate poly(ethylene glycol) (PEG) [56], poly(vinyl alcohol) [57], and poly(2-hydroxy ethyl methacrylate) [58]. Moreover, hydrogel material can be used to be biodegradable by fusing PLA [59] units or by using cleavable enzyme peptide groupings into the polymer backbone structure. The biocompatibility is requisite to uses in which the requirement of cells is an absolute necessity, for example, tissue building and regenerative drugs [60]. Manufactured frameworks have a few clear preferences over Matrigel or other regular gels for 3D societies.

To begin with, the utilization of engineered materials can limit the generally poor reproducibility of natural ECMs among groups and the subsequent absence of consistency between societies, as they are frequently just prepared and produced. Second, these platforms take into account the presence of various mechanical and protein structures, so it is conceivable to improve both mechanical and substance signs for 3D cell societies. Third, these hydrogels have high water content, empowering transport of oxygen, supplements, and solvent factors, which are all critical to cell capacities. However, these hydrogels have certain issues in proper gaseous distribution, heterogeneities present in the engineered cell microenvironment, and lopsided circulation of solvent development factors inside the framework and entanglement in imaging and cell investigation [61]. This leads to the creation of imperfect models.

There are numerous different kinds of frameworks accessible, which have been created to look like the working of the human body. The interest for frameworks that enough copy the *in vivo* condition of cells prompted the improvement of "body-on-a-chip" frameworks [62]. Their particular similarity with utilitarian tissue is accomplished by imbedding the most recent age small scale fluidics so as to upgrade the perfusion of the microenvironment [63]. The motivation behind the framework is to fill in as a brief supporting structure permitting not just 3D support but also allow formation of vessels for exchange of gases and nutrition.

4.1 Arrangement and qualities of a framework for a scaffold

The 3D scaffold platform is essential to fill in as a transitory complacent and supportive frame permitting the 3D growth of tissue development as well as the arrangement. Also, the additional support is to give the nutrition expected for cell development, separation and tissue development [64]. The fundamental necessities to accomplish utilitarian tissue are the right kind of cells, an appropriate environment and legitimate frameworks. The perfect framework ought to have the option to manage cell bond, multiplication, actuation, and separation. For example, mesenchymal stem cells (MSCs) were appeared to indicate ancestry and focus on phenotypes with extraordinary susceptibility to the difference in stress or elasticity of the scaffold. The importance of the structure of the framework has major importance in the field of pharmaceutical applications. Depending upon the pore size, the spread of cells around the structure, and the formation of the vessels, the medication capacity and delivery technique may vary. The drug delivery technique can be perfected and adapted based on the type of scaffolds or tissues that are representing the organ systems of the body. A 2D culture has only few methods of direct drug delivery, which thus cannot help to understand the various challenges that drugs face while being administered to the human body. Fig. 3A shows show cells can be embedded in a scaffold for further usage in tissue engineering and disease research. These scaffolds can be embedded with various drugs and supplements to help in growth and differentiation of the cells and formation of vessels.

The ideal framework embedded with cells and furnished with all other development factors is to make a biological structure conductive to cell

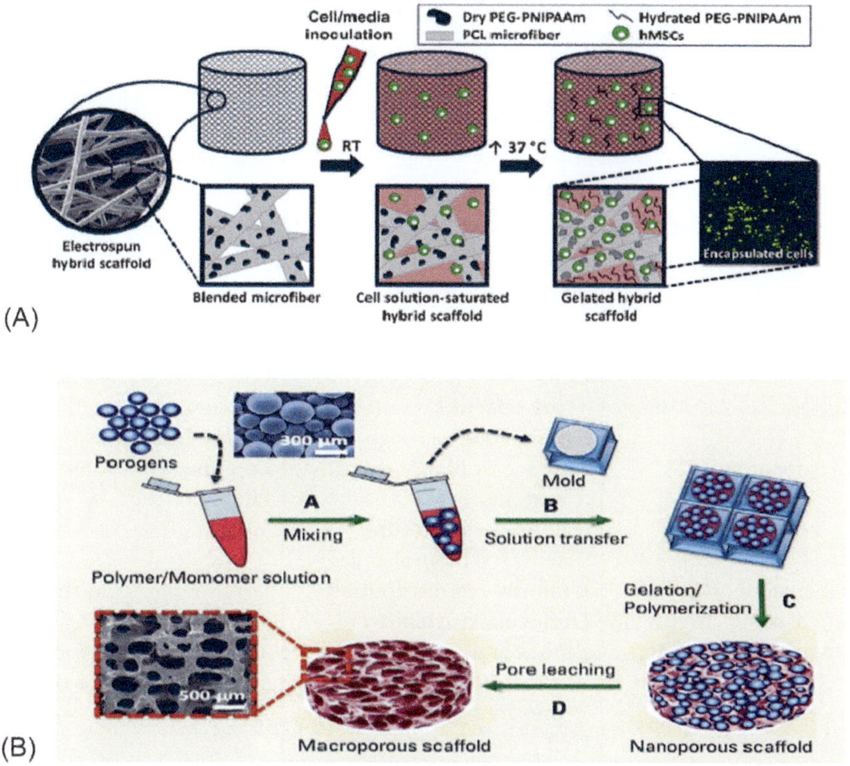

Fig. 3 (A) Example of an electrospun scaffold embedded with cells [65] (B) Preparation of porous scaffold using melt-molding and salt leaching [66].

development and adhesion [67]. The prepared tissue should develop within the encompassing original tissues. Tissue engineering helps to create a similar structure which mimics the biophysical characteristics of the local ECM. Table 2 has an assortment of the different techniques used to prepare scaffolds and platforms suitable for usage in any biological systems.

The sorts of platforms acquired are subject to the manufacturing procedures. It additionally relies on the sort of materials utilized, regardless of whether in single or in compound structure. By and large, particulate draining or dissolvable throwing can be utilized to create permeable frameworks, though electrospinning is helpful for manufacturing strong platforms [77], and various bio-printing could be utilized to deliver platforms with characterized structures (Table 2). These kinds of frameworks have been acknowledged for 3D culture.

Table 2 Platforms and scaffolds can be made utilizing an assortment of techniques.

Technique	Advantages	Disadvantages
3D printing [68]	Very specific technique Different shapes can be prepared Highly controlled and easy to handle	Expensive technique Only specific materials can be used
Salt draining [69]	Cheaper and very effective Simple method of pore formation	Pores may not be evenly spread
Electrospinning [70]	Nanofibres prepared easily Cells can be embedded in the structure Easy to use	Expensive technique Low mechanical resistance Highly brittle
Gas foaming [71]	Absence of highly caustic solvents Can be used to utilize sensitive biomolecules	Low mechanical resistance Highly brittle
Fiber networks [72]	Uniform cell distribution High cellular viability	Low strength and stiffness
Phase division [73]	Porous and interconnected scaffolds prepared Cell growth on different parts	Temperature dependent Very specific materials required
Melt molding [74]	Less amount of polymer required Easy to prepare desired shape	Long process
Emulsion-freeze drying [75]	Simple technique Wide range of pores formed	Volatile salts may be lost from structure
Solution casting [76]	Better uniform thickness No defects	Must be soluble in a particular solvent Minimum solid content and viscosity required

In order to prevent the de-differentiation of the cancer cells, the cell lines were straightforwardly seeded inside the 3D permeable hydroxyapatite clay. Of note, cells developed on scaffold platforms are frequently not thought of as related to 3D culture, as cells commonly follow and prolong as per the presence of the scaffold. Moreover, permeable frameworks still may not be able to provide a uniform rate of dispersion and nutrient supply to all the tissues around it, thus leaving a lot to be desired when it comes to preparing complicated organs like the heart or lungs [78]. But permeable frameworks are necessary and are a better improvement from the previous structures that were designed in the 1970s. Fig. 3B is an adequate representation of the formation of porous scaffolds and how different measures of porosity are obtained by different techniques.

The main requirement of 3D cultures was to emulate the microenvironment of the human body for testing the efficacy and dosage of various kinds of medications. The three-dimensional models are prepared by seeding various layers of the cells on a scaffold or microcarrier beads and then either using them alone or separately [79]. The effectiveness of drug delivery to all cells on the scaffold, the prevention of further growth or increase in growth of tissues will help to understand the pharmacokinetics and pharmacodynamics related to the tissues.

Many other uses of multi-dimensional cell models incorporate the examination of malignant growth, pharmaceutical dynamics, tissue arrangement and organ designing. A ground-breaking advantage of the circular organoid models above the customary monolayer cells is the increased impact of the similarity of the oncogenic tissues as well as their nutrient supply that can impact pharmacological viability. Many circular organoid cultures of different cell types that arrive at sizes of 400–600 μm or all the more regularly experience focal necrosis [80]. This loss of cell viability in the middle of the organoid develops in light of the fact that nutrients and gaseous exchange is constrained in the middle, and the environment becomes more acidic and there is an aggregation of waste. Therefore, an important factor is to ensure that the organoids prepared receive proper transfer of nutrients and gases in order to have thriving cells even in the middle.

Likewise, strong tumors frequently have locales with various expansion rates and areas with gentle to serious oxygen inadequacies, because of the absence of blood supply to developing tumor structures [81]. Along these lines, the circular cell cultures can imitate the effect and physiology of large tumors(without the presence of capillaries) and can be utilized to anticipate the importance and dosage of various forms of cancer therapy and to foresee the fast growing capability of specific cancers.

5 Case study of 3D culture models in pharmaceutical applications

The benefit of going through actual applications for multi-dimensional carcinoma models was featured by the research group [82]. The efficacy of a promising cancer drug can be effectively evaluated on the 3D culture as compared to a monolayer 2D model. Two different anti-cancer drugs, 5-fluorouracil (5-FU) and tirapazamine were used in interacting with various methods of activity that created significant reactions in 3D spheroids as compared to regular monolayer cell cultures of human epithelial carcinoma cells. The prepared spheroids stayed feasible after high dosage 5-FU treatment, while cells developed as simple monolayer culture did not endure the treatment. Since normal cultured cells first multiply at a moderately uniform rate over the plastic substrate, 5-FU can affect the DNA of the cells. Rather, the organoids prepared have a lower growth rate, prompting a decreased effectiveness against the drug 5-FU. Thus, this experiment proved whether the drug was not quite effective for a tumor like structure. This saved the research from enduring the valley of death and allowed the researchers to go back to the drawing board with better modifications.

At less concentration of O_2, tirapazamine releases free radicals which affect the monolayer cells as well as spherical cultures. Moreover, hypoxia occurs at the inside of the 3D spherical cultures slow down the cellular multiplication and making the spheroids less susceptible to anti-neoplastic drugs, for example, cisplatin [81]. Another research had exhibited that the diseased HCT116 cells developed in 3D are in fact not affected to cisplatin than the monolayer culture of same cell line [83]. The target mechanism of the antineoplastic drugs present and mentioned here can be understood more clearly by using the organoid culture. The previous outcomes show that the advancement of such systems can conceivably be improved utilizing spherical cultures.

6 Significance of 3D co-cultures

With an end goal to all the more likely improvement of the *in vivo* model, 3D simultaneous multi-cultures have been created. The co-cultures can be made by either seeding two different kinds of monolayers with each other, or seeding one layer on one type of scaffold and allowing them to interact with another type of cells [84]. Moreover, in such model the vascular architecture can be incorporated. Actually executing a vascularized segment in multidimensional tissues is significant since veins can make oxygen inclinations that can prompt growth of endothelial cells and advance the cell relocation by means of chemotaxis [85,86]. Sometime the drugs trial on 2D model, fails to reflect the same efficacy when trial on 3D models. Such results do not only imply the failure of 2D models but exhibit the pharmacokinetic inconstancy. Such outcomes also reflect the necessity of the 3D co-culture models, to mimic the biological situation as close as possible.

Another run of the mill highlight of numerous tumors is their capacity to experience a carcinogenic phase known as EMT. This system actuates metastatic spread of cancer cells that could make similar expressions of the actual tumor model. In the year of 2012, approximately 8 million individuals passed away due to malignant growth and their metastasis had all the earmarks of being the significant reason for death [87]. Despite the fact that metastatic movement enormously affects the rest of the tissues, the procedure is as yet not completely comprehended. In this manner, it is essential to create better replications of the carcinogenic mechanisms in 3D framework.

7 Preferences of utilizing 3D culture frameworks

During three-dimensional printing, biocompatible compounds seeded with cells and other bioactive composite scaffolds are appropriated in normal designs and combined together to shape multi-dimensional scaffolds that can contain distinctive kind of cells. As such, an increasingly exact portrayal of the *in vivo* circumstance is accomplished contrasted with customary 2D and 3D cell societies. Taking everything into account, various methodologies exist that consider spheroid arrangement.

It is imperative to have the option to reproduce the homeostatic condition of the living life form and for that 3D co-cultures or the "lab on a chip" platform can be utilized which also omit the necessity of animal experiments to some extent. There are a wide range of circumstances, hereditary contrasts and environmental cues (smoking, corpulence, and different illnesses) that would expect you to take a huge gathering of various types of individuals for a uniform outcome, which can be both very costly and present numerous moral difficulties.

Hence a optimum frameworks is required for developing organoids, spheroids or other form of 3D cultures at lower cost. Since the spheroid's structure, form and condition firmly rely upon the refined framework, it is significant to create institutionalized protocols before spheroid frameworks can be executed and standardized in the public pipeline. These conventions will restrict the fluctuation between various trial arrangements and hence will build the equivalence of the outcomes. Top to bottom investigations must be performed to figure out which refined framework is most appropriate for a specific application. To figure out which framework is best for assessing antitumor restorative adequacy, multiple factors must be thought through, including the yield, cost, spheroid size, size dissemination, and shape [88]. At the point when autonomous examinations utilize the equivalent trial arrangement, it turns out to be progressively important to analyze spheroid parameters, for example, volume development, reasonability, endurance, relocation, and patho-physiological status [89]. When an institutionalized 3D framework gets generally acknowledged for remedial viability assurance, it can enhance the pharmaceutical drug optimization, as it provides practical quantitative data that is regularly missed by customary *in vitro* and *in vivo* medication screening models.

8 Challenges in 3D culture frameworks

3D culture frameworks are increasingly costly. Furthermore, they should be kept up more cautiously and carefully. All platforms based refined frameworks, just as the HDP method and using the fluid overlay framework, have one normal downside: the main nutritional medium can become stale which can cause issues identified with supplement nutritional transfer. It is particularly disadvantageous when developing spheroids over longer time frames [90].

For long term development, use of non-dormant culture frameworks, for example, the spinner jar can be a reasonable alternate. Inside the spinner carafe, cells do not have a solid substrate and subsequently they associate with one another and absorb nutrients from the surrounding liquid culture. Spherical cultures can likewise be shaped utilizing a separate commencement dish from an alleged progenitor dish, which is intended for transient development [91]. After commencement, the spheroids are moved into the spinning jar. The spinner flasks existed for more than 30 years however it stays as one of the most proficient platform free frameworks for getting a lot of spheroids inside a specific size range. However there can be no doubt that they are extremely useful and better and more efficient methods are required to prepare them in order to bridge the gap between hypothetical pharmaceutical concepts and actual live saving drugs.

9 Conclusion

Both 2D and 3D culture systems are used in pharmaceutical applications for drug trial and development. Beside widespread application of 2D culture system, 3D methods are now emerging as promising tool to overcome the inherent limitations of 2D system. The 3D system shows great promises as it can maintain the cell-cell and cell-ECM interaction in a way resemble to the *in vivo* condition. However, to establish the 3D system as a replacement of 2D model, more experiments need to be conducted to ensure the reproducibility of the system. Already various types of promising techniques and processes are being used to show how the 3D cultures are fast becoming the norm for serious assays to check the efficacy of the selected drugs. Various drugs can also be produced in a large scale basis without having to utilize animals; instead different kinds of 3D cultures can be used to produce these products. Thanks to 3D culture, human mortality due to drug testing may reduce considerably. Also, the valley of death which costs close to 1 billion dollars loss will be avoided for future pharmaceutical applications. Various different types of drugs are already being processed *via* 3D culture, and the FDA has shown great promise in accepting such assays for consideration before proceeding to clinical trials [92]. A better standardized set of protocols and techniques can help improving this global industry the right way forward and help alleviate human suffering for the greater good.

References

[1] I. Kola, J. Landis, Can the pharmaceutical industry reduce attrition rates?. Nat. Rev. Drug Discov. 3 (8) (2004) 711–715, https://doi.org/10.1038/nrd1470.

[2] Z. Tan, T. Liu, J. Zhong, Y. Yang, W. Tan, Control of cell growth on 3D-printed cell culture platforms for tissue engineering. J. Biomed. Mater. Res. A 105 (12) (2017) 3281–3292, https://doi.org/10.1002/jbm.a.36188.

[3] E.L.S. Fong, T.B. Toh, H. Yu, E.K.-H. Chow, 3D culture as a clinically relevant model for personalized medicine. SLAS Technol. 22 (3) (2017) 245–253, https://doi.org/10.1177/2472630317697251.

[4] Biocompare, Drug Discovery in 3D | Biocompare: The Buyer's Guide for Life Scientists; n.d. https://www.biocompare.com/Editorial-Articles/359035-Drug-Discovery-in-3D/.

[5] J.H. Sung, Y.I. Wang, N.N. Sriram, M. Jackson, C. Long, J.J. Hickman, M.L. Shuler, Recent advances in body-on-a-chip systems. Anal. Chem. (2019), https://doi.org/10.1021/acs.analchem.8b05293. American Chemical Society.

[6] Y. Fang, R.M. Eglen, Three-dimensional cell cultures in drug discovery and development. SLAS Discov. (2017), https://doi.org/10.1177/1087057117696795. Oxford University Press.

[7] S.J. Jitcy, S.T. Malindisa, M. Ntwasa, Two-dimensional (2D) and three-dimensional (3D) cell culturing in drug discovery. in: Cell Culture, IntechOpen, 2019https://doi.org/10.5772/intechopen.81552.

[8] D. van den Brand, C. Veelken, L. Massuger, R. Brock, Penetration in 3D tumor spheroids and explants: adding a further dimension to the structure-activity relationship of cell-penetrating peptides. Biochim. Biophys. Acta Biomembr. 1860 (6) (2018) 1342–1349, https://doi.org/10.1016/j.bbamem.2018.03.007.

[9] N. Gupta, J.R. Liu, B. Patel, D.E. Solomon, B. Vaidya, V. Gupta, Microfluidics-based 3D cell culture models: utility in novel drug discovery and delivery research. Bioeng. Transl. Med. 1 (1) (2016) 63–81, https://doi.org/10.1002/btm2.10013.

[10] D. Anton, H. Burckel, E. Josset, G. Noel, Three-dimensional cell culture: a breakthrough in vivo. Int. J. Mol. Sci. (2015)https://doi.org/10.3390/ijms16035517.

[11] D. Lv, Z. Hu, L. Lu, H. Lu, X. Xiuli, Three-dimensional cell culture: a powerful tool in tumor research and drug discovery. Oncol. Lett. (2017), https://doi.org/10.3892/ol.2017.7134. Spandidos Publications.

[12] R. Edmondson, J.J. Broglie, A.F. Adcock, L. Yang, Three-dimensional cell culture systems and their applications in drug discovery and cell-based biosensors. Assay Drug Dev. Technol. (2014), https://doi.org/10.1089/adt.2014.573. Mary Ann Liebert Inc.

[13] C. Giaccotto, R.E. Santerre, J.A. Vernon, Drug prices and research and development investment behavior in the pharmaceutical industry. J. Law Econ. 48 (1) (2005) 195–214, https://doi.org/10.1086/426882.

[14] J. Avorn, The $2.6 billion pill—methodologic and policy considerations, N. Engl. J. Med. 372 (2015) 1877–1879. Massachussetts Medical Society. https://doi.org/10.1056/NEJMp1500848.

[15] Medicinenet, Drug Approvals—From Invention to Market...12 Years!; n.d. https://www.medicinenet.com/script/main/art.asp?articlekey=9877.

[16] C. Zuppinger, 3D culture for cardiac cells. BBA-Mol. Cell. Res. 1863 (7) (2016) 1873–1881, https://doi.org/10.1016/j.bbamcr.2015.11.036.

[17] Labskin, How to Improve on the 90% Failure Rate in Drug Discovery—Labskin; n.d. https://www.labskin.co.uk/how-to-improve-on-the-90-failure-rate-in-drug-discovery/.

[18] T. Hartung, Food for thought; look back in anger - what clinical studies tell us about preclinical work. ALTEX 30 (3) (2013) 275–291, https://doi.org/10.14573/altex.2013.3.275.

[19] J. Hoarau-Véchot, A. Rafii, C. Touboul, J. Pasquier, Halfway between 2D and animal models: are 3D cultures the ideal tool to study cancer-microenvironment interactions?. Int. J. Mol. Sci. (2018)https://doi.org/10.3390/ijms19010181. MDPI AG.

[20] F.M. Chen, X. Liu, Advancing biomaterials of human origin for tissue engineering. Prog. Polym. Sci. (2016)https://doi.org/10.1016/j.progpolymsci.2015.02.004. Elsevier Ltd.

[21] M.J. Waring, J. Arrowsmith, A.R. Leach, P.D. Leeson, S. Mandrell, R.M. Owen, G. Pairaudeau, et al., An analysis of the attrition of drug candidates from four major pharmaceutical companies. Nat. Rev. Drug Discov. (2015)https://doi.org/10.1038/nrd4609. Nature Publishing Group.

[22] D. Cunningham, Z. You, In vitro and in vivo model systems used in prostate cancer research. J. Biol. Methods 2 (1) (2015) 17, https://doi.org/10.14440/jbm.2015.63.

[23] L. Falzone, S. Salomone, M. Libra, Evolution of cancer pharmacological treatments at the turn of the third millennium. Front. Pharmacol. (2018)https://doi.org/10.3389/fphar.2018.01300. Frontiers Media S.A.

[24] R. Sever, J.S. Brugge, Signal transduction in cancer. Cold Spring Harb. Perspect. Med. 5 (4) (2015)https://doi.org/10.1101/cshperspect.a006098.

[25] A. Gregor, E. Filová, M. Novák, J. Kronek, H. Chlup, M. Buzgo, V. Blahnová, et al., Designing of PLA scaffolds for bone tissue replacement fabricated by ordinary commercial 3D printer. J. Biol. Eng. 11 (1) (2017) 31, https://doi.org/10.1186/s13036-017-0074-3.

[26] E. Fennema, N. Rivron, J. Rouwkema, C. van Blitterswijk, J. De Boer, Spheroid culture as a tool for creating 3D complex tissues. Trends Biotechnol. (2013)https://doi.org/10.1016/j.tibtech.2012.12.003. Elsevier.

[27] A. Abugomaa, M. Elbadawy, M. Yamanaka, et al., Establishment of 2.5D organoid culture model using 3D bladder cancer organoid culture. Sci. Rep. 10 (2020) 9393, https://doi.org/10.1038/s41598-020-66229-w.

[28] S.R. Caliari, J.A. Burdick, A practical guide to hydrogels for cell culture. Nat. Methods (2016), https://doi.org/10.1038/nmeth.3839. Nature Publishing Group.
[29] M. Liu, X. Zeng, C. Ma, et al., Injectable hydrogels for cartilage and bone tissue engineering. Bone Res. 5 (2017) 17014, https://doi.org/10.1038/boneres.2017.14.
[30] I. Matai, G. Kaur, A. Seyedsalehi, A. McClinton, C.T. Laurencin, Progress in 3D bioprinting technology for tissue/organ regenerative engineering. Biomaterials (2020) https://doi.org/10.1016/j.biomaterials.2019.119536. Elsevier Ltd.
[31] R.M. Sutherland, B. Sordat, J. Bamat, H. Gabbert, B. Bourrat, W. Mueller-Klieser, Oxygenation and differentiation in multicellular spheroids of human colon carcinoma, Cancer Res. 46 (10) (1986) 5320–5329.
[32] S.B. Lee, S.H. Han, S. Park, Long-term culture of intestinal organoids, Methods Mol. Biol. 1817 (2018) 123–135. Humana Press Inc. https://doi.org/10.1007/978-1-4939-8600-2_13.
[33] M. Ravi, V. Paramesh, S.R. Kaviya, E. Anuradha, F.D. Paul Solomon, 3D cell culture systems: advantages and applications. J. Cell. Physiol. 230 (1) (2015) 16–26, https://doi.org/10.1002/jcp.24683.
[34] P.S. Sankar, M.F.C. Mat, K. Muniandy, B.L.S. Xiang, P.S. Ling, S.L.L. Hoe, A.S.B. Khoo, N. Mohana-Kumaran, Modeling nasopharyngeal carcinoma in three dimensions (review), Oncol. Lett. (2017) Spandidos Publications, doi:10.3892/ol.2017.5697.
[35] B. Li, X. Wang, W. Yu, W. Gou, X. Yuan, J. Peng, Q. Guo, S. Lu, Past, present, and future of microcarrier-based tissue engineering. J. Orthop. Transl. (2015), https://doi.org/10.1016/j.jot.2015.02.003. Elsevier (Singapore) Pte Ltd.
[36] M. Shri, H. Agrawal, P. Rani, D. Singh, S.K. Onteru, Hanging drop, a best three-dimensional (3D) culture method for primary buffalo and sheep hepatocytes. Sci. Rep. 7 (1) (2017) 1–14, https://doi.org/10.1038/s41598-017-01355-6.
[37] A. Dhaliwal, Three dimensional cell culture : a review. Mater. Methods 2 (2012) 162, https://doi.org/10.13070/mm.en.2.162.
[38] P. Benien, A. Swami, 3D tumor models: history, advances and future perspectives. Future Oncol. (2014), https://doi.org/10.2217/fon.13.274. Future Medicine Ltd.
[39] M. Simian, M.J. Bissell, Organoids: a historical perspective of thinking in three dimensions. J. Cell Biol. 216 (1) (2017) 31–40, https://doi.org/10.1083/jcb.201610056.
[40] J.Z. Tong, P. De Lagausie, V. Furlan, T. Cresteil, O. Bernard, F. Alvarez, Long-term culture of adult rat hepatocyte spheroids. Exp. Cell Res. 200 (2) (1992) 326–332, https://doi.org/10.1016/0014-4827(92)90179-C.
[41] G. Kaushik, M.P. Ponnusamy, S.K. Batra, Concise review: current status of three-dimensional organoids as preclinical models. Stem Cells (2018), https://doi.org/10.1002/stem.2852. Wiley-Blackwell.
[42] X. Yin, B.E. Mead, H. Safaee, R. Langer, J.M. Karp, O. Levy, Engineering stem cell organoids. Cell Stem Cell (2016), https://doi.org/10.1016/j.stem.2015.12.005. Cell Press.
[43] K.I.W. Kane, E.L. Moreno, S. Hachi, M. Walter, J. Jarazo, M.A.P. Oliveira, T. Hankemeier, et al., Automated microfluidic cell culture of stem cell derived dopaminergic neurons. Sci. Rep. 9 (1) (2019) 1–12, https://doi.org/10.1038/s41598-018-34828-3.
[44] Q. Lv, L. Nair, C.T. Laurencin, Fabrication, characterization, and in vitro evaluation of poly(lactic acid glycolic acid)/nano-hydroxyapatite composite microsphere-based scaffolds for bone tissue engineering in rotating bioreactors. J. Biomed. Mater. Res. A 91 (3) (2009) 679–691, https://doi.org/10.1002/jbm.a.32302.
[45] I.M. El-Sherbiny, M.H. Yacoub, Hydrogel scaffolds for tissue engineering: progress and challenges. Glob. Cardiol. Sci. Pract. 2013 (3) (2013) 38, https://doi.org/10.5339/gcsp.2013.38.
[46] S. Ahmed, S. Ikram, Chitosan based scaffolds and their applications in wound healing. Achiev. Life Sci. 10 (1) (2016) 27–37, https://doi.org/10.1016/j.als.2016.04.001.

[47] J. Glowacki, S. Mizuno, Collagen scaffolds for tissue engineering. Biopolymers 89 (5) (2008) 338–344, https://doi.org/10.1002/bip.20871.
[48] C. Choong, J.T. Triffitt, Z.F. Cui, Polycaprolactone scaffolds for bone tissue engineering: effects of a calcium phosphate coating layer on osteogenic cells. Food Bioprod. Process. 82 (2) (2004) 117–125, https://doi.org/10.1205/0960308041614864.
[49] J.L. Ryszkowska, M. Auguścik, A. Sheikh, A.R. Boccaccini, Biodegradable polyurethane composite scaffolds containing bioglass® for bone tissue engineering. Compos. Sci. Technol. 70 (13) (2010) 1894–1908, https://doi.org/10.1016/j.compscitech.2010.05.011.
[50] Q.L. Loh, C. Choong, Three-dimensional scaffolds for tissue engineering applications: role of porosity and pore size. Tissue Eng. Part B Rev. (2013), https://doi.org/10.1089/ten.teb.2012.0437. Mary Ann Liebert, Inc.
[51] A.M. Sitarski, H. Fairfield, C. Falank, M.R. Reagan, 3D tissue engineered in vitro models of cancer in bone. ACS Biomater Sci. Eng. (2018), https://doi.org/10.1021/acsbiomaterials.7b00097. American Chemical Society.
[52] R.D. Kamm, R. Bashir, N. Arora, R.D. Dar, M.U. Gillette, L.G. Griffith, M.L. Kemp, et al., Perspective: the promise of multi-cellular engineered living systems. APL Bioeng. 2 (4) (2018) 040901https://doi.org/10.1063/1.5038337.
[53] D. Sun, J. Lu, Z. Chen, Y. Yu, Y. Li, A novel three-dimensional microfluidic platform for on chip multicellular tumor spheroid formation and culture. Microfluid. Nanofluid. (2014)https://doi.org/10.1007/s10404-014-1373-3.
[54] J.C. Ashworth, J.L. Thompson, J.R. James, C.E. Slater, S. Pijuan-Galitó, K. Lis-Slimak, R.J. Holley, et al., Peptide gels of fully-defined composition and mechanics for probing cell-cell and cell-matrix interactions in vitro. Matrix Biol. 85–86 (2019) 15–33, https://doi.org/10.1016/j.matbio.2019.06.009.
[55] G. Turnbull, J. Clarke, F. Picard, P. Riches, L. Jia, F. Han, B. Li, W. Shu, 3D bioactive composite scaffolds for bone tissue engineering. Bioact. Mater. (2018), https://doi.org/10.1016/j.bioactmat.2017.10.001. KeAi Communications Co.
[56] A. Escudero-Castellanos, B.E. Ocampo-García, M.V. Domínguez-García, J. Flores-Estrada, M.V. Flores-Merino, Hydrogels based on poly(ethylene glycol) as scaffolds for tissue engineering application: biocompatibility assessment and effect of the sterilization process. J. Mater. Sci. Mater. Med. 27 (12) (2016), https://doi.org/10.1007/s10856-016-5793-3.
[57] E. Dattola, E.I. Parrotta, S. Scalise, G. Perozziello, T. Limongi, P. Candeloro, M. L. Coluccio, et al., Development of 3D PVA scaffolds for cardiac tissue engineering and cell screening applications. RSC Adv. 9 (8) (2019) 4246–4257, https://doi.org/10.1039/C8RA08187E.
[58] O. Janoušková, M. Přádný, M. Vetrík, E.C. Krumbholcová, J. Michálek, M. D. Smrčková, Biomimetic modification of dual porosity poly(2-hydroxyethyl methacrylate) hydrogel scaffolds-porosity and stem cell growth evaluation. Biomed. Mater. 14 (5) (2019) 055004, https://doi.org/10.1088/1748-605X/ab2856.
[59] A. Basu, K.R. Kunduru, S. Doppalapudi, A.J. Domb, W. Khan, Poly(lactic acid) based hydrogels. Adv. Drug Deliv. Rev. (2016), https://doi.org/10.1016/j.addr.2016.07.004. Elsevier B.V.
[60] A.S. Mao, D.J. Mooney, Regenerative medicine: current therapies and future directions. Proc. Natl. Acad. Sci. U. S. A. 112 (47) (2015) 14452–14459, https://doi.org/10.1073/pnas.1508520112.
[61] S.H. Park, T.G. Kim, H.C. Kim, D.Y. Yang, T.G. Park, Development of dual scale scaffolds via direct polymer melt deposition and electrospinning for applications in tissue regeneration. Acta Biomater. 4 (5) (2008) 1198–1207, https://doi.org/10.1016/j.actbio.2008.03.019.
[62] B. Zhang, A. Korolj, B.F.L. Lai, M. Radisic, Advances in organ-on-a-chip engineering. Nat. Rev. Mater. (2018), https://doi.org/10.1038/s41578-018-0034-7. Nature Publishing Group.

[63] G. Štumberger, B. Vihar, Freeform perfusable microfluidics embedded in hydrogel matrices. Materials 11 (12) (2018), https://doi.org/10.3390/ma11122529.
[64] K.P. Krafts, Tissue repair: the hidden drama. Organogenesis (2010), https://doi.org/10.4161/org.6.4.12555. Taylor & Francis.
[65] A.R. Brunelle, C.B. Horner, K. Low, G. Ico, J. Nam, Electrospun thermosensitive hydrogel scaffold for enhanced chondrogenesis of human mesenchymal stem cells. Acta Biomater. 66 (2018) 166–176, https://doi.org/10.1016/j.actbio.2017.11.020.
[66] S.A. Bencherif, T.M. Braschler, P. Renaud, Advances in the design of macroporous polymer scaffolds for potential applications in dentistry. J. Periodontal Implant Sci. (2013), https://doi.org/10.5051/jpis.2013.43.6.251.
[67] I. Elgali, O. Omar, C. Dahlin, P. Thomsen, Guided bone regeneration: materials and biological mechanisms revisited. Eur. J. Oral Sci. (2017), https://doi.org/10.1111/eos.12364. Blackwell Munksgaard.
[68] J. Liu, C. Yan, 3D printing of scaffolds for tissue engineering. in: 3D Printing, InTech, 2018, https://doi.org/10.5772/intechopen.78145.
[69] R.G.J.C. Heijkants, T.G. Van Tienen, J.H. De Groot, A.J. Pennings, P. Buma, R.P. H. Veth, A.J. Schouten, Preparation of a polyurethane scaffold for tissue engineering made by a combination of salt leaching and freeze-drying of dioxane. J. Mater. Sci. 41 (8) (2006) 2423–2428, https://doi.org/10.1007/s10853-006-7065-y.
[70] I. Jun, H.S. Han, J.R. Edwards, H. Jeon, Electrospun fibrous scaffolds for tissue engineering: viewpoints on architecture and fabrication. Int. J. Mol. Sci. (2018), https://doi.org/10.3390/ijms19030745. MDPI AG.
[71] W. Chen, H. Zhou, M. Tang, M.D. Weir, C. Bao, H.H.K. Xu, Gas-foaming calcium phosphate cement scaffold encapsulating human umbilical cord stem cells. Tissue Eng. Part A 18 (7–8) (2012) 816–827, https://doi.org/10.1089/ten.tea.2011.0267.
[72] A. Tamayol, M. Akbari, N. Annabi, A. Paul, A. Khademhosseini, D. Juncker, Fiber-based tissue engineering: progress, challenges, and opportunities. Biotechnol. Adv. (2013), https://doi.org/10.1016/j.biotechadv.2012.11.007. NIH Public Access.
[73] R. Pisano, A.A. Barresi, D. Fissore, Innovation in monitoring food freeze drying. Drying Technol. 29 (16) (2011) 1920–1931, https://doi.org/10.1080/07373937.2011.596299.
[74] Q. Hou, D.W. Grijpma, J. Feijen, Porous polymeric structures for tissue engineering prepared by a coagulation, compression moulding and salt leaching technique. Biomaterials 24 (11) (2003) 1937–1947, https://doi.org/10.1016/S0142-9612(02)00562-8.
[75] T. Garg, O. Singh, S. Arora, R.S.R. Murthy, Scaffold: a novel carrier for cell and drug delivery. Crit. Rev. Ther. Drug Carrier Syst. (2012), https://doi.org/10.1615/CritRevTherDrugCarrierSyst.v29.i1.10. Begell House Inc.
[76] A. Sola, J. Bertacchini, D. D'Avella, L. Anselmi, T. Maraldi, S. Marmiroli, M. Messori, Development of solvent-casting particulate leaching (SCPL) polymer scaffolds as improved three-dimensional supports to mimic the bone marrow niche. Mater. Sci. Eng. C 96 (2019) 153–165, https://doi.org/10.1016/j.msec.2018.10.086.
[77] V. Pande, A. Kharde, P. Bhawar, V. Abhale, Scaffolds : porous scaffold for modulated drug delivery, Austin Ther. 3 (1) (2016) 1027.
[78] C. Mandrycky, K. Phong, Y. Zheng, Tissue engineering toward organ-specific regeneration and disease modeling. MRS Commun. 7 (3) (2017) 332–347, https://doi.org/10.1557/mrc.2017.58.
[79] P.H. Jakob, J. Kehrer, P. Flood, C. Wiegel, U. Haselmann, M. Meissner, E.H. K. Stelzer, E.G. Reynaud, A 3-D cell culture system to study epithelia functions using microcarriers. Cytotechnology 68 (5) (2016) 1813–1825, https://doi.org/10.1007/s10616-015-9935-0.
[80] M.A. Lancaster, J.A. Knoblich, Generation of cerebral organoids from human pluripotent stem cells. Nat. Protoc. 9 (10) (2014) 2329–2340, https://doi.org/10.1038/nprot.2014.158.

[81] G. Mehta, A.Y. Hsiao, M. Ingram, G.D. Luker, S. Takayama, Opportunities and challenges for use of tumor spheroids as models to test drug delivery and efficacy. J. Control. Release 164 (2) (2012) 192–204, https://doi.org/10.1016/j.jconrel.2012.04.045.

[82] Y.C. Tung, A.Y. Hsiao, S.G. Allen, Y.S. Torisawa, M. Ho, S. Takayama, High-throughput 3D spheroid culture and drug testing using a 384 hanging drop array. Analyst 136 (3) (2011) 473–478, https://doi.org/10.1039/c0an00609b.

[83] W. Fayad, et al., Restriction of cisplatin induction of acute apoptosis to a subpopulation of cells in a three-dimensional carcinoma culture model. Int. J. Cancer (2009), https://doi.org/10.1002/ijc.24627. In press.

[84] M. Carter, J. Shieh, Cell Culture Techniques, Guide to Research Techniques in Neuroscience, second ed., Academic Press, 2015, .pp. 295–310. http://www.sciencedirect.com/science/article/pii/B9780128005118000149. (Chapter 14), https://doi.org/10.1016/B978-0-12-800511-8.00014-9.

[85] B. Mosadegh, G. Xiong, S. Dunham, J.K. Min, Current progress in 3D printing for cardiovascular tissue engineering. Biomed. Mater. 10 (3) (2015) 034002https://doi.org/10.1088/1748-6041/10/3/034002.

[86] S.S. Verbridge, A. Chakrabarti, P. Delnero, B. Kwee, J.D. Varner, A.D. Stroock, C. Fischbach, Physicochemical regulation of endothelial sprouting in a 3D microfluidic angiogenesis model. J. Biomed. Mater. Res. A 101 (10) (2013) 2948–2956, https://doi.org/10.1002/jbm.a.34587.

[87] L.A. Torre, R.L. Siegel, E.M. Ward, A. Jemal, Global cancer incidence and mortality rates and trends—an update. Cancer Epidemiol. Biomarkers Prev. (2016), https://doi.org/10.1158/1055-9965.EPI-15-0578. American Association for Cancer Research Inc.

[88] C. Te Kuo, J.Y. Wang, Y.F. Lin, A.M. Wo, B.P.C. Chen, H. Lee, Three-dimensional spheroid culture targeting versatile tissue bioassays using a PDMS-based hanging drop Array. Sci. Rep. 7 (1) (2017), https://doi.org/10.1038/s41598-017-04718-1.

[89] A. Accardo, M.C. Blatché, R. Courson, I. Loubinoux, C. Thibault, L. Malaquin, C. Vieu, Multiphoton direct laser writing and 3D imaging of polymeric freestanding architectures for cell colonization. Small 13 (27) (2017), https://doi.org/10.1002/smll.201700621.

[90] M. Dr RichardEglen, Dr Feng Li, Dr Tony Frutos, https://www.ddw-online.com/drug-discovery/p322782-increasing-drug-discovery-success-with-3d-cell-culture-systems.html.

[91] A. Abbott, Biology's new dimension. Nature (2003), https://doi.org/10.1038/424870a.

[92] The Guardian, FDA Approves First Prescription Drug Made through 3D Printing | Science | The Guardian, https://www.theguardian.com/science/2015/aug/04/fda-first-prescription-drug-3d-printing.

CHAPTER 10

Polysaccharide-based polymeric gels as drug delivery vehicles

Dilshad Qureshi[a], Amit Kumar Nayak[b], Doman Kim[c], Samarendra Maji[d], Arfat Anis[e], Biswaranjan Mohanty[f], and Kunal Pal[a]

[a]Department of Biotechnology and Medical Engineering, National Institute of Technology Rourkela, Rourkela, Odisha, India

[b]Department of Pharmaceutics, Seemanta Institute of Pharmaceutical Sciences, Mayurbhanj, Odisha, India

[c]Department of International Agricultural Technology & Institute of Green BioScience and Technology, Seoul National University, Gwangwon-do, Republic of Korea

[d]Department of Chemistry, Faculty of Engineering and Technology, SRM Institute of Science and Technology, SRM Nagar, Kattankulathur, Tamil Nadu, India

[e]SABIC Polymer Research Center, Department of Chemical Engineering, King Saud University, Riyadh, Saudi Arabia

[f]Department of Pharmaceutics, Institute of Pharmacy and Technology, Salipur, Cuttack, Odisha, India

1 Introduction

The hydrogels are made from water-soluble polymers. The water-soluble polymers are crosslinked, either by physical or chemical methods, to form a three-dimensional network [1]. The polymeric network structure imbibes and hold water within itself and does not release the water under normal circumstances [2]. Since water-soluble polymers are available in a wide range of chemistry (chemical or physical), various types of hydrogels can be synthesized. It has been possible to synthesize hydrogels in different physical forms, e.g., nanoparticles, microparticles, films, composites, phase-separated systems, and slabs [3, 4]. Due to the availability of the hydrogels in various types, hydrogels have extensively been harnessed for controlled drug delivery, tissue engineering, regenerative medicine [5], wound dressings [6], tissue adhesives [7], and cell immobilization (both microbial and mammalian cells) applications [8, 9]. Among the various applications, the use of hydrogels for the development of drug delivery systems has gained much attention, which is due to the unique properties that are presented by the hydrogels [10]. The hydrogels have been reported to be inherently highly biocompatible, i.e., the hydrogels do not elicit toxic reactions when placed in contact with the tissues. Due to the inherent biocompatibility of the hydrogels, the hydrogels have found applications in implantable and injectable drug delivery systems apart from oral delivery systems and suppositories [11]. The biocompatible nature of the hydrogels has been explained by

Advances and Challenges in Pharmaceutical Technology
https://doi.org/10.1016/B978-0-12-820043-8.00013-X

the presence of water molecules within the hydrogels, which have physicochemical properties that are similar to the extracellular matrices [12]. It is noteworthy that due to the availability of the polymers in comprehensive chemistries, it is feasible to design hydrogels that are similar to the composition of the extracellular matrices and have matching mechanical properties [13]. Many of the hydrogels are non-biodegradable, i.e., the hydrogels do not degrade. Non-degradable hydrogels are problematic in case the same is to be implanted within the human body. Such hydrogels are to be explanted out once they have performed their duty. The patients face difficulty as they have to undergo surgery both during implantation and explantation procedures [14]. Hence, researchers are looking forward to synthesizing the biodegradable hydrogels or hydrogels that undergo dissolution under biological environment [15]. The hydrogels that are meant to biodegrade or undergo dissolution are intended to change their physical forms under the influence of enzymes, hydrolysis, or variations in the environmental conditions (e.g., pH, temperature, ultrasound, light, or electric field) [10]. Another essential property of the hydrogels is their easy deformability. Since the hydrogels can undergo easy deformation, the hydrogels have been reported to conform to the shape of the surface over which they have been applied. Under such conditions, the adhesive properties (bioadhesive or mucoadhesive) of the hydrogels can help to immobilize and securing the hydrogels over the application site (e.g., wound surface and buccal cavity) even if the surface of the application site is not horizontal [16].

As mentioned earlier, hydrogels are three-dimensional network structures and form a depot formulation. The inter-polymeric spaces appear as pores within the polymeric architecture. The nature of these pores plays an essential role in the diffusion of drugs through the hydrogel matrices [17]. One of the easiest ways to tailor the porous structures is by varying the crosslinking density of the polymer matrix [18] (Fig. 1). In general, an increase in the crosslinking density decreases the pore size, which in turn increases the interactions between the polymer architecture and the drug molecules. The increased interaction reduces the drug diffusion and delays the release of the drug molecules from the delivery systems [19]. Further, the affinity of the polymer chains that have been used for the synthesis of the hydrogels towards the biological fluids also plays a vital role in manipulating the porous structure of the hydrogels. A hydrogel may become swollen in a specific liquid while deswell in another liquid. The fluid-specific swelling and deswelling property of the hydrogels have been employed to load the drugs into the hydrogel matrices and subsequently release the drugs in a controlled manner from the hydrogel matrices [20].

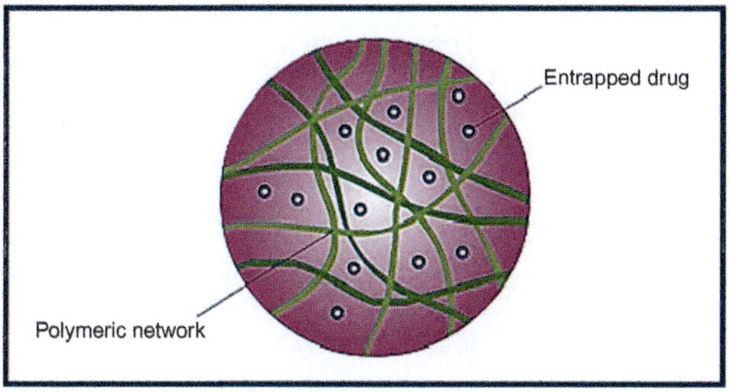

Fig. 1 Porous polymer-based hydrogel. *(Reproduced with permission from T.G. Barclay, C.M. Day, N. Petrovsky, S. Garg, Review of polysaccharide particle-based functional drug delivery Carbohydr. Polym. (2019).)*

Like the two sides of a coin, hydrogels too have several limitations despite the aforesaid advantages. One of the significant disadvantages of the hydrogels is their poor mechanical properties. This makes them unsuitable for many applications, including load-bearing applications [21]. However, this can be overcome by employing various techniques like crosslinking and synthesis of composite hydrogels, to make the hydrogels suitable for drug delivery applications, including depot formulations [22–24]. A summary of the crosslinking strategies for the synthesis of the hydrogels has been given in Fig. 2. Further, many of the hydrogels are not injectable, which makes them unsuitable to be used as injectable depot formulations. The addition of the drugs within the hydrogel matrices have been found to alter the properties of the hydrogels [25]. Also, the drug molecules may not be homogeneously dispersed within the hydrogel matrices that may result in the heterogeneous features of the matrices and batch variations. These limitations are more pronounced, especially when the hydrophobic drugs are to be incorporated within the hydrogels [26]. The drug-polymer interactions can lead to the differential drug release. Hence, proper care has to be taken while designing drug delivery systems to avoid erratic drug releases [14]. Since the hydrogels are constituted mostly of water, controlling the release of the hydrophilic drug molecules from the delivery vehicles is a tedious task for the formulation scientists. The drugs may be released as quickly as a few minutes. Considering the aforesaid limitations, the formulation scientists have to be very careful while designing hydrogel-based drug delivery systems [25].

In this chapter, an attempt has been made to discuss the various release mechanisms that are associated with the drug release from the hydrogel

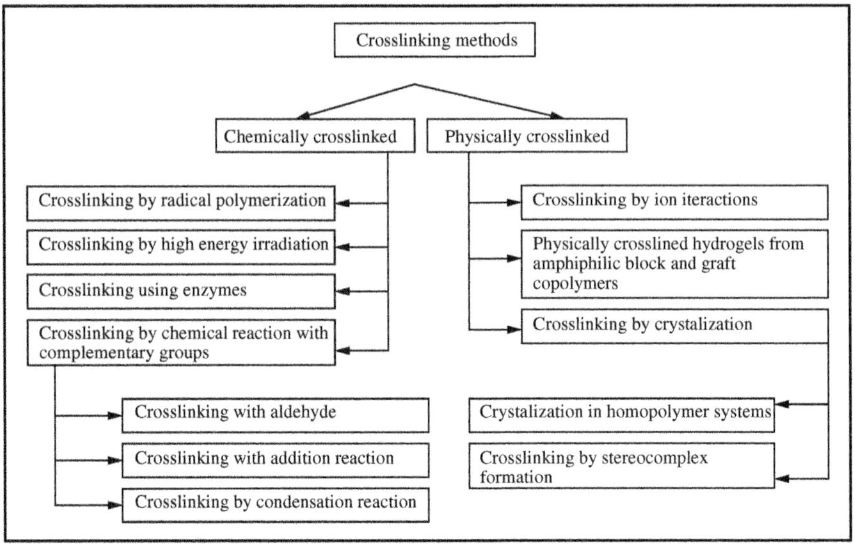

Fig. 2 Crosslinking strategies used in the synthesis of the hydrogels. *(Reproduced with permission from M. Hamidi, A. Azadi, P. Rafiei, Hydrogel nanoparticles in drug delivery. Adv. Drug Deliv. Rev. 60 (15) (2008) 1638–1649.)*

matrices, types of hydrogel systems, and applications of the polysaccharide (biopolymer)-based hydrogels as drug delivery systems.

2 Release mechanism from hydrogel matrices

The release of the drugs from the hydrogel-based drug delivery systems with respect to time can be explained either by sophisticated or straightforward mechanisms. All the models consider the rate-limiting step in the release process to understand the release phenomena. In this section, various release mechanisms will be discussed.

2.1 Diffusion-controlled release

This mechanism is frequently used for explaining the drug release mechanics from the hydrogels. Fick's first law of diffusion has been used by the researchers to describe the diffusion-controlled release process from the hydrogel-based delivery systems [27]. As mentioned earlier, the diffusion of the drugs is dependent on the pore size (also known as mesh size) of the polymer network of the hydrogels [4, 28]. The simplest way to alter the pore size is by manipulating the polymer concentration. An increase in the polymer concentration in the hydrogels results in a decrease in the pore size, while the same increases when the polymer concentration is decreased. However, care should be taken while manipulating the pore size by altering the polymer concentration. A significant reduction of the polymer concentration may compromise the mechanical stability of the hydrogels and promote the degradation of the hydrogel matrices. The pore size can also be altered by tailoring the crosslinking density of the hydrogels [29]. An increase in the crosslinking density reduces the pore size and vice versa [30, 31]. The diffusion-controlled delivery systems do not change their physical form (volume), either through swelling or degradation, throughout the release process. Further, the drug diffusion process can be manipulated by applying an external stimulus. One such example is the application of electric current to promote the diffusion of the drugs across the skin and/or ocular tissue [32]. Increasing drug diffusion using electric current is regarded as "iontophoresis." The typical pore size within the swollen hydrogel matrices lies in the range of 5 nm and 100 nm. Since the pore size is much bigger than most of the small-molecule drugs, the diffusion of the drugs occurs in an unhindered fashion. However, in the case of the macromolecular drug molecules (e.g., oligonucleotides, peptides, and proteins), there will be a retardation factor exerted by the smaller pores. This would lead to the sustained release of the macromolecular drugs from the hydrogel matrices. If unhindered diffusion is required for the macromolecular drugs (so as to have a quicker release), the hydrogels have to be designed such that the pore size of the hydrogels is more significant than the hydrodynamic radii of the macromolecular drugs [14].

The diffusion layer model (also known as film theory) has been used for estimating the drug release properties. This model proposes that the release of the drug is a two-step process. Initially, a thin saturated layer of the drug solution is formed just outside the delivery matrices when the dissolution media reacts with the drug delivery matrices. Herein, the concentration of the drug remains more-or-less constant. Hence, the layer is also sometimes regarded as the stagnant layer. In the second step, the drug moves away from the stagnant layer via diffusion (from higher concentration to lower concentration) into the bulk dissolution media [33, 34]. The diffusion-controlled drug delivery systems are generally categorized into reservoir and matrix type delivery systems (Fig. 3). The reservoir type delivery systems are also regarded as the laminated matrix type delivery systems. These delivery systems have been developed in various shapes like cylinders, spheres and discs. Such delivery vehicles have a layer of an insoluble polymeric semipermeable membrane that encapsulates the drugs that are present either as

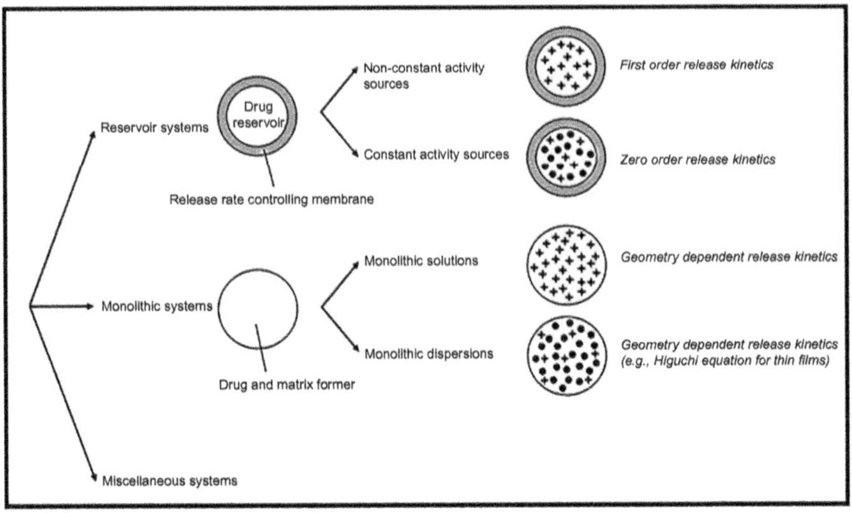

Fig. 3 The diffusion-controlled drug delivery systems. Stars represent individual drug molecules, black circles drug crystals and/or amorphous aggregates. Only spherical dosage forms are illustrated, but the classification system is applicable to any type of geometry. *(Reproduced with permission from J. Siepmann, F. Siepmann, Mathematical modeling of drug delivery. Int. J. Pharm. 364 (2) (2008) 328–343.)*

powder or liquid [33, 35]. If the size of the delivery vehicles is within 100 nm, then the delivery vehicles are regarded as the nano-reservoir systems. On the other hand, if the size is a few microns, then the delivery systems are considered as micro-reservoir systems [14]. The reservoir type delivery systems are produced using non-biodegradable polymers. This ensures that the drug is slowly diffused out of the delivery vehicle. The rate-limiting step during the release process in these types of delivery systems is the partitioning of the drug molecules within the polymeric membrane. The drug molecules are then subsequently diffused into the dissolution medium [36]. On the other hand, matrix type delivery systems consist of a homogenous dispersion of the drug particles in a continuum polymer matrix. The dissolution media initially imbibes into the outer layer of the polymer matrix. The drug in the outer layer gets dissolved and diffuses out of the delivery system. As time progresses, the dissolution media front moves to the core of the delivery system. During this process, the drug particles that are present in the matrix system get dissolved and starts diffusing outside. The release of the drug has been reported to be uniform in the case of matrix type delivery system [37, 38]. The synthesis of the matrix type delivery systems is more straightforward than the reservoir type delivery system. The high-molecular-weight drugs can be delivered conveniently using a matrix type delivery system. However, the reservoir type delivery system can release drugs following zero-order kinetics. Also, the rate of the drug release in these systems can be achieved just using polymers with different physicochemical properties.

2.2 Swelling-controlled release

The swelling-controlled release systems are categorized under the broad classification of solvent-activated release systems (Fig. 4). The swelling-controlled delivery is exhibited by the hydrogels that can imbibe a large amount of water. Hydrogels are generally cross-linked structures and visco-elastic materials. The release from these systems is affected by three mechanisms, namely, water concentration gradient, polymer stress gradient [39], and osmotic forces [40, 41]. The release of the drug from the swelling-controlled drug delivery systems occurs due to the permeability of the drug molecules when the hydrogel matrices start swelling. The release is mainly governed by anomalous transport, where Fickian diffusion and case II drug transport plays a significant role [42, 43]. Case II transport allows the release of drugs at a constant rate. In these systems, the choice of a proper polymer matrix can hinder the Fickian diffusion, thereby delaying the release of the drug molecules from the delivery system. Hence, Case II transport may be the predominant release method that would result in the formation of a zero-order drug release vehicle [39].

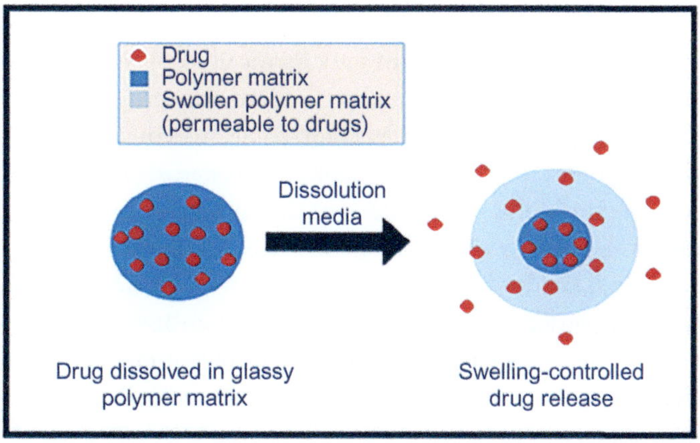

Fig. 4 Mechanics of drug release from swelling-controlled drug delivery systems.

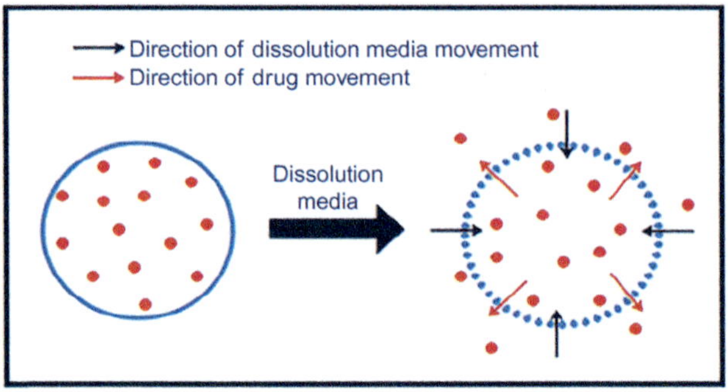

Fig. 5 Mechanics of drug release from dissolution and diffusion-controlled drug delivery systems.

2.3 Dissolution and diffusion-controlled release

In these types of drug delivery systems, dissolutions of the polymer matrices and diffusion of the drug molecules are both involved simultaneously. The drug molecules are compartmentalized within a thin polymeric membrane. The said membrane is partially soluble in nature. When placed within the biological fluids (dissolution media), the polymeric membranes are partially solubilized. This results in the formation of the pores that allow the dissolution media in gaining entry to the core, which houses the drug molecules (Fig. 5). The dissolution media then dissolves the drug particles and initiate the release process by diffusion [38].

2.4 Chemically controlled release

The mechanics of drug release from the chemically-controlled drug delivery systems involve chemical reactions to initiate the drug-release from the delivery vehicles. Bioerosion or biodegradable polymers are usually used for the development of these drug delivery systems. The erosion of the drug-containing polymer matrix starts at the surface of the delivery system. This results in the release of the drugs into the biological fluid. As the outer layer is degraded, the bulk matrix is exposed to the biological fluid, which then forms the outer layer. The bioerodable delivery systems are not required to be removed from the body. The release rate of the drug from such delivery systems is primarily dependent on the degradation rate of the polymer matrix [44, 45].

3 Types of hydrogel systems

The hydrogel systems are broadly classified as traditional hydrogels, filled hydrogels, stimuli-sensitive hydrogels, and self-assembled hydrogels. In this section, the different types of hydrogels will be discussed in brief.

3.1 Traditional hydrogels

The traditional hydrogels can be produced either using natural or synthetic polymers [46]. The hydrogels produced using natural polymers are usually weaker than hydrogels prepared using synthetic polymers. On the contrary, the biodegradability of the synthetic polymers is quite low as compared to the natural polymers. Many researchers have proposed different design criteria to obtain hydrogels with suitable degradation properties [1]. These hydrogels are produced by conventional crosslinking methodologies. The hydrogels prepared using physical interactions and/or secondary interactions are regarded as reversible or physical hydrogels. These hydrogels are formed when the polymeric chains undergo molecular entanglements and/or are involved in the formation of secondary forces (e.g., ionic, H-bonding or hydrophobic) that promote the formation of polymeric networks [47]. However, the hydrogels formed via physical interactions are not homogeneous. This may result in the batch-to-batch variations in the properties of the physical hydrogels. On the other hand, hydrogels that are produced by the help of chemical reactions are regarded either as permanent or synthetic hydrogels [48]. In these hydrogels, new covalent bonds are formed that results in the formation of a network structure [49]. Here, multifunctional reagents called "crosslinkers" are used. The crosslinkers are small molecules as compared to the polymer chains. The multifunctional groups, present on the crosslinkers, react with the polymer chains to form a network structure that can hold water inside it [50]. Several physical and chemical methods of hydrogel formation have been described by Pal et al. [51].

3.2 Filled hydrogels

Filled hydrogels are the hydrogels which encapsulate another phase within its structures. In general, the internal phases of the filled hydrogels may either be hydrophobic (oils and fats) or hydrophilic. In either of the cases, the external phase is the hydrophilic hydrogel matrices. The oil containing filled hydrogels are also regarded as emulgels [52]. The oil/fat-filled hydrogels may contain emulsifiers (e.g., polyglycerol polyricinoleate, sorbitan monooleate, sorbitan monopalmitate, Tween 80, and Tween 20). The addition of the emulsifiers helps to stabilize the oil/fat droplets during the synthesis stage and also help in obtaining oil/fat droplets of uniform sizes [53, 54]. If the internal phase is semi-solid in nature, then the filled hydrogels are regarded as "bigels." The bigels with dispersed hydrophobic phase are formed when oleogels are mixed with hydrogels [55]. On the contrary, the bigels with dispersed hydrophilic phases are formed due to the phase-separation process. Such systems are created by the protein and polysaccharide mixtures. A typical example of phase-separated hydrogels includes the starch-gelatin phase-separated hydrogels. Segregative phase separation happens when the net charge over the polysaccharide is either neutral or similar to that of proteins. If the protein and the polysaccharide bear opposite charges, then phase separation occurs via associative interactions. The properties of these systems are dependent on the pH and the ionic strength of the immediate environment [56, 57]. Filled hydrogels can also be prepared by entrapping nano- or microparticles in a hydrophilic hydrogel matrix. The particles may either be polymeric or inorganic particles [58, 59]. The size and the distribution of the internal phases can alter the properties of the filled hydrogels. Fig. 6 summarizes the different types of filled hydrogels.

3.3 Stimuli-sensitive hydrogels

The evaluation of the hydrogels that can alter their physicochemical properties and phase states with the change in the external conditions has gained

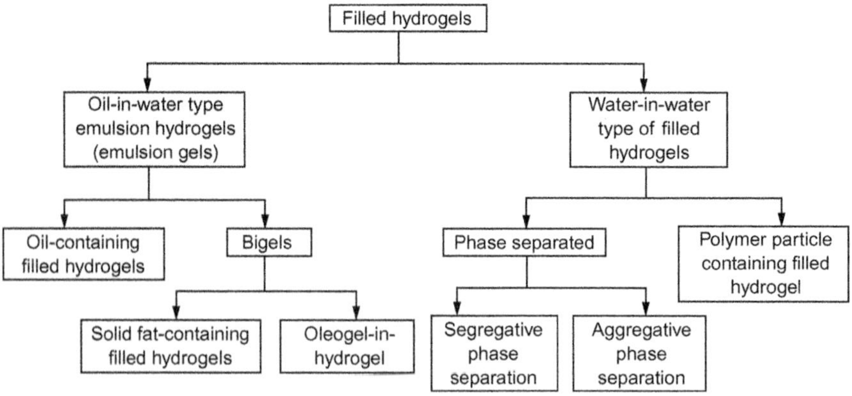

Fig. 6 Classification of filled hydrogels. *(Reproduced with permission from B. Choudhary, S.R. Paul, S.K. Nayak, D. Qureshi, K. Pal, Synthesis and biomedical applications of filled hydrogels Polymeric Gels, Elsevier, 2018, pp. 283–302.)*

importance in the last couple of decades. These hydrogels can also display phase transition at body temperatures [60]. The external conditions (also known as stimuli) that induce the changes are broadly categorized into (i) physical [61], (ii) chemical, and (iii) biological stimuli [62, 63]. The physical stimuli include light, electric field, magnetic, temperature, mechanical forces, and ultrasound [64]; the chemical stimuli include ionic strength, pH, and solvent [64]; and the biological stimuli include the changes in the environment due to the activity of the biological entities (e.g., enzymes) [64]. The mechanics of the various types of stimuli-responsive hydrogels have been summarized in Fig. 7. Nowadays, hydrogels that are sensitive to multiple stimuli are also being investigated [62, 65]. The stimuli-sensitive hydrogels are developed either using stimuli-responsive polymers or by grafting stimuli-responsive polymers on a polymer backbone [66]. These hydrogels have been explored in various areas including tissue engineering, drug delivery, microfluidic devices [67], gene delivery, and sensors [68].

The alteration in the stimuli alters the phase volume, microstructures, surface charge, and/or the combination of these properties of the stimuli-sensitive hydrogels [69]. The effect on the stimuli-responsive hydrogels is usually reversible. In other words, as the stimuli are altered to the usual conditions, the hydrogels regain their original shape, size, and properties [70].

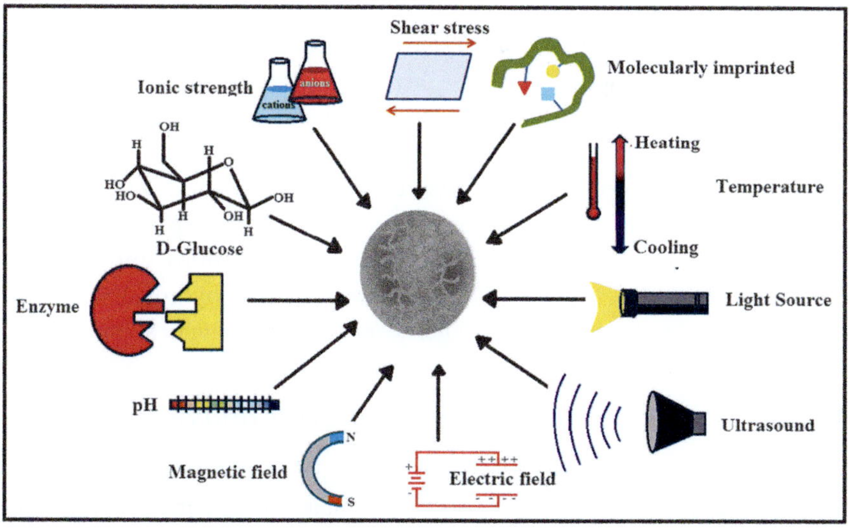

Fig. 7 Summary of the mechanics of stimuli-responsive hydrogels. *(Reproduced with permission from D. Qureshi, S.K. Nayak, S. Maji, A. Anis, D. Kim, K. Pal, Environment sensitive hydrogels for drug delivery applications, Eur. Polym. J. (2019) 109220.)*

Since the stimuli-responsive hydrogels can respond very quickly to the stimuli, they have been explored to synthesize controlled drug delivery systems [70]. These hydrogel systems allow the delivery of the drug molecules at a specific spatiotemporal environment [71, 72]. Qureshi et al. [64] have written a comprehensive review of the environment-sensitive hydrogels for controlled drug delivery applications [64].

3.4 Self-assembled hydrogels

The interest in the self-assembled hydrogels has exponentially increased in the last decade. Such formulations have found applications in drug and cell delivery due to the advantages they offer as compared to the conventional delivery systems. Many stimuli-responsive polymers that are soluble in water can be manipulated to form gels under the influence of stimuli, e.g., pH, light, and temperature [41]. In other words, these polymer solutions in water undergo a sol-to-gel transition in the presence of the stimuli. These polymers undergo self-assembly under physiological conditions to form gelled structures (Fig. 8). Various amphiphilic block copolymers and peptide oligomers with specific sequences have been reported to show this property [73, 74]. Water-soluble polymers can be made to display self-assembly by introducing functional groups within the polymer structures, which can induce physical crosslinking either by hydrophobic interactions, change in pH, and promoting the formation of the stereocomplexed crystallizable domains [75]. The synthesis of the self-assembled hydrogels has also been implemented by photo-irradiation induced chemical crosslinking methods. In this case, the polymers are grafted with photo-sensitive functional groups (e.g., azide group) [50]. However, the synthesis of self-assembled hydrogels by the physical crosslinking method is more preferred over the chemical method. The physically crosslinked hydrogels undergo spontaneous interactions (e.g., ionic interactions, hydrophobic interactions, Van der Waals interactions, and hydrogen bonding) to form the hydrogels. Due to this reason, these hydrogels have often been employed to formulate in situ gel-forming drug delivery systems [76].

4 Biopolymer-based hydrogels in drug delivery

Delivery of therapeutic agents from the hydrogel matrices can be controlled spatiotemporally. They mitigate the inefficiencies of the conventional delivery systems owing to their easily tunable nature and controllable degradable characteristics. In the present scenario, bio-based polymers are recognized as attractive materials for the synthesis of delivery systems because of their

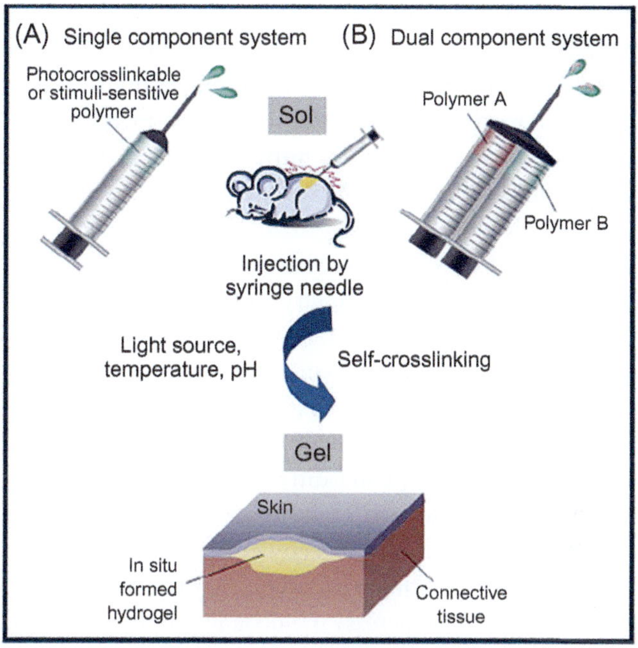

Fig. 8 Schematic illustration for in situ forming hydrogels as an injectable matrix. Hydrogels are formed in situ, after injection through a syringe needle, by (A) photo-polymerization or stimuli-sensitive sol-gel transition of a single polymer, and (B) stereocomplexation or self-crosslinking of two polymer components. *(Reproduced with permission from H.J. Chung, T.G. Park, Self-assembled and nanostructured hydrogels for drug delivery and tissue engineering, Nano Today 4 (5) (2009) 429–437.)*

excellent biocompatible, renewable, non-toxic, economical, biodegradable, and stable nature. Additionally, their structural and compositional features closely resemble the macromolecules that are present in the ECM. Many biopolymers have been explored for designing the drug delivery systems to date [14, 77]. The properties of the most common biopolymers of polysachharide-origin have been comprehensively reviewed in this section. Moreover, the recent applications of polysaccharide-based hydrogels in the field of sustained and controlled drug delivery have also been reviewed.

4.1 Chitosan

Chitosan (CS) is a cationic polysaccharide that is exclusively composed of monomers, namely β-(1 → 4)-linked glucosamine and *N*-acetyl-D-glucosamine, which are randomly distributed along the backbone [78]. It is the deacetylated derivative of chitin and serves as a structural component of the crustacean shells, and cell walls of insects and fungi [79]. The degree of acetylation and molar mass of CS has an essential impact on its applications [80]. On the structural basis, CS contains three different functional chemical groups, an amino group (C-2), and a primary (C-3) and a secondary (C-6) hydroxyl groups [81]. The modification of these reactive groups (—NH_2 and —OH) can easily alter the properties (e.g., solubility, mucoadhesive nature, and stability) of CS [82]. CS exhibits specific biological characteristics such as biodegradability, biocompatibility, and non-toxicity, which are responsible for its immense popularity in biomedical applications [83]. Moreover, it also possesses a range of medicinal properties, e.g., antioxidant, antimicrobial, and low immunogenicity [83]. This polycationic polymer demonstrates solubility in dilute organic and inorganic acidic solutions, which carries a pH lower than its pK_a (around 6.3) and form shear-thinning fluid [84]. The amenability and non-toxic nature of CS have made it a convenient material to be utilized in gastrointestinal drug delivery. The cationic moieties of the CS interact with the negatively charged mucous membrane, thereby providing excellent mucoadhesion and increase the drug residence time [83]. It also acts as a permeation/penetration enhancer for hydrophilic drugs with low oral bioavailability. The protonated and soluble form of CS reversibly opens the tight epithelial junctions ultimately facilitating the delivery of the hydrophilic compounds [85, 86]. Due to its versatile nature, CS has gained potential importance in various drug delivery applications.

4.1.1 Drug delivery applications

Potiwiput et al. [87] prepared carboxymethyl chitosan (CMC) and alginate (Alg) based hydrogels encapsulated with in situ synthesized calcium phosphate (CaP) particles. Dual crosslinking of the hydrogels was achieved using two physical crosslinking techniques, namely, electrostatic interactions (between amine groups of CMC and carboxyl moieties of Alg) and ionic crosslinking (internal gelation of Alg with $CaCO_3$). The authors reported that dual-crosslinking enhanced the swelling behavior and rheological properties of the hydrogels. The hydrogels demonstrated the diffusion-controlled release of two model drugs (e.g., silver sulfadiazine (SSD) and tetracycline hydrochloride (TH)). The hydrogels were also used to encapsulate human adipose-derived stem cells (ASCs). The authors concluded that the hydrogels could be explored as injectable scaffolds and drug delivery devices [87]. The preparation of robust and mechanically stable hydrogel networks can be achieved by using the covalent crosslinking of polymer chains.

Martínez-Martínez et al. [88] prepared an oral delivery system based on CS hydrogel that was covalently crosslinked with tetrakis ((hydroxymethyl) phosphonium chloride). The prepared hydrogels were observed to be biocompatible, mucoadhesive, and less-toxic. The release of camptothecin (CPT, an anti-tumor drug) showed a pH-dependent release. The authors reported that the mucoadhesive nature of the hydrogels prolonged the release of the CPT at the target site, which ultimately reduced its intestinal toxicity [88]. An alternative approach to crosslinking technique, for preparing CS hydrogels, is the polyelectrolyte complexation (PEC) [89]. CS is a cationic polymer in acidic conditions ($pH < pK_a$), and therefore, it interacts spontaneously with the polyanions to form polyelectrolyte complexes [90]. Motiei et al. [91] reported the preparation of polyelectrolyte nanoparticles (PENs) with a CS-based inner core surrounding with an outer shell of polycationic polyethylene amine (PEI). The core was synthesized by the ionic gelation method using dextran sulfate and tripolyphosphate at pH 8 (bPENs) and pH 3 (aPENs), respectively. These polyanions enhanced the interactions inside the core and furthermore assisted in the assembly of the polycationic PEI shell. Transmission electron microscopy (TEM) results demonstrated that the core/shell structure was distinctively visible in bPENs while aPENs appeared structurally more compact and did not have any distinguished shell. The release profiles of nicotinic acid (INN)-incorporated PENs demonstrated pH-dependent behavior. Both PENs released INN in a sustained manner at their specified pH. The release profile of the drugs could be best analyzed using Korsmeyer-Peppas model [91]. Shi et al. [41, 92] reported the development of an injectable thermosensitive hydrogel system of n-hexanoyl glycol chitosan (H-GCS) for ophthalmic drug delivery applications. H-GCS aqueous sol transitioned into a gelled network at 32°C (sol-gel transition temperature). H-GCS-based hydrogels released the drug levofloxacin in a sustained and controlled manner. It was also observed that the hydrogels caused minimal cytotoxicity to L-929 cells and human corneal epithelial cells (HCEC) at concentrations below 0.8 mg/mL. Ocular irritation test on the rabbit eyes confirmed the ocular biocompatibility of the prepared hydrogels. Single topical instillation of H-GCS hydrogels in rabbit eyes demonstrated prolonged precorneal retention of fluorescein sodium in comparison to its aqueous solution [92]. Wang et al. [63, 93] prepared smart injectable UV-cross-linkable and thermoresponsive CS-based hydrogels. The hydrogels could be used for the NIR-responsive on-demand and localized delivery of doxorubicin (DOX) (anti-cancer drug). The thermoresponsive nature was imparted to CS by grafting poly *N*-isopropyl acrylamide

(PNIPAM, a temperature-sensitive polymer) units on CS. The incorporation of the UV-cross-linkable unit (methacryloyl groups) imparted the UV-crosslinkablility to CS. Furthermore, the incorporation of photothermal carbon (PTC) facilitated copolymer hydrogel to undergo volume shrinkage upon NIR-irradiation. The authors reported that NIR-irradiation triggered the elevation in temperature of the hydrogels due to the photothermal effect of PTC and caused 42% shrinkage of the hydrogels. NIR-irradiation also boosted the release of doxorubicin from the hybrid hydrogels to approximately 40% [93]. A comprehensive tabulation of some recent applications of CS in the arena of drug delivery has been presented in Table 1.

4.2 Cellulose

Cellulose has been regarded as the most abundant biopolymer. Its chemical structure is composed of repeating D-glucose residues that are linked together with β-(1→4)-glycosidic bonds. The linkage causes the formation of anhydroglucose units (six-carbon sugar) with a total of three hydroxyl (—OH) groups at different carbon positions [104]. The high crystallinity and water-insoluble nature of cellulose can be attributed to (i) the above-mentioned glycosidic linkages [105], and (ii) intense intra-/intermolecular hydrogen bonding among the hydroxyl groups [106]. This linear polymer imparts high mechanical strength to the cell walls of plants, fungi, bacteria, and algae [104]. Cellulose encompasses various biological features such as biocompatibility, mechanical strength, biodegradability, and hydrophilic nature. These attributes are responsible for its massive utilization in multiple fields including pharmaceuticals, medical sciences, and industries. Limited water-solubility of cellulose impedes the exploitation of this biopolymer in the pharmaceutical and biomedical industries. However, the —OH functional groups present in cellulose structure can undergo chemical modifications (including esterification or etherification) to produce cellulose-derivatives named as "cellulosics" [106]. These water-soluble derivatives of cellulose obtained either as esters (e.g., cellulose acetate (CA), cellulose acetate butyrate (CAB)) [107] or ethers (e.g., carboxymethyl cellulose (CMC), ethyl cellulose (EC), methylcellulose (MC), hydroxypropyl cellulose (HPC), and hydroxypropyl methylcellulose (HPMC)) have become potential materials to be utilized in drug delivery applications [108]. Hydrogel's preparation can be done by physically or chemically crosslinking the

Table 1 Drug delivery applications of chitosan-based hydrogel systems.

S. No.	Polymer used	Preparation technique	Drug	Application	References
1.	Catechol modified-chitosan (Cat-CS)	Genipin mediated crosslinking	Sulfasalazine (SSZ)	Treatment of ulcerative colitis	[94]
2.	Citral and low molecular weight CS	–	5-Fluorouracil	Local anticancer therapy	[95]
3.	CS functionalized with furfural (CF) and *N*-maleoyl alanine functionalized hydroxypropyl β-cyclodextrin (HPCD-AMI)	Diels-Alder reaction	Methyl Orange (MO)	Drug delivery system	[96]
4.	Lotus root amylopectin (LRA) and CS	–	Curcumin	Encapsulation and sustained release of curcumin	[97]
5.	Oxidized pullulan (OP) and CS-grafted-dihydrocaffeic acid (CS-DA)	Schiff base reaction	DOX and amoxicillin (Amox)	pH-responsive localized drug delivery	[98]
6.	CS and PVA	Water-in-oil (W/O) emulsion method and physical crosslinking	Levonorgestrel	Long-term and controlled release system for contraceptive delivery	[99]
7.	CS and nitrosalicylaldehyde	In situ hydrogelation	Diclofenac sodium salt	Local chemotherapy	[100]
8.	Carboxymethyl cellulose, β-cyclodextrin, and CS, and Fe_3O_4 nanoparticles	Precipitation chemical method	Methotrexate	Drug delivery system	[101]
9.	Carboxymethyl chitosan (CMCS) and gelatin	Ortho ester linking	DOX	Local implantable drug delivery system	[102]
10.	CMCS and dialdehyde-β-cyclodextrin (DA-β-CD)	Sodium periodate oxidation	Phenolphthalein (PhP)	Drug delivery system	[103]

aqueous solution of these cellulosics. Moreover, the fabrication of the composite hydrogels with specific attributes is also plausible by combining various natural or synthetic polymers with cellulose or its derivatives [109].

4.2.1 Drug delivery applications

Liu et al. [86, 110] prepared pH/NIR-light responsive hydrogel comprised of cellulose nanofibrils (TOCNFs) (prepared using 2,2,6,6-tetramethylpiperidine-1-oxyl(TEMPO)-mediated oxidation method) and polydopamine (PDA). The aqueous solution of the components was physically crosslinked using Ca^{2+} ions as a crosslinker to synthesize the TOCNFs/PDA hydrogels. The composite hydrogels demonstrated the on-demand release of tetracycline hydrochloride (TCH) (model drug) under NIR-light irradiation or acidic environment. Moreover, the in vivo skin defect test performed in a rabbit model revealed the potentiality of the TOCNFs/PDA hydrogels as a wound dressing [110]. Treesuppharat et al. [111] developed composite hydrogels based on bacterial cellulose (BC), which was copolymerized with gelatin using glutaraldehyde as a crosslinking agent. The hydrogels were observed to be thermally stable, chemical resistant, and mechanically robust. Atomic force microscopy (AMF) revealed that cellulose filled the cavities of the gelatin network, which resulted in the sustained release of the model drug (methylene blue) [111]. Xu et al. [112, 113] prepared cellulose nanocrystal (CNC) and chitosan (CS), based nanocomposite hydrogels, for the sustained delivery of theophylline. Periodate-oxidation of CNC was performed to obtain dialdehyde nanocellulose (DACNC), which served as both matrix and crosslinker. CNC/CS composites were fabricated by crosslinking DACNC and CS in different ratios. The characterization of the hydrogels revealed that increment in the CS content increased the swelling ratio and isoelectric point of the hydrogels. As depicted from the drug release studies, the cumulative drug release percentage of theophylline was higher in gastric fluid (pH-1.5) as compared to intestinal fluid (pH-7.4). The authors concluded that the CNC/CS hydrogels hold promising potential to be used as gastric-specific drug delivery vehicles [112]. Mukherjee et al. [114] reported the preparation of organogels-in-hydrogel type bigel matrix based on soybean oil (SO) and hydroxypropyl methylcellulose (HPMCK4M) for topical drug delivery applications. Soybean oil-SPAN 60 (organogelator) mixtures were used to prepare organogel formulations. The aforesaid organogel formulations were mixed with HPMCK4M hydrogels for fabricating four batches ofbigels. Rheological characterization of the formulations revealed their pseudoplastic nature. In vitro drug release study showed that bigels comprising of 3% w/v HPMCK4M hydrogel and soybean oil-based organogel with 20% w/v SPAN 60 showed improved release of paracetamol in comparison to its corresponding control organogels formulation. Moreover, non-fickian type drug diffusion was observed in the case of bigels [114]. Park et al. [115] synthesized pH-responsive hydrogels by radical polymerization of carboxymethyl cellulose (CMC) and 2-hydroxyethyl acrylate (HEA) as a transdermal drug delivery system. The swelling of the hydrogels was higher at pH 7.5 and 8.5 as compared to pH 5.5. The rheological analysis suggested the formation of stable network structure. However, grafting and crosslinking density had a profound impact on mechanical properties. The release behavior of naringenin (NRG, model drug) from the hydrogels was observed to be Fickian diffusion mediated. In vivo skin permeation study confirmed that the synthesized hydrogel systems could enhance the transdermal delivery of the model drug [115]. A comprehensive tabulation of some recent applications of cellulose in the arena of drug delivery has been presented in Table 2.

4.3 Pullulan

Pullulan is one of the biopolymers which has been explored and exploited for biomedical applications. It is a non-ionic water-soluble polysaccharide that is obtained from the aerobically fermented culture of the fungus *Aureobasidium pullulans* [126]. Its linear structure is composed of maltotriose units (structurally defined as α-(1$\rightarrow$4)-linked glucose units) as the building blocks, which are linked together by α-(1$\rightarrow$6) glycosidic linkages [127]. The physiological activity and extreme flexibility of pullulan to chemical modifications are attributed to the presence of various hydroxyl groups [128]. This polysaccharide has attained considerable attention from researchers in the biomedical arena that can be attributed to its biocompatible, non-immunogenic, non-carcinogenic, non-mutagenic, biodegradable, and non-toxic nature [129]. Moreover, it possesses excellent adhesive properties that enable its utilization in the production of oxygen-resistant films [130]. The water-soluble nature of pullulan allows it to be used as a carrier system for therapeutic agents [131]. However, its structure can be chemically modified in order to fabricate various pullulan-derivatives, which possess partial solubility or complete insolubility in water [131]. The modifications are achieved by the derivatization of the hydroxyl groups present on the

Table 2 Drug delivery applications of cellulose-based hydrogel systems.

S. No.	Polymer used	Preparation technique	Drug	Application	References
1.	Gelatin and cellulose nanocrystals (CNCs)	Glutaraldehyde based crosslinking	Theophylline	pH-sensitive controlled drug delivery systems	[116]
2.	β-Cyclodextrin (βCD) and hydroxypropyl-methylcellulose (HPMC)	Citric acid	Ketoconazole	Delivery of poorly soluble drugs	[117]
3.	Carboxymethyl cellulose, poly(lactic acid), and itaconic acid	Facile graft copolymerization	Amoxicillin	Gastric drug delivery	[118]
4.	Carboxy-ethyl cellulose (CEC), dithiodipropionate dihydrazide, and dibenzaldehyde-terminated poly(ethylene glycol)	4-amino-DL-phenylalanine (4a-Phe) catalysis	Doxorubicin (DOX)	pH/redox dual responsive and self-healable hydrogels for sustained drug release and 3D cell culture	[119]
5.	Cellulose (CEL), carboxymethyl-β-cyclodextrin/cellulose (cmβCD), carboxymethyl cellulose (CMC), β-cyclodextrin (β-CD)	–	Tetracycline	Superabsorbent hydrogels for sustained and effective drug delivery	[120]
6.	Carboxymethyl cellulose and graphene quantum dot	Copper(II) acetate solution crosslinking	Naproxen (NPX)	Oral drug delivery	[121]

7.	Hyaluronic acid (HA) and carboxymethyl cellulose (CMC)	–	Bovine Serum albumin (BSA)	Controlled drug delivery, encapsulation of cells, tissue engineering, and regenerative medicine	[122]
8.	Hydroxypropyl methyl cellulose and sodium alginate	Polyelectrolyte complexation	Bovine serum albumin (BSA), metformin hydrochloride (MH), and indomethacin (IDM)	Drug carrier	[123]
9.	Carboxymethyl cellulose (CMC) and graphene quantum dot nanocomposite	–	DOX	Anticancer drug delivery system	[124]
10.	Carboxymethyl cellulose (CMC) and graphene quantum dots (GQDs)	–	Doxorubicin (DOX)	pH-oral sensitive drug delivery	[125]

pyranose rings of pullulan. Various chemical reactions such as esterification, amidation, etherification, copolymerization, oxidation, and sulfation can lead to the synthesis of a series of pullulan-derivatives [132]. Derivatized pullulan (e.g., cholesterol-bearing pullulan, pullulan acetate, carboxymethyl pullulan, and pullulan succinylate) have found numerous applications in biomedical applications [132]. Interestingly, for extending its applications in targeted and sustained drug delivery, pullulan can be modified into various forms such as hydrogels, nanogels, microspheres, and nanoparticles [132]. It serves as a promising candidate for oral drug delivery applications due to its non-attackable nature to the digestive enzymes in the gut [133].

4.3.1 Drug delivery applications

Cinay et al. [134] reported the fabrication of pH-responsive composite hydrogels comprising of poly (ethylene glycol) (PEG) grafted methacrylic acid (P(MAA-g-EG)) and acryloyl group modified-cholesterol-bearing pullulan (CHPOA) nanogels for controlled drug delivery applications. Two different techniques, namely, bulk and surface-initiated photopolymerization methods were used for synthesizing nanocarrier-incorporated hydrogels. The photopolymerization step in both the aforementioned approaches utilized UV radiation and visible light for the preparation of the hydrogels. The hybrid hydrogels were characterized for their swelling and drug release behavior under acidic and basic conditions. A higher swelling ratio was observed in the nanogel-incorporated hydrogels as compared to control P(MAA-g-EG) hydrogels. Moreover, the conjugation of CHPOA-nanogel into the hydrogels led to the enhancement in the loading and release percentages of pregabalin (PGB, model drug). In vitro cytotoxic analysis revealed the non-toxic nature of the prepared hydrogels [134]. Cheng et al. [135] developed a hydrogel system based on aldehyded pullulan (A-Pul), ε-poly-L-lysine (ε-PL) and branched polyethyleneimine (BPEI) using Schiff-base reaction. The hydrogels were loaded with two anticancer drugs, namely, Cisplatin (CDDP) and DOX, for combinational drug therapy. The hydrogels exhibited a macroporous network structure with excellent swelling behavior. In vitro drug release profiles showed controlled and pH-triggered release of the drugs. Additionally, injecting the hydrogels subcutaneously into the dorsal areas of the Spraguee-Dawley (SD) rats in the sol state displayed fast in situ gelation followed by controlled in vivo degradation. Moreover, the in vivo antitumor efficacy of the dual-drug loaded hydrogels assessed in H_{22} liver tumor-bearing mice demonstrated their effective performance owing to the synergistic effect of CDDP and

DOX. Furthermore, the hydrogels showed low in vivo toxicity and excellent cytocompatibility with Human hepatoma cells (Huh)-7 cells [135]. Zhang et al. [136–138] prepared gelatin-based hydrogels using pullulan dialdehyde (PDA) as the crosslinker. It was observed that PDA-crosslinked gelatin hydrogels possessed excellent mechanical properties owing to the synergistic presence of both; covalent (short-range) and physical (long-range) crosslinking. Moreover, MTT assay confirmed the non-cytotoxic nature of the hydrogels against the MC3T3 cells [136]. Lin et al. [139] developed glucose-responsive hydrogels comprising of covalently modified pullulan and concanavalin A (Con-A) for the smart insulin delivery. Carboxylation of pullulan with succinic anhydride was performed to yield pullulan derivative with —COOH groups (CPUL). EDC/NHS (1-ethyl-3-(3-dimethylamino propyl) carbodiimide/*N*-hydroxysuccinimide) crosslinking method was used to prepare covalently crosslinked ConA-CPUL hydrogel that exhibited glucose-responsive behavior. Different amounts of insulin were loaded in ConA-CPUL hydrogels in situ. Insulin/ConA-CPUL hydrogels showed elevated levels of insulin release with an increase in glucose concentration. The authors concluded that these shear-thinning pullulan-based glucose-sensitive hydrogels could potentially be used as closed-loop systems in diabetes therapy [139]. Lachowicz et al. [140] prepared electrostatically crosslinked nanohydrogels, comprising of pullulan and dextran derivatives. Grafting of the pullulan backbone with trimethylammonium groups yielded a cationic derivative of (C-PUL). Spontaneous interaction of C-PUL with anionic dextran sulfate (DEX-A) in water caused the formation of nanohydrogels. It was observed that the alterations in the polycation: polyanion ratio could easily tune the size and stability of the hydrogels. The hydrophobic model drug, piroxicam (PIX), entrapped within the nanohydrogels showed a gradual release with no "burst effect." PIX-loaded hydrogels were biocompatible and demonstrated lower cytotoxic behavior as compared to the free drug [140]. A comprehensive tabulation of some recent applications of pullulan in the arena of drug delivery has been presented in Table 3.

4.4 Hyaluronic acid

Hyaluronic acid (HA) or hyaluronan is a non-sulfated glycosaminoglycan that is typically distributed in various tissues and physiological fluids of the body [148]. As a major constituent of the extracellular matrix (ECM), HA has been reported to accomplish a variety of physiological functions

Table 3 Drug delivery applications of pullulan-based hydrogel systems.

S. No.	Polymer used	Preparation technique	Drug	Application	References
1.	Pullulan and PVA	Glutaraldehyde assisted water-in-oil emulsion crosslinking	Pirfenidone (PFD)	Treatment of Idiopathic Pulmonary Fibrosis (IPF).	[141]
2.	Pemulen and Pullulan	–	Clotrimazole (CTZ)	Treatment of vulvovaginal candidiasis	[142]
3.	Hydroxypropyl cyclosophoraose-pullulan	Emulsion-crosslinking method	Naproxen	Controlled drug delivery	[143]
4.	Pullulan and silanized hydroxypropyl methylcellulose (Si-HPMC)	Sodium trimetaphosphate (STMP)-based crosslinking	Transforming growth factor-b1 (TGF-b1) and growth and differentiation factor-5 (GDF-5)	Intervertebral disc regenerative medicine	[144]
5.	Silk fibroin and pullulan	Horseradish peroxidase and hydrogen peroxide-based crosslinking	Rabbit bone marrow-derived mesenchymal stem cells	Mesenchymal stem cells delivery	[145]
6.	Pullulan	–	*Rhus verniciflua* extract	Atopic dermatitis treatment	[146]
7.	Poly(acrylamide) (PAAm), pullulan (PUL) and PVA	Radical polymerization	Rivastigmine tartarate	Electrically-responsive transdermal drug delivery systems	[147]

such as (i) protection of tissues by maintaining the viscoelastic cushion, (ii) regulation of water transport and hydration in tissues, (iii) creation of lubricious regions among biointerfaces, (iv) interaction with proteins and proteoglycans to make large assemblies in ECM, and (v) cell-receptor mediated signaling to control cellular physiological processes [149]. Chemically, the structure of HA consists of alternating repeating units of D-glucuronic acid (β-(1 → 4)-linked) and *N*-acetylglucosamine (β-(1 → 3)-linked) [150]. The cellular synthesis of HA on the plasma membrane is mediated by three transmembrane enzymes, which are collectively named as hyaluronan synthases (HAS1, HAS2, HAS3). The synthesized HA is then ultimately released in the ECM [151]. HA is a versatile biomolecule with various biological features including but not limited to biocompatibility, biodegradability, non-immunogenicity, viscoelasticity [152] and non-toxicity [151]. Owing to these features, HA has become a potential biomaterial for medical, cosmetics, and pharmaceutical applications [153]. However, this polyanionic natural polymer is rapidly degraded in vivo and exhibits low mechanical properties, which restricts its much anticipated clinical applications [154, 155]. A key strategy for the production of mechanically robust hyaluronan is the chemical modification of the hydroxyl and carboxyl groups that are present on the D-glucuronic acid and *N*-acetylglucosamine sugar residues. Although HA and HA-derived materials show significant differences in their physicochemical properties, both exclusively retain the biocompatible, biodegradable, therapeutic, and pharmacologically-active nature of the pristine polymer [155]. HA has extensively been utilized for delivering the therapeutic agents from different administration routes such as ophthalmic, nasal, pulmonary, and topical [153].

4.4.1 Drug delivery applications

Non-invasive therapies such as photodynamic therapy (PDT) for the treatment of cancer and tumor growth have attracted considerable attention. PDT relies on the NIR-light triggered generation of cytotoxic reactive oxygen species (ROS) from the photosensitizers. Xu et al. [156] combined chemo-photodynamic therapy and NIR-light stimulation to induce the on-demand release of the therapeutic agents from the delivery matrices. The authors prepared an in situ-forming HA-based hydrogel, which exhibited NIR-light triggered ROS-degradable behavior for the delivery of protoporphyrin IX (PpIX) and doxorubicin (DOX). Adipic dihydrazide modified HA (HA-ADH) was conjugated with PpIX to synthesize water-soluble HA-ADH-PpIX conjugate. The conjugate was mixed with ROS-cleavable thioketal linker (TK) to prepare pH-responsive and self-healing hydrogels. The prepared hydrogels were safe for human use. NIR-light irradiation induced the generation of ROS in HA-ADH-PpIX hydrogels, which in turn, triggered the cleavage of TK-CHO that triggered the degradation of the hydrogel matrix. In vitro release kinetics of DOX was observed to be dependent on the light-induced degradation of the hydrogels. Moreover, the treatment of MCF-7 and 4T1 cancer cells with HA-ADH-PpIX hydrogels demonstrated decrement in cell viabilities upon NIR-light irradiation. The authors concluded that the designed hydrogels could be utilized as a promising localized delivery carrier for chemo-photodynamic cancer therapy [156]. Ma et al. [157] developed injectable hydrogels that comprised of aldehyde hyaluronic acid (HA-CHO) and hydrazide-modified poly (γ-glutamic acid) (γ-PGA-ADH) for sustained protein delivery. The hydrogels were biocompatible and showed fast in situ gelation with excellent mechanical properties. HA/γ-PGA hydrogels demonstrated a controlled release of BSA (model drug) with the release rate being dependent on the crosslinking density. Modeling of the drug release kinetics revealed that the hydrogels followed diffusion or Case-II relaxation release mechanism [157]. Lee and coworkers reported the preparation of silane-modified HA hydrogels using a one-pot synthesis technique. BDDE (1,4-butanediol diglycidyl ether) was used as the crosslinker and GPTMS (3-glycidyloxypropyl-trimethoxysilane) was used as the drug-capping agent. The injectable framework of the hydrogels was obtained due to the sol-to-gel transition nature of the silane (self-condensation reaction). The silane-modified hydrogels were observed to be mechanically and biochemically stable with no biocompatibility issues. The drug (vancomycin drug) was loaded within the hydrogel using the sol-gel transition method. This allowed avoiding the use of harsh chemical agents for the encapsulation process. Antibacterial tests confirmed that the hydrogels released vancomycin in a sustained manner and good efficacy [158]. Lu et al. [159] fabricated pH-responsive injectable hydrogels comprising of carboxymethyl hexanoyl chitosan (CHC) and low molecular weight HA (LMWHA). Förster resonance energy transfer (FRET) fluorescence spectra and imaging of the hydrogels revealed that CHC and LMWHA assembled into colloidal particles, which further self-aggregated to form supramolecular hydrogels. CHC-LMWHA composite hydrogels showed excellent injectability, biocompatibility, cytocompatibility, shape-persistent ability, and adhesive property. The hydrogels demonstrated sustained and pH-responsive release of berberine (naturally occurring anti-apoptotic and anti-arthritis alkaloid). Berberine-loaded CHC-LMWHA hydrogels prevented cartilage degeneration by protecting the sodium nitroprusside-induced apoptosis of human chondrocyte primary cells [159]. A comprehensive tabulation of some recent applications of HA in the arena of drug delivery has been presented in Table 4.

4.5 Alginate

Alginate is a versatile and functional biomaterial that has been well-studied in the field of pharmaceuticals, tissue, and biomedical engineering. Alginate (or alginic acid) and its salts (Na, Ca, K, and Mg) can be extracted from the cell walls of the brown algae, which belong to the genera *Laminaria, Eisenia, Ascophyllum*, and *Cystoseira* [168, 169]. The linear heteropolysaccharide alginate comprises of β-(1→4)-linked D-mannuronic acid (M) and α-(1→4)-linked L-guluronic acid (G). The polymer consists of homopolymeric MMMM and GGGG blocks interspersed with alternating MGMGMG residues [169, 170]. On the basis of the extraction sources, the ratio of these blocks in the alginate polymer backbone varies significantly [171]. This biopolymer and its salts are recognized as biocompatible and non-toxic in nature [172]. Divalent cations (e.g., Ca^{2+}, Mg^{2+}) induce the formation of crosslinks among the carboxylate groups of guluronate residues and form a gel-like structure [173]. According to the egg-box model, four carboxylate groups of G-blocks interacts with only one divalent cation to produce these physically crosslinked alginate hydrogels [174]. High guluronate content in the alginate backbone causes the formation of rigid and strong gels while increment in the mannuronate residues imparts softness and elasticity to the hydrogels [175]. Crosslinking of alginate can also be achieved at low pH levels (<3.5) owing to the intermolecular hydrogen bonding and is referred to as "acid gelation" [174]. However, in an attempt to modify the physicochemical (e.g., solubility, hydrophobicity) and biological properties of the alginate, chemical functionalization of its hydroxyl and carboxyl groups have been reported to be an optimal strategy [171]. Alginate hydrogels serve as the matrix for the encapsulation and delivery of a varied range of therapeutics such as cells, DNA, and proteins. In addition, the porous nature of the alginate hydrogel networks allows the sustained and controlled release of the entrapped molecules [175].

Table 4 Drug delivery applications of HA-based hydrogel systems.

S. No.	Polymer used	Preparation technique	Drug	Application	References
1.	Oxidized hyaluronic acid (oxi-HA)/ 3,30-dithiobis (propionohydrazide) (DTP)	Schiff base reaction	BSA	Redox responsive drug delivery system	[160]
2.	PNIPAAm, poly(ethylene glycol) diacrylate (PEGDA), tyrosine (Tyr), and HA	Enzymatic and chemical crosslinking	Infliximab (INF) antibody	Intraocular release of proteins	[161]
3.	PNIPAM and HA	Radical polymerization and Michael addition	Luteolin	Transdermal delivery system for psoriasis skin relief	[162]
4.	Carboxymethyl chitosan (CC), aldehyde hyaluronic acid (AHA), and poly (lactic-*co*-glycolic acid) (PLGA)	–	Vancomycin and vascular endothelial growth factor (VEGF)	Management of the infected wounds	[163]
5.	PNIPAM, HA, CS, acrylic acid Polylactide-*co*-glycolid (PLGA)	–	Melatonin	Cartilage tissue engineering	[164]

6.	Thiol and hydrazide functionalized hyaluronic acid (HA-SH/ CDH) and oxidized sodium alginate (oxi-Alg)	Dual covalent crosslinking via hydrazone and disulphide bonds	BSA	Drug delivery system	[138]
7.	Pluronic F-127 (PF127), *N*,*N*,*N*-trimethyl chitosan (TMC), and polyethylene glycolated hyaluronic acid (PEG-HA)	–	Gallic acid	Textile-based transdermal therapy	[165]
8.	Hyaluronic acid and self-nanoemulsion of Miconazole nitrate (MZ-NE)	Microwave-assisted technique	Miconazole nitrate (MZ)	Treatment of oral thrush	[166]
9.	HA-modified with glycidyl methacrylate and cross-linkable block polymer (CBC)	Atom transfer radical polymerization (ATRP)	Simvastatin	Treatment of atherosclerosis	[167]

4.5.1 Drug delivery applications

Singh and Kumar [176] prepared polymeric hydrogel films based on sodium alginate (SA), sterculia gum (SG), and PVA using radiation polymerization method for the controlled delivery of citicoline (nerve regenerating agent) in the brain. The characterization of the films revealed their antioxidant, non-hemolytic, non-thrombogenic, and mucoadhesive nature. SA-SG-PVA hydrogels showed sustained release of the regenerating agent without any "burst effect." The release profile of citicoline from the hydrogels followed the Fickian diffusion mechanism. The authors suggested the potentiality of the SA-SG-PVA films as a brain drug delivery system and wound dressing [176]. Chen et al. [23, 177] developed self-healing carboxyethyl chitosan (CEC)/ oxidized alginate (OAlg) based hydrogels incorporated with magnetic gelatin microspheres (MGMs) as a platform for drug delivery and soft tissue engineering. Schiff-base reaction was used to prepare the hydrogels by crosslinking the aldehyde groups of OAlg and amino gropus of CEC. Furthermore, to enhance the mechanical and biological capabilities, the composite hydrogels were reinforced with 5-fluorouracil (5-FU)-loaded MGMs. It was observed that MGM-loaded ECE/OAlg hydrogels demonstrated quicker gelation, lower swelling and degradation rates as compared to the control hydrogel. MGMs had a significant impact on the performance of the hydrogels. It was noticed from the compressive and rheological tests that 5-Fu/MGMs/Gels possessed high compressive modulus, stable network structure, and self-healing properties. Furthermore, in vitro performed drug release studies revealed that the prepared composite hydrogels loaded with MGMs could efficiently release 5-FU delivery in a steady and sustained manner [177]. Bahadori et al. [178] fabricated core-shell nano-drug delivery hydrogel systems based on SA and inulin. The hydrogels were tested for the treatment of intestinal disorders. The inner-core of the nanoparticles was prepared by crosslinking SA with $CaCl_2$. The nanoparticles were loaded with the model inflammatory drug, 5-aminosalicylic acid (5-ASA). The negatively charged 5-ASA-loaded alginate cores were subsequently coated with quaternized inulin (QIn) via Coulomb interactions. The core-shell nanoparticles showed excellent bio-adhesion property due to the presence of the inulin protective shell, whereas the inner core of alginate impeded the degradation of 5-ASA from acidic pH in the stomach. The particles appeared as homogenous spheres. However, the size of the particles was enlarged when the SA inner core was coated with QIn. The drug was released from the hydrogels in the sustained and controlled fashion from the nanoparticles (at intestine pH) owing to the slow degradation of the inulin by the microflora in the bowel [178]. Delivery of lipophilic therapeutics using emulsion-based carriers has been tremendously explored. However, the practical applications of these systems are limited owing to their thermodynamically unstable nature. It has been reported that filling the emulsions in the three-dimensional network of hydrogels provide immense stability to these emulsion systems. Xu et al. [112, 113] reported the preparation of emulsion-filled ALG hydrogel beads for the encapsulation and release of a hydrophobic active component (curcumin, Cur). Tea seed oil-based nanoemulsions loaded with curcumin w1ere prepared using spontaneous emulsification method. A dispersion of ALG solution and Cur-loaded nanoemulsions was added dropwise in $CaCl_2$ solution for the preparation of alginate hydrogel beads. The beads had a spherical morphology. In vitro drug release studies showed that the release behavior of Cur from the filled hydrogels was pH-dependent. The release of Cur from the filled hydrogels was analyzed using Hixson-Crowell's model, which indicated a diffusion-controlled release behavior of the active agent [112]. Modification of the functional moieties of alginate (i.e., hydroxyl and carboxyl groups) can be performed to fabricate covalently-crosslinked hydrogels. These hydrogels possess higher mechanical stability as compared to alginate gels that are crosslinked using the ionic gelation method. Anugrah et al. [179] fabricated diselenide-crosslinked NIR-light responsive alginate-based hydrogels for on-demand and controlled therapeutic release. Norbornene-functionalized alginates (Alg-Nb) were linked together with diselenide-tetrazine (Se-Tz) crosslinkers using an inverse electron demand Diels-Alder (IEDDA) click chemistry. During the gelation process, the hydrogels were incorporated with doxorubicin and NIR-light sensitive dye (indocyanine green, ICG). It was observed that the degree of crosslinking varied the microstructural, gelation, swelling, and mechanical attributes of the hydrogels. The covalent crosslinks of the alginate hydrogels were destroyed when irradiated with NIR-light. Under physiological conditions, the hydrogels demonstrated a sustained release profile of doxorubicin. However, NIR-light irradiation resulted in the ROS generation (from ICG), which was responsible for the degradation of the diselenide bonds. Hence, the NIR-light irradiation triggered the rapid release of doxorubicin [179]. A comprehensive tabulation of some recent applications of alginate in the arena of drug delivery has been presented in Table 5.

Table 5 Drug delivery applications of alginate-based hydrogel systems.

S. No.	Polymer used	Preparation technique	Drug	Application	References
1.	Methacrylated alginate (Alg-MA)	Ionic crosslinking using calcium chloride ($CaCl_2$) and covalent crosslinking using photoinitiators and visible (i.e., green) or UV light irradiation	Doxorubicin hydrochloride (DOX-HCl)	Chemotherapeutic delivery carrier for the treatment of cancer	[180]
2.	Carbon quantum dots (CQDs) and calcium alginate (CA)	Ionic gelation and solution casting method	Vancomycin	pH-responsive gastrointestinal drug delivery	[181]
3.	Dopamine, polydopamine, and alginate (ALG)	Self-gelation	Gatifloxacin (GFLX)	Drug delivery system	[182]
4.	Dexamethasone Sodium Phosphate (DEXP) and alginate	Calcium ion crosslinking	DEXP	Drug delivery system	[137]
5.	SA and PNIPAAm	Free radical polymerization	Plasmid DNA (pDNA), complexed with an amphipathic cell-penetrating peptide termed RALA	Drug delivery system for prostate cancer	[183]

6.	Alginate, iron oxide nanoparticles, and gelatin particles	Calcium ion crosslinking	DOX	Magnetic-field responsive drug delivery system	[184]
7.	Sodium alginate, hydroxypropyl methylcellulose (HPMC), Carbapol 934, liquid paraffin, clove oil, and mentha oil	–	Ornidazole	Topical drug delivery	[185]
8.	*Terminalia arjuna* gum and sodium alginate	Simulated tear-fluid induced gelation	Moxifloxacin HCl (MOX-HCl)	Ophthalmic drug delivery	[186]
9.	Poly (2-dimethylamino ethyl methacrylate), poly(*N*-isopropylacrylamide), sodium alginate, and iron oxide nanoparticles	–	Doxorubicin hydrochloride	Cancer treatment	[187]
10.	Itaconic acid grafted starch and alginate, graphene sheets, Fe_3O_4 nanoparticles	Graft copolymerization	Guaifenesin (GFN)	Magnetically triggered drug delivery	[188]

5 Conclusion

Hydrogels have tremendously been explored as appealing drug delivery systems owing to their biocompatible nature and ECM-like properties. The release of the therapeutic agents from the hydrogel matrices occurs through different mechanisms such as diffusion, polymeric matrix swelling, and material degradation. With their potential amenability, hydrogels can be tailored into different architectural forms such as films, nanogels, microparticles, nanoparticles, and slabs. For achieving a precisely predicted and sustained drug release pattern, hydrogels have been synthesized in the form of filled, self-assembled, and stimuli-responsive hydrogels. The risks associated with the surgical implantation of the non-degradable hydrogels paved the way for the utilization of natural biopolymer-based biodegradable hydrogels (biocompatible and non-toxic). The recent developments of these hydrogels in targeted drug delivery have been tremendous. It can be predicted that with subtle modifications, natural polymer-based hydrogels can be used as promising and superlative drug delivery vehicles.

Acknowledgments

The funds leveraged from the Indo-Korea joint research program of Department of Science and Technology, Government of India (Sanction order # INT/Korea/P-37, June 15, 2017) and International Cooperation Program managed by the NRF(2016K1A3A1A19945059), Republic of Korea, are hereby acknowledged.

References

[1] E.M. Ahmed, Hydrogel: preparation, characterization, and applications: a review, J. Adv. Res. 6 (2) (2015) 105–121.
[2] K.G. Subramanian, V. Vijayakumar, Hydrogels: classification, synthesis, characterization, and applications, in: Encyclopedia of Biomedical Polymers and Polymeric Biomaterials, vol. 11, CRC Press, 2015, pp. 3879–3892. Volume Set.
[3] A. Chyzy, M. Tomczykowa, M.E. Plonska-Brzezinska, Hydrogels as potential nano-, micro- and macro-scale systems for controlled drug delivery, Materials 13 (1) (2020) 188.
[4] T.R. Hoare, D.S. Kohane, Hydrogels in drug delivery: progress and challenges, Polymer 49 (8) (2008) 1993–2007.
[5] B.-S. Kim, C.-S. Cho, Injectable hydrogels for regenerative medicine, Tissue Eng. Regen. Med. 15 (5) (2018) 511–512.
[6] Y. Budama-Kilinc, R. Cakir-Koc, B. Aslan, B. Özkan, H. Mutlu, E. Üstün, Hydrogels in regenerative medicine, Biomaterials in Regenerative Medicine, IntechOpen, 2017, pp. 277–301.
[7] M. Rahimnejad, W. Zhong, Mussel-inspired hydrogel tissue adhesives for wound closure, RSC Adv. 7 (75) (2017) 47380–47396.
[8] T. Andersen, P. Auk-Emblem, M. Dornish, 3D cell culture in alginate hydrogels, Microarrays 4 (2) (2015) 133–161.

[9] I. Kwiecień, M. Kwiecień, Application of polysaccharide-based hydrogels as probiotic delivery systems, Gels 4 (2) (2018) 47.
[10] A. Onaciu, R.A. Munteanu, A.I. Moldovan, C.S. Moldovan, I. Berindan-Neagoe, Hydrogels based drug delivery synthesis, characterization and administration, Pharmaceutics 11 (9) (2019) 432.
[11] K.Y. Lee, D.J. Mooney, Hydrogels for tissue engineering, Chem. Rev. 101 (7) (2001) 1869–1880.
[12] J. Saroia, W. Yanen, Q. Wei, K. Zhang, T. Lu, B. Zhang, A review on biocompatibility nature of hydrogels with 3D printing techniques, tissue engineering application and its future prospective, Bio Des. Manuf. 1 (4) (2018) 265–279.
[13] P. Ducheyne, Comprehensive Biomaterials, vol. 1, Elsevier, 2015.
[14] J. Li, D.J. Mooney, Designing hydrogels for controlled drug delivery, Nat. Rev. Mater. 1 (12) (2016) 1–17.
[15] K.R. Kamath, K. Park, Biodegradable hydrogels in drug delivery, Adv. Drug Deliv. Rev. 11 (1–2) (1993) 59–84.
[16] S.P. Shetye, A. Godbole, S. Bhilegaokar, P. Gajare, Hydrogels: introduction, preparation, characterization and applications, Human J. 1 (2015) 47–71.
[17] A. Fahr, Voigt's Pharmaceutical Technology, John Wiley & Sons, 2018.
[18] E. Jabbari, Hydrogels in Tissue Engineering, MDPI, 2018.
[19] M.K. Nguyen, E. Alsberg, Bioactive factor delivery strategies from engineered polymer hydrogels for therapeutic medicine, Prog. Polym. Sci. 39 (7) (2014) 1235–1265.
[20] A.H. Karoyo, L.D. Wilson, Physicochemical properties and the gelation process of supramolecular hydrogels: a review, Gels 3 (1) (2017) 1.
[21] Q. Chai, Y. Jiao, X. Yu, Hydrogels for biomedical applications: their characteristics and the mechanisms behind them, Gels 3 (1) (2017) 6–21.
[22] S.J. Buwalda, T. Vermonden, W.E. Hennink, Hydrogels for therapeutic delivery: current developments and future directions, Biomacromolecules 18 (2) (2017) 316–330.
[23] D.X. Chen, Chen, Glaser, Extrusion Bioprinting of Scaffolds for Tissue Engineering Applications, Springer, 2019.
[24] W. Hu, Z. Wang, Y. Xiao, S. Zhang, J. Wang, Advances in crosslinking strategies of biomedical hydrogels, Biomater. Sci. 7 (3) (2019) 843–855.
[25] T.R.R. Singh, G. Laverty, R. Donnelly, Hydrogels: Design, Synthesis and Application in Drug Delivery and Regenerative Medicine, CRC Press, 2018.
[26] M. McKenzie, D. Betts, A. Suh, K. Bui, L.D. Kim, H. Cho, Hydrogel-based drug delivery systems for poorly water-soluble drugs, Molecules 20 (11) (2015) 20397–20408.
[27] M.S. Hasnain, A.K. Nayak, Alginates: Versatile Polymers in Biomedical Applications and Therapeutics, CRC Press, 2019.
[28] M. Jawaid, F. Mohammad, Nanocellulose and Nanohydrogel Matrices: Biotechnological and Biomedical Applications, John Wiley & Sons, 2017.
[29] S. Mohapatra, S. Ranjan, N. Dasgupta, R. Kumar, S. Thomas, Characterization and Biology of Nanomaterials for Drug Delivery: Nanoscience and Nanotechnology in Drug Delivery, Elsevier, 2018.
[30] J.M. Anderson, S.W. Kim, Recent Advances in Drug Delivery Systems, Springer Science & Business Media, 2012.
[31] S.M. LaNasa, I.T. Hoffecker, S.J. Bryant, Presence of pores and hydrogel composition influence tensile properties of scaffolds fabricated from well-defined sphere templates, J. Biomed. Mater. Res. B Appl. Biomater. 96 (2) (2011) 294–302.
[32] A.L.M. Ruela, A.G. Perissinato, M.E.S. de Lino, P.S. Mudrik, G.R. Pereira, Evaluation of skin absorption of drugs from topical and transdermal formulations, Braz. J. Pharm. Sci. 52 (3) (2016) 527–544.
[33] R. Narayanaswamy, V.P. Torchilin, Hydrogels and their applications in targeted drug delivery, Molecules 24 (3) (2019) 603.

[34] J. Siepmann, N. Peppas, Hydrophilic matrices for controlled drug delivery: an improved mathematical model to predict the resulting drug release kinetics (the "sequential layer" model), Pharm. Res. 17 (10) (2000) 1290–1298.
[35] J. Siepmann, F. Siepmann, Modeling of diffusion controlled drug delivery, J. Control. Release 161 (2) (2012) 351–362.
[36] S.A. Stewart, J. Domínguez-Robles, R.F. Donnelly, E. Larrañeta, Implantable polymeric drug delivery devices: classification, manufacture, materials, and clinical applications, Polymers 10 (12) (2018) 1379.
[37] H. Patel, D.R. Panchal, U. Patel, T. Brahmbhatt, M. Suthar, Matrix type drug delivery system: a review, J. Pharm. Sci. Biosci. Res. 1 (3) (2011) 143–151.
[38] A. Reddy, R. Karthikeyan, R. Vejandla, P. Babu, Controlled release matrix drug delivery system—a review, Int. J. Allied Med. Sci. Clin. Res. 5 (2017) 2017–2384.
[39] D. Caccavo, S. Cascone, G. Lamberti, A.A. Barba, A. Larsson, Swellable hydrogel-based systems for controlled drug delivery, in: A. Demir Sezer (Ed.), Smart Drug Delivery System, IntechOpen, 2016, pp. 237–303.
[40] K.J. Buschow, R.W. Cahn, M.C. Flemings, B. Ilschner, E.J. Kramer, S. Mahajan, Encyclopedia of materials, Sci. Technol. 1 (2001) 11.
[41] Q. Shi, H. Liu, D. Tang, Y. Li, X. Li, F. Xu, Bioactuators based on stimulus-responsive hydrogels and their emerging biomedical applications, NPG Asia Mater. 11 (1) (2019) 1–21.
[42] K.R. Rao, K.P. Devi, Swelling controlled-release systems: recent developments and applications, Int. J. Pharm. 48 (1–3) (1988) 1–13.
[43] V.K. Thakur, M.K. Thakur, Handbook of Polymers for Pharmaceutical Technologies, Wiley Online Library, 2015.
[44] C. Engineer, J. Parikh, A. Raval, Review on hydrolytic degradation behavior of biodegradable polymers from controlled drug delivery system, Trends Biomater. Artif. Organs 25 (2) (2011).
[45] C.R. Rabin, S.J. Siegel, Delivery systems and dosing for antipsychotics, in: Current Antipsychotics, Springer, 2012, pp. 267–298.
[46] Y.N. Sudhakar, M. Selvakumar, D.K. Bhat, Methods of preparation of biopolymer electrolytes (Chapter 2), in: Y.N. Sudhakar, M. Selvakumar, D.K. Bhat (Eds.), Biopolymer Electrolytes, Elsevier, 2018, pp. 35–52.
[47] N. Chirani, L. Gritsch, F.L. Motta, S. Fare, History and applications of hydrogels, J. Biomed. Sci. 4 (2) (2015).
[48] S.K. Gulrez, S. Al-Assaf, G.O. Phillips, Hydrogels: methods of preparation, characterisation and applications, in: A. Carpi (Ed.), Progress in Molecular and Environmental Bioengineering—From Analysis and Modeling to Technology Applications, IntechOpen, 2011, pp. 117–150.
[49] C. Migliaresi, A. Motta, Scaffolds for Tissue Engineering: Biological Design, Materials, and Fabrication, Jenny Stanford Publishing, 2014.
[50] R. Parhi, Cross-linked hydrogel for pharmaceutical applications: a review, Adv. Pharm. Bull. 7 (4) (2017) 515–530.
[51] K. Pal, B. Behera, S. Roy, S. Sekhar Ray, G. Thakur, Chitosan based delivery systems on a length scale: nano to macro, Soft Mater. 11 (2) (2013) 125–142.
[52] K. Ashara, M. Soniwala, K. Shah, Emulgel: a novel drug delivery system, J. Pak. Assoc. Dermatol. 26 (3) (2017) 244–249.
[53] D.J. McClements, Emulsion design to improve the delivery of functional lipophilic components, Annu. Rev. Food Sci. Technol. 1 (2010) 241–269.
[54] S.K. Sah, A. Badola, B.K. Nayak, Emulgel: magnifying the application of topical drug delivery, Indian J. Pharm. Biol. Res. 5 (01) (2017) 25–33.
[55] A. Shakeel, F.R. Lupi, D. Gabriele, N. Baldino, B. De Cindio, Bigels: a unique class of materials for drug delivery applications, Soft Mater. 16 (2) (2018) 77–93.

[56] B. Choudhary, S.R. Paul, S.K. Nayak, D. Qureshi, K. Pal, Synthesis and biomedical applications of filled hydrogels, in: Polymeric Gels, Elsevier, 2018, pp. 283–302.
[57] C. Schmitt, C. Sanchez, S. Desobry-Banon, J. Hardy, Structure and technofunctional properties of protein-polysaccharide complexes: a review, Crit. Rev. Food Sci. Nutr. 38 (8) (1998) 689–753.
[58] A. Boonmahitthisud, L. Nakajima, K.D. Nguyen, T. Kobayashi, Composite effect of silica nanoparticle on the mechanical properties of cellulose-based hydrogels derived from cottonseed hulls, J. Appl. Polym. Sci. 134 (10) (2017) 1–12.
[59] P. Thoniyot, M.J. Tan, A.A. Karim, D.J. Young, X.J. Loh, Nanoparticle–hydrogel composites: concept, design, and applications of these promising, multi-functional materials, Adv. Sci. 2 (1–2) (2015) 1400010.
[60] R. Ghizal, G.R. Fatima, S. Srivastava, Smart polymers and their applications, Int. J. Eng. Technol. Manag. Appl. Sci. 2 (4) (2014) 104–115.
[61] F. Greco, V. Mattoli, Introduction to active smart materials for biomedical applications, in: Piezoelectric Nanomaterials for Biomedical Applications, Springer, 2012, pp. 1–27.
[62] E. Cabane, X. Zhang, K. Langowska, C.G. Palivan, W. Meier, Stimuli-responsive polymers and their applications in nanomedicine, Biointerphases 7 (1) (2012) 9.
[63] D. Wang, Y. Jin, X. Zhu, D. Yan, Synthesis and applications of stimuli-responsive hyperbranched polymers, Prog. Polym. Sci. 64 (2017) 114–153.
[64] D. Qureshi, S.K. Nayak, S. Maji, A. Anis, D. Kim, K. Pal, Environment sensitive hydrogels for drug delivery applications, Eur. Polym. J. (2019) 109220.
[65] P. Schattling, F.D. Jochum, P. Theato, Multi-stimuli responsive polymers—the all-in-one talents, Polym. Chem. 5 (1) (2014) 25–36.
[66] A.P. Mathew, K.-H. Cho, S. Uthaman, C.-S. Cho, I.-K. Park, Stimuli-regulated smart polymeric systems for gene therapy, Polymers 9 (4) (2017) 152.
[67] N.A. Pattanashetti, G.B. Heggannavar, M.Y. Kariduraganavar, Smart biopolymers and their biomedical applications, Proc. Manuf. 12 (2017) 263–279.
[68] M.R. Aguilar, C. Elvira, A. Gallardo, B. Vázquez, J. Román, Smart polymers and their applications as biomaterials, Top. Tissue Eng. 3 (6) (2007).
[69] J.M. Knipe, N.A. Peppas, Multi-responsive hydrogels for drug delivery and tissue engineering applications, Regen. Biomater. 1 (1) (2014) 57–65.
[70] M. Ebara, Y. Kotsuchibashi, R. Narain, N. Idota, Y.-J. Kim, J.M. Hoffman, et al., Smart Biomaterials, Springer, 2014.
[71] S. Gao, G. Tang, D. Hua, R. Xiong, J. Han, S. Jiang, et al., Stimuli-responsive bio-based polymeric systems and their applications, J. Mater. Chem. B 7 (5) (2019) 709–729.
[72] R. Kulkarni, S. Biswanath, Electrically responsive smart hydrogels in drug delivery: a review, J. Appl. Biomater. Biomech. 5 (3) (2007) 125–139.
[73] A. Dehsorkhi, V. Castelletto, I.W. Hamley, Self-assembling amphiphilic peptides, J. Pept. Sci. 20 (7) (2014) 453–467.
[74] D. Lombardo, M.A. Kiselev, S. Magazù, P. Calandra, Amphiphiles self-assembly: basic concepts and future perspectives of supramolecular approaches, Adv. Condens. Matter Phys. 2015 (2015) 1–22.
[75] H.J. Chung, T.G. Park, Self-assembled and nanostructured hydrogels for drug delivery and tissue engineering, Nano Today 4 (5) (2009) 429–437.
[76] E. Chung, N. Shah, R.N. Shah, Nanomaterials for cartilage tissue engineering, in: Nanomaterials in Tissue Engineering, Elsevier, 2013, pp. 301–334.
[77] S. Gopi, A. Amalraj, S. Thomas, Effective drug delivery system of biopolymers based on nanomaterials and hydrogels—a review, Drug Des. 5 (129) (2016). 2169-0138. 100012.
[78] S. Nimesh, Gene Therapy: Potential Applications of Nanotechnology, Elsevier, 2013.

[79] R. Cheung, T. Ng, J. Wong, W. Chan, Chitosan: an update on potential biomedical and pharmaceutical applications, Mar. Drugs 13 (8) (2015) 5156–5186.
[80] R. Parhi, Drug delivery applications of chitin and chitosan: a review, Environ. Chem. Lett. (2020), https://doi.org/10.1007/s10311-020-00963-5.
[81] V. Manigandan, R. Karthik, S. Ramachandran, S. Rajagopal, Chitosan applications in food industry, in: Biopolymers for Food Design, Elsevier, 2018, pp. 469–491.
[82] M.A. Mohammed, J. Syeda, K.M. Wasan, E.K. Wasan, An overview of chitosan nanoparticles and its application in non-parenteral drug delivery, Pharmaceutics 9 (4) (2017) 53.
[83] C. Saikia, P. Gogoi, T. Maji, Chitosan: a promising biopolymer in drug delivery applications, J. Mol. Genet. Med. 4 (006) (2015).
[84] E. Szymańska, K. Winnicka, Stability of chitosan—a challenge for pharmaceutical and biomedical applications, Mar. Drugs 13 (4) (2015) 1819–1846.
[85] M.-C. Chen, F.-L. Mi, Z.-X. Liao, C.-W. Hsiao, K. Sonaje, M.-F. Chung, et al., Recent advances in chitosan-based nanoparticles for oral delivery of macromolecules, Adv. Drug Deliv. Rev. 65 (6) (2013) 865–879.
[86] H. Liu, C. Wang, C. Li, Y. Qin, Z. Wang, F. Yang, et al., A functional chitosan-based hydrogel as a wound dressing and drug delivery system in the treatment of wound healing, RSC Adv. 8 (14) (2018) 7533–7549.
[87] S. Potiwiput, H. Tan, G. Yuan, S. Li, T. Zhou, J. Li, et al., Dual-crosslinked alginate/carboxymethyl chitosan hydrogel containing in situ synthesized calcium phosphate particles for drug delivery application, Mater. Chem. Phys. 241 (2019).
[88] M. Martínez-Martínez, G. Rodríguez-Berna, M. Bermejo, I. Gonzalez-Alvarez, M. Gonzalez-Alvarez, V. Merino, Covalently crosslinked organophosphorous derivatives-chitosan hydrogel as a drug delivery system for oral administration of camptothecin, Eur. J. Pharm. Biopharm. 136 (2019) 174–183.
[89] H. Hamedi, S. Moradi, S.M. Hudson, A.E. Tonelli, Chitosan based hydrogels and their applications for drug delivery in wound dressings: a review, Carbohydr. Polym. 199 (2018) 445–460.
[90] J.H. Hamman, Chitosan based polyelectrolyte complexes as potential carrier materials in drug delivery systems, Mar. Drugs 8 (4) (2010) 1305–1322.
[91] M. Motiei, V. Sedlařík, L.A. Lucia, H. Fei, L. Münster, Stabilization of chitosan-based polyelectrolyte nanoparticle cargo delivery biomaterials by a multiple ionic cross-linking strategy, Carbohydr. Polym. 231 (2020) 115709.
[92] H. Shi, Y. Wang, Z. Bao, D. Lin, H. Liu, A. Yu, et al., Thermosensitive glycol chitosan-based hydrogel as a topical ocular drug delivery system for enhanced ocular bioavailability, Int. J. Pharm. 570 (2019) 118688.
[93] L. Wang, B. Li, F. Xu, Z. Xu, D. Wei, Y. Feng, et al., UV-crosslinkable and thermo-responsive chitosan hybrid hydrogel for NIR-triggered localized on-demand drug delivery, Carbohydr. Polym. 174 (2017) 904–914.
[94] J. Xu, M. Tam, S. Samaei, S. Lerouge, J. Barralet, M.M. Stevenson, M. Cerruti, Mucoadhesive chitosan hydrogels as rectal drug delivery vessels to treat ulcerative colitis, Acta Biomater. 48 (2017) 247–257.
[95] D. Ailincai, L. Tartau Mititelu, L. Marin, Drug delivery systems based on biocompatible imino-chitosan hydrogels for local anticancer therapy, Drug Deliv. 25 (1) (2018) 1080–1090.
[96] M. Zhang, J. Wang, Z. Jin, Supramolecular hydrogel formation between chitosan and hydroxypropyl β-cyclodextrin via Diels-Alder reaction and its drug delivery, Int. J. Biol. Macromol. 114 (2018) 381–391.
[97] K. Liu, R.-L. Huang, X.-Q. Zha, Q.-M. Li, L.-H. Pan, J.-P. Luo, Encapsulation and sustained release of curcumin by a composite hydrogel of lotus root amylopectin and chitosan, Carbohydr. Polym. 232 (2019) 115810.
[98] Y. Liang, X. Zhao, P.X. Ma, B. Guo, Y. Du, X. Han, pH-responsive injectable hydrogels with mucosal adhesiveness based on chitosan-grafted-dihydrocaffeic acid

and oxidized pullulan for localized drug delivery, J. Colloid Interface Sci. 536 (2019) 224–234.
[99] J. Long, A. Etxabide Etxeberria, C. Kornelsen, A. Nand, S. Ray, C. Bunt, A. Seyfoddin, Development of a long-term drug delivery system with levonorgestrel loaded chitosan microspheres embedded in poly (vinyl alcohol) hydrogel, ACS Appl. Bio Mater. 2 (7) (2019) 2766–2779.
[100] A.M. Craciun, L.M. Tartau, M. Pinteala, L. Marin, Nitrosalicyl-imine-chitosan hydrogels based drug delivery systems for long term sustained release in local therapy, J. Colloid Interface Sci. 536 (2019) 196–207.
[101] Z. Naderi, J. Azizian, E. Moniri, N. Farhadyar, Synthesis and characterization of carboxymethyl cellulose/β-cyclodextrin/chitosan hydrogels and investigating the effect of magnetic nanoparticles (Fe_3O_4) on a novel carrier for a controlled release of methotrexate as drug delivery, J. Inorg. Organomet. Polym. Mater. (2019) 1–13.
[102] J. Gao, Y. Xu, Y. Zheng, X. Wang, S. Li, G. Yan, et al., pH-sensitive carboxymethyl chitosan hydrogels via acid-labile ortho ester linkage as an implantable drug delivery system, Carbohydr. Polym. 225 (2019) 115237.
[103] C. Lou, X. Tian, H. Deng, Y. Wang, X. Jiang, Dialdehyde-β-cyclodextrin-crosslinked carboxymethyl chitosan hydrogel for drug release, Carbohydr. Polym. 231 (2020) 115678.
[104] N. Kayra, A. Aytekin, Synthesis of cellulose-based hydrogels: preparation, formation, mixture, and modification, in: Cellulose-Based Superabsorbent Hydrogels, 2019, pp. 407–434.
[105] A. Sannino, C. Demitri, M. Madaghiele, Biodegradable cellulose-based hydrogels: design and applications, Materials 2 (2) (2009) 353–373.
[106] D.E. Ciolacu, D.M. Suflet, Cellulose-based hydrogels for medical/pharmaceutical applications, in: Biomass as Renewable Raw Material to Obtain Bioproducts of High-Tech Value, Elsevier, 2018, pp. 401–439.
[107] S.F. Kabir, P.P. Sikdar, B. Haque, M.R. Bhuiyan, A. Ali, M. Islam, Cellulose-based hydrogel materials: chemistry, properties and their prospective applications, Prog. Biomater. 7 (3) (2018) 153–174.
[108] B. Sun, M. Zhang, J. Shen, Z. He, P. Fatehi, Y. Ni, Applications of cellulose-based materials in sustained drug delivery systems, Curr. Med. Chem. 26 (14) (2019) 2485–2501.
[109] M. Mishra, Handbook of Encapsulation and Controlled Release, CRC Press, 2015.
[110] Y. Liu, Y. Sui, C. Liu, C. Liu, M. Wu, B. Li, Y. Li, A physically crosslinked polydopamine/nanocellulose hydrogel as potential versatile vehicles for drug delivery and wound healing, Carbohydr. Polym. 188 (2018) 27–36.
[111] W. Treesuppharat, P. Rojanapanthu, C. Siangsanoh, H. Manuspiya, S. Ummartyotin, Synthesis and characterization of bacterial cellulose and gelatin-based hydrogel composites for drug-delivery systems, Biotechnol. Rep. 15 (2017) 84–91.
[112] Q. Xu, Y. Ji, Q. Sun, Y. Fu, Y. Xu, L. Jin, Fabrication of cellulose nanocrystal/chitosan hydrogel for controlled drug release, Nanomaterials 9 (2) (2019) 253.
[113] W. Xu, L. Huang, W. Jin, P. Ge, B.R. Shah, D. Zhu, J. Jing, Encapsulation and release behavior of curcumin based on nanoemulsions-filled alginate hydrogel beads, Int. J. Biol. Macromol. 134 (2019) 210–215.
[114] S. Mukherjee, S. Majee, G.R. Biswas, Formulation and in vitro characterisation of soybean oil-HPMCK4M based bigel matrix for topical drug delivery, Int. J. Appl. Pharm. 11 (5) (2019), https://doi.org/10.22159/ijap.2019v11i5.33987.
[115] S.H. Park, H.S. Shin, S.N. Park, A novel pH-responsive hydrogel based on carboxymethyl cellulose/2-hydroxyethyl acrylate for transdermal delivery of naringenin, Carbohydr. Polym. 200 (2018) 341–352.
[116] S.Y. Ooi, I. Ahmad, M.C.I.M. Amin, Cellulose nanocrystals extracted from rice husks as a reinforcing material in gelatin hydrogels for use in controlled drug delivery systems, Ind. Crop. Prod. 93 (2016) 227–234.

[117] V.S. Ghorpade, A.V. Yadav, R.J. Dias, Citric acid crosslinked cyclodextrin/hydroxypropylmethylcellulose hydrogel films for hydrophobic drug delivery, Int. J. Biol. Macromol. 93 (2016) 75–86.
[118] S. Sood, V.K. Gupta, S. Agarwal, K. Dev, D. Pathania, Controlled release of antibiotic amoxicillin drug using carboxymethyl cellulose-cl-poly (lactic acid-co-itaconic acid) hydrogel, Int. J. Biol. Macromol. 101 (2017) 612–620.
[119] X. Yang, G. Liu, L. Peng, J. Guo, L. Tao, J. Yuan, et al., Highly efficient self-healable and dual responsive cellulose-based hydrogels for controlled release and 3D cell culture, Adv. Funct. Mater. 27 (40) (2017) 1703174.
[120] D. Jeong, S.-W. Joo, Y. Hu, V.V. Shinde, E. Cho, S. Jung, Carboxymethyl cellulose-based superabsorbent hydrogels containing carboxymehtyl β-cyclodextrin for enhanced mechanical strength and effective drug delivery, Eur. Polym. J. 105 (2018) 17–25.
[121] S. Javanbakht, N. Nazari, R. Rakhshaei, H. Namazi, Cu-crosslinked carboxymethylcellulose/naproxen/graphene quantum dot nanocomposite hydrogel beads for naproxen oral delivery, Carbohydr. Polym. 195 (2018) 453–459.
[122] S. Deng, X. Li, W. Yang, K. He, X. Ye, Injectable in situ cross-linking hyaluronic acid/carboxymethyl cellulose based hydrogels for drug release, J. Biomater. Sci. Polym. Ed. 29 (13) (2018) 1643–1655.
[123] Y. Hu, S. Zhang, D. Han, Z. Ding, S. Zeng, X. Xiao, Construction and evaluation of the hydroxypropyl methyl cellulose-sodium alginate composite hydrogel system for sustained drug release, J. Polym. Res. 25 (7) (2018) 148.
[124] S. Javanbakht, H. Namazi, Doxorubicin loaded carboxymethyl cellulose/graphene quantum dot nanocomposite hydrogel films as a potential anticancer drug delivery system, Mater. Sci. Eng. C 87 (2018) 50–59.
[125] R. Rakhshaei, H. Namazi, H. Hamishehkar, M. Rahimi, Graphene quantum dot cross-linked carboxymethyl cellulose nanocomposite hydrogel for pH-sensitive oral anticancer drug delivery with potential bioimaging properties, Int. J. Biol. Macromol. 150 (2019) 1121–1129.
[126] A.G. Grigoras, Drug delivery systems based on pullulan polysaccharides and their derivatives, in: Pharmaceuticals from Microbes, Springer, 2019, pp. 99–141.
[127] R. Lochhead, The use of polymers in cosmetic products, in: K. Sakamoto, R. Lochhead, H. Maibach, Y. Yamashita (Eds.), Cosmetic Science and Technology: Theoretical Principles and Applications, Elsevier, 2017, pp. 171–221.
[128] A. Grenha, S. Rodrigues, Pullulan-based nanoparticles: future therapeutic applications in transmucosal protein delivery, Ther. Deliv. 4 (11) (2013) 1339–1341.
[129] G. Patel, C. Dalwadi, 19—Polymeric nanobiomaterials for tumor targeting, in: A.M. Holban, A.M. Grumezescu (Eds.), Nanoarchitectonics for Smart Delivery and Drug Targeting, William Andrew Publishing, 2016, pp. 541–564.
[130] A. Grumezescu, A. Butu, Materials for Biomedical Engineering: Organic Micro and Nanostructures, Elsevier Science, 2019.
[131] M. Abhilash, D. Thomas, Biopolymers for biocomposites and chemical sensor applications, in: Biopolymer Composites in Electronics, Elsevier, 2017, pp. 405–435.
[132] R.S. Singh, N. Kaur, J.F. Kennedy, Pullulan and pullulan derivatives as promising biomolecules for drug and gene targeting, Carbohydr. Polym. 123 (2015) 190–207.
[133] V. Grumezescu, A. Grumezescu, Materials for Biomedical Engineering: Inorganic Micro-and Nanostructures, Elsevier, 2019.
[134] G.E. Cinay, P. Erkoc, M. Alipour, Y. Hashimoto, Y. Sasaki, K. Akiyoshi, S. Kizilel, Nanogel-integrated pH-responsive composite hydrogels for controlled drug delivery, ACS Biomater. Sci. Eng. 3 (3) (2017) 370–380.
[135] C. Cheng, X. Zhang, Y. Meng, L. Chen, Q. Zhang, Development of a dual drug-loaded hydrogel delivery system for enhanced cancer therapy: in situ formation, degradation and synergistic antitumor efficiency, J. Mater. Chem. B 5 (43) (2017) 8487–8497.

[136] L. Zhang, J. Liu, X. Zheng, A. Zhang, X. Zhang, K. Tang, Pullulan dialdehyde crosslinked gelatin hydrogels with high strength for biomedical applications, Carbohydr. Polym. 216 (2019) 45–53.
[137] R. Zhang, L. Lei, Q. Song, X. Li, Calcium ion cross-linking alginate/dexamethasone sodium phosphate hybrid hydrogel for extended drug release, Colloids Surf. B: Biointerfaces 175 (2019) 569–575.
[138] Y. Zhang, X. Li, N. Zhong, Y. Huang, K. He, X. Ye, Injectable in situ dual-crosslinking hyaluronic acid and sodium alginate based hydrogels for drug release, J. Biomater. Sci. Polym. Ed. 30 (12) (2019) 995–1007.
[139] K. Lin, J. Yi, X. Mao, H. Wu, L.-M. Zhang, L. Yang, Glucose-sensitive hydrogels from covalently modified carboxylated pullulan and concanavalin A for smart controlled release of insulin, React. Funct. Polym. 139 (2019) 112–119.
[140] D. Lachowicz, P. Mielczarek, R. Wirecka, K. Berent, A. Karewicz, M. Szuwarzyński, S. Zapotoczny, Nanohydrogels based on self-assembly of cationic pullulan and anionic dextran derivatives for efficient delivery of Piroxicam, Pharmaceutics 11 (12) (2019) 622.
[141] S.R. Soni, A. Ghosh, Exploring pullulan-poly (vinyl alcohol) interpenetrating network microspheres as controlled release drug delivery device, Carbohydr. Polym. 174 (2017) 812–822.
[142] J.A. de Lima, T.C. Paines, M.H. Motta, W.B. Weber, S.S. dos Santos, L. Cruz, da Silva, C. d. B., Novel Pemulen/pullulan blended hydrogel containing clotrimazole-loaded cationic nanocapsules: evaluation of mucoadhesion and vaginal permeation, Mater. Sci. Eng. C 79 (2017) 886–893.
[143] J.M. Choi, B. Lee, D. Jeong, K.H. Park, E.-J. Choi, Y.-J. Jeon, et al., Characterization and regulated naproxen release of hydroxypropyl cyclosophoraose-pullulan microspheres, J. Ind. Eng. Chem. 48 (2017) 108–118.
[144] N. Henry, J. Clouet, A. Fragale, L. Griveau, C. Chédeville, J. Véziers, et al., Pullulan microbeads/Si-HPMC hydrogel injectable system for the sustained delivery of GDF-5 and TGF-β1: new insight into intervertebral disc regenerative medicine, Drug Deliv. 24 (1) (2017) 999–1010.
[145] T. Li, X. Song, C. Weng, X. Wang, J. Wu, L. Sun, et al., Enzymatically crosslinked and mechanically tunable silk fibroin/pullulan hydrogels for mesenchymal stem cells delivery, Int. J. Biol. Macromol. 115 (2018) 300–307.
[146] J.H. Jeong, S.K. Back, J.H. An, N.S. Lee, D.K. Kim, C.S. Na, et al., Topical film prepared with Rhus verniciflua extract-loaded pullulan hydrogel for atopic dermatitis treatment, J. Biomed. Mater. Res. B Appl. Biomater. 107 (7) (2019) 2325–2334.
[147] S.B. Patil, S.Z. Inamdar, K.K. Das, K.G. Akamanchi, A.V. Patil, A.C. Inamadar, et al., Tailor-made electrically-responsive poly(acrylamide)-graft-pullulan copolymer based transdermal drug delivery systems: synthesis, characterization, in-vitro and ex-vivo evaluation, J. Drug Deliv. Sci. Technol. (2020) 101525, https://doi.org/10.1016/j.jddst.2020.101525.
[148] C. Chircov, A.M. Grumezescu, L.E. Bejenaru, Hyaluronic acid-based scaffolds for tissue engineering, Romanian J. Morphol. Embryol. 59 (2018) 71–76.
[149] M.K. Cowman, T.A. Schmidt, P. Raghavan, A. Stecco, Viscoelastic properties of hyaluronan in physiological conditions, F1000Research 4 (2015) 1–13.
[150] M. Bantzi, S. Rigol, A. Giannis, Synthesis of a hexasaccharide partial sequence of hyaluronan for click chemistry and more, Beilstein J. Org. Chem. 11 (1) (2015) 604–607.
[151] Y. Zhu, Z. Pang, Hyaluronic acid in drug delivery applications, in: Natural Polysaccharides in Drug Delivery and Biomedical Applications, Elsevier, 2019, pp. 307–325.
[152] G. Huang, H. Huang, Application of hyaluronic acid as carriers in drug delivery, Drug Deliv. 25 (1) (2018) 766–772.
[153] A. Fallacara, E. Baldini, S. Manfredini, S. Vertuani, Hyaluronic acid in the third millennium, Polymers 10 (7) (2018) 701.

[154] A.A.M. Shimojo, A.M.B. Pires, R. Lichy, A.A. Rodrigues, M.H.A. Santana, The crosslinking degree controls the mechanical, rheological, and swelling properties of hyaluronic acid microparticles, J. Biomed. Mater. Res. A 103 (2) (2015) 730–737.
[155] X.Z. Shu, G.D. Prestwich, Therapeutic biomaterials from chemically modified hyaluronan, in: H.G. Garg, C.A. Hales (Eds.), Chemistry and Biology of Hyaluronan, Elsevier, 2004, pp. 475–504.
[156] X. Xu, Z. Zeng, Z. Huang, Y. Sun, Y. Huang, J. Chen, et al., Near-infrared light-triggered degradable hyaluronic acid hydrogel for on-demand drug release and combined chemo-photodynamic therapy, Carbohydr. Polym. 229 (2020) 115394.
[157] X. Ma, T. Xu, W. Chen, H. Qin, B. Chi, Z. Ye, Injectable hydrogels based on the hyaluronic acid and poly (γ-glutamic acid) for controlled protein delivery, Carbohydr. Polym. 179 (2018) 100–109.
[158] H.-Y. Lee, H.-E. Kim, S.-H. Jeong, One-pot synthesis of silane-modified hyaluronic acid hydrogels for effective antibacterial drug delivery via sol–gel stabilization, Colloids Surf. B: Biointerfaces 174 (2019) 308–315.
[159] K.-Y. Lu, Y.-C. Lin, H.-T. Lu, Y.-C. Ho, S.-C. Weng, M.-L. Tsai, F.-L. Mi, A novel injectable in situ forming gel based on carboxymethyl hexanoyl chitosan/hyaluronic acid polymer blending for sustained release of berberine, Carbohydr. Polym. 206 (2019) 664–673.
[160] R. Zhang, X. Li, K. He, X. Sheng, S. Deng, Y. Shen, et al., Preparation and properties of redox responsive modified hyaluronic acid hydrogels for drug release, Polym. Adv. Technol. 28 (12) (2017) 1759–1763.
[161] R. Egbu, S. Brocchini, P.T. Khaw, S. Awwad, Antibody loaded collapsible hyaluronic acid hydrogels for intraocular delivery, Eur. J. Pharm. Biopharm. 124 (2018) 95–103.
[162] A.R. Kim, S.L. Lee, S.N. Park, Properties and in vitro drug release of pH-and temperature-sensitive double cross-linked interpenetrating polymer network hydrogels based on hyaluronic acid/poly (N-isopropylacrylamide) for transdermal delivery of luteolin, Int. J. Biol. Macromol. 118 (2018) 731–740.
[163] J. Huang, J. Ren, G. Chen, Z. Li, Y. Liu, G. Wang, X. Wu, Tunable sequential drug delivery system based on chitosan/hyaluronic acid hydrogels and PLGA microspheres for management of non-healing infected wounds, Mater. Sci. Eng. C 89 (2018) 213–222.
[164] Z. Atoufi, S.K. Kamrava, S.M. Davachi, M. Hassanabadi, S.S. Garakani, R. Alizadeh, et al., Injectable PNIPAM/hyaluronic acid hydrogels containing multipurpose modified particles for cartilage tissue engineering: synthesis, characterization, drug release and cell culture study, Int. J. Biol. Macromol. 139 (2019) 1168–1181.
[165] S. Chatterjee, P.C.-L. Hui, C.-W. Kan, W. Wang, Dual-responsive (pH/temperature) Pluronic F-127 hydrogel drug delivery system for textile-based transdermal therapy, Sci. Rep. 9 (1) (2019) 1–13.
[166] K.M. Hosny, H.M. Aldawsari, R.H. Bahmdan, A.M. Sindi, M. Kurakula, M.M. Alrobaian, et al., Preparation, optimization, and evaluation of hyaluronic acid-based hydrogel loaded with miconazole self-nanoemulsion for the treatment of oral thrush, AAPS PharmSciTech 20 (7) (2019) 297.
[167] S.Y. Yao, M.L. Shen, S.J. Li, X.D. Wu, M.M. Zhang, L.N. Ma, Y.P. Li, Application of a mechanically responsive, inflammatory macrophage-targeted dual-sensitive hydrogel drug carrier for atherosclerosis, Colloids Surf. B: Biointerfaces 186 (2020) 110718.
[168] N.K. Sachan, S. Pushkar, A. Jha, A. Bhattcharya, Sodium alginate: the wonder polymer for controlled drug delivery, J. Pharm. Res. 2 (8) (2009) 1191–1199.
[169] P.N. Sudha, Industrial Applications of Marine Biopolymers, CRC Press, 2017.
[170] L. Grøndahl, G. Lawrie, A. Jejurikar, Alginate-based drug delivery devices, in: Biointegration of Medical Implant Materials, Elsevier, 2010, pp. 236–266.

[171] D. Jain, D. Bar-Shalom, Alginate drug delivery systems: application in context of pharmaceutical and biomedical research, Drug Dev. Ind. Pharm. 40 (12) (2014) 1576–1584.
[172] S. Haider, A. Haider, Hydrogels, BoD–Books on Demand, 2018.
[173] J. Roberts, P. Martens, Engineering biosynthetic cell encapsulation systems, in: Biosynthetic Polymers for Medical Applications, Elsevier, 2016, pp. 205–239.
[174] K. Chaturvedi, K. Ganguly, U.A. More, K.R. Reddy, T. Dugge, B. Naik, et al., Sodium alginate in drug delivery and biomedical areas, in: Natural Polysaccharides in Drug Delivery and Biomedical Applications, Elsevier, 2019, pp. 59–100.
[175] K. Nahar, M.K. Hossain, T.A. Khan, Alginate and its versatile application in drug delivery, J. Pharm. Sci. Res. 9 (5) (2017) 606.
[176] B. Singh, A. Kumar, Synthesis and characterization of alginate and sterculia gum based hydrogel for brain drug delivery applications, Int. J. Biol. Macromol. 148 (2020) 248–257.
[177] X. Chen, M. Fan, H. Tan, B. Ren, G. Yuan, Y. Jia, et al., Magnetic and self-healing chitosan-alginate hydrogel encapsulated gelatin microspheres via covalent cross-linking for drug delivery, Mater. Sci. Eng. C 101 (2019) 619–629.
[178] F. Bahadori, S.A. Buse, S. Akyıl, M.S. Eroğlu, Synthesis and engineering of sodium alginate/inulin core-shell nano-hydrogels for controlled-release oral delivery of 5-ASA, Org. Commun. 12 (3) (2019) 142.
[179] D.S.B. Anugrah, K. Ramesh, M. Kim, K. Hyun, K.T. Lim, Near-infrared light-responsive alginate hydrogels based on diselenide-containing cross-linkage for on demand degradation and drug release, Carbohydr. Polym. 223 (2019) 115070.
[180] S.L. Fenn, T. Miao, R.M. Scherrer, R.A. Oldinski, Dual-cross-linked methacrylated alginate sub-microspheres for intracellular chemotherapeutic delivery, ACS Appl. Mater. Interfaces 8 (28) (2016) 17775–17783.
[181] N. Sarkar, G. Sahoo, R. Das, G. Prusty, S.K. Swain, Carbon quantum dot tailored calcium alginate hydrogel for pH responsive controlled delivery of vancomycin, Eur. J. Pharm. Sci. 109 (2017) 359–371.
[182] B. Gao, L. Chen, Y. Zhao, X. Yan, X. Wang, C. Zhou, et al., Methods to prepare dopamine/polydopamine modified alginate hydrogels and their special improved properties for drug delivery, Eur. Polym. J. 110 (2019) 192–201.
[183] M. Chalanqui, S. Pentlavalli, C. McCrudden, P. Chambers, M. Ziminska, N. Dunne, H. McCarthy, Influence of alginate backbone on efficacy of thermo-responsive alginate-gP (NIPAAm) hydrogel as a vehicle for sustained and controlled gene delivery, Mater. Sci. Eng. C 95 (2019) 409–421.
[184] C. Kim, H. Kim, H. Park, K.Y. Lee, Controlling the porous structure of alginate ferrogel for anticancer drug delivery under magnetic stimulation, Carbohydr. Polym. 223 (2019) 115045.
[185] K. Raju, G. Sneha, R. Khatoon, M. Ashwini, G. Shirisha, B. Ajay, B.R. Narender, Formulation and evaluation of ornidazole topical emulgel, World J. Pharm. Pharm. Sci. 8 (7) (2019) 1179–1197.
[186] S. Noreen, S.A. Ghumman, F. Batool, B. Ijaz, M. Basharat, S. Noureen, et al., Terminalia arjuna gum/alginate in situ gel system with prolonged retention time for ophthalmic drug delivery, Int. J. Biol. Macromol. 152 (2019) 1056–1067.
[187] M. Hayati, G. Rezanejade Bardajee, M. Ramezani, S.S. Hosseini, F. Mizani, Temperature/pH/magnetic triple-sensitive nanogel–hydrogel nanocomposite for release of anticancer drug, Polym. Int. 69 (2) (2020) 156–164.
[188] S. Forouzandehdel, S. Forouzandehdel, M.R. Rami, Synthesis of a novel magnetic starch-alginic acid-based biomaterial for drug delivery, Carbohydr. Res. 487 (2020) 107889.

CHAPTER 11

Role of targeted drug delivery in cancer therapeutics

Fiona Concy Rodrigues, N.G. Devi, and Goutam Thakur
Department of Biomedical Engineering, Manipal Institute of Technology, MAHE, Manipal, India

1 Introduction

Cancer is broadly characterized by the abnormal uncontrolled proliferation of malignant cells and is one of the top causes of mortality worldwide. According to statistics, there are an estimated 13 million new cases of cancer globally out of which, 2.5 million new cases of cancer occur annually in India at any given point of time. There is a report which states that there will be an estimated 13.1 million cancer-related deaths by the year 2030 [1]. With the advancement in medical technology, it is now possible to detect and screen tumors at earlier stages and suggest a suitable treatment. However, many times there are cases in which the disease is so advanced that the treatment remains ineffective. Many approaches have been studied and new strategies have been adopted to understand and treat this deadly disease.

The treatment of cancer depends on the origin, type, stage, and spread of cancer. In usual cases, a multimodal approach is adopted with a combination treatment of surgery, chemotherapy, radiation, immunotherapy, and hormone therapy. Chemotherapy and radiation are usually adopted to reduce the incidence of further recurrence of cancer and to limits its spread. The aim of chemotherapy is to target the DNA synthesis and mitosis to inhibit the growth and rapidly dividing cells [2]. Many agents have been discovered with their respective modes of action by targeting different pathways or targets in cancer. Some of the most commonly adopted antineoplastic agents approved or under development have been depicted below (Fig. 1) [3].

One of the major setbacks associated with this treatment is the lack of selectivity between normal and diseased tissue. This nonselective treatment causes a number of adverse side effects, for example, loss of immunity, appetite, hair and feeling nausea and fatigue [4]. Despite undergoing dreadful treatments, patients often develop cases of multi drug resistance owing to the large doses of such highly toxic drugs [5]. This treatment strategy, despite inflicting harm, has been one of the most effective strategies to combat

Advances and Challenges in Pharmaceutical Technology
https://doi.org/10.1016/B978-0-12-820043-8.00008-6

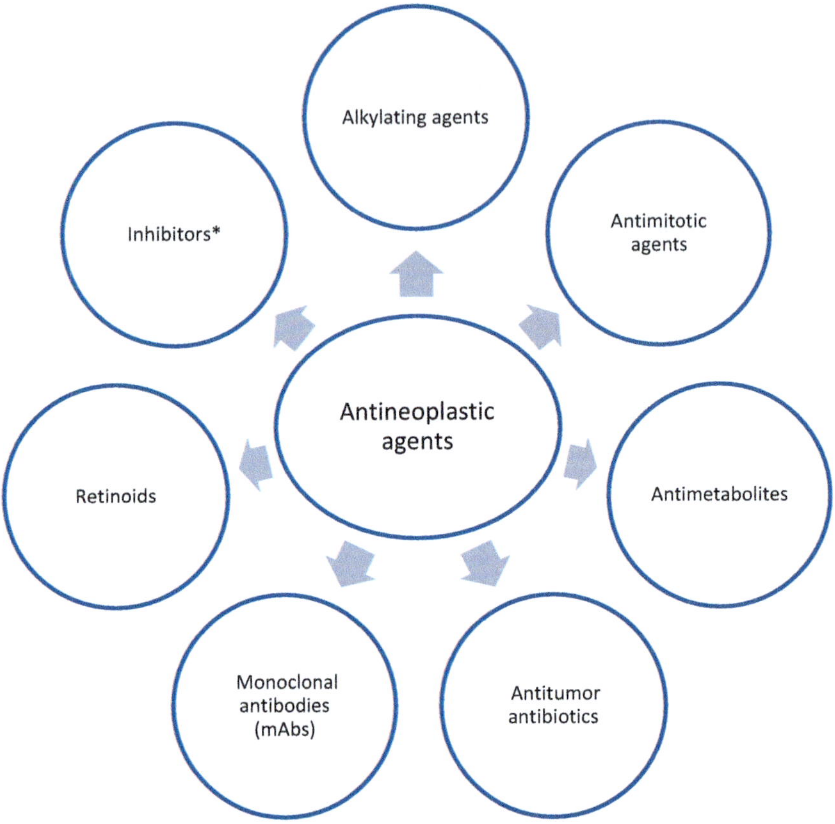

Fig. 1 Schematic representation of the different types of antineoplastic agents. *Inhibitors - Cyclin dependent kinase (CDK) inhibitors, Tyrosine kinase inhibitors, Histone deacetylase (HDAC) inhibitors, Poly(ADP-ribose) polymerase (PARP) inhibitors, Mitogen-activated protein kinase (MAPK) inhibitors, Serine/threonine-protein kinase B-Raf inhibitors, Mammalian target of rapamycin (mTOR) inhibitors, Phosphoinositide 3-kinase (PI3K) inhibitors, Ribonucleotide reductase (RNR) inhibitors, DNA methyltransferase inhibitors.

cancer. To address this issue, there is a pressing need to develop strategies to ensure that the effective dosages of the drug are carried only to the non-healthy tissue. This calls for the development of efficient targeted drug delivery systems [6, 7].

2 Need for targeted drug delivery

There is a surge in the advancements in pharmacological medicine and oncology as separate domains over the last fifty years. However, over the last decade, interdisciplinary research with respect to both the fields has called for the development of carrier-based drug delivery specific to cancer. The quest to deliver the most effective dosage of the drug to the precise site of action has driven research in this field. Here the importance of site-specific drug delivery is of even paramount importance due to the lack of specificity demonstrated by chemotherapy and radiation therapy. The aim of targeted drug delivery in the case of cancer therapy is to preferentially kill cancerous cells. Another challenge posed by the cancer cells is the transport and penetration of the therapeutically effective drug concentration to the solid tumor to facilitate the desired effect. The administration of high doses of such toxic drugs could further cause adverse effects to the patients. Hence, there is a requirement for the design and development of suitable carriers to not only improve the pharmacological effectiveness of the drug but to also reduce the systemic toxicity in these patients [8,9].

3 Targeting

Targeting cancer cells and tumors are challenging due to the uniqueness of different types of cancers. An inclination in understanding the molecular makeup of the switches of cancer progression has led to specific drugs and treatment techniques. Designing drugs that regulate molecular targets and pathways exclusive to cancer has been identified as the most suitable treatment [10]. The transporting of the drug to the site can either be direct or within a suitable drug delivery system depending on the cancer and its cell surface. The design of drug delivery systems must take into account about the type of ligand, the specific target, the microenvironment of the tumor, the changing dynamics of the tumor, the mode of delivering the ligand, and the facilitation of the conjugation of the ligand and target [11]. Drug targeting can be achieved by the exploitation of the altered pathophysiology observed in the cancerous cells. This type of targeting can be broadly classified into three types—passive targeting, active targeting and physical targeting (Fig. 2) [9,12].

3.1 Passive targeting

Passive targeting exploits the perforated vasculature demonstrated by the tumor cells and ensures the accumulation of the drug around such leaky vasculature which is termed as the EPR effect. This effect was first recognized and reported by Matsumura and Maeda in 1986 and stands for enhanced permeation and retention [13]. It relies on the low molecular weight of the drug

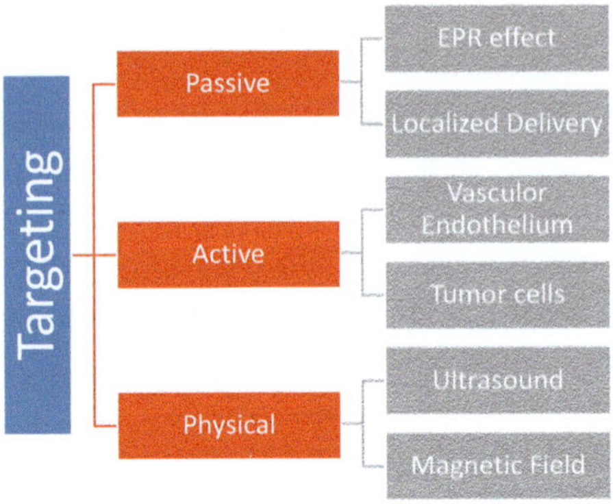

Fig. 2 Schematic representation of the different approaches to targeting tumor cells.

being able to surpass through the blood vessel owing to its small size. It has been noted that the size of the gaps usually ranges from 100 to 780nm depending on the type of tumors [14]. Hence, the drug model should be designed in a way to allow the passage through this barrier along with the understanding of the elevated and suppressed levels of molecular targets. The delivery systems must be designed to sustain therapeutic activity during circulation. Other factors such as the size of the tumor, the surface of the tumor, angiogenesis, degree of tumor vascularization and degree of tumor penetration must also be taken into account while designing suitable drug delivery systems [15,16]. As depicted in the figure (Fig. 3A), where small molecules and nanocarriers are able to escape through the perforated vasculature. Another approach to EPR has been the conjugation of drugs to an appropriate polymer so that the drug can only be released at the site of action [17]. Localized delivery, which is also a mode of passive targeting, is the accumulation of the drug agents around the localized tumor site. However, this approach cannot be adopted for every type of cancer but has been effective for selected types of cancer such as prostate cancer [9,18].

3.2 Active targeting

Active targeting achieves unique interactions between the drug/drug carrier and target cells by specific ligand-target interactions. This sort of interaction can be facilitated only when the ligand and the target can be in close proximity to each other. The drug delivery systems are carried by the blood circulation and are aided to reach and attach to the target site with the help of some ligand-receptor interactions [12]. This has been illustrated by Fig. 3B. which depicts the presence of surface receptors that are specific

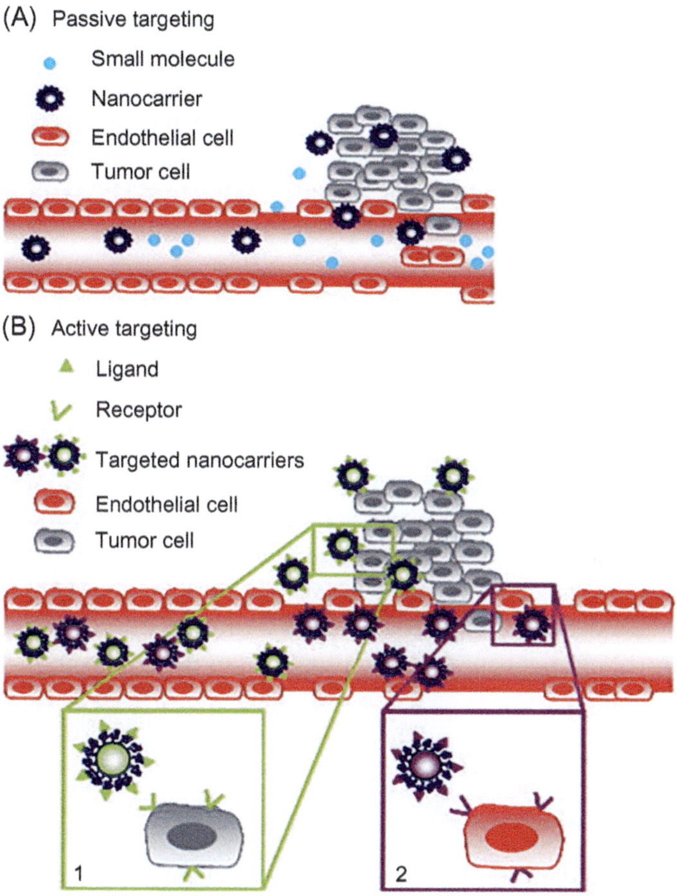

Fig. 3 Schematic representation of the mode of action in (A) Active targeting and (B) Passive targeting. *Source: Reproduced from Danhier, Fabienne, Olivier Feron, and Véronique Préat, To exploit the tumor microenvironment: passive and active tumor targeting of nanocarriers for anti-cancer drug delivery, J. Contr. Release 148 (2) (2010) 135–46 with permission from Elsevier.*

to the ligand. During the progression of cancer due to the abnormal growth and proliferation of cells, the tumor vasculature differs from the normal vasculature. Also, the existing vasculature gives rise to blood vessels—angiogenesis, which plays a crucial role in the elevation of targets and enzymes. Due to this, it is important that the drug carrier must be able to retain the blood circulation for a prolonged period of time to achieve the desired binding [9]. Many targets such as the antibodies, Endoglin (CD105), vascular endothelial growth factor (VEGF), fibroblast growth factor (FGF), endothelial growth factor (EGF), and their receptors have been exploited for this purpose [19–22]. With the advances in molecular biology, different protein biomarkers for tumor cells have been also been identified. Many such TAA—tumor-associated antigens are potential targets and can be found on the surface of tumor cells and can be selectively targeted [23].

3.3 Physical targeting

Physical targeting is a novel strategy and it employs an external stimulus to serve as a carrier and release the drug. Two of the most commonly utilized techniques are the ultrasound and the magnetic field. In micellar formulations, focussing ultrasound can trigger the release of the drug from its polymer system. In some cases, ultrasound has caused the degradation of the micelle to release the drug. This is an advantageous strategy as it is noninvasive and is simple to control the drug release at the specific site [24,25]. With a similar approach, magnetic fields can also be used for the same purpose. The formulation of the therapeutic agent is encapsulated within a magnetic drug carrier and is injected intravenously. This formulation is then externally guided by a magnetic field to accumulate locally at the site of the tumor [26,27]. These types of targeting are being extensively investigated and many examples such as iron-oxide nanoparticles as well as radioisotopes for administration are being studied [28].

4 Biomaterials for 3D tumor models

The battle with the prognosis of cancer, to develop new and efficient techniques for its early detection and treatment is ongoing. Different research groups are trying every day to make improvements in this field not only in terms of machinery and therapeutics but also in terms of new culturing systems, diagnostics as well as biomarkers and cellular delivery devices. Biomaterials are one such aide which has proved to develop cancer therapy [29]. First biomaterials were only limited to prosthetics and devices but now they are applied as sophisticated systems that interact with cells and tissues. Biomaterial devices have also been able to mimic the tumor models in vitro with the well-defined microenvironment and serve as a template to design therapeutic agents and alter the immune responses [30]. Conventional two-dimensional culture models for studying cancer biology have been employed to develop anticancer therapeutics. However, the results observed in vitro model have not been replicated in the in vivo model and have indicated a poor predictive drug response [31,32]. To build better models, the three-dimensional models have proved to be more efficient as they bear a closer resemblance to the tumor models and are able to mimic the micro-environment of the tumor [33]. For the development of better therapeutic entities, bio-medically relevant results are achieved by stimulating the phenotype and recapitulating the extracellular matrix of the tumor [34,35]. Several synthetic and natural-origin biomaterials, for instance, collagen, laminin, HA (hyaluronic acid), and reconstituted basement membrane (rBM or Matrigel) have been used regularly to imitate cancer models as they are compatible and can be remodeled [36,37]. Developments are being made to achieve clinically impactful responses, for example, the first tailored biomaterial-based cancer immunotherapy (WDVAX) has lately progressed to a clinical trial [38]. This period marks the emergence of further developments and a better understanding of biomaterials in cancer therapies and the design of new and advanced 3D in vitro models.

5 Types of targeted drug delivery

The different types of targeted drug carriers that have been extensively and most commonly been adopted for cancer therapy have been selected and discussed below. These systems have been discussed in detail, elaborating about their composition, mode of action, clinically approved formulations and finally the advances and challenges that have been identified.

5.1 Antibody-drug conjugates

Antibody-drug conjugates (ADCs) are a newly developed approach for targeted delivery of anticancer drugs, which provides a site-specific and sustained delivery of the cytotoxic drug to targeted malignant cells [2]. Investigation on the "targeted small-molecule therapeutics" for suppressing tumors by inhibiting specific molecules involved in the progression of cancer cells was initiated in the early 1970s. The origin of the ADC concept is from the "magnetic bullet" idea proposed by Paul Ehrlich [39]. ADC is the most desirable way to facilitate efficient site-specific targeted delivery of cytotoxic drug which minimizes chemo-associated side effects [40]. Currently, approximately 50 ADCs are known to be in different phases of clinical trials for cancer therapy. Over the past 20 years, four ADCs have been approved by the US Food and Drug Administration (FDA) for clinical use. The first ADC to gain FDA approval was gemtuzumab ozogamicin (Mylotarg®) in 2000 for treating acute myelogenous leukemia. The other three ADCs approved by FDA are brentuximab vedotin (Adcetris®) in

2011, and inotuzumab ozogamicin (Besponsa®) in 2017, was approved for hematologic malignancies, and trastuzumab emtansine (Kadcyla®) in 2013, which was approved for metastatic breast cancer [40,41].

The three key parts of an ADC targeted delivery system are the antibody (monoclonal antibodies), the cytotoxic drug (payload) and the linker. In ADC the monoclonal antibodies (mAbs) are covalently attached to the cytotoxic drug via the linker (Fig. 4A) [42]. The mAb should be specific for targeting an antigen that is overexpressed on tumor cells relative to normal cells. An antibody molecule in an ADC can be linked only to 2–4 payloads without varying its pharmacokinetic properties, thus the need to use highly potent cytotoxic drugs have the ability to kill the tumor cells. If the antibody molecule is attached to too many payloads, the conjugate might get recognized by the body as a damaged protein form which leads to quick clearance of the conjugate from the system [43,44]. Upon binding of the ADC with the target antigen, the ADC-antigen complex undergoes rapid internalization by receptor-mediated endocytosis, followed by releasing a sufficient quantity of the drug to kill the cell (Fig. 4B). The freely circulating cytotoxic drug induces cell apoptosis by either binding to the minor groove of DNA resulting in strand breakage, or binding to tubulin resulting in microtubule disruption [43].

The most important steps in the development of an ADC include appropriate target (antigen-binding site) selection, development of stable linkers and development of highly cytotoxic drugs [44]. The appropriate selection

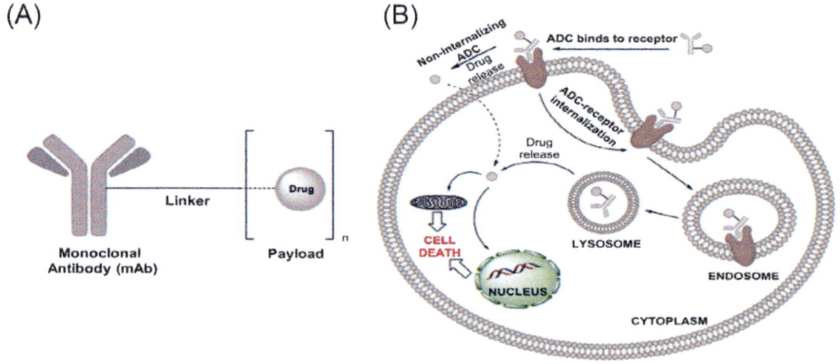

Fig. 4 (A) Schematic representation of an antibody-drug conjugate. (B) Mechanism of action of ADC. *Source: Reproduced from del Solar, Virginia, and María Contel, Metal-based antibody drug conjugates. Potential and challenges in their application as targeted therapies in cancer, J. Inorg. Biochem. 199 (2019) 110780 with permission from Elsevier.*

of a target or an antigen binding site is an important step in ADC development. In order to achieve site-specific targeted delivery of anticancer therapeutics, the targeted antigen should have a high expression on tumor cells compared to normal cells. In the development of ADC, the mAbs are selected on the basis of their internalization ability and binding affinity. The mAb selected should have high affinity and avidity for its target tumor antigen, increasing the potential for internalization of the cytotoxic drug [40,45].

The exact conjugation of the antibody and the payload has a profound influence on the selectivity, pharmacokinetics, and therapeutic efficiency of the ADC. The selection of linker and the site of conjugation plays a vital role in the overall success of an ADC. The linker is the central part of ADCs that connects the cytotoxic payload to the mAbs. In particular, the linkers must be stable in the systemic circulation (pH 7.3–7.5) to limit early payload release which prevents damage to healthy tissues while maximizing payload release inside the targeted tumor cell [41]. Once the ADC-antigen complex got internalized due to the intracellular conditions such as low pH or the mildly acidic environment in lysosomes (pH 4.5–5.0) the linker undergoes hydrolysis resulting in destabilization of the linker or mAb backbone, causing the dissociation of the cytotoxic drug [43,44]. The linkers are generally divided based on their release mechanism as, cleavable and noncleavable. Cleavable linkers are sub-classified as acid-sensitive and protease-sensitive. Acid-liable linkers remain stable in the systemic circulation and require the low-pH environment in endosomes and lysosomes to trigger hydrolysis and release of payloads. An example of the acid-sensitive cleavable linker is hydrazone, which was used as the linker in the inotuzumzb and ozogamicin ADCs. The protease-sensitive peptide linkers are more stable in systemic circulation than the acid-sensitive linkers. Peptide linkers, such as valine-citrulline are superior to hydrazone. Many studies on linker design have shown that non-cleavable linkers have improved plasma stability than cleavable linkers, and hence reduced risk of non-specific drug release [46].

The site of conjugation, where the linker-payload connects to the antibody molecule is a major factor determining the therapeutic efficacy of the ADC. In order to increase the targeted delivery and therapeutic index of the system, two conjugation methods called side chain cysteine (SH group) and lysine (NH_2 group) are used for attaching the cytotoxic drug with the mAb. The conjugation process should sustain the mAb and the drug in their original forms, without altering the biological activity of the drug and, integrity and binding capacity of the antibody to the antigen. Lysine is the most commonly used conjugation method in ADCs. They are highly expressed on the antibody surface, thus are easily accessible. Mylotarg® had utilized this conjugation method for conjugating the drug calicheamicin with the humanized anti-CD33 mAb via bifunctional acid-sensitive hydrazine linkers. Conjugation via lysines results in heterogeneous mixtures. In contrast to lysines conjugation via cysteine produces more uniform products were as they are available in much lower number in human immunoglobulin G (IgG) scaffolds. This conjugation method was used in Adcetris® to conjugate monomethyl auristatin E (MMAE) with the anti-CD30 mAb via cleavable dipeptide linkers [40].

The selection of a highly potent drug is another important step in the development of an ADC since only a limited number of payloads can be attached to the antibody. Most of the drugs used in ADCs are highly potent and are commonly targeted to tubulin or DNA. The most commonly used clinically approved chemotherapeutics are doxorubicin, methotrexate, and 5-fluorouracil. The payloads used in ADCs have an IC_{50} value in the sub-nanomolar range in the cell culture [41,47]. The payloads used in ADCs can be classified into two main classes on the basis of their mechanism of action as DNA damaging agents and tubulin inhibitors. Clicheamicins, duocarmycins, and pyrrolobenzodiazepines are examples of chemotherapeutics in this class. The cytotoxic payloads maytansinoid and auristatins are the widely used tubulin inhibiting agents. The two derivatives of auristatin payload monomethyl auristatin E (MMAE) and monomethyl auristatin F (MMAF) linked via cysteine residues of the mAb are used in several ADCs [45,48].

5.1.1 Advances and challenges

Currently, more than 80 ADCs have received FDA approval in the United States. Extensive researches are ongoing to develop ADCs with novel linkers providing a well-sustained half-life while reducing the toxicity by the effective release of the payload at the target site. Recently, Li et al. [49] developed an ADC with humanized CD-19 mAb (RC58) conjugated to oxide dumycin-7, a DNA damaging agent via maleimide-PEG-based linker for treating B-cell lymphoma. The results showed that the developed ADC had a significant effect on CD-19 positive lymphoma cells. Singh et al. [50] have designed a maytansinoid loaded ADCs with peptide linkers for the effective release of drug within targeted tumor cells. The ADCs consist of antibodies anti-EpCAM and anti-EGFR was prepared using a triglycyl peptide linker (CX) and non-cleavable SMCC linker. The results of both ADCs anti-EpCAM-CX and anti-EGFR-CX showed improved antitumor activity over SMCC ADC. Recently, Nagaoka et al. [51] studied the antitumor effect of an ADC with human monoclonal antibody (SNS-622) and emtansine (DM1) payload in targeting aspartyl-β-hydroxylase (ASPH), a type II transmembrane protein overexpressed in pancreatic ductal adenocarcinoma (PDACs) and patient-derived xenograft (PDX). The developed SNA-622-DM1 ADC showed significant inhibition of PDX and optimum effect on PDAC cells.

The major challenge associated with ADCs is payload related toxicity and low therapeutic index. Several studies on improving these limitations are going on and some of those studies have shown that the development of homogeneous ADCs, using high potent payloads and bispecific mAbs with enzyme activated linkers can address those limitations. Although ADCs are highly selective, due to the small number of antigens on the tumor cell surface and the limited DAR ratio, only a small amount of the drug reaches the targeted tumor site. It has also been noted that very limited studies in this domain have been reported, hence, research regarding ADC could be further investigated and could open up doors to experimentation with different combinations of drug and antibodies.

5.2 Liposomal drug delivery systems

Liposomes are a dominant class of targeted drug delivery systems for cancer treatment, which was first reported by Alee Bangham in 1963. Liposomes are highly biocompatible phospholipids based spherical vesicles, composed of one or several concentric lipid bilayers surrounding discrete aqueous spaces (Fig. 5A). Liposomes have several advantages as a drug carrier, such as high biocompatibility, low immunogenicity, ability to load both hydrophilic and hydrophobic agents and enhancing their solubility, protection of the encapsulated drug from the external media, increases drug half-life, ability to reduce toxicity and increase efficiency and so on. The size of liposome vesicle is an important factor in determining their circulation half-life, also the number and size of bilayers are crucial parameters that determine the amount of drug loading in the liposomes. The size of the liposome vesicles varies from nanometers (30 nm) to several micrometers (2.5 μm) [52]. On the basis of their number of bilayers, liposomes are categorized into two types: unilamellar vesicles and multilamellar vesicles. Unilamellar liposomes consist of a single phospholipid bilayer and are further divided into small unilamellar vesicles (20–100 nm) and large unilamellar vesicles (>100 nm). One

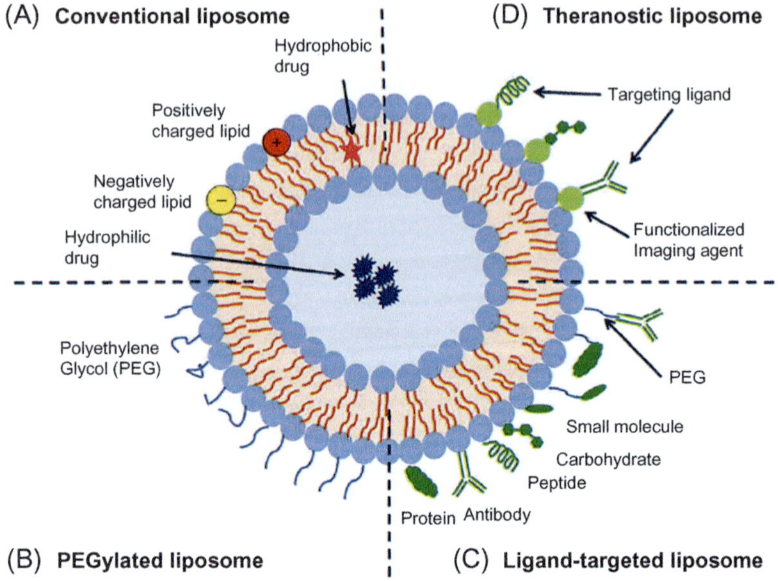

Fig. 5 (A) Conventional liposome, (B) PEGylated (stealth or long-circulating liposomes), (C) Active targeted liposomes with specific targeting moiety (ligands); (D) Theranostic liposome. *Source: Reproduced from Hossen, Sarwar et al., Smart nanocarrier-based drug delivery systems for cancer therapy and toxicity studies: a review, Adv. Res. 15 (2019) 1–18 with permission from Elsevier.*

or more unilamellar vesicles, which are separated by water combines together to form multilamellar liposomes, resulting in an onion-like structure (>500 nm). It was observed that medium-sized liposomes (40–200 nm) have longer circulation time, low renal clearance rate and more specific accumulation at tumor sites. Findings showed that liposomes with size bigger than 300 nm or smaller than 70 nm are easily recognized by the Mononuclear Phagocyte System (MPS), resulting in rapid clearance from the blood circulation and eventually accumulate in the liver, spleen and other organs [53,54]. The encapsulation of the chemotherapeutic drug in liposomes can limit drug-mediated damage to normal tissues thus improve its therapeutic potency. The most commonly used methods for targeting liposomal anticancer drugs to cancerous cells are passive targeting and active targeting.

In passive targeting, the delivery of the loaded drug is achieved by the EPR effect of the tumor vasculature. However, for effective drug delivery, the liposomes need to have prolonged systemic circulation. Longer circulation time allows the liposome to reach an effective concentration within the tumor interstitial space, resulting in the accumulation of the drug in tumor tissue through the EPR effect. The longer circulation time of liposomes can be achieved by modifying the surface of the liposomal drug delivery system with biocompatible, inert and hydrophilic polymers such as polysaccharides, synthetic polymers, etc. These surface-modified liposomes are known as stealth or long-circulating liposomes [54,55] (Fig. 5B). PEGylation is one such surface modification technique in which the surface of the carrier is coated with polyethylene glycol (PEG). Surface coating with PEG prevents opsonization and liposome uptake by macrophages. The development of PEGylated long-circulating liposomes was a major process took place in the 1990s. Liposomes functionalized with PEG were first studied by Blume et al. [56] and Klibanov et al. [57] in 1990. Their findings showed that the steric hindrance effect offered by PEG has improved surface properties of the liposome, which prevents the rapid clearance of liposomes by RES. Some of the examples for passive targeting liposome formulated drugs in the market are liposomal encapsulated doxorubicin, vincristine and cisplatin. Doxil® a stealth PEGylated liposomal doxorubicin (DOX) was the first nano-drug approved by FDA in 1995 for treating patients with AIDS-related Kapos' sarcoma. Doxil® was also used in treating ovarian cancer in 1999, and in 2007 for treating multiple myeloma [53].

Active targeting overcomes all the limitations of passive targeting by using a targeting moiety (ligands) on the surface of the carrier (Fig. 5C). In active targeting the surface of the liposome carrier is manipulated with certain antibodies, so-called immunoliposomes (ILP) [52]. In active targeting the ligands attached to the liposome surface binds to the overexpressed receptors or antigens at the tumor site, improving the tumor-selective delivery of the liposomal encapsulated drug. Thus, active targeting can be described as ligand and receptor interaction. Findings showed that some ligands can trigger the release of the liposomal encapsulated drug inside the tumor cell by receptor-mediated endocytosis process [58]. Target selection is a significant factor in the active targeted liposome drug delivery system in order to protect normal cells from damage. Upon binding of the ligand with the specific receptor on the tumor cell surface, the ligand-receptor complex undergoes rapid internalization by receptor-mediated endocytosis. Once internalized the ligand-receptor complex will be coated with the plasma membrane forming an endocytic vesicle. These endocytic vesicles fuse with an early endosome resulting in the maturation of endosome and subsequent decrease in pH (~5). At lower pH, the ligands will get dissociated from its receptors, resulting in the release of loaded drug form the liposomes by partial degradation [59]. Rituximab (Rituxan®) in 1997 and Ibritumomab tiuxetan (Zevalin®) in 2002 are two FDA approved anti-CD20 antibody used for treating CD-20 positive B-cell nonHodgkin's lymphoma. Gemtuzumab ozoamicin (Mylotarg®) is another FDA approved anti-CD33 antibody in 2000 for treating relapsed and refractory acute myelogenous leukemia [60].

An additional targeting method has been developed to increase the efficiency of passive and active targeting in site-specific drug delivery, by using an external trigger or stimulus called stimulus-responsive liposomal anticancer drug targeting. This targeting approach allows the nanoparticles to release the entrapped drug in the presence of their stimuli. In stimulus-responsive targeting the release of entrapped drug by liposome occurs by two ways: due to the exposure of the liposomes to dynamic microenvironmental changes (endogenous or exogenous stimuli) such as a change in pH, temperature, etc. present at the tumor site or by employing artificial extracorporeal physical stimulus such as heat, ultrasound, magnetic effect, etc. [52,54]. pH-sensitive liposomes are the most widely used stimulus-responsive liposome-based drug delivery system. Compared to normal tissues tumor tissues exhibit lower pH levels, this decrease in pH level triggers the release of cytotoxic drug from the nanoparticle [55]. In addition to all the above-mentioned liposomes, there are theranostic liposomes that can carry both therapeutic drugs and imaging agents (Fig. 5D).

5.2.1 Advances and challenges

The recent advancements in the liposomal drug delivery system for cancer therapy have been done by several researchers in the past decade. Tian et al. [61] synthesized a novel dual-functional paclitaxel-loaded liposome by modifying with hyaluronic acid. The results showed that hyaluronic acid-coated liposomes had a significant effect on A559 and A549/Taxol cell inhibition and could overcome multidrug resistance (MDR). Deshpande et al. [62] constructed a PEGylated dual-functional liposome for the targeted delivery of DOX, by modifying with transferrin and octa-arginine, an arginine-rich cell-penetrating peptide. The results showed that the developed liposome had a significant inhibitory effect against A2780 ovarian cancer in vitro and in vivo and with efficient DOX delivery reducing in vitro cytotoxicity. In another study, Lakkadwala et al. [63] developed transferrin and penetrating peptides modified 5-fluorouracil (5-FU) loaded liposome to evaluate their antitumor effect against brain cancer. The in vitro studies showed that the dual functionalized liposome could transport the 5-FU across the brain endothelial barrier and deliver the anticancer drug to tumor cells. Sesarman et al. [64] prepared PEGylated DOX liposome co-encapsulated with curcumin (CUR) for treating colorectal cancer. The in vivo studies of this PEG-modified co-loaded liposome system showed increased antitumor activity on C26 colon carcinoma. A study by Xiao et al. [65] aimed to develop a gadolinium modified liposome to improve the solubility of sorafenib (SF), an anticancer drug with low water solubility. They studied the in vivo and in vitro activity of the system against hepatocellular carcinoma, demonstrating that this liposome system could be used as a promising approach for the efficient delivery of SF in HCC treatment. Ramadas et al. [66] found that matrix metalloproteinases (MMPs) inhibitor, epigallocatechin gallate (EGCG) and an anticancer drug, paclitaxel co-loaded liposomal system is an effective approach for treating invasive breast cancer. In vivo studies showed that the developed liposome system significantly induced tumor cell death on MDA-MB-231 breast cancer cells. Recently, Saraswat et al. [67] developed a stealth liposome for quercetin (QC) using egg-PC, cholesterol and DSPE-PEG2000 lipids for evaluating the anticancer effects of QC. The liposome formulation significantly increased the aqueous solubility and cytotoxic efficacy of QC in cervical cancer cells in vitro, and relevant antitumor activities were observed. Li et al. [68] constructed DOX thermosensitive peptide modified liposome, to improve controlled drug release for the chemotherapy of ovarian carcinoma. The studies in vitro and in vivo showed that the peptide modified liposomes exhibited thermally controlled release of DOX by selectively accumulating in the tumor region, which significantly increases the local drug concentration. Dunne et al. [69] designed a novel thermosensitive liposome system for co-loading DOX and alvespimycin. The in vivo studies of this system showed effective heat-mediated triggering of drug release from liposomes within breast cancer tumor vasculature.

Even though antibody-based liposomal drug delivery systems showed good results in vitro, their use in in vivo studies has been obstructed due to several factors such as the sensitivity of antibodies to environmental changes (temperature, enzyme, etc.) and are not stable in organic solvents. Also due to the large size of antibodies after conjugation into the liposome surface increases the overall size of the liposomal system after conjugation resulting in rapid clearance by RES. Another major challenge in developing a liposomal drug delivery system is the physical and chemical instability of the system besides its cost. The physical instability of the system may lead to fusion and aggregation of liposomes, which results in increasing vesicle size and leakage of the loaded drug during their storage [52]. Fusion, aggregation, and leakage of the liposomal loaded drug can also happen due to the chemical stability of the loaded drug substance or by the chemical stability of the lipid component in the liposomes. Since the liposomes undergo chemical degradation by hydrolysis and peroxidation, when designed need to conduct stress and accelerated stability testing to evaluate possible degradation of the liposomes [53].

5.3 Polymer drug conjugates

Polymer drug conjugates (PDG) are a promising novel anticancer drug delivery strategy to enhance the bioavailability of drug molecules and reduce systemic toxicity by delivering the drug at specific sites. The concept of polymer-drug conjugates (PDC) was first proposed by Helmut Ringsdorf in 1975 for the targeted delivery of hydrophobic drugs. The three main components of the Ringsdorf proposed model were a hydrophilic component, drug and a targeting protein [70]. From then on, extensive studies have been carried out on developing PDC based chemotherapeutic drug delivery system. PDCs offer many advantages over conventional chemotherapy such as improve aqueous solubility, protection from systemic degradation, site-specific drug delivery and increase circulation half-life of the loaded hydrophobic drug. In PDCs based drug delivery systems the drug molecules are covalently attached to the functional groups of the hydrophilic polymer/copolymer molecules with the help of linkers (or spacer). The linkers used in PDCs are mostly designed to facilitate preferable drug release from the polymer system, they are also bio-responsive, thus it undergoes dissociation in specific biological conditions (example: change in temperature, pH). The administration of polymer conjugated with chemotherapeutic agents into the body primarily aims in localized drug delivery to the tumor cells. Once the PDC got internalized into the tumor cells, due to the intracellular conditions such as a shift in pH or exposure to cleaving enzymes (esterase, lipase or protease) the bio-responsive linkers undergo hydrolysis resulting in destabilization of the linker or polymer backbone. The destabilization of the linker causes the preferable release of anticancer drug at the tumor site over a substantial period, making the treatment an effective and safer option for treating cancer [71]. The main advantage of using PDCs in targeted anticancer drug delivery is that the efficiency and undesired drug leakage can be controlled by altering the chemical and physical properties of polymers.

Compared to other targeted drug delivery strategies in cancer therapy, PDCs have higher drug loading capacity, controlled release and enhanced stability with minimum or no damage to normal tissues.

The poor specificity of the conjugate system in reaching tumor cells may lead to the early release of conjugated agents within the body and only a very small amount reaches the tumor site. Moreover, in some cases the amount of drug reaching the tumor site will not be sufficient to carry out a strong therapeutic effect or drug may undergo rapid clearance from the tumor site. These limitations of conjugated systems can be addressed by developing targeted PDC systems. Tumor targeting by PDC can be achieved by two approaches passive and active targeting. The passive approach involves the EPR effect and active targeting is achieved by the incorporation of targeting moiety (ligands) into polymer backbone like galactosamine [72]. The ligands used for targeting PDC should have a higher affinity towards the target site. Water-soluble polymers are the most widely used carriers for conjugation with biologically active molecules (chemotherapeutic drugs), to improve the biocompatibility, water-solubility, extended systemic circulation time and pharmacokinetics of hydrophobic drugs. Various natural and synthetic polymers like polyethylene glycol (PEG), N-(2-hydroxypropyl) methacrylamide (HPMA), polyglutamic acid (PLGA), dextran, carboxymethylcellulose, and others are used as carriers to prepare PDCs for cancer therapy. Among all these polymers PEG and HPMA are most extensively investigated by researchers for anticancer drug delivery. Water-soluble targeted PDCs generally consist of a hydrophilic polymer, low molecular weight drug, the biodegradable linker between polymer chain and drug, and a targeting moiety (ligands) [73]. PEG is the most commonly used carrier in PDCs, which have shown enhanced in vivo stability and longer circulation time of the therapeutic agents. PEG-asparaginase (Oncaspar®), PEG-adenosine deaminase (Adagen®), PEG-interferon α-2b (PEG-Intron®), PEG-granulocyte colony-stimulating factor (Neulasta®) and PEG-growth hormone receptor antagonist (Somavert®) are some of the FDA approved PEG-drug conjugates [74].

Polymeric micelles are another polymeric drug conjugation approach in targeted anticancer drug delivery. Micelles are colloidal particles that consist of self-assembled aggregates of amphiphilic copolymer molecules. Most formulations of polymeric micelles utilize block-copolymers with hydrophilic and hydrophobic domains. These polymers can also self-assemble into an inner core and an outer shell. The inner core is mainly composed of hydrophobic blocks whereas the outer shell is usually composed of hydrophilic blocks [74]. The hydrophobic therapeutic agents are contained within the hydrophobic core of the micelles. The drug is covalently bound to the hydrophobic chains in the micelles. PEG is the most extensively used hydrophilic polymer, due to its ability to protect the drug from enzymatic degradation and resist uptake by the reticuloendothelial system (RES). Polymeric micelles based on PEG has been used for the delivery of several hydrophobic therapeutic agents such as doxorubicin (DOX), paclitaxel (PXT), camptothecin (CPT), β lapachone, etc. NK105, a PEG-b-poly(aspartate-4-phenyl-1-butanolate) is an example of polymeric micelles based on PEG. Several polymer micelles are currently in various phases of clinical trials. The major advantage of polymeric micelles over all other polymer conjugates is their ease of fabrication, due to their innate self-assemble properties [2]. Poly (β-benzyl-L-aspartate) (PBLA), PLGA, polycaprolactone (PCL), etc. are some of the commonly used polymers as the hydrophobic core of polymeric micelles [74].

5.3.1 Advances and challenges

Dextran methotrexate (MTX) conjugate a soluble polymer conjugate, was the first reported PDC for cancer treatment. A large number of different polymer conjugate systems with chemotherapeutics such as doxorubicin, paclitaxel, etc. have been developed and clinically tested. Over the past few decades, several types of research have been done on the development of a novel targeted drug delivery system for cancer therapy using the diversified properties of polymers that have remained one of the core research areas. Luo et al. [75] synthesized water-soluble PEG-PXT conjugates with two different molecular weights 6 and 20 kDa for pulmonary application in lung cancer. The results showed that conjugation to PEG improved the solubility of hydrophobic chemotherapeutic agents with improved stability and sustained PTX release in the lung. PTX was also conjugated to trimethyl chitosan (TMC-PTX) and modified with folic acid (FA) by He and Yin [76] for oral and intravenous delivery of PTX. The in vivo studies of the developed polymer-drug-conjugate showed improved PTX solubility, controlled release of PTX and enhanced antitumor efficacy due to the increased accumulation in the tumor. Wallat et al. [77] have synthesized alkyne decorated DOX loaded pH-responsive fluorous polymer conjugates for treating breast and ovarian cancer. The in vitro study results showed that the developed system had an insufficient cytotoxic effect on A2780 and OVCAR3 cells compared to free DOX. The cellular uptake studies showed that the internalization of free DOX and F-DOX by the cancer cell lines was similar to a limited extend. Islam et al. [78] developed a pH-responsive polymer-drug-conjugate system consists of pirarubicin (THP), an anticancer prodrug conjugated to HPMA (PTHP). The system showed increased THP cellular uptake at pH 7.6. A study by Pantshwa et al. [79] aimed to develop stimuli-responsive methotrexate (MTX) entrapped amphiphilic poly (N-isopropyl acrylamide)-b-poly (aspartic acid) nano-micelles using thermal ring-cleavage polymerization of L-Aspartic acid-N-carboxy anhydride (L-Asp-NCA) conjugated to poly(N-isopropyl acrylamide) (PNIPAmeNH2). The MTX-loaded nano-micelles showed increased cellular uptake, sustained MTX release from the micelles and had a significant effect on NIH: OVCAR-5 ovarian cancer cells with minimum or no systemic toxicity. Han et al. [80] constructed a polymeric micelles system for inhibiting breast cancer metastasis by targeting the *E*-selectin (an antigen like family only expressed on tumor endothelial cells) and hyaluronic acid (HA) receptor on tumor cells. The polymeric micelles conjugate consists of HA conjugated to PTX engineered with E-selectin binding peptide-polyethylene glycol-1-octadecylamine (Esbp-PEG-OA), targeting moiety. The cellular uptake studies showed increased internalization of the Esbp-micelle by E-selectin mediated endocytosis. Esbp modified micelles significantly suppressed breast cancer metastasis compared to unmodified micelles and free PTX.

Despite all the unique properties demonstrated by PDC as drug carriers, two major issues to be addressed is the lack of commercial acceptability and clinical success. Several studies on PDC based anticancer drug delivery have demonstrated the low cytotoxicity of the conjugates compared to the free drug [77]. Another challenge associated with PDC based anticancer delivery system is their physicochemical heterogeneity. Even though the small molecular weight cancer therapeutics have a well-defined chemical structure, the individual PDC molecules vary in molecular weight and drug loading. Due to the heterogeneity reproducibility of PDC synthesis for clinical usage is a major issue. Extensive Studies on precise synthetic approaches such as drug-initiated, living-in situ polymerization and living polymerization of prodrug monomers for the synthesis of well-defined PDCs are being developed.

5.4 Nanoparticulate-based drug delivery

Nanotechnology-based drug delivery system is a rapidly developing area for targeted delivery of anticancer drugs in cancer therapy. Nanoparticles (NPs) have been precedent in solving several limitations of conventional chemotherapy, including non-specific biodistribution, rapid drug clearance, biodegradation, limited targeting, and unwanted adverse effects. The special properties of nanosized drug delivery system such as hyperpermeability to tumor vasculature, and potential to protect the therapeutic agents from biodegradation has broadened research on developing the nanoparticle-based system as drug carriers for treating cancer [81]. Moreover, the unique biological properties such as small size (1–100 nm) and high surface to volume ratio of NPs, allow the binding, absorption, transport and sustained release of anticancer drugs to the targeted tumor site. Surface characteristics, size, and shape of NPs are very crucial in regulating the systems therapeutic efficacy and biodistribution of an anticancer drug in the targeted site [82]. Surface characteristics of NPs play a crucial role in determining its systemic circulation time relating to their capture to macrophages. Ideally, the nanoparticulate-based drug carriers should have a hydrophilic surface to break free from the capture by macrophages. This can be achieved by minimizing opsonization, which can be carried out by either surface coating the NP with hydrophilic polymers such as Polyethylene glycol (PEG), called PEGylation or by constructing nanoparticles from block copolymers with hydrophilic and hydrophobic units [81,83]. The PEGylation of NPs prolong their circulation in the blood and also improves its bioavailability. The NPs with hydrophobic or unmodified surface often leads to opsonization events resulting in rapid clearance of particles by the reticuloendothelial system (RES) from the body. The size of the nanocarrier system is another important factor which determines the long-term fate, bioavailability, cellular toxicity and target-specific delivery of the loaded drug. NPs used in targeted drug delivery applications should have an optimum size (7–100 nm), which can prevent its leakage into blood capillaries and to cease their rapid clearance by macrophage capture of the RES, such as the liver. NPs with hydrodynamic diameter less than 7 nm will undergo renal filtration and urinary excretion, while NPs bigger than 100 nm are often cleared from the system by phagocytic cells. In addition to surface characteristics and size, the shape of NPs is another important factor in determining circulation time, biodistribution, and cellular uptake of the nanoparticle-based anticancer drug delivery system. The majority of the nanoparticle-based anticancer drug delivery systems are generated in a spherical shape [81].

The targeted delivery of cytotoxic drugs into tumor cells by NPs aims in maximizing the killing effect on cancer cells by controlled release of the encapsulated drug from NPs while minimizing off-target toxic effects.

Compared to other delivery systems, nanoparticulate based drug delivery systems shows several advantages as NPs can be targeted directly to the tumor site using several strategies. Due to the smaller size of the NPs, they can get internalized into the targeted tumor site easily by endocytosis mechanism [84]. The commonly used strategies for transporting the drug-loaded NPs to the targeted site are passive, active and stimuli-responsive targeting. The passive targeting of NPs to tumor sites mainly depends on the physiological properties of the tumor. The EPR effect allows the drug-loaded NPs to passively accumulate in the targeted tumor cells. Although passive targeting via the EPR effect solves several problems of conventional cancer therapy by improving target-specific drug delivery to tumor cells, it exhibits some of the limitations of conventional cancer therapy such as inconsistent accumulation of NPs within the tumor site. Genexol-PM™, ProLindac™, and Opaxio™ are examples for the passively targeted NPs currently in clinical use. In active targeting approach, the accumulation of the NPs inside the tumor cell is achieved by conjugating the drug-loaded NPs with specific targeting moiety that binds to overexpressed antigens or receptors on the targeted tumor cell surface [82] (Fig. 6). Actively targeted NPs can be designed to target specific cells, tissues or organs by conjugating the surface of carrier with a specific targeting moiety. Currently, NPs based on liposomes and polymers are known to be in different clinical development phases [81]. The advantage of targeting NPs actively is the target-specific delivery of loaded drugs at the targeted tumor sites. Also, due to prolonged circulation time they remain at the targeted tumor site for a longer time, thereby increasing the NP accumulation. In addition to passive and active targeting, the targeted delivery of anticancer drugs at tumor sites can be achieved by using stimuli-responsive (e.g. pH, temperature changes) NPs. This targeting approach allows the NPs to release their payload exactly at their action site in the presence of specific stimuli resulting in enhanced bioavailability.

Nanoparticulate based drug delivery system in cancer therapy of submicron size (3–200 nm), can be made from a variety of materials. On the basis of the material used in the fabrication, nanocarriers are classified into two types: organic nanocarriers and inorganic nanocarriers. Organic nanocarriers are comprised of liposomes, lipids, dendrimers, carbon nanotubes, emulsions, and synthetic polymers. Among all of these organic nanocarriers, studies on polymeric nanocarriers for targeted drug delivery have significantly evolved over the year. In cancer therapy, polymeric nanocarriers have been extensively investigated due to their permeability, biocompatibility and

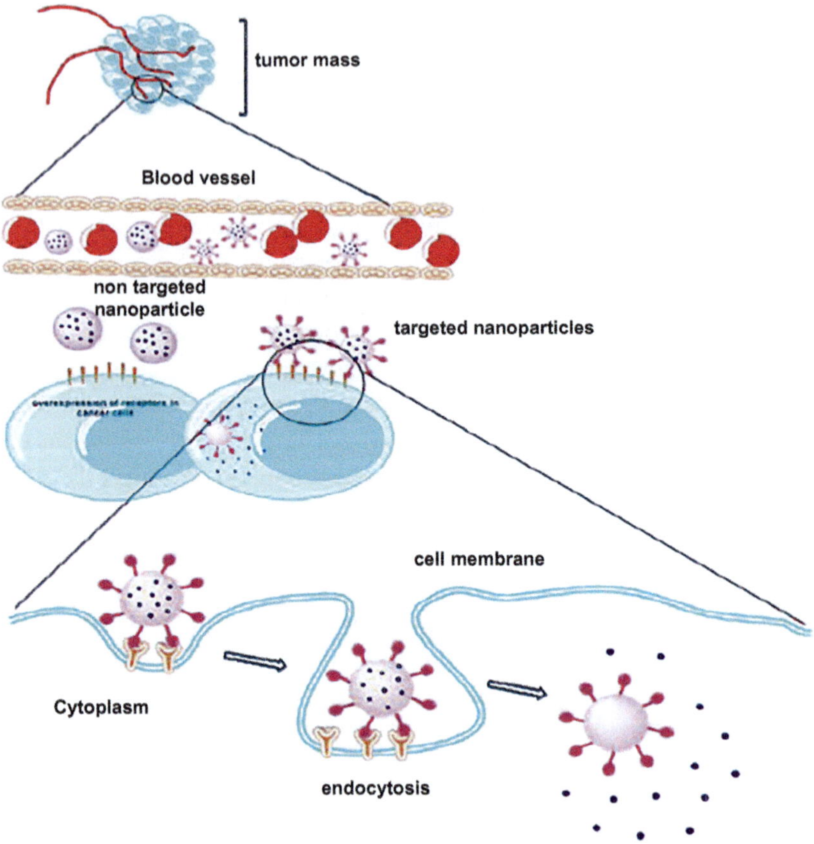

Fig. 6 Mechanism of action of active targeted nanoparticles in the tumor site. *Source: Reproduced from Bahrami, Behdokht et al., Nanoparticles and targeted drug delivery in cancer therapy, Immunol. Lett. 190 (April) (2017) 64–83 with permission from Elsevier.*

prolonged systemic circulation half-life by conjugating the surface with sterically amphiphilic polymers such as PEG. In recent years several studies have been extensively investigated the use of inorganic nanocarrier platforms for the delivery of anticancer therapeutics and imaging. Inorganic materials used for nanocarrier fabrication in cancer treatment are usually metals such as quantum dots, gold, carbon nanotubes, layered double hydroxides, mesoporous silica, and magnetic nanoparticles. In nanoparticulate based drug delivery systems small organic molecules such as liposomes, micelles, dendrimers, nanospheres, and nanocapsules are the most commonly employed nanocarrier materials due to their relative ease of preparation, and stability [1]. Depending on the type of nanocarrier used, the release of drug from the system differ. The release of encapsulated anticancer drugs from nanocarriers depends on many factors including, pH, temperature, drug solubility, erosion of nanoparticle matrix, etc. [85].

5.4.1 Advances and challenges

The nanoparticulate based treatment and diagnosis of cancer are largely still under the developmental phase. Nab-paclitaxel (Abraxane®) and Doxil® are the two US FDA approved nanotechnology-based drugs available in the market. Several studies from the past few years have shown the promising application of nanoparticles as efficient drug delivery systems in cancer therapy. A study by Nima et al. [86] showed the potential of nanoparticles as targeting systems for doxorubicin delivery to breast cancer and prostate cancer cells. Recently, Changizi et al. [87] developed theranostic nanoparticles by conjugating polymeric gold-photoactive nanoparticles with folic acid (PGPNPs-FA) for active targeting and fluorescence tracking in LNCaP prostate cancer cells. The results showed significant cellular uptake of PGPNPs-FA NPs by LNCaP prostate cancer cells. Tao et al. [88] designed a novel photothermal agent (PTA) using PEGylated two-dimensional quantum dots (AMQDs) for cancer photothermal therapy (PTT). The in vivo therapeutic and biocompatibility studies of the PEG-coated AMQDs on MCF-7 cells showed excellent stability in the physiological environment and high accumulation at the tumor sites. One of the major recent advancements in nanoparticulate based drug delivery is the development of curcumin nanoparticles in cancer treatment. Curcumin (Cur) is a hydrophobic phytochemical derived from the rhizome of the medicinal plant turmeric, with excellent anticancer potential has known to have poor aqueous solubility which decreases their bioavailability. Recently, Elbialy et al. [89] aimed to develop a multifunctional curcumin-loaded mesoporous silica NPs surface coated with PEG for improving the bioavailability of Cur for treating cancer. The in vitro and in vivo studies results showed that the developed Cur loaded NPs had a significant effect on tumor cell inhibition and could be used as a promising nanocarrier system for enhancement of Cur bioavailability. Cur was also encapsulated in a phenylboronic acid (PBA) conjugated pH-responsive ZnO by Kundu et al. [90]. The PBA conjugated Cur loaded nanoparticles showed increased cellular uptake and had significant inhibition effect on breast cancer cells (MCF-7).

The nanoparticle-based system plays a vital role in biomedical applications, especially in anticancer drug delivery and has proven to be a good alternative for conventional chemotherapy. However, NPs are not free from limitations. In the nanoparticulate based anticancer drug delivery system, due to the submicron micron size of the NPs the number of particles increases, resulting in increased inter-particle friction, further rising colloidal instability. The colloidal instability of NPs can dramatically alter their in vivo performance, thus storage of prepared NPs under appropriate conditions is also critical. Also, the large surface area of NPs increases its chemical reactivity resulting in excessive reactive oxygen species (ROS) production. This further results in oxidative stress, inflammation, DNA and protein damage, thereby causing gene toxicity. Reproducible, large-scale synthesis of homogeneous nanoparticles for anticancer drug delivery is still an obstacle. The reproducibility of the synthesis and efficacy of NPs can be only ensured by the thorough characterization of these nanocarriers at every stage of the production process. Nanoparticle-based drug delivery system is considered a promising alternative to chemotherapy for cancer treatment. Therefore, nanoparticulate based drug delivery is a topic under continuous research that provides a lot of scopes and challenges for researchers all around the globe.

6 Conclusion and future perspectives

Cancer is a highly variable disease and with the research, ever-evolving to keep pace with the disease, targeted drug delivery has been a key motivator in research. There is a constant need to improve site-specific drug delivery as it has not been truly achieved yet. Even though there are a variety of combinational strategies adopted to achieve the ideal carrier for delivery, sometimes the drug delivered at the site is of very poor concentrations. There have been a number of liposomal formulations as well as nanoparticle formulations that are in line for clinical approval as well as some that have been used in the market such as Doxil and Abraxane, etc. This chapter has touched upon the essential targeting strategies that have been adopted for cancer therapy agents as well as different types of formulations that have been studied and marketed. Research should be directed towards the identification of compatible biomaterials, biomarkers and translational medicine as the advancement in the field of molecular biology has excelled over the last decade. Investigations for better three-dimensional models to mimic the tumor models must also be undertaken to facilitate the testing of immunotherapy in vitro which could lead to a better predictive study in vivo and this paves the way for better site-specific delivery. Also, the future lies in combinational therapy and different conjugations should be tested to serve as potential platforms and therapeutic agents for cancer treatment.

References

[1] S. Senapati, A.K. Mahanta, S. Kumar, P. Maiti, Controlled drug delivery vehicles for cancer treatment and their performance, Signal Transduct. Target. Ther. 3 (1) (2018) 7.

[2] Z.-R. Lu, P. Qiao, Drug delivery in cancer therapy, quo vadis? Mol. Pharm. 15 (9) (2018) 3603–3616.

[3] T. Feng, Y. Zhao, Nanomaterial-Based Drug Delivery Carriers for Cancer Therapy, Springer, Singapore, 2017, pp. 15–54.

[4] A. Coates, et al., On the receiving end- -patient perception of the side-effects of cancer chemotherapy, Eur. J. Cancer Clin. Oncol. 19 (2) (1983) 203–208.

[5] M.J.A. de Jonge, J. Verweij, Renal toxicities of chemotherapy, Semin. Oncol. 33 (1) (2006) 68–73.

[6] Z. Li, et al., Cancer drug delivery in the nano era: an overview and perspectives, Oncol. Rep. 38 (2) (2017) 611–624.

[7] A.G. Tzakos, et al., Novel oncology therapeutics: targeted drug delivery for cancer, J. Drug Deliv. 2013 (2013) 1–5.

[8] V.P. Torchilin, Drug targeting, Eur. J. Pharm. Sci. 11 (Suppl. 2) (2000) S81–S91.

[9] J.K. Vasir, V. Labhasetwar, Targeted drug delivery in cancer therapy, Technol. Cancer Res. Treat. 4 (4) (2005) 363–374.

[10] M. Mimeault, R. Hauke, S.K. Batra, Recent advances on the molecular mechanisms involved in the drug resistance of cancer cells and novel targeting therapies, Clin. Pharmacol. Ther. 83 (5) (2008) 673–691.

[11] D.E. Gerber, Targeted therapies: a new generation of cancer treatments, Am. Fam. Physician 77 (3) (2008) 311–319.

[12] Y.H. Bae, K. Park, Targeted drug delivery to tumors: myths, reality and possibility, J. Control. Release 153 (3) (2011) 198–205.

[13] Y. Matsumura, H. Maeda, A new concept for macromolecular therapeutics in cancer chemotherapy: mechanism of tumoritropic accumulation of proteins and the antitumor agent smancs, Cancer Res. 46 (12 Pt 1) (1986) 6387–6392.

[14] S.K. Hobbs, et al., Regulation of transport pathways in tumor vessels: role of tumor type and microenvironment, Proc. Natl. Acad. Sci. U. S. A. 95 (8) (1998) 4607–4612.

[15] H. Maeda, The enhanced permeability and retention (EPR) effect in tumor vasculature: the key role of tumor-selective macromolecular drug targeting, Adv. Enzyme Regul. 41 (1) (2001) 189–207.

[16] S.M. Moghimi, A.C. Hunter, J.C. Murray, Long-circulating and target-specific nanoparticles: theory to practice, Pharmacol. Rev. 53 (2) (2001) 283–318.

[17] R. Duncan, Polymer conjugates for tumour targeting and intracytoplasmic delivery. The EPR effect as a common gateway? Pharm. Sci. Technol. Today 2 (11) (1999) 441–449.

[18] R. Bazak, et al., Passive targeting of nanoparticles to cancer: a comprehensive review of the literature, Mol. Clin. Oncol. 2 (6) (2014) 904–908.

[19] A. Béduneau, et al., Design of targeted lipid nanocapsules by conjugation of whole antibodies and antibody fab' fragments, Biomaterials 28 (33) (2007) 4978–4990.

[20] P.M. Deckert, Current constructs and targets in clinical development for antibody-based cancer therapy, Curr. Drug Targets 10 (2) (2009) 158–175.

[21] E. Fonsatti, M. Altomonte, P. Arslan, M. Maio, Endoglin (CD105): a target for anti-angiogenetic cancer therapy, Curr. Drug Targets 4 (4) (2003) 291–296.

[22] P.E. Thorpe, Vascular targeting agents as cancer therapeutics, Clin. Cancer Res. 10 (2) (2004) 415–427.

[23] S. Stevanovic, Identification of tumour-associated t-cell epitopes for vaccine development, Nat. Rev. Cancer 2 (7) (2002) 514–520.

[24] N. Rapoport, Combined cancer therapy by micellar-encapsulated drug and ultrasound, Int. J. Pharm. 277 (1–2) (2004) 155–162.

[25] N.Y. Rapoport, et al., Ultrasound-triggered drug targeting of tumors in vitro and in vivo, Ultrasonics 42 (1–9) (2004) 943–950.
[26] A.S. Lübbe, C. Alexiou, C. Bergemann, Clinical applications of magnetic drug targeting, J. Surg. Res. 95 (2) (2001) 200–206.
[27] U.O. Häfeli, Magnetically modulated therapeutic systems, Int. J. Pharm. 277 (1–2) (2004) 19–24.
[28] T.K. Jain, et al., Iron oxide nanoparticles for sustained delivery of anticancer agents, Mol. Pharm. 2 (3) (2005) 194–205.
[29] D.W. Hutmacher, Biomaterials offer cancer research the third dimension, Nat. Mater. 9 (2) (2010) 90–93.
[30] L. Gu, D.J. Mooney, Biomaterials and emerging anticancer therapeutics: engineering the microenvironment, Nat. Rev. Cancer 16 (1) (2016) 56–66.
[31] B.J. Gill, J.L. West, Modeling the tumor extracellular matrix: tissue engineering tools repurposed towards new frontiers in cancer biology, J. Biomech. 47 (9) (2014) 1969–1978.
[32] D.W. Infanger, M.E. Lynch, C. Fischbach, Engineered culture models for studies of tumor-microenvironment interactions, Annu. Rev. Biomed. Eng. 15 (2013) 29–53.
[33] K.L. Schmeichel, M.J. Bissell, Modeling tissue-specific signaling and organ function in three dimensions, J. Cell Sci. 116 (Pt 12) (2003) 2377–2388.
[34] D.W. Hutmacher, et al., Translating tissue engineering technology platforms into cancer research, J. Cell. Mol. Med. 13 (8a) (2009) 1417–1427.
[35] C. Feder-Mengus, et al., New dimensions in tumor immunology: what does 3D culture reveal? Trends Mol. Med. 14 (8) (2008) 333–340.
[36] T. Elsdale, J. Bard, Collagen substrata for studies on cell behavior, J. Cell Biol. 54 (3) (1972) 626–637.
[37] F. Sabeh, R. Shimizu-Hirota, S.J. Weiss, Protease-dependent versus -independent cancer cell invasion programs: three-dimensional amoeboid movement revisited, J. Cell Biol. 185 (1) (2009) 11–19.
[38] E. Dolgin, Cancer vaccines: material breach, Nature 504 (7480) (2013) S16–S17.
[39] K. Strebhardt, A. Ullrich, Paul Ehrlich's magic bullet concept: 100 years of progress, Nat. Rev. Cancer 8 (6) (2008) 473–480.
[40] N. Dan, et al., Antibody-drug conjugates for cancer therapy: chemistry to clinical implications, Pharmaceuticals 11 (2) (2018) 32.
[41] A.C. Parslow, et al., Antibody—drug conjugates for cancer therapy, Biomedicine 4 (3) (2016) 14.
[42] J.M. Lambert, C.Q. Morris, Antibody–drug conjugates (ADCs) for personalized treatment of solid tumors: a review, Adv. Ther. 34 (5) (2017) 1015–1035.
[43] E.L. Sievers, P.D. Senter, Antibody-drug conjugates in cancer therapy, Annu. Rev. Med. 64 (1) (2013) 15–29.
[44] L. Ducry, B. Stump, Antibody-drug conjugates: linking cytotoxic payloads to monoclonal antibodies, Bioconjug. Chem. 21 (1) (2010) 5–13.
[45] A. Thomas, B.A. Teicher, R. Hassan, Antibody—drug conjugates for cancer therapy, Lancet Oncol. 17 (6) (2016) 254–262.
[46] P.M. Drake, D. Rabuka, An emerging playbook for antibody-drug conjugates: lessons from the laboratory and clinic suggest a strategy for improving efficacy and safety, Curr. Opin. Chem. Biol. 28 (2015) 174–180.
[47] N. Diamantis, U. Banerji, Antibody-drug conjugates—an emerging class of cancer treatment, Br. J. Cancer 114 (4) (2016) 362–367.
[48] M. Abdollahpour-Alitappeh, et al., Antibody-drug conjugates (ADCs) for cancer therapy: strategies, challenges, and successes, J. Cell. Physiol. 234 (5) (2019) 5628–5642.
[49] Z. Li, et al., Development of novel anti-CD19 antibody-drug conjugates for B-cell lymphoma treatment, Int. Immunopharmacol. 62 (June) (2018) 299–308.

[50] R. Singh, et al., A new, triglycyl peptide linker for antibody – drug conjugates (ADCs) with improved targeted killing of cancer cells, Molecular Cancer Ther. 15 (6) (2016) 1311–1320.
[51] K. Nagaoka, et al., Anti-tumor activity of antibody drug conjugate targeting aspartate-β-hydroxylase in pancreatic ductal adenocarcinoma, Cancer Lett. 449 (2019) 87–98.
[52] T.O.B. Olusanya, et al., Liposomal drug delivery systems and anticancer drugs, Molecules 23 (4) (2018) 1–17.
[53] Z. Vahed, R.S. Sepideh, S. Davaran, S. Sharifi, Liposome-based drug co-delivery systems in cancer cells, Mater. Sci. Eng. C 71 (2017) 1327–1341.
[54] M. Li, et al., Composition design and medical application of liposomes, Eur. J. Med. Chem. 164 (2019) 640–653.
[55] M. Abri Aghdam, et al., Recent advances on thermosensitive and PH-sensitive liposomes employed in controlled release, J. Control. Release 315 (2019) 1–22.
[56] G. Blume, G. Cevc, Liposomes for the sustained drug release in vivo, BBA-Biomembranes 1029 (1) (1990) 91–97.
[57] A.L. Klibanov, K. Maruyama, V.P. Torchilin, L. Huang, Amphipathic polyethyleneglycols effectively prolong the circulation time of liposomes, FEBS Lett. 268 (1) (1990) 235–237.
[58] P. Tosha, P. Kaushika, P. Rudree, S. Shreeraj, Liposomal formulations in cancer therapy: passive versus active targeting, Asian J. Pharm. Res. Dev. 7 (5) (2019) 92–102.
[59] P. Yingchoncharoen, D.S. Kalinowski, D.R. Richardson, Lipid-based drug delivery systems in cancer therapy: what is available and what is yet to come, Pharmacol. Rev. 68 (3) (2016) 701–787.
[60] S. Fathi, A.K. Oyelere, Liposomal drug delivery systems for targeted cancer therapy: is active targeting the best choice? Future Med. Chem. 8 (17) (2016) 2091–2112.
[61] Y. Tian, et al., Overcoming drug-resistant lung cancer by paclitaxel-loaded hyaluronic acid-coated liposomes targeted to mitochondria, Drug Dev. Ind. Pharm. 44 (12) (2018) 2071–2082.
[62] P. Deshpande, et al., Transferrin and octaarginine modified dual-functional liposomes with improved csancer cell targeting and enhanced intracellular delivery for the treatment of ovarian Cancer, Drug Deliv. 25 (1) (2018) 517–532.
[63] S. Lakkadwala, J. Singh, Dual functionalized 5-fluorouracil liposomes as highly efficient nanomedicine for glioblastoma treatment as assessed in an in vitro brain tumor model, J. Pharm. Sci. 107 (11) (2018) 2902–2913.
[64] A. Sesarman, et al., Co-delivery of curcumin and doxorubicin in PEGylated liposomes favored the antineoplastic C26 murine Colon carcinoma microenvironment, Drug Deliv. Transl. Res. 9 (1) (2019) 260–272.
[65] Y. Xiao, et al., Sorafenib and gadolinium co-loaded liposomes for drug delivery and MRI-guided HCC treatment, Colloids Surf. B Biointerfaces 141 (2016) 83–92.
[66] S.K. Ramadass, et al., Paclitaxel/epigallocatechin gallate coloaded liposome: a synergistic delivery to control the invasiveness of MDA-MB-231 breast cancer cells, Colloids Surf. B Biointerfaces 125 (2015) 65–72.
[67] A.L. Saraswat, T.J. Maher, Development and optimization of stealth liposomal system for enhanced in vitro cytotoxic effect of quercetin, J. Drug Deliv. Sci. Technol. 55 (June 2019) (2020) 101477.
[68] S. Li, et al., A novel tumor-targeted thermosensitive liposomal cerasome used for thermally controlled drug release, Int. J. Pharm. 570 (24) (2019) 118660.
[69] M. Dunne, et al., Heat-activated drug delivery increases tumor accumulation of synergistic chemotherapies, J. Control. Release 308 (March) (2019) 197–208.
[70] J. Khan, A. Alexander, S. Saraf, S. Saraf, Exploring the role of polymeric conjugates toward anti-cancer drug delivery: current trends and future projections corresponding author: Dr. Shailendra Saraf, Int. J. Pharm. 58 (1) (2018) 500–514.

[71] H. Xu, et al., Targeted polymer-drug conjugates: current progress and future perspective, Colloids Surf. B Biointerfaces 136 (2015) 729–734.
[72] G. Jeswani, S.D. Paul, A.K. Jha, Advances in the delivery of cancer therapeutics, Curr. Drug Deliv. 15 (1) (2018) 21–36.
[73] M.F. Seifu, L.K. Nath, Polymer-drug conjugates: novel carriers for cancer chemotherapy, Polymer Plast. Technol. Mater. 58 (2) (2019) 158–171.
[74] N.L. Ghandeharia, Hamidreza, Polymeric conjugates for drug delivery, Chem. Mater. 24 (5) (2013) 840–853.
[75] T. Luo, et al., Synthesis and in vitro evaluation of polyethylene glycol-paclitaxel conjugates for lung cancer therapy, Pharm. Res. 33 (7) (2016) 1671–1681.
[76] R. He, C. Yin, Trimethyl chitosan based conjugates for oral and intravenous delivery of paclitaxel, Acta Biomater. 53 (2017) 355–366.
[77] J.D. Wallat, J.K. Harrison, J.K. Pokorski, PH responsive doxorubicin delivery by fluorous polymers for cancer treatment, Mol. Pharm. 15 (8) (2018) 2954–2962.
[78] A.W. Islam, J. Fang, T. Etrych, P. Chytil, HPMA copolymer conjugate with pirarubicin: in vitro and ex vivo stability and drug release study, Int. J. Pharm. 536 (108–115) (2017).
[79] J. Pantshwa, et al., Journal of drug delivery science and technology synthesis of novel amphiphilic poly (N-isopropylacrylamide)—b—poly (aspartic acid) nanomicelles for potential targeted chemotherapy in ovarian cancer, J. Drug Deliv. Sci. Technol. 39 (2017) 308–323.
[80] X. Han, et al., Free paclitaxel-loaded E-selectin binding peptide modified micelle self-assembled from hyaluronic acid-paclitaxel conjugate inhibit breast cancer metastasis in a murine model, Int. J. Pharm. 528 (1–2) (2017) 33–46.
[81] S. Raj, et al., Specific targeting Cancer cells with nanoparticles and drug delivery in cancer therapy, Semin. Cancer Biol. (2019). S1044-579X(19)30216-0.
[82] S. Tran, P.-J. DeGiovanni, B. Piel, P. Rai, Cancer nanomedicine: a review of recent success in drug delivery, Clin. Transl. Med. 6 (1) (2017).
[83] A. Dadwal, A. Baldi, R.K. Narang, Nanoparticles as carriers for drug delivery in cancer, Artif. Cells Nanomed. Biotechnol. 46 (Suppl. 2) (2018) 295–305.
[84] B. Bahrami, et al., Nanoparticles and targeted drug delivery in cancer therapy, Immunol. Lett. 190 (April) (2017) 64–83.
[85] S.A.A. Rizvi, A.M. Saleh, Applications of nanoparticle systems in drug delivery technology, Saudi Pharm. J. 26 (1) (2018) 64–70.
[86] Z.A. Nima, et al., Targeting nano drug delivery to cancer cells using tunable, multilayer, silver-decorated gold nanorods, J. Appl. Toxicol. 37 (12) (2017) 1370–1378.
[87] O. Changizi, et al., Enhanced radiosensitivity of LNCaP prostate cancer cell line by gold-photoactive nanoparticles modified with folic acid, Photodiagnosis Photodyn. Ther. 29 (2020) 101602.
[88] W. Tao, et al., Antimonene quantum dots : synthesis and application as near-infrared photothermal agents for effective cancer therapy angewandte, Cancer Ther. 100049 (56) (2017) 11896–11900.
[89] N.S. Elbialy, S. Faisal, B. Fahad, A. Noorwali, Multifunctional curcumin-loaded mesoporous silica nanoparticles for cancer chemoprevention and therapy, Microporous Mesoporous Mater. 291 (109540) (2020) 1387–1811.
[90] M. Kundu, et al., PH-responsive and targeted delivery of curcumin via phenylboronic acid-functionalized ZnO nanoparticles for breast cancer therapy, J. Adv. Res. 18 (2019) 161–172.

CHAPTER 12

Nanopolymeric systems to improve brain cancer treatment outcomes

Luiza Steffens[a], Mabilly Cox Holanda de Barros Dias[b], Pablo Ricardo Arantes[a], Jeferson Gustavo Henn[a], Michael Nugent[b], and Dinara Jaqueline Moura[a]

[a]Laboratory of Genetic Toxicology, Federal University of Health Sciences of Porto Alegre—UFCSPA, Porto Alegre, Rio Grande do Sul, Brazil

[b]Athlone Institute of Technology, Athlone, County Westmeath, Ireland

1 Introduction

Cancer is a complex disease and one of the main reasons for human morbidity. The projection of new GLOBOCAN 2012 cases worldwide was 14.1 million per year, with an estimated increase to 19.3 million by 2025 [1]. Surgery, radiotherapy (RT), chemotherapy (CT) and immunotherapy are the main approaches used for cancer therapy [2]. To eradicate localized cancers, surgery and RT works relatively well, however, when metastasis is associated, CT and immunotherapy are the main functional treatments [3]. However, high toxicity, an insufficient amount of therapeutic agents released, water insolubility of some drugs, erratic circulation, non-specific biodistribution and drug release for both cancerous and non-tumor cells are the main limitations of CT [2, 4]. Consequently, intelligent and efficient strategies are needed to decrease these mortality statistics [1, 5] through a more specific delivery to the cancer targets.

2 Glioblastoma overview

Brain tumors develop from several different types of cells. Gliomas, for example, are brain tumors primarily similar to typical glial cells, such as astrocytes, oligodendrocytes, and ependymal cells [6]. In general, these tumors are incurable and highly resistant to CT and RT. Gliomas can infiltrate through brain regions and are essentially categorized according to their morphological similarity with corresponding glial cell natures, cytostructure and

Advances and Challenges in Pharmaceutical Technology
https://doi.org/10.1016/B978-0-12-820043-8.00001-3

immunohistology [6]. The glioma classification system, according to the World Health Organization (WHO), distinguishes astrocytomas in four degrees (I, II, III and IV) and oligodendrogliomas, in two degrees (II and III). Grade IV astrocytoma, known as glioblastoma (GB) is the most prevalent and poorly forebode glioma [6].

The National Cancer Institute (INCA) estimates that for each year of the 2018/2019 biennium, there were 11,320 new cases of brain tumors/central nervous system (CNS), 5810 in men and 5510 in women in Brazil [7]. These numbers correspond to an estimated risk of 5.62 new cases per 100,000 men and 5.17 per 100,000 women. GB is one of the most frequent and aggressive types of CNS cancer in adults [8], comprising 16% of primary brain tumors. Because it is a predominantly recurrent primary malignant brain cancer, it accounts for 77% of all malignant brain cancers [9, 10].

Clinical features and malignancy are used to categorize gliomas, with GB having histopathology comprising cell pleomorphism, high mitotic rate, and cellularity with unusual nuclei. Furthermore, GB can be distinguished by the heterogeneity of cells, the resistance of cancer cells to treatment with chemotherapeutics, development of necrosis and angiogenesis [11]. Although GBs essentially develop only in the brain, they can occur in some cases in the spinal cord, cerebellum, and brain STEM (American Association of Neuroscience Nurses [12]). Initially, it was believed that GB was exclusively derived from glial cells, but more recent studies suggest that GBs can develop from several cells with neural trunk characteristics [13].

GBs develop primarily in patients with a mean age of 64 years [10]. The mean survival time in patients with GB is less than 50 weeks, even with exhaustive management which includes surgical resection, CT and RT [14]. The current clinical protocol for GB is a combinatory procedure of RT doses and systematic delivery of anticancer drugs [15]. The BC Cancer Treatment Management Guidelines (protocol: CNAJ12TZRT) for GB treatment includes concomitant treatment of temozolomide (TMZ) and RT for 5 days, with 2 days of RT, totaling 7 days of exposure. This cycle on the clinical protocol of patients must be repeated for 6 weeks [16], although combinatorial treatment is most effective, it leads to frustrating side effects. Given the complex features of GB cells and the presence of physiological barriers, particularly the blood-brain barrier (BBB), the cure of this cancer remains a challenge [17, 18].

2.1 Molecular aspects of glioblastoma

GBs have a complex genetic profile. The Atlas Cancer Genome project analyzed the GB genomic profile of 200 tumor specimens and approximately

600 genes were overexpressed [19]. Based on this analysis, three major protein signaling pathways were established: (a) tumor protein 53 (p53), (b) receptor tyrosine kinase/Ras/phosphoinositide 3-kinase and (c) pathways related to retinoblastoma [20]. Due to modifications in these protein pathways, primary and secondary tumors predominantly have uncontrolled cell proliferation. This allows malignant cells to avoid cell cycle checkpoints and cell death processes, markedly senescence and apoptosis [19].

Primary and secondary GBs exhibit complex patterns of gene expression: epidermal growth factor receptor (EGFR) overexpression, phosphatase and tensin homolog (PTEN) mutations, and loss of the long arm of chromosome 10 are examples of genetic and molecular characteristics of primary GB. On the other hand, mutations in isocitrate dehydrogenase 1 (IDH1) and p53, as well as the loss of the long arm of chromosome 19 are commonly observed in the secondary GB [21–23].

Nek1 is a protein that is associated with tumor resistance in gliomas [24]. It has recently been reported that Nek1 is overexpressed in different gliomas and, interestingly, the level of expression is directly related to the degree of tumor severity, proliferation rate, and TMZ resistance. In light of analyzing patient samples, the association of Nek1 expression with decreased survival prognosis ($p = 0.03$) was observed [24]. Inhibitors of the kinase activity of this protein have already been tested [25] to inhibit the DNA damage response signaled by Nek1. Although the results indicate that Nek1 is somehow related to the tumor process, much information is still missing to identify its true role in tumor resistance.

Given all the complex changes in the signaling pathways of these tumor cells, GBs have become a major obstacle in oncology with several progression and survival outcomes. These molecular changes are extremely important to understand GBs and could potentially be used for targeted therapy [26–28].

2.2 The challenge imposed by the blood brain barrier (BBB)

The blood brain barrier (BBB) is a special tissue that is crucial in the biodistribution of drugs from the general organism and the CNS (central nervous system), comprising a challenge on GB treatment. Due to its lipophilic character, cells peculiarity and permeability limitations, the BBB constitutes a challenge on the delivery of medicines through it. The four main reasons can be listed as:

(a) The cerebral vasculature of the BBB cell composition is formed by narrow endothelial lines. These endothelial cells can develop a strong connection between neighboring cells. Therefore, the paracellular transport of the brain is restricted by this physical limit [29];

(b) The presence of specific enzymes that degrade some drugs, preventing them from reaching specific sites in the brain;

(c) The active efflux transporters, which sends the drugs back to the blood, resulting in their metabolism and consequent inability to reach their destination [30];

(d) The astrocyte extensions form a lamella network being close to and adhered to the outer surface of the BBB endothelium [31].

The narrowed junctions between the cells in the BBB are essentially composed of occludins and claudins [32]. The latter group is critical for the restriction of small molecules. For example, claudin 5 restricts molecules smaller than 800 Da. Also, impairment in the functioning of some claudins, for example, claudin 3 is associated with increased penetration of BBB in the vasculature of cancer [33]. However, in infiltrating and micrometastatic gliomas, the barrier remains intact. This important aspect indicates that the modulation of permeability at these sites is critical and can be used to increase the effectiveness of the treatment.

The two main forms of transport through the BBB are carrier-mediated transport and receiver-mediated transport. Plus, molecular transport through BBB is specifically limited to particles larger than 12 nm [30]. It is estimated that almost all large molecules cannot cross the barrier, and that is why several promising drugs do not reach the tumor site in sufficient concentrations [34].

Because of this, BBB prevents the route of numerous antitumor drugs to brain sites, making CT in GB one of the biggest challenges among current cancer treatments. Intending to improve drug transport or avoid the BBB, many research groups have been developing new technologies to overcome these obstacles. An option to transport drugs to the brain bypassing the cited limitations is the use of nanocarriers [35]. Biochemical changes in drugs and the use of nanocarriers have been proposed by researchers, allowing the local release of drugs at sufficient concentrations, avoiding exposure to the system and overcoming BBB limitations [3].

2.3 Current strategies in GB treatment

The treatment of recently diagnosed GB comprises multidisciplinary tactics. The possible tactics include surgery, RT, CT, Immunotherapy, and hyperthermia. Currently, the standard therapy for GBs uses surgical resection followed by simultaneous RT and CT with TMZ, known as Temodal®

(National Comprehensive Cancer Network [36]). Remaining and infiltrating malignant cells persist consistently within the brain, leading to advanced tumor progression and even the relapse of the disease [22]. There are several complications related to the management of GB, including insufficient drug delivery, damage to healthy tissues and resistance to CT. Among the existing approaches, CT, immunotherapy, and RT can, to a certain extent, assist patients with GB. However, beyond their benefits, these strategies have unfavorable characteristics [37].

2.3.1 Surgery

GBs are usually invasive tumors and are located regularly in important places in the brain; in particular, at the areas responsible for sensory control, motor function and speech, hence surgical resection is risky though still used. To say nothing of, these tumors are highly invasive; making complete resection of the tumor mass unfeasible.

2.3.2 Immunotherapy

In the last decade, interest in the development of immunotherapeutic ways to deal with tumor malignancy has increased. Among the various immunotherapeutic methods, stimulated lymphocytes, viral drugs, peptide-based vaccines, cytokines, and dendritic cell treatment appear to be successful approaches for use in GB therapy [38]. A large number of studies focus on the administration of bevacizumabumab, a humanized monoclonal anti-vascular endothelial growth factor antibody (anti-VEGF), which acts on angiogenesis [39]. Even so, when combined with TMZ in newly diagnosed patientsit was not able to increase overall patient survival [39].

Another treatment strategy is the use of an anti-EGFRvIII vaccine. This bioproduct, called rindopepimute, was demonstrated satisfactory in the phase II study combined with TMZ with a survival of 21.8 months [40], but the phase III study was discontinued in 2016 after the second partial analysis demonstrate failure to increase survival compared to standard (20.1 *vs* 20 months) [41]. Another immunological treatment is the use of an antibody that inhibits the programmed T cell death receptor 1 (PD-1) or its linker (PD-L1) [42]. The PD-1 inhibitor (pembrolizumab) has already led to satisfactory responses in patients with advanced melanoma, non-small cell lung cancer and renal carcinoma [43]. In gliomas, a clinical study was started in 2014, with an estimated end date for 2018, but still has no partial results (NCT02017717).

Although immunotherapies are interesting options for GBs treatment, there are some complications, frequently patient selection to develop clinical trials, monitoring of body response, evaluation of side effects and clinical consequences [38]. Equally important, some issues need to be assessed: first, effective preclinical outcomes should be carefully considered, as cross-cutting variations are crucial; second, determination of the maximum acceptable dosage is not achievable in most cases, and finally, immune responses during *in vivo* monitoring are very complex.

2.3.3 Radiotherapy

Radiotherapy is the most common initial treatment option for GB patients is RT. However, there are numerous drawbacks regarding the use of this method. An important aspect is that the reaction of the tumor to RT is influenced by its size. Given the multifaceted biology of the tumor, large ones almost do not respond to RT. In contrast, with smaller samples, the earlier the treatment begins, the better the result, given that RT is most effective in the first 5 years after resection [14]. The side effects of RT include both acute and chronic effects. Acute side effects include sterility, gastrointestinal abscesses, and damage to the epithelial surfaces of the mouth and throat. Late effects include heart disease, hair loss, fibrosis and lymphedema [37].

2.3.4 Chemotherapy

TMZ is the main drug of choice for the treatment of GB [44], consisting of a second-generation Imitatrazazine, and its toxic effect is due to DNA methylation [45]. TMZ is a prodrug, which means that it is spontaneously hydrolyzed at physiological pH and converted into the active drug, 5-(3-methyltriazen-1-yl) imidazole-4-carboxamide (MTIC) [45].

TMZ is administered orally, being regularly used in the early stages of GB treatment. Even though TMZ has benefits in GB therapy, there are numerous complications. To be cited, the existence of BBB which limits the delivery of drugs to specific sites, as well as O6-methylguanine DNA methyltransferase (MGMT) which can lead to DNA repairing. The latter consequence can modify the cellular phenotype and, consequently, increase the cellular resistance to TMZ [46]. These phenotypes can alter the expression of key proteins, to enumerate p53, PTEN, EGFR, Mouse double minute 2 homolog (Mdm-2) and galectin-1. To overcome these complications, new drugs that can target these proteins are needed. For example, to inhibit MGMT it is possible to use O6-benzylguanine as well, some tyrosine kinase inhibitors may suppress EGFR and finally, nutlin-3 can act as an inhibitor of Mdm2 [47, 48].

Dacarbazine (DTIC) is another antitumor agent that despite having erratic absorption and low solubility [49], it has been studied in recurrent GBs, proving to be effective [50]. However, DTIC causes severe toxicity in healthy tissues [49]. Thus, new strategies and formulations are required to decrease systemic toxicity.

CT has numerous harmful effects, notably nausea and vomiting, rash, diarrhea, hair loss, insomnia, infertility, and nerve damage. Some reports infer that CT in association with different strategies, especially RT, offers the most efficient GB treatment strategy [51, 52].

Immunotherapy, RT and CT have both beneficial and disadvantageous results. It is believed that the combination of these treatments may positively improve the effects. van Linde et al. [53] applied a neoadjuvant management approach to the newly identified GB, which comprises a chimeric monoclonal antibody, bevacizumab, followed by RT together with TMZ. Their results revealed that this combinatorial treatment was safe; however, patients presented a low response rate [54].

2.3.5 Hyperthermia

Hyperthermia is an alternative technique for GB management [55] based on the elimination of malignant cells dependent on heat production at the site of the tumor. This treatment, therefore, induces functional changes in the cells and this can lead to cell death [56]. Hyperthermia occurs at a temperature of approximately 45 °C and stimulates various intra- and extracellular mechanisms including cell degradation, cellular organelle aggregation, and protein unfolding, which may lead to apoptosis [57].

Treatment parameters, chiefly the temperature at the site of the tumor and the duration of exposure, can significantly alter the efficiency of this treatment [58]. Among the conventional methods of hyperthermia induction, the most common are infrared irradiation, microwave, and ultrasound [59]. However, all of these methods have limitations, notably low heat diffusion for tumor metastases. Particularly in tumors that are well vascularized, heat can be dissipated by the blood and this can be a problem because healthy tissues can be affected. To solve these problems, in 1957, magnetic particles began to be used for hyperthermia [60], making it known as magnetic hyperthermia therapy (THM). The limitations that come with using this method are the preparation of the materials and their cost [60]. Yet, over the years, much research has been done to prove that hyperthermia is the only strategy lacking for the success of GB therapy. However, very few combinatorial studies have been reported. Some of the clinical studies conducted so far have described that hyperthermia could be safely applied with minimal side effects [61].

To develop targeted treatment complexes with low toxicity to healthy tissues, several different techniques are being investigated, such as radio-immunotherapy, stereotactic radiosurgery, hyperfractionation, and iodine-125 brachytherapy. However, these methods have not significantly increased patient survival [62]. Due to some deficiencies, expressively low tissue accumulation of drugs, low specificity, and toxicity, the conventional management of GB has restricted its use in patients. Consequently, the use of new nanotechnologies is an encouraging approach, since they present promising results in the treatment of GB.

3 Drug delivery systems (DDS)

Numerous obstacles in brain cancer therapy are being addressed in this work, such as low absorption, bioavailability, and low drug biodistribution since proper clinical results are not found using conventional DDS [63]. To overcome these problems, the development of new DDS capable of improving the drugs pharmacological properties is an urgent need. At this point, DDS-associated nanotechnology could be the best technology in which the created systems should prevent side effects in healthy cells [64] since upgraded DDS are capable of improving compliance with patient treatment. Furthermore, they can increase the useful life of the drug and, consequently, reduce the costs associated with medical care [65]. Subsequently, the development of DDS which exhibit controlled properties is demonstrating great potential and various nano-DDS formulations are being offered for cancer treatment [66–72].

To develop an efficient DDS, some features are essential, to enumerate:

(a) a specific target;
(b) a useful drug carrier;
(c) a suitable route of administration.

First of all, the physical-chemical characteristics of the DDS system and the drug should be clearly understood as the goal is to target the cancer cells and avoid harming normal cells. In this topic, another important aspect is drug concentration, nano-DDS may contain minor amounts of drugs, decreasing toxicity, and consequently side effects [3].

As outlined by Wen and Reardon [73], despite all efforts to treat GBs, no new approach has modified the prognosis of the disease, justifying the need for studies on promising systems using nanocarriers.

3.1 Electrospinning technique for DDS production

There are several methodologies associated with the production of DDS, notably the electrospinning technique. With this method, there is the removal of the usual emulsion step that can trigger loss of bioactivity of sensitive drugs, a point that differs and highlights it from other methods used to produce nanocarriers [74].

This technique allows the production of polymeric NP and NF with diameters varying from 3 nm to 5 μm [75]. These nanoproducts have several applications in nanocatalysis, tissue engineering, protective clothing and filter membranes [75, 76].

Electrospinning is an interesting technique given the robustness of the equipment. It consists of a support with the polymer solution, a syringe connected to a needle, a pressure pump whose purpose is to control the flow of the solution, an electric source of high voltage and a collector carrier produced from a conductive material [77]. The development of the Nano product occurs when the solution is pumped by the system and a high voltage electric field is applied to induce repulsion between the charges in the polymer solution. Due to the applied voltage, deformation occurs in the drop formed at the end of the needle, changing its shape to a cone-like structure called the Taylor Cone [78]. When the electrostatic force exceeds the polymer surface tension at the needle end, a jet (spray or spin) is formed toward the collector plate while the solvent evaporates allowing the material to solidify [79].

The final characteristics of the fibers and particles produced, such as their diameter, morphology, degree of alignment and mechanical strength may be interfered by various adjustable parameters in the art. To clarify these effects, several studies have been carried out elucidating the correlation between the parameters used and the resulting fiber. Pham et al. [79] in their review indicated variations in fiber morphology and diameter as a function of changes in electrophilic parameters [79]. Lee [80] studied the influence of molecular weight on the structure and properties of polyvinyl alcohol (PVA) nanofibers and demonstrated optimal conditions for obtaining uniform fibers [80]. Among the parameters that can affect the process are: the polymer used and the conformation of its chain, solvent, elasticity, conductivity, polarity and solvent surface tension, applied an electric field, the distance between capillary, viscosity and solution concentration [76].

The resulting structure of the nanoproduct, is usually related to the polymer concentration. Spherical particles are fabricated at lower concentrations of polymers. To find the ideal mixture, it is essential to modify the viscosity of the solution. If its concentration is extremely low, there is no suitable chain entanglement within a drop to stabilize the NP development during electrospray, nevertheless, if the concentration is too elevated there is an increased amount of chain webs, creating polymer networks and fiber production [81].

4 Nanotechnological approaches

The category of technology that engages in employment, production, and analysis of nanometer-sized materials is known as nanotechnology. This category uses nanomaterials for practically all aspects of use [63, 65, 82–88]. Drug delivery systems (DDS) based on nanotechnology have improved pharmacokinetic characteristics. Among them, it is possible to mention high clearance rate, distribution of a large volume of drugs and high bioavailability of drugs in the tumor, due to the greater effect of enhanced permeability and retention (EPR). Coupled with it, nanomedicine-based methods can transport drugs through biological barriers [89].

The most searched nanocarrier formulations are emulsions, dendrimers, micelles, liposomes, and polymers. Polymeric nanoproducts can be divided into different systems, including nanospheres, nanocapsules, nanoparticles (NP) and nanofibers (NF) [90]. The first group is composed of solid colloidal particles in which drugs can be dispersed or adsorbed onto a matrix made of one or a mix of polymers. The size of nanospheres varies from 100 to 200 nm [91]. Nanocapsules, on the other hand, are vesicular systems. In this type of systems, the drug is encapsulated by a polymer capsule. The nanocapsules are a common system to be used in the entrapment of hydrophobic substances and, consequently, some homopolymers that may prevent the opsonization of these systems can be used in the formulation, including polylactic acid (PLA) [18]. Finally, NP and NF are solid nanoproducts (size between 1 and 1000 nm) in which it is possible to encapsulate or dissolve drugs [92].

NP and NF have interesting qualities since it is possible to use these carriers for various purposes, for instance, diagnosis and imaging, gene delivery and drug release. Due to the incorporation of the drug into the material, the common side effects of the drugs can be decreased and also the rapid enzymatic degradation of the drug.

Additionally, these biofunctionalized products can point the target tissues in higher concentrations, being a good choice in the delivery of biological substances proteins, peptides, nucleic acids and antibiotics [93]. Further, depending on the composition developed, such products, especially NP, may be administered by various routes, including intravascular, oral, nasal, transdermal and intraocular [94, 95]. Therefore, nanotechnology-based DDSs appear as useful approaches in therapies related to CNS diseases.

Polymer properties, chiefly crystallinity, solubility, composition, molecular weight, polydispersity, and hydrophobicity are important features for developing a DDS [2]. Therefore, studies of physicochemical characterization and bio interactivity considering the different fluids, barriers and environments of the human body are crucial for building advanced knowledge at this matter. Altogether, NPs is the most commonly approached system when talking about GBs treatment, with a rising up area on NF for controlled and local delivery.

4.1 PNPS in the treatment of glioblastoma

Considering the numerous benefits of polymeric NP (PNP), markedly targeted drug administration, improved bioavailability, decreased drug side effects, drug protection against degradation, improved drug solubility and sustained release of the drug, its use in the diagnosis and treatment of CNS diseases is worthwhile [96]. The targeted DDS can be described as a material capable of delivering a medicament in a precise place with an appropriate dose. This method can be separated into passive and active [97]. In the first system, the release depends on differences between healthy and unhealthy tissues. It is known that damaged tissues present several changes in a physiological condition through the EPR effect [98, 99]. However, tumor sites usually have high concentrations of drugs when compared to normal sites [100]. The second system, using nanotechnology techniques, can be performed by adding specific ligands on the surface of the PNPs [101]. Conjugation of specific nucleic acids, peptides, and antibodies coating PNP leads to specific targeting (Fig. 1). Not to mention, this enables the modification of the tumor cells signaling pathways [15]. The ability to readily modify the surface is a supplementary outstanding property of the PNPs [87, 102]. However, to accomplish the desired delivery of the drug, it is important to understand the nature of the PNP and the properties of the drug.

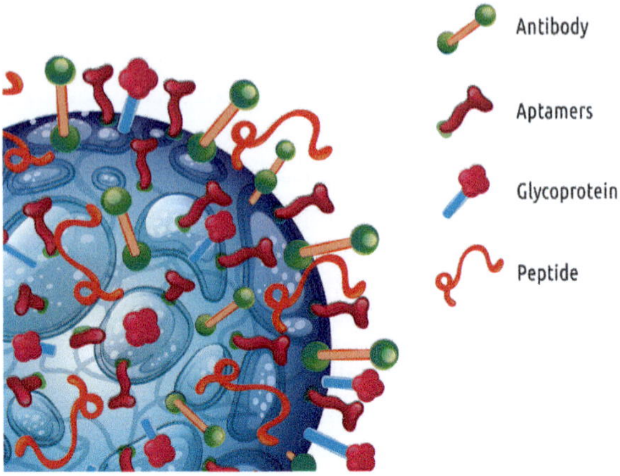

Fig. 1 Schematic representation of polymeric nanoparticles and biofunctionalization options.

Polymers such as poly (lactic-*co*-glycolic acid) (PLGA), poly (methyl methacrylate) (PMMA), poly (ethyleneimines) (PEI), polyesters, poly (methylene malonates) (PMM) and poly (alkyl cyanoacrylates) (PACA) were used as drug delivery carriers [103]. The correct selection of the polymer will depend on the purpose of the treatment. For example, PEG is a biocompatible, inert and hydrophilic polymer used to design PNP, increasing its stability [104]. However, oxidative damage avoids its long-term uses [105]. PLA and PLGA are the most effective polymers used to date since their metabolites are eliminated by the Krebs cycle and are therefore non-toxic [106]. PVA, a non-toxic hydrophilic polymer approved by the Food and Drug Administration (FDA), is an interesting option for designing nanoproducts since this polymer has low cytotoxicity and is biodegradable. The PVA also has an excellent history in biomedical applications, mainly hydrogels [107–110].

The NP could be efficient in GB treatment due to its various properties, including biocompatibility, small size and simple, fast and low-cost production [18]. Furthermore, NPs can enhance various aspects of the drug, including stability and release [111]. There are numerous assortments of NP construction and different design strategies. The need to develop effective DDS, both safe and biodegradable systems, has made PNPs one of the most encouraging technology, achieving wide recognition in the field of DDS, particularly in cancer therapies [88, 94, 95].

PNPs are submicron-sized colloidal particles that can encapsulate or adsorb drugs, imaging substances and biological agents within or on their surface. Considering that the structure of the PNP matrix allows the addition of drugs, the surface facilitates its distribution in the body. Besides, PNPs offer resolutions to CT limitations, including resistance to usual medications and undesirable effects [18].

PNPs can be synthesized from synthetic polymers. To point out, PVA, (poly (ethylene glycol)-PEG, poly (acrylamide)-AMP, ε-poly (caprolactone)-PCL and poly (acrylate)-PA) or natural compounds like chitosan, gelatin, polysaccharides and albumin [112]. Synthetic PNPs such as PLA, PLGA and PGA are biocompatible and degrade hydrolytically [113]. Among the biocompatible PNPs, the FDA approved PVA, PLGA, and PCL. In order to deliver agents to brain tumors, mainly cationic polymers have been used, especially for transporting nucleic acids, since they have positive charges that allow them to interact with DNA and RNA [94, 95, 114].

PNPs can be allocated in biodegradable or non-biodegradable PNPs. Due to the low immunogenicity, high bioavailability and stability, low cytotoxicity and administrable release of the drug, biodegradable PNPs are generally used for DDS and are the appropriate option for the encapsulation of hydrophobic drugs [18]. Moreover, these PNPs show better stability in biological fluids [115], and their biodegradable versions are often employed for the purpose of improving the solubility and bioavailability of bioactive substances [100].

Although PNPs are extensively studied for GB drug delivery, BBB remains a challenge [116]. Significant results were obtained when an intravenous delivery of poly (*n*-butyl-2-cyanoacrylate) NP (PBCA) was tested [117]. Moreover, different polymers, notably PGA and PLA, were similarly investigated as carriers increasing drug delivery efficiency [118]. Using an unconventional route, intranasal administration that is able to bypass the BBB, Sekerdag et al. [119] analyzed the lipid-fused pseudo-thiosalicylic acid-FEG-PEG-PLGA hybrid NPs in rats and these PNPs achieved a substantial reduction of the tumor area.

Recently, Baghirov et al. [120] produced a new platform with focused ultrasound (FuS) and poly (2-ethyl-butyl cyanoacrylate) microbubbles that could space the BBB and accumulate in the cerebral parenchyma. In this area, Mead et al. [121] studied in rats a combined method with the delivery of genetic material with a continuous release. Remarkably, this system achieved important results without inducing toxicity or activation of astrocytes.

The first PNP used for drug delivery to the CNS was PBCA PNP [122]. Subsequently, the NP design has been based on the ability of PNPs to target BBB and to deliver drugs to the tumor site using receptor-mediated endocytosis (Table 1). This mechanism causes endogenous materials to pass through the BBB. Besides, changes in the surface of PNPs, for instance, surfactant coverage or ligand binding, significantly increase cell uptake, especially when the linker is associated with receptor-related proteins, which enhances receptor binding to CNS cells [3].

Surface changing with surfactants is a very common modification. In 2010, Gelperina et al. demonstrated that NP of PLGA with a poloxamer-188-modified surface showed more substantial results than PNP without modification. In another study using glioblastomas from 101/18 rats, Wohlfart et al. [135] showed that PLGA NP with poloxamer-188 and doxorubicin (DOX) were able to reach the brain and cross the BBB. By the same token, polysorbate-80 (P-80) modified poly (isohexyl cyanoacrylate) (PIHC) was tested *in vitro* against GB cells, showing a promising use as a non-invasive treatment of GB. An alternative to NP coating is glutathione-coated PNP. Geldenhuys et al. [136] presented PLGA NPs with paclitaxel (PTX) and coated with glutathione, which resulted in an increased capacity to reach the passage through the BBB.

Additional modifications, namely protein binding, can lead to better results. Transferrin (TF), a brain-directed glycoprotein ligand which has a receptor on endothelial cells, was used by Ren et al. [127, 134] to functionalize PEG-PLA NP with results demonstrating that the formulation reaches C6 cells and crosses the BBB. Lactoferrin glycoprotein (LF) used in PEG-PLGA NP was also effective in reaching the brains of mice [128]. Son et al. [147] studied NP dissociates of disulfide modified with rabies virus glycoprotein (RVG) (SS-PEI) that were used to deliver microRNA to GB.

A red blood cell membrane-coated NP with the neurotoxin candoxin (CDX), a peptide that shows high affinity for receptors expressed on CNS endothelial cells, including nicotinic acetylcholine receptors, have also been shown to be efficient [148]. Because these NPs are very flexible, this methodology system could be used for several types of membranes, serving different medical purposes.

Fang et al. [149] used chlorotoxin (CTX), a neurotoxin that binds to cell membranes, on the surface of a PNP that contained a chitosan nucleus functionalized with biotin and PEG and was covalently conjugated to TMZ. TMZ-CTX-PNP showed higher uptake by GB cells and were more efficient in treatment than PNP without CTX. They were also able to cross

Table 1 Polymeric nanoparticles in the treatment of glioblastomas.

NP	Surface coating	Ligand	Drug	Results	References
MPEG	PLA	Folate	PTX	Folic acid induced increased cell absorption	Wang et al. [123]
PLGA	PEG	AS1411, a DNA aptamer	PTX	Extended drug circulation and accumulation of PTX in tumor tissue. Increased inhibition of glioma growth in rats and increased animal survival	Guo et al. [98, 99]
PLGA	PEG	F3 and tLyp-1 peptides	PTX	Drug accumulation in tumor tissue, increased tissue penetration and increased animal survival	Hu et al. [124]
PEG-*co*-PCL	MMP-2/9		PTX	Increased PTX efficacy. Increased animal survival	Gu et al. [125]
PLGA	Poloxamer 188/ polysorbate 80		DOX/ loperamide	Increased treatment effectiveness due to formulation parameters such as surfactants, drug and stabilizer	Gelperina et al. [126]
PLA	PEG	Transferrin		Functionalized NP were able to penetrate the tumor tissue	Ren et al. [127]
PLGA	PEG	Lactoferrin		Viability assays demonstrated low systemic toxicity	Hu et al. [128]
GNRs	PEG	Arg-Gly-Asp		Increased toxicity due to avβ3 binding	Verma et al. [44]

Continued

Table 1 Polymeric nanoparticles in the treatment of glioblastomas—cont'd

NP	Surface coating	Ligand	Drug	Results	References
PEG-PCL			PTX	Increased toxicity due to microtubule stabilization	Xin et al. [129]
PLGA			Celecoxib	COX inhibition	Vera et al. [130]
PEG-SLN			c-Met siRNA	C-Met inhibition and decreased cell proliferation	Jin et al. [131]
Liposomal PEG			DOX	Increase of DOX infusion through BBB	Birngruber et al. [132]
PLGA	Poloxamer 188		DOX	DOX accumulation in glioma	Agarwal et al. [133]
PAMAM dendrimer			mir-21 5-FU	Increased efficacy of 5-FU and apoptosis in glioblastoma	Ren et al. [134]
PLGA	Poloxamer 188		DOX	Effective DOX delivery through BBB	Wohlfart et al. [135]
PLGA	Glutathione		PTX	Increased transport ability through BBB	Geldenhuys et al. [136]
SS-PEI		RVG	micro-RNA	Increased transport ability through BBB	Jin et al. [131]
PIHC	Polysorbate 80		DOX	Powerful vector for use in noninvasive therapies	Wohlfart et al. [135]
MNP	PEG	CTX		Increase of drug efficacy	Sun et al. [137]
SPION	PEG	CTX		Increased cell absorption and decreased tumor invasion	Veiseh et al. [138]
MNP	PEG and chitosan	CTX	siRNA	Ideal system for delivery of genetic material in the presence of a magnetic field	Veiseh et al. [139]
PLGA			Celecoxib	Increase of antitumor activity	Suzuki et al. [140]
PEG-*co*-PTMC	2-Deoxy-D-glucose			Increase of cellular absorption	Jiang et al. [141]

MNP	PEG, chitosan and PEI	CTX	GFP	Increased cell absorption and GFP expression	Kievit et al. [142]
MNP	PEG	Arg-Gly-Asp		Target-specific in glioblastoma cells	Cabada et al. [143]
MNP	PEG		DOX	Drug release promoted by infrared light	Agarwal et al. [133]
AuNP	PEG			Increased DNA damage to glioblastoma cells after radiotherapy	Joh et al. [144]
AuNP	PEG		DOX	Increased cell absorption and drug half-life	Ruan et al. [145]
PEG-PCL			PTX Angiopep	Increased transport ability through BBB	Dilnawaz et al. [146]

NP: nanoparticles; MPEG: methoxypolyethylene glycol; PLA: polylactic acid; PTX: paclitaxel; PLGA: poly(lactic-*co*-glycolic acid); PEG: polyethylene glycol; PCL: polycaprolactone; COX: cyclooxygenase; MMP-2/9: matrix metalloproteinase-2/9; DOX: doxorubicin; GNRs: gold nanorods; SLN: solid lipid nanoparticle; c-Met: tyrosine-protein kinase Met; BBB: blood-brain barrier; 5-FU: fluorouracil; PAMAM: poly(amidoamine); SS-PEI: disulfide–polyethylenimine; RVG: rabies virus glycoprotein; PIHC: polymer-infiltrated hybrid ceramic; MNP: magnetic nanoparticles; SPION: super-paramagnetic iron oxide nanoparticles; CTX: chlorotoxin; PTMC: trimethylene carbonate; GFP: green fluorescent protein; Au: gold.

the BBB and transport the drug to the tumor site, demonstrating a promising option to transport TMZ in significant concentrations and avoid toxicity in healthy cells [149].

Comparatively, Liu et al. [150,151] constructed a liposome containing TF (TF-CPP-SSL) in the nucleus *via* PEG. These TF-CPP-SSL PNPs showed efficient BBB penetration, increased drug circulation *in vivo* and were able to target the tumor. Coupled with it, they showed high cellular uptake and were able to avoid lysosomal degradation.

An important aspect to analyze the prognosis in GB patients is the necessary resection. However, it is difficult to establish a safe level of resection due to GB malignancy. In a recent study, the authors developed fluorescence-guided surgery, allowing a safer resection of GB. Tang et al. conjugated an aptamer 32 to a pegylated NP producing an aptamer probe capable of specifically binding to cancer cells [152].

The discovery of promising molecular targets to treat GB is a research challenge; however, it has stimulated the development of genetic treatments (Table 1). The low success rate of the available therapies is related to the inability of the vectors to reach the brain [3].

PTX is used to prevent mitosis because it promotes congregation and stabilization of microtubules. However, PTX is a highly hydrophobic substance and is not able to cross the BBB. To solve this problem, PTX can be absorbed into a PNP [153]. Angiopep (ANG1005) is a peptide with 19 amino acids that is conjugated with the taxane, a mitotic inhibitor. PEG-PCL-NP conjugated to angiopep and PTX were tested in glioma therapy, reducing GB cell proliferation and increasing cellular apoptosis [129]. Furthermore, using PLA NP, Wang et al. [123] produced a monomethoxy-PEG-PLA-PTX-NP modified folate target which increased cytotoxicity against HeLa and C6 cell lines.

In a study using AS1411 (Ap), a DNA aptamer, the authors produced an NP for siRNA delivery that significantly decreased cell proliferation due to prolonged drug release and showed better biodistribution of the drug at the tumor site compared with Taxol® [98, 99]. In a different study with PEG-PLA-NP, the F3 peptide, which similarly to Ap, binds to nucleolin, was used to target tumor cells by promoting an increase in cell uptake. These PNPs exhibited increased accumulation, tumor site infiltration and survival of the animal model [124].

Gu et al. [125] conjugated the low molecular weight activated protamine MMP-2/9 with PEG-*co*-PCL-NP, these NP improved uptake into cells, while *in vivo* results showed accumulation of NP at the tumor site. Also, mice with C6 implants treated with the formulation had a longer survival time when compared to the control groups.

A cyclooxygenase inhibitor (COX-2), celecoxib, presents its action promoting cell death and decreasing cell proliferation [140]. In a dose-dependent manner, PLGA-NP with celecoxib showed activity against U87MG and C6 cell lines. These results reveal these NP as an efficient choice for drug delivery in GB [154].

Another promising strategy is polyamidoamine dendrimers (PAMAMs) used as transporters of 5-fluorouracil (5-FU) and antisense oligonucleotide miR-21 (as-miR-21) to specific sites in GB [155]. When tested on U251 cells, co-delivery of 5-FU and as-miR-21 significantly increased cytotoxicity and cell death. Therefore, co-administration of drugs and molecular vectors, especially if the tumor has a known gene overexpression, is an interesting and effective approach in the treatment of GB [127, 134].

Equally important is the down-regulation of the protein tyrosine kinase Met (c-Met) which is a tyrosine kinase receptor that can interact with hepatocyte growth factor and this interaction leads to an increase in cell proliferation, corroborating with invasiveness and resistance to CT. Therefore, with the use of siRNA it is possible to suppress c-Met and this allows therapy improvement. The PEG/c-Met-siRNA formulation significantly inhibited tumor drug resistance and cell proliferation [131].

Liu et al. [150,151] presented a PNP that successfully transports and protects siRNAs. The formulated PNPs involved PEGylated AuNPs coated with PEI and chitosan. The siRNA for Ape1 was added to the formulation and the NPs were tested on brain cancer cells showing that the reduction of Ape1 expression is related to an enhancement of DNA damage after exposure to radiation.

Recently, Wadajkar et al. [156] formulated particulate nanocarriers using PLGA and PEG binding the Fn14 receptor, generally overexpressed in GB invasive, on the surface. The increase in observed efficacy suggests that these PNPs optimized the supply of drugs to treat invasive GBs. In another study, the researchers developed a PNP using polystyrene (PS) with a surface functionalized with an antibody that binds to Fn14. The results showed that this PNP was able to penetrate tumor tissue by selectively targeting GB cells [157]. Although these studies with PNP appear promising, the usual methods of preparation are sometimes inconsistent and complex, which impairs serial production. On the other hand, electro-spinning is an interesting method to produce NP without the use of surfactants and on a large scale.

4.2 PNFS in the treatment of glioblastoma

Recurrent GBs are less responsive to CT than the original tumor and are extremely invasive [158]. Despite the application of all treatment strategies, GB remains an incurable disease and the overall survival of the patient between 12 and 15 months after the initial diagnosis [159]. Autopsy studies suggest that recurrent GB are mainly local and appear within 2 cm of the initial site of the tumor [160]. Therefore, localized CT by using polymeric NF that can be administered during surgery provides an alternative DDS for the treatment of GB, significantly reducing systemic side effects and preventing tumor recurrence.

Although *in loco* treatment option is extremely promising, few studies have been performed using NF in the treatment of GB (Table 2). As previously mentioned, PTX presents several difficulties in its application [161], so a local application of the drug is an interesting choice to improve its effectiveness. In one study, PLGA and PTX were used to make submicron and microfibers. Both formulations showed a prolonged release of PTX *in vitro* over 80 days, in which submicrofibers presented augmented release rates matched to microfibers due to the higher polymeric degradation rate and surface area. The superiority of NF over the systemic administration of Taxol® in terms of apoptosis has also been reported. Not to mention, inhibition of the tumor growth in mice disclosed that the animals treated with the fiber presented smaller tumors after 24 days and 32 days post tumor inoculation compared to animals treated with Taxol®, indicating significant advantages of prolonged release of PTX [162]. In a parallel report, PTGA NF loaded with PTX with a high drug release showed a drug penetration into the rat brain at 5 mm from the implant site, even after 42 days post-implant. Furthermore, NF demonstrated tumor inhibition and a low rate of tumor proliferation after 41 days of treatment in animal models compared to placebo [163].

Recently, Ramachandran et al. [167] developed PLZ-PLA-PCL NF containing TMZ implanted in an orthotopic model of GB, which demonstrated a continuous drug release of 116 mg per day with a small extravasation to the peripheral blood (less than 100 ng), demonstrating 1000-fold higher drug concentration at the tumor site compared to peripheral blood.

BCNU [1,3-bis (2-chloroethyl)-1-nitrosourea] (carmustine) is one of the most used antitumor drugs for GB treatment that is capable of penetrating BBB [180]. The mechanism of action of BCNU is based on the formation of DNA and RNA crosslinks leading to inhibition of DNA synthesis,

Table 2 Polymeric nanofibers in the treatment of glioblastomas.

NF	Drug	Results	References
PLGA	PTX	Extended PTX release up to 80 days in which sub-microfibers demonstrated faster release than microfibers	Ranganath and Wang [162]
PLGA	PTX	Increased drug penetration (5 mm) observed after 42 days of implantation	Ranganath et al. [163]
PPC-Ca-alginate MPs	PTX, TMZ	Synergistic effect of drugs	Ni et al. [164]
PCL-Diol-*b*-PU	TMZ	Extended TMZ release up to 30 days	Irani et al. [165]
PCL	TMZ	NF upregulated p53 and Bax more than only TMZ	Tavakoli et al. [166]
PLGA, PLA, PCL	TMZ	Animals treated with extended release NF showed longer survival (> 4 months) than those treated with short release NF (74 days)	Ramachandran et al. [167]
PCL	TMZ	Inhibition of tumor growth and favoring neuron differentiation for tissue reconstruction	Huang et al. [168]
PCL-Diol-b-PU-CS NP	TMZ and Au NP	Increased antitumor activity	Irani et al. [169,170]
PCL-Diol-*b*-PU	TMZ and Au NP	Increased antitumor activity	Irani et al. [169,170]
PEG-PLLA	BCNU	BCNU showed toxicity only after 72 h due to its controlled release	Xu et al. [171]
PLGA	BCNU	NF showed BCNU release for more than 6 weeks in rat brains	Tseng et al. [172]

Continued

Table 2 Polymeric nanofibers in the treatment of glioblastomas—cont'd

NF	Drug	Results	References
PEO, PLA	Rapamycin	Decreased cell viability after treatment	Wang et al. [173]
PC, PVP	Mycophenolic acid	Fibers prepared with coaxial electrospinning demonstrated gradual drug release	Han et al. [174]
PCL	Daunorubicin	Controlled and extended drug release	Lian and Meng [175]
PLGA	BCNU, irinotecan, cisplatin, combretastatin	Decreased tumor malignancy and tumor growth, and increased survival of the animal model	Tseng et al. [176]
PCL-GT	SN-38	Increased cytotoxicity after 72h	Zhu et al. [177]
PLGA,PEI	MMP-2 mRNA and PTX	Significant regression of tumor growth in drug and gene fiber treatment	Lei et al. [178]
PLA	TRAIL	NF released TRAIL protein, leading to reduced tumor volume in mice	Bagó et al. [179]
PCEC	Curcumin	Blockage of cell proliferation signaling pathways	Guo et al. [98, 99]

NF: nanofibers; PLGA: poly(lactic-*co*-glycolic acid); PTX: paclitaxel; PPC: polypropylene copolymer; Ca: calcium; MPs: magnetic nanoparticles; TMZ: temozolomide; PCL: polycaprolactone; PU: polyurethane; p53: tumor protein 53; Bax: Bcl-2-associated X protein; PLA: polylactic acid; CS: chitosan; NPs: nanoparticles; PEG: polyethylene glycol; PLLA: poly-L-lactide; Au: gold; BCNU: bis-chloroethylnitrosourea; PEO: polyethylene oxide; PC: polycarbonate; PVP: polyvinylpyrrolidone; GT: glutathione; PEI: polyethylenimine; PCEC: polycaprolactone/poly (ethylene glycol)/polycaprolactone.

RNA production, and translation of RNA [171]. Gliadel® is a biodegradable poly (carboxyphenoxy-propane/sebacic acid) wafer that releases BCNU; it can be implanted into the tumor cavity after surgical resection [181]. This wafer has been approved by the FDA for clinical use and has shown some improvements in the survival of patients up to 2 months longer than those who did not receive the treatment [181]. Nevertheless, fast release of the drug and wafer stiffness are major disadvantages. In view of this, Tseng et al. [172] produced PLGA NFs releasing high amounts of BCNU for a prolonged time (>6 weeks) in the rat brain, coupled with no observation of inflammatory reaction in the brain tissue.

In one study, MMP-2 was chosen as a therapeutic target. An RNA plasmid suppressing the expression of MMP-2 in cancer cells was designed and complexed with PEI. This plasmid together with PTX were encapsulated in PLGA NF aiming to achieve prolonged release of both. It demonstrated substantial tumor growth suppression evaluated in intracranial tumor models and compared to NF with PTX and commercial PTX, demonstrating synergistic therapeutic effects of the gene and PTX [182].

In another research, different CT drugs such as bis-chloroethyl nitrosourea, irinotecan, cisplatin and an antiangiogenic drug, combretastatin, were loaded into PLF NF. CT drugs were rapidly released from NF, while combretastatin exhibited a prolonged release up to 2 weeks. This formulation was considered efficient in decreasing tumor progression and extending survival in GB rats compared to NF without combretastin [176].

Although there has been an increase in the interest of researchers in developing NF that are able to provide gradual release of drugs over a long period to improve the effects thereof, the concentrations used in the formulations are often toxic. In addition, the experimental procedure tested in the animals does not reflect what happens in the clinic, considering that surgical resections are not performed.

5 Adverse effects of nanocarriers

Typically, most studies that develop nanoproducts use biocompatible and biodegradable materials. Therefore, the harmful effects of nanocarriers should be minimal. However, some described adverse effects might be related to the surface area of the product, because this surface can result in the formation of reactive oxygen species (ROS), markedly superoxide anion and hydrogen peroxide [183]. Due to the existence of electron receptor sites and donors on the transport surface that tend to react with oxygen molecules, this consequently leads to the oxidation of nearby chemical compounds or even organelles [183]. Consequently, an essential factor to analyze when evaluating the side effects of nanoproducts is their surface area.

In like manner, a common side effect is the hypersensitivity reactions that some patients describe following intravenous injection. This adverse effect is common not only for NP formulations but also for several others drugs and can be avoided by administering different medication to patients or by decreasing the infusion rate [184]. Generally, hypersensitivity reactions are greater in the administration of free drugs when compared to DDS formulations. An interesting example is the administration of DOX. It is known that this drug induces a cardiotoxicity effect which can be decreased when a DDS formulation is applied. Equally important, the evaluation of the possible toxicity related to the remaining carrier substance needs to be highlighted so the production of biodegradable polymers with a known life expectancy is aimed to produce safe formulations [3,185].

6 Clinical trials

Although several nanopolymeric systems are available in the market, only a few clinical trials focusing on brain cancers are being developed, all in phase I or II (Table 3). One of them was a phase I completed the trial (NCT00734682) of intravenous liposomal irinotecan in 34 patients with recurrent high-grade gliomas (HGG). Dosing was given IV every 3 weeks, the maximum tolerated doses were established between 120 and 150 mg/m^2 and the toxicity profile was felt acceptable, guaranteeing the continuation of this study to the phase II [186]. Also, a phase II study (NCT01663012) using PEGylated liposomes containing etirinotecan pegol (NKTR-102; the active compound of irinotecan) to treat bevacizumab-resistant HGG was carried out. Among the 20 patients treated for 3 cycles, three of them had partial MRI responses and 10 patients had stable disease and, once this study did not explore the efficacy of NKTR-102 over other therapies, it might warrant further investigation of this molecule against HGG [187].

Other two studies, NCT01386580 and NCT01818713, were based on a glutathione PEGylated liposomal doxorubicin (2B3-101/targeted NC), which demonstrated a 5-fold higher brain delivery in rats as compared to Caelyx® and Doxil®, two commercial pegylated liposomal DOX formulations. Phase I studies were carried out in 37 patients with HGG ($n = 13$) or solid tumors in advanced stages and brain metastases ($n = 24$), and the results

Table 3 Clinical trials related to nanoparticles in the treatment of GBM.

Clinical trials identifier number	Treatment name/platform	Intervention	Phase
NCT01386580	2B3-101/targeted NC (PEGylated liposomes)	DOX	I/IIa
NCT01818713	2B3-101/targeted NC (PEGylated liposomes)	DOX	II
NCT01663012	NKTR-102 In Bevacizumab-Resistant High Grade Glioma (PEGylated liposomes)	Etirinotecan pegol	II
NCT00734682	Nanoliposomal CPT-11 (NL CPT-11) (PEGylated liposomes)	Irinotecan	I
NCT00944801	RNOP-09: Pegylated Liposomal Doxorubicin	DOX	II
NCT01806675	18F FPPRGD2 PET/CT or PET/MRI Imaging of αvβ3 Integrins Expression as a Biomarker of Angiogenesis	2-Fluoropropionyl-labeled pegylated dimeric RGD peptide	II

indicated a good tolerability up to a dose of 15 mg/m^2/week with or without a combination with trastuzumab and 16 patients showed stable tumor progression after 2 cycles of treatment. In phase IIa, the safety, tolerability and intracranial anti-tumor activity of the same liposomal formulation were evaluated in patients with HGG ($n = 18$) and breast cancer brain metastases ($n = 10$). Results up to March 15th 2014 revealed that in both groups the side effects were similar to the treatment with Caelyx® and Doxil®, and 31% of patients with HGG presented stable disease [188]. In the same way, the phase II trial NCT00944801 investigated the outcomes of the administration of RNOP-09 (PEGylated liposomal doxorubicin) and prolonged TMZ coupled with radiotherapy in 63 patients at the newly diagnosed GBs. Altogether, results indicated that the toxicity of the three methods combination was well tolerated, and the progression-free survival and median global survival did not result in an expressive improvement at the outcome of the patient in comparison with databases [189].

Finally, the phase II study NCT01806675 consists in a prospective trial using a new tracer (18F FPPRGD2 PET/CT or PET/MRI) for prognosis and early assessment of response to anti-angiogenic therapy in patients with GBs and other types of cancer-based on $\alpha_v\beta_3$-positive cells. This trial involves 25 patients and is still in progress. Although new data have not been released, some studies concerning other types of cancer revealed that this approach could provide comparable diagnostic information as compared to conventional methods. Thus, a similar outcome might be observed in the diagnosis of GBs cases [190].

7 How to study drug and polymer interactions with multiscale computational methods

Polymers present particular aspects that differ from the angstrom level of a bond between atoms to nanometers of the chain, micrometers, millimeters, and polymeric nanofibers. The different time scales for each material properties may vary from femtoseconds to seconds or hours (phase separation in blends). On the literature, there are many examples of multiscale nature of polymer systems [191–198]. Because of this, several computational methods were developed in order to address these issues [199–207]. These new methods introduce new possibilities to construct, optimize and predict the properties and structures of polymers. Not only that, these methods allowed the study of these materials with other molecules [208–212]. Therefore, it is possible to predict polymer-drug interactions in nanosystems by using *in silico* methods.

Until the present moment, no computational method can cover different size scales of polymers [213]. Thus, the multiscale simulation approach is one of the best choices to deal with this issue. The multiscale strategy combines various methods and is considered one of the most important subjects in computational chemistry. To perform a multiscale method, different theories and models are combined. Following the scales that are divided the different methods (Fig. 2):

- The quantum scale: In this method, the nuclei and electrons are part of the calculation and quantum mechanics (QM) calculations are used to model the system. This allows the study of several events associated with chemical reaction, in particular, the disruption and formation of chemical bonds between atoms, the transitions in electrons in the nucleus, and others important phenomena on polymers material that need to be modeled at the quantum scale.
- The atomistic scale: In the atomistic calculations, all atoms are present, defined and treated by single spheres. The force field, a typical system

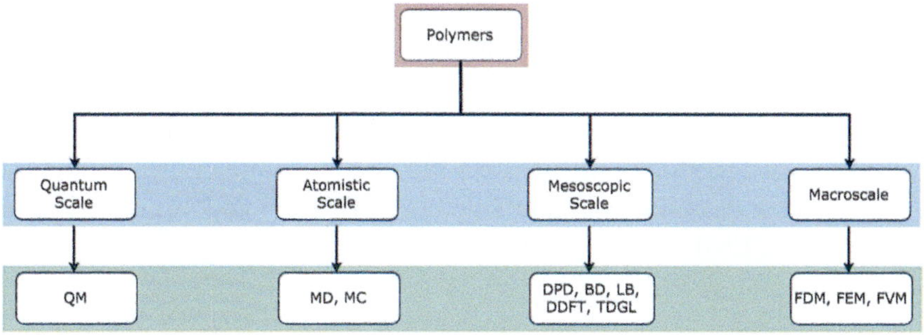

Fig. 2 Different theories and models to study polymeric materials with multiscale approach.

interaction, is responsible for the potential energy and these interactions include the bonded interactions such as the bond angle, the bond length, and the dihedral angle potentials between atoms. Moreover, force fields also contain non-bonded interactions. These interactions act between atoms in other molecules and in the same molecule. Within force fields, non-bonded interactions can be divided into two types: Van der Waals interactions and electrostatic interactions. Monte Carlo (MC) and Molecular dynamics (MD) simulations are examples used at this level to study molecular processes covering a larger system (high number of atoms), for example, proteins, membranes, and nucleic acids.

- The mesoscopic scale: In this approach, the molecules are defined as particles, known as beads. Some details of the system are implemented implicitly, allowing the possibility to simulate some systems on higher time scales when compared to atomistic simulations. An example is a coarse-grained model, where the particles are accumulated in beads. During the calculation, for definition of the system, interactions between beads are used. Several techniques were made to investigate the structure of polymers at mesoscopic scale, among them lattice Boltzmann (LB), Brownian dynamics (BD) dissipative particle dynamics (DPD), time-dependent Ginzburg-Landau (TDGL), and dynamic density functional theory (DDFT).
- The macroscale: In this methodology, the system is considered as continuous and the details of atoms and molecules are ignored. The constitutive laws are responsible for the behavior of the system. The conservation laws are associated with constitutive laws to simulate the systems. The main functions, namely stress and velocity are continuous. On the other hand, a finite number of locations which separate

continuity regions is considered discontinuous on these calculations. An excellent analogy to describe this scale would be to replace a heterogeneous material (details of atoms and molecules) with an equivalent homogeneous model (macroscale). To perform molecular simulations at this scale, the following methods can be used: finite difference method (FDM) and finite volume method (FVM), and finite element method (FEM).

8 Final considerations

GB is an extremely aggressive disease, however, current therapy to treat GBs does not suggest efficient results and cure is a distant reality. The use of nanotechnology can be seen as a promising strategy to GB therapy; at the same time, several unknown responses need to be further investigated. Researchers are not yet aware of all possible effects when using DDS, in general, toxicity and efficacy parameters. Nowadays, cancers such as GB, which are particularly malignant and very resistant to CT, can be treated with nanocarriers that improve the efficacy of existing and new medicines. Currently, several vectors are being investigated, in contrast, only some formulations using this innovative approach are commercially available. Even so, nanocarriers are a promising strategy for the effective delivery of stable release of drugs at tumor sites for a sustained period.

Because GBs can invade functional areas of the brain, the second surgical resection is jeopardy. Autopsy studies suggest that recurrent GBs are mainly local and appear within two centimeters of the initial site of the tumor. Therefore, direct and controlled CT *in loco* can be provided with the alternative DDS for the treatment of GB. To emphasize, it can significantly reduce systemic side effects and can prevent tumor recurrence.

While many studies like this have established good results, further research is essential to the developing of successful DDS. In this regard, nanocarriers with biofunctionalized surfaces and molecular targets are necessary to improve delivery *in vivo*. In addition, developing functionalized nanocarriers can provide multiple targets and overcome BBB problems. Other important aspects that need to be better understood during the fabrication of DDS are:

(a) The microenvironment of GB, because it is important to elucidate the molecular implications that can affect the carrier;

(b) Responses of the immune system.

Clinical trials must be conducted to analyze the pharmacological characteristics and body response of formulations. Only with these types of assessments, it is possible to obtain optimized therapies. In general, it is also essential to develop intelligent methods that can evaluate and follow up on the results of the DDS formulation. It is extremely necessary to understand cell interactions after transporter uptake and changes in the tumor microenvironment.

The development of polymers structure and the study of their interactions with other biomolecules, as anticancer drugs, require a comprehensive knowledge of the phenomena at different time and length scales. Because of this need, the theoretical and computational methods present great progress, allowing the study of these systems. The development of multiscale methods enables the design of polymeric materials simultaneously on many scales instead of trial-and-error experimentation.

Multiscale simulations approaches may be useful strategies for assessing the behavior of molecules and their molecular alterations after carrier treatment since such technologies can predict dynamic changes on important molecules for cancer therapy. Finally, approaches based on nanotechnology can be used to improve and direct the delivery of new and known drugs for the management of various human diseases; the days are prosperous for strategies based on nanotechnology.

References

[1] J. Ferlay, et al., Cancer incidence and mortality worldwide: sources, methods and major patterns in GLOBOCAN 2012, Int. J. Cancer 136 (2015) 359–386, https://doi.org/10.1002/ijc.29210.

[2] F. Masood, Polymeric nanoparticles for targeted drug delivery system for cancer therapy, Mat. Sci. Eng. C 60 (2016) 569–578, https://doi.org/10.1016/j.msec.2015.11.067.

[3] S. Aftab, et al., Nanomedicine: an effective tool in cancer therapy, Int. J. Pharm. 540 (2018) 132–149, https://doi.org/10.1016/j.ijpharm.2018.02.007.

[4] K. Cho, et al., Therapeutic nanoparticles for drug delivery in cancer, Clin. Cancer Res. 14 (2008) 1310–1316, https://doi.org/10.1158/1078-0432.CCR-07-1441.

[5] L. Steffens, et al., Natural polysaccharides for the delivery of anticancer therapeutics, in: M.S. Hasnain, A.K. Nayak (Eds.), Natural Polysaccharides in Drug Delivery and Biomedical Applications, Elsevier, 2019, https://doi.org/10.1016/B978-0-12-817055-7.00019-4.

[6] T. Glaser, et al., Targeted nanotechnology in glioblastoma multiforme, Front. Pharm. 8 (2017) 1–14, https://doi.org/10.3389/fphar.2017.00166.

[7] Instituto Nacional de Câncer José Alencar Gomes da Silva, Estimativa 2018: Incidência de Câncer no Brasil. Rio de Janeiro, 2017, Available from: https://www.inca.gov.br/sites/ufu.sti.inca.local/files//media/document//estimativa-incidencia-de-cancer-no-brasil-2018.pdf. (Accessed 27 March 2019).

[8] A.A. Brandes, et al., Glioblastoma in adults, Crit. Rev. Oncol. Hematol. 67 (2008) 139–152, https://doi.org/10.1016/j.critrevonc.2008.02.005.
[9] J. Ching, et al., A novel treatment strategy for glioblastoma multiforme and glioma associated seizures: increasing glutamate uptake with PPARγ agonists, J. Clin. Neurosci. 22 (2015) 21–28, https://doi.org/10.1016/j.jocn.2014.09.001.
[10] S. Thakkar, M. Misra, Electrospun polymeric nanofibers: new horizons in drug delivery, Eur. J. Pharm. Sci. 107 (2017) 148–167, https://doi.org/10.1016/j.ejps.2017.07.001.
[11] Q.T. Ostrom, et al., CBTRuS statistical report: primary brain and central nervous system tumors diagnosed in the United States in 2008–2012, Neuro Oncol. 16 (Suppl. 4) (2015) iv1–v62, https://doi.org/10.1093/neuonc/nov189.
[12] American Association of Neurosciences Nurses, Care of the Adult Patient With a Brain Tumor, 2014, Available from: https://www.abta.org/healthcare-professionals/clinical-practice-guidelines/. (Accessed 27 March 2019).
[13] M.E. Davis, Glioblastoma: overview of disease and treatment, Clin. J. Oncol. Nurs. 20 (2016) S2–S8, https://doi.org/10.1188/16.CJON.S1.2-8.
[14] R. Stupp, et al., Radiotherapy plus concomitant and adjuvant temozolomide for glioblastoma, N. Engl. J. Med. 10 (2005) 987–996, https://doi.org/10.1056/NEJMoa043330.
[15] S.S. Kim, et al., A nanoparticle carrying the p53 gene targets tumors including cancer stem cells, sensitizes glioblastoma to chemotherapy and improves survival, ACS Nano 8 (2014) 5494–5514, https://doi.org/10.1021/nn5014484.
[16] BC Cancer Agency Management Guidelines, BC Cancer Protocol Summary for Concomitant (Dual Modality) and Adjuvant Temozolomide for Newly Diagnosed Malignant Gliomas with Radiation, 2018, Available from: http://www.bccancer.bc.ca/health-professionals/clinical-resources/chemotherapy-protocols/neuro-oncology. (Accessed 27 March 2019).
[17] S. Lapointe, et al., Primary brain tumours in adults, Lancet 392 (2018) 432–446.
[18] F. Pourgholi, et al., Nanoparticles: novel vehicles in treatment of glioblastoma, Biomed. Pharmacother. 77 (2016) 98–107, https://doi.org/10.1016/j.biopha.2015.12.014.
[19] J. Chen, R.M. McKay, L.F. Parada, Malignant glioma: lessons from genomics, mouse models, and stem cells, Cell 30 (2012) 36–47, https://doi.org/10.1016/j.cell.2012.03.009.
[20] D.W. Parsons, et al., An integrated genomic analysis of human glioblastoma multiforme, Science 26 (2008) 1807–1812, https://doi.org/10.1126/science.1164382.
[21] C. Alifieris, D.T. Trafalis, Glioblastoma multiforme: pathogenesis and treatment, Pharmacol. Ther. 152 (2015) 63–82, https://doi.org/10.1016/j.pharmthera.2015.05.005.
[22] T.A. Wilson, M.A. Karajannis, D.H. Harter, Glioblastoma multiforme: state of the art and future therapeutics, Surg. Neurol. Int. 8 (2014) 64, https://doi.org/10.4103/2152-7806.132138.
[23] R.M. Young, A. Jamshidi, G. Davis, J.H. Sherman, Current trends in the surgical management and treatment of adult glioblastoma, Ann. Transl. Med. 3 (2015) 121, https://doi.org/10.3978/j.issn.2305-5839.2015.05.10.
[24] J. Zhu, et al., Frequent Nek1 overexpression in human gliomas, Biochem. Biophys. Res. Commun. 476 (4) (2016) 522–527.
[25] T.D. Melo-Hanchuk, et al., NEK1 kinase domain structure and its dynamic protein interactome after exposure to cisplatin, Sci. Rep. 7 (2017) 5445.
[26] C.W. Brennan, et al., The somatic genomic landscape of glioblastoma, Cell 10 (2013) 462–477, https://doi.org/10.1016/j.cell.2013.09.034.

[27] R.G. Verhaak, et al., Integrated genomic analysis identifies clinically relevant subtypes of glioblastoma characterized by abnormalities in PDGFRA, IDH1, EGFR, and NF1, Cancer Cell 17 (2010) 98–110, https://doi.org/10.1016/j.ccr.2009.12.020.
[28] H. Wang, et al., The challenges and the promise of molecular targeted therapy in malignant gliomas, Neoplasia 17 (2015) 239–255, https://doi.org/10.1016/j.neo.2015.02.002.
[29] H. Yang, Nanoparticle-mediated brain-specific drug delivery, imaging, and diagnosis, Pharm. Res. 27 (2010) 1759–1771, https://doi.org/10.1007/s11095-010-0141-7.
[30] I.T. Papademetriou, T. Porter, Promising approaches to circumvent the blood-brain barrier: progress, pitfalls and clinical prospects in brain cancer, Ther. Deliv. 6 (2015) 989–1016, https://doi.org/10.4155/tde.15.48.
[31] F.L. Cardoso, et al., Looking at the blood-brain barrier: molecular anatomy and possible investigation approaches, Brain Res. Rev. 64 (2010) 328–363.
[32] R.F. Haseloff, et al., Transmembrane proteins of the tight junctions at the blood-brain barrier: structural and functional aspects, Semin. Cell Dev. Biol. 38 (2015) 16–25, https://doi.org/10.1016/j.semcdb.2014.11.004.
[33] H. Wolburg, et al., Localization of claudin-3 in tight junctions of the blood-brain barrier is selectively lost during experimental autoimmune encephalomyelitis and human glioblastoma multiforme, Acta Neuropathol. 105 (2003) 586–592, https://doi.org/10.1007/s00401-003-0688-z.
[34] W.M. Pardridge, The blood-brain barrier: bottleneck in brain drug development, NeuroRx 2 (2005) 3–14, https://doi.org/10.1602/neurorx.2.1.3.
[35] T. Patel, J. Zhou, J.M. Piepmeier, W.M. Saltzman, Polymeric nanoparticles for drug delivery to the central nervous system, Adv. Drug Deliv. Rev. 15 (2012) 701–705, https://doi.org/10.1016/j.addr.2011.12.006.
[36] National Comprehensive Cancer Network, Clinical Practice Guidelines in Oncology: Central Nervous System Cancers, 2015, p. 1. Available from: https://www.nccn.org/professionals/physician_gls/pdf/cns.pdf. (Accessed 27 March 2019).
[37] M.D. Piroth, et al., Postoperative radiotherapy of glioblastoma multiforme: analysis and critical assessment of different treatment strategies and predictive factors, Strahlenther. Onkol. 183 (2007) 695–702, https://doi.org/10.1007/s00066-007-1739-5.
[38] S. Polyzoidis, et al., Active dendritic cell immunotherapy for glioblastoma: current status and challenges, Br. J. Neurosurg. 29 (2015) 197–205, https://doi.org/10.3109/02688697.2014.994473.
[39] M.R. Gilbert, et al., A randomized trial of bevacizumab for newly diagnosed glioblastoma, N. Engl. J. Med. 370 (2014) 699–708.
[40] B.D. Choi, et al., EGFRvIII-targeted vaccination therapy of malignant glioma, Brain Pathol. 19 (2009) 713–723.
[41] A.A. Elsamadicy, Prospect of rindopepimut in the treatment of glioblastoma, Expert. Opin. Biol. Ther. 17 (2017) 507–513.
[42] S. Xue, et al., Blocking the PD-1/PD-L1 pathway in glioma: a potential new treatment strategy, J. Hematol. Oncol. 10 (2017) 81.
[43] J. Polivka Jr., et al., Advances in experimental targeted therapy and immunotherapy for patients with glioblastoma multiforme, Anticancer Res. 37 (2017) 21–33.
[44] J. Verma, S. Lal, C.J. Van Noorden, Nanoparticles for hyperthermic therapy: synthesis strategies and applications in glioblastoma, Int. J. Nanomedicine 10 (2014) 2863–2877, https://doi.org/10.2147/IJN.S57501.
[45] F. Baumann, et al., Combined thalidomide and temozolomide treatment in patients with glioblastoma multiforme, J. Neuro-Oncol. 67 (2004) 191–200.
[46] M.E. Hegi, et al., Correlation of O6-methylguanine methyltransferase (MGMT) promoter methylation with clinical outcomes in glioblastoma and clinical strategies to modulate MGMT activity, J. Clin. Oncol. 1 (2008) 4189–4199, https://doi.org/10.1200/JCO.2007.11.5964.

[47] P. Bischoff, A. Altmeyer, F. Dumont, Radiosensitising agents for the radiotherapy of cancer: advances in traditional and hypoxia targeted radiosensitisers, Expert Opin. Ther. Pat. 19 (2009) 643–662, https://doi.org/10.1517/13543770902824172.
[48] J.C. Sok, et al., Mutant epidermal growth factor receptor (EGFRvIII) contributes to head and neck cancer growth and resistance to EGFR targeting, Clin. Cancer Res. 1 (2006) 5064–5073, https://doi.org/10.1158/1078-0432.CCR-06-0913.
[49] J.M. Di Bei, C.Y. Bi-Botti, Formulation of dacarbazine-loaded cubosomes—part I: influence of formulation variables, AAPS PharmSciTech 10 (3) (2009) 1032–1039.
[50] B. Fazeny-Dorner, et al., Survival with dacarbazine and fotemustine in newly diagnosed glioblastoma multiforme, Br. J. Cancer 88 (2003) 496–501.
[51] N. Charnley, C. West, P. Price, Assessment of Drug Resistance in Anticancer Therapy by Nuclear Imaging in Drug Resistance in Cancer Cells, Springer, 2009, pp. 295–313.
[52] H.L. Wong, et al., Chemotherapy with anticancer drugs encapsulated in solid lipid nanoparticles, Adv. Drug Deliv. Rev. 10 (2007) 491–504, https://doi.org/10.1016/j.addr.2007.04.008.
[53] M.E. van Linde, et al., Bevacizumab in combination with radiotherapy and temozolomide for patients with newly diagnosed glioblastoma multiforme, Oncologist 20 (2015) 107–108, https://doi.org/10.1634/theoncologist.2014-0418.
[54] K. Messaoudi, A. Clavreul, F. Lagarce, Toward an effective strategy in glioblastoma treatment. Part I: resistance mechanisms and strategies to overcome resistance of glioblastoma to temozolomide, Drug Discov. Today 20 (2015) 899–905, https://doi.org/10.1016/j.drudis.2015.02.011.
[55] W. Gong, et al., Improving efficiency of adriamycin crossing blood brain barrier by combination of thermosensitive liposomes and hyperthermia, Biol. Pharm. Bull. 34 (2011) 1058–1064.
[56] L. Yang, et al., Shape-controlled synthesis of protein-conjugated silver sulfidenanocrystals and study on the inhibition of tumor cell viability, Chem. Commun. 26 (2008) 2995–2997, https://doi.org/10.1039/b804274h.
[57] B. Hildebrandt, et al., The cellular and molecular basis of hyperthermia, Crit. Rev. Oncol. Hematol. 43 (2002) 33–56.
[58] G. Baronzio, A. Gramaglia, G. Fiorentini, Current role and future perspectives of hyperthermia for prostate cancer treatment, In Vivo 23 (2009) 143–146.
[59] S. Mornet, et al., Magnetic nanoparticle design for medical diagnosis and therapy, J. Mater. Chem. 14 (2004) 2161–2175, https://doi.org/10.1039/B402025A.
[60] R. Gilchrist, et al., Selective inductive heating of lymph nodes, Ann. Surg. 146 (1957) 596.
[61] R. Grupta, D. Sharma, Evolution of magnetic hyperthermia for glioblastoma multiforme therapy, ACS Chem. Neurosci. 10 (2019) 1157–1172.
[62] I.J. Barani, D.A. Larson, Radiation therapy of glioblastoma, Cancer Treat. Res. 163 (2015) 49–73, https://doi.org/10.1007/978-3-319-12048-5_4.
[63] X.X. Xia, et al., Hydrophobic drug-triggered self assembly of nanoparticles from silk-elastin-like protein polymers for drug delivery, Biomacromolecules 15 (2014) 908–914.
[64] G. Zhu, et al., Self-assembled, aptamer-tethered DNA nanotrains for targeted transport of molecular drugs in cancer theranostics, Proc. Natl. Acad. Sci. 110 (2013) 7998–8003.
[65] T. Aminabhavi, S. Dharupaneedi, U. More, The role of nanotechnology and chitosan-based biomaterials for tissue engineering and therapeutic delivery, in: J.A. Jennings, J.D. Bumgardner (Eds.), Chitosan Based Biomaterials, Tissue Engineering and Therapeutics, Vol. 2, Elsevier, 2017, pp. 1–29.
[66] H.B. Man, et al., Synthesis of nanodiamond–daunorubicin conjugates to overcome multidrug chemoresistance in leukemia, nanomedicine: nanotechnology, Biol. Med. 10 (2014) 359–369.

[67] W.K. Oh, H. Yoon, J. Jang, Size control of magnetic carbon nanoparticles for drug delivery, Biomaterials 31 (2010) 1342–1348.
[68] L. Pan, Overcoming multidrug resistance of cancer cells by direct intranuclear drug delivery using TAT-conjugated mesoporous silica nanoparticles, Biomaterials 34 (2013) 2719–2730.
[69] W. She, et al., Dendronized heparin–doxorubicin conjugate based nanoparticle as pH-responsive drug delivery system for cancer therapy, Biomaterials 34 (2013) 2252–2264.
[70] B. Sun, et al., Robust, active tumor-targeting and fast bioresponsive anticancer nanotherapeutics based on natural endogenous materials, Acta Biomater. 45 (2016) 223–233.
[71] T. Sun, et al., Engineered nanoparticles for drug delivery in cancer therapy, Angew. Chem. Int. Ed. 53 (2014) 12320–12364.
[72] G. Xi, et al., Convection-enhanced delivery of nanodiamond drug delivery platforms for intracranial tumor treatment, Nanomedicine 10 (2014) 381–391.
[73] P.Y. Wen, D.A. Reardon, Neuro-oncology in 2015: progress in glioma diagnosis, classification and treatment, Nat. Rev. Neurol. 12 (2016) 69–70, https://doi.org/10.1038/nrneurol.2015.242.
[74] Y. Cao, et al., Drug release from core-shell PVA/silk fibroin nanoparticles fabricated by one-step electrospraying, Sci. Rep. 7 (2017) 11913.
[75] T.B. Subbiah, S. Pararneswaran, S. Ramkumar, et al., Electrospinning of nanofibers, J. App. Polym. Sci. 96 (2) (2005) 557–569.
[76] A. Frenot, I.S. Chronakis, Polymer nanofibers assembled by electrospinning, Curr. Opin. Colloid Interface Sci. 8 (2003) 64–75.
[77] B. Ghorani, N. Tucker, Fundamentals of electrospinning as a novel delivery vehicle for bioactive compounds in food nanotechnology, Food Hydrocolloids 51 (2015) 227–240.
[78] G. Taylor, Disintegration of water drops in an electric field, Proc. R. Soc. Lond. A Math. Phys. Sci. 280 (1964) 383–397.
[79] W. Pham, U. Sharma, A.G. Mikos, Electrospinning of polymeric nanofibers for tissue Enginnering applications: a review, Tissue Eng. 12 (2006) 1197–1211.
[80] J.S. Lee, Role of molecular weight of atactic poly(vinyl alcohol) (PVA) in the structure and properties of PVA nanofabric prepared by electrospinning, J. Appl. Polym. Sci. 93 (2004) 1638–1646.
[81] B. Felice, et al., Electrosprayed poly(vinyl alcohol) particles: preparation and evaluation of their drug release profile, Polym. Int. 64 (2015) 1722–1732.
[82] T.M. Aminabhavi, A.S. Deshmukh, Polysaccharide-based hydrogels as biomaterials, in: Polymeric Hydrogels as Smart Biomaterials, Springer, 2016, pp. 45–71.
[83] K. Ganguly, et al., Polysaccharide-based micro/nanohydrogels for delivering macromolecular therapeutics, J. Control. Release 193 (2014) 162–173.
[84] N.K. Jain, V. Mishra, N.K. Mehra, Targeted drug delivery to macrophages, Expert Opin. Drug Deliv. 10 (2013) 353–367.
[85] A. Lamprecht, et al., Nanomedicines in gastroenterology and hepatology, Nat. Rev. Gastroenterol. Hepatol. 12 (2015) 195–204.
[86] D. Peer, et al., Nanocarriers as an emerging platform for cancer therapy, Nat. Nanotechnol. 2 (2007) 751–760.
[87] M.F. Sohail, et al., Folate grafted thiolated chitosan enveloped nanoliposomes with enhanced oral bioavailability and anticancer activity of docetaxel, J. Mater. Chem. B 4 (2016) 6240–6248.
[88] S. Wilhelm, et al., Analysis of nanoparticle delivery to tumors, Nat. Rev. Mater. 1 (2016) 16014.

[89] Y. Cheng, et al., Deep penetration of a PDT drug into tumors by noncovalent drug-gold nanoparticle conjugates, J. Am. Chem. Soc. 133 (2011) 2583–2591.
[90] K. Letchford, H. Burt, A review of the formation and classification of amphiphilic block copolymer nanoparticulate structures: micelles, nanospheres, nanocapsules and polymersomes, Eur. J. Pharm. Biopharm. 65 (2007) 259–269, https://doi.org/10.1016/j.ejpb.2006.11.009.
[91] C.P. Reis, et al., Nanoencapsulation I. Methods for preparation of drug-loaded polymeric nanoparticles, Nanomedicine 2 (2006) 8–21, https://doi.org/10.1016/j.nano.2005.12.003.
[92] M. Teixeira, et al., Development and characterization of PLGA nanospheres and nanocapsules containing xanthone and 3-methoxyxanthone, Eur. J. Pharm. Biopharm. 59 (2005) 491–500, https://doi.org/10.1016/j.ejpb.2004.09.002.
[93] A. Misra, et al., Drug delivery to the central nervous system: a review, J. Pharm. Pharm. Sci. 6 (2003) 252–273.
[94] S.Y. Tzeng, J.J. Green, Subtle changes to polymer structure and degradation mechanism enable highly effective nanoparticles for siRNA and DNA delivery to human brain cancer, Adv. Healthc. Mater. 2 (2013) 468–480, https://doi.org/10.1002/adhm.201200257.
[95] S.Y. Tzeng, J.J. Green, Therapeutic nanomedicine for brain cancer, Ther. Deliv. 4 (2013) 687–704, https://doi.org/10.4155/tde.13.38.
[96] J. Zhang, et al., Multifunctional envelope-type mesoporous silica nanoparticles for tumor triggered targeting drug delivery, J. Am. Chem. Soc. 135 (2013) 5068–5073.
[97] E. Gullotti, Y. Yeo, Extracellularly activated nanocarriers: a new paradigm of tumor targeted drug delivery, Mol. Pharm. 6 (2009) 1041–1051, https://doi.org/10.1021/mp900090z.
[98] G. Guo, et al., Preparation of curcumin loaded poly (e-caprolactone)–poly (ethylene glycol)–poly (e-caprolactone) nanofibers and their *in vitro* antitumor activity against glioma 9L cells, Nanoscale 3 (2011) 3825–3832.
[99] J. Guo, et al., Aptamer-functionalized PEG–PLGA nanoparticles for enhanced anti-glioma drug delivery, Biomaterials 32 (2011) 8010–8020.
[100] A. Kumari, S.K. Yadav, S.C. Yadav, Biodegradable polymeric nanoparticles based drug delivery systems, Colloids Surf. B Biointerfaces 1 (2010) 1–18, https://doi.org/10.1016/j.colsurfb.2009.09.001.
[101] R. Bazak, et al., Cancer active targeting by nanoparticles: a comprehensive review of literature, J. Cancer Res. Clin. Oncol. 141 (2015) 769–784, https://doi.org/10.1007/s00432-014-1767-3.
[102] C.F. Greineder, et al., Molecular engineering of high affinity single-chain antibody fragment for endothelial targeting of proteins and nanocarriers in rodents and humans, J. Control. Release 226 (2016) 229–237.
[103] P.V. Kulkarni, et al., Quinoline-n-butylcyanoacrylate-based nanoparticles for brain targeting for the diagnosis of Alzheimer's disease, Wiley Interdiscip. Rev. Nanomed. Nanobiotechnol. 2 (2010) 35–47.
[104] A.J. Cole, et al., Magnetic brain tumor targeting and biodistribution of long-circulating PEG-modified, cross-linked starchcoated iron oxide nanoparticles, Biomaterials 32 (2011) 6291–6301.
[105] J.V. Jokerst, et al., Nanoparticle PEGylation for imaging and therapy, Nanomedicine 6 (2011) 715–728.
[106] F. Danhier, et al., PLGA-based nanoparticles: an overview of biomedical applications, J. Control. Release 161 (2012) 505–522.
[107] M. Canillas, et al., Bioactive composites fabricated by freezing-thawing method for bone regeneration applications, Polym. Phys. 54 (2016) 761–773.
[108] W. Chen, et al., Charge-conversional and reduction-sensitive poly(vinyl alcohol) nanogels for enhanced cell uptake and efficient intracellular doxorubicin release, J. Control. Release 205 (2015) 15–24.

[109] G.G. De Lima, et al., A novel pH-sensitive ceramic-hydrogel for biomedical applications, Polym. Adv. Technol. 26 (2015) 1439–1446.
[110] L.S. Reinhardt, et al., Freeze-thaw electrospun PVA-dacarbazine nanoparticles: preparation, characterization and anticar evaluation, Int. J. Polym. Mater. Polym. Biomater. (2019) 1–12.
[111] A. Dikpati, et al., Targeted drug delivery to CNS using nanoparticles, J. Adv. Pharm. Sci. 2 (2012) 179–191.
[112] A.Z. Wilczewska, et al., Nanoparticles as drug delivery systems, Pharmacol. Rep. 64 (2012) 1020–1037.
[113] P. Couvreur, et al., Nanocapsule technology: a review, Crit. Rev. Ther. Drug Carrier Syst. 19 (2002) 99–134.
[114] H. He, et al., PEGylated poly(amidoamine) dendrimer-based dual-targeting carrier for treating brain tumors, Biomaterials 32 (2011) 478–487.
[115] C.Y. Lee, Strategies of temozolomide in future glioblastoma treatment, OncoTargets Ther. 10 (2017) 265–270, https://doi.org/10.2147/OTT.S120662.
[116] J. Ma, et al., Nano-enabled drug delivery systems for brain cancer and Alzheimer's disease: research patterns and opportunities, nanomedicine: nanotechnology, Biol. Med. 11 (2015) 1763–1771.
[117] Y. Lin, et al., Delivery of large molecules via poly(butyl cyanoacrylate) nanoparticles into the injured rat brain, J. Nanotechnol. 23 (2012) 165101.
[118] R.C. Mundargi, et al., Nano/micro technologies for delivering macromolecular therapeutics using poly (D,L-lactide-co-glycolide) and its derivatives, J. Control. Release 125 (2008) 193–209.
[119] E. Sekerdag, et al., A potential non-invasive glioblastoma treatment: nose-to-brain delivery of farnesylthiosalicylic acid incorporated hybrid nanoparticles, J. Control. Release 261 (2017) 187–198, https://doi.org/10.1016/j.jconrel.2017.06.032.
[120] H. Baghirov, et al., Ultrasound-mediated delivery and distribution of polymeric nanoparticles in the normal brain parenchyma and melanoma metastases, Cancer Res. 77 (2017) 1–18, https://doi.org/10.1158/1538-7445.AM2017-3109.
[121] B.P. Mead, et al., Targeted gene transfer to the brain via the delivery of brain-penetrating DNA nanoparticles with focused ultrasound, J. Control Release 223 (2016) 109–117, https://doi.org/10.1016/j.jconrel.2015.12.034.
[122] J. Kreuter, et al., Passage of peptides through the blood-brain barrier with colloidal polymer particles (nanoparticles), Brain Res. 674 (1995) 171–174.
[123] J. Wang, et al., Folate decorated hybrid polymeric nanoparticles for chemically and physically combined paclitaxel loading and targeted delivery, Biomacromolecules 12 (2010) 228–234.
[124] Q. Hu, et al., F3 peptide-functionalized PEG-PLA nanoparticles co-administrated with tLyp-1 peptide for anti-glioma drug delivery, Biomaterials 34 (2013) 1135–1145.
[125] G. Gu, et al., PEG-co-PCL nanoparticles modified with MMP-2/9 activatable low molecular weight protamine for enhanced targeted glioblastoma therapy, Biomaterials 34 (2013) 196–208.
[126] S. Gelperina, et al., Drug delivery to the brain using surfactant-coated poly(lactide-co-glycolide) nanoparticles: influence of the formulation parameters, Eur. J. Pharm. Biopharm. 74 (2) (2010) 157–163, https://doi.org/10.1016/j.ejpb.2009.09.003.
[127] W.H. Ren, et al., Development of transferrin functionalized poly(ethylene glycol)/poly (lactic acid) amphiphilic block copolymeric micelles as a potential delivery system targeting brain glioma, J. Mater. Sci. 21 (2010) 2673–2681.
[128] K. Hu, et al., Lactoferrin conjugated PEGPLGA nanoparticles for brain delivery: preparation, characterization and efficacy in Parkinson's disease, Int. J. Pharm. 415 (2011) 273–283.

[129] H. Xin, et al., Anti-glioblastoma efficacy and safety of paclitaxel-loading Angiopep-conjugated dual targeting PEG-PCL nanoparticles, Biomaterials 33 (2012) 8167–8176, https://doi.org/10.1016/j.biomaterials.2012.07.046.
[130] M. Vera, et al., New celecoxib multiparticulate systems to improve glioblastoma treatment, Int. J. Pharm. 473 (2014) 518–527, https://doi.org/10.1016/j.ijpharm.2014.07.028.
[131] J. Jin, et al., *In vivo* specific delivery of c-Met siRNA to glioblastoma using cationic solid lipid nanoparticles, Bioconjug. Chem. 21 (2011) 2568–2572, https://doi.org/10.1021/bc200406n.
[132] T. Birngruber, et al., Enhanced doxorubicin delivery to the brain administered through glutathione PEGylated liposomal doxorubicin (2B3-101) as compared with generic Caelyx,®/Doxil®—a cerebral open flow microperfusion pilot study, J. Pharm. Sci. 103 (7) (2014) 1948–1954, https://doi.org/10.1002/jps.23994.
[133] A. Agarwal, et al., Remote triggered release of doxorubicin in tumors by synergistic application of thermosensitive liposomes and gold nanorods, ACS Nano 5 (2011) 4919–4926.
[134] Y. Ren, et al., Co-delivery of as-miR-21 and 5-Fu by poly(amidoamine) dendrimer attenuates human glioma cell growth *in vitro*, J. Biomater. Sci. Polym. Ed. 21 (2010) 303–314, https://doi.org/10.1163/156856209X415828.
[135] S. Wohlfart, et al., Efficient chemotherapy of rat glioblastoma using doxorubicin loaded PLGA nanoparticles with different stabilizers, PLoS ONE 6 (2011), e19121.
[136] W. Geldenhuys, et al., Brain-targeted delivery of paclitaxel using glutathione-coated nanoparticles for brain cancers, J. Drug Target. 19 (2011) 837–845.
[137] C. Sun, et al., Tumor-targeted drug delivery and MRI contrast enhancement by chlorotoxin-conjugated iron oxide nanoparticles, Nanomedicine 3 (4) (2008) 495–505.
[138] O. Veiseh, et al., Inhibition of tumor-cell invasion with chlorotoxin-bound superparamagnetic nanoparticles, Small 5 (2) (2009) 256–264.
[139] O. Veiseh, et al., Chlorotoxin bound magnetic nanovector tailored for cancer cell targeting, imaging, and siRNA delivery, Biomaterials 31 (31) (2010) 8032–8042.
[140] K. Suzuki, et al., Celecoxib enhances radiosensitivity of hypoxic glioblastoma cells through endoplasmic reticulum stress, Neuro-Oncology 15 (2013) 1186–1199, https://doi.org/10.1093/neuonc/not062.
[141] X. Jiang, et al., Nanoparticles of 2-deoxy-D-glucose functionalized poly (ethylene glycol)-co-poly(trimethylene carbonate) for dual-targeted drug delivery in glioma treatment, Biomaterials 35 (2014) 518–529.
[142] F.M. Kievit, et al., Chlorotoxin labeled magnetic nanovectors for targeted gene delivery to glioma, ACS Nano 4 (2010) 4587–4594.
[143] T.F. Cabada, et al., Induction of cell death in a glioblastoma line by hyperthermic therapy based on gold nanorods, Int. J. Nanomed. 7 (2012) 1511.
[144] D.Y. Joh, et al., Selective targeting of brain tumors with gold nanoparticle-induced radiosensitization, PLoS ONE 8 (2013), e62425.
[145] S. Ruan, et al., Tumor microenvironment sensitive doxorubicin delivery and release to glioma using angiopep-2 decorated gold nanoparticles, Biomaterials 37 (2015) 425–435.
[146] F. Dilnawaz, et al., The transport of non-surfactant based paclitaxel loaded magnetic nanoparticles across the blood brain barrier in a rat model, Biomaterials 33 (2012) 2936–2951, https://doi.org/10.1016/j.biomaterials.2011.12.046.
[147] S. Son, et al., A brain-targeted rabies virus glycoprotein-disulfide linked PEI nanocarrier for delivery of neurogenic microRNA, Biomaterials 32 (2011) 4968–4975.
[148] Z. Chai, et al., A facile approach to functionalizing cell membrane-coated nanoparticles with neurotoxin-derived peptide for brain-targeted drug delivery, J. Control Release 264 (2017) 102–111, https://doi.org/10.1016/j.jconrel.2017.08.027.

[149] C. Fang, et al., Temozolomide nanoparticles for targeted glioblastoma therapy, HHS Public Access 7 (12) (2016) 6674–6682, https://doi.org/10.1021/am5092165.
[150] C. Liu, et al., A dual-mediated liposomal drug delivery system targeting the brain: rational construction, integrity evaluation across the blood–brain barrier, and the transporting mechanism to glioma cells, Int. J. Nanomedicine 12 (2017) 2407–2425, https://doi.org/10.2147/IJN.S131367.
[151] Z. Liu, H. Yan, H. Li, Silencing of DNA repair sensitizes pediatric brain tumor cells to γ-irradiation using gold nanoparticles, Environ. Toxicol. Pharmacol. 53 (2017) 40–45, https://doi.org/10.1016/j.etap.2017.04.017.
[152] J. Tang, et al., Aptamer-conjugated PEGylated quantum dots targeting epidermal growth factor receptor variant III for fluorescence imaging of glioma, Int. J. Nanomedicine 12 (2017) 3899–3911, https://doi.org/10.2147/IJN.S133166.
[153] Q. Hu, et al., Glioma therapy using tumor homing and penetrating peptide-functionalized PEG-PLA nanoparticles loaded with paclitaxel, Biomaterials 34 (2013) 5640–5650, https://doi.org/10.1016/j.biomaterials.2013.04.025.
[154] T.H. Kim, et al., Preparation of polylactide-co-glycolide nanoparticles incorporating celecoxib and their antitumor activity against brain tumor cells, Int. J. Nanomedicine 6 (2011) 2621–2631.
[155] H. Yan, et al., Two-order targeted brain tumor imaging by using an optical/paramagnetic nanoprobe across the blood brain barrier, ACS Nano 6 (2012) 410–420.
[156] A.S. Wadajkar, et al., Decreased non-specific adhesivity, receptor targeted (DART) nanoparticles exhibit improved dispersion, cellular uptake, and tumor retention in invasive gliomas, J. Control. Release 267 (2017) 144–153, https://doi.org/10.1016/j.jconrel.2017.09.006.
[157] C.S. Schneider, et al., Minimizing the non-specific binding of nanoparticles to the brain enables active targeting of Fn14-positive glioblastoma cells, Biomaterials 42 (2015) 42–51, https://doi.org/10.1016/j.biomaterials.2014.11.054.
[158] R. Stupp, et al., Effects of radiotherapy with concomitant and adjuvant temozolomide versus radiotherapy alone on survival in glioblastoma in a randomised phase III study: 5-year analysis of the EORTC-NCIC trial, Lancet Oncol. 10 (2009) 459–466.
[159] M. Weller, T. Cloughesy, J.R. Perry, W. Wick, Standards of care for treatment of recurrent glioblastoma—are we there yet? Neuro-Oncology 15 (2013) 4–27.
[160] B. Campos, L.R. Olsen, T. Urup, H.S. Poulsen, A comprehensive profile of recurrent glioblastoma, Oncogene (2016) 1–7.
[161] Q. Hu, et al., Biomaterials peptide for anti-glioma drug delivery, Biomaterials 34 (4) (2013) 1135–1145, https://doi.org/10.1016/j.biomaterials.2012.10.048.
[162] S.H. Ranganath, C. Wang, Biodegradable microfiber implants delivering paclitaxel for post-surgical chemotherapy against malignant glioma, Biomaterials 29 (2008) 2996–3003.
[163] S.H. Ranganath, et al., The use of submicron/nanoscale PLGA implants to deliver paclitaxel with enhanced pharmacokinetics and therapeutic efficacy in intracranial glioblastoma in mice, Biomaterials 31 (2010) 5199–5207.
[164] S. Ni, et al., Biodegradable implants efficiently deliver combination of paclitaxel and temozolomide to glioma C6 cancer cells *in vitro*, Ann. Biomed. Eng. 42 (2014) 214–221.
[165] M. Irani, et al., Electrospun biocompatible poly (e-caprolactonediol) based polyurethane core/shell nanofibrous scaffold for controlled release of temozolomide, Int. J. Polym. Mater. Polym. Biomater. 67 (2018) 361–366.
[166] R. Tavakoli, et al., Prolonged drug release using PCL–TMZ nanofibers induce the apoptotic behavior of U87 glioma cells, Int. J. Polym. Mater. Polym. Biomater. 67 (15) (2018) 873–878.
[167] R. Ramachandran, et al., Theranostic 3-dimensional nano brain-implant for prolonged and localized treatment of recurrent glioma, Sci. Rep. 7 (2017) 43271.

[168] D. Huang, et al., A potential nanofiber membrane device for filling surgical residual cavity to prevent glioma recurrence and improve local neural tissue reconstruction, PLoS One 11 (2016), e0161435.
[169] M. Irani, et al., A novel biocompatible drug delivery system of chitosan/temozolomide nanoparticles loaded PCL-PU nanofibers for sustained delivery of temozolomide, Int. J. Biol. Macromol. 97 (2017) 744–751.
[170] M. Irani, et al., The sustained delivery of temozolomide from electrospun PCL-diol-b-PU/gold nanocomposite nanofibers to treat glioblastoma tumors, Mater. Sci. Eng. C 75 (2017) 165–174.
[171] X. Xu, et al., BCNU-loaded PEG-PLLA ultrafine fibers and their *in vitro* antitumor activity against glioma C6 cells, J. Control. Release 114 (2006) 307–316.
[172] Y. Tseng, et al., Sustainable release of carmustine from biodegradable poly [(D,L)-lactide-co-glycolide] nanofibrous membranes in the cerebral cavity: *in vitro* and *in vivo* studies, Expert Opin. Drug Deliv. 10 (2013) 879–888.
[173] B. Wang, et al., Local *in vitro* delivery of rapamycin from electrospun PEO/ PDLLA nanofibers for glioblastoma treatment, Biomed. Pharmacother. 83 (2016) 1345–1352.
[174] D. Han, et al., In-vitro evaluation of MPA-loaded electrospun coaxial fiber membranes for local treatment of glioblastoma tumor cells, J. Drug Deliv. Sci. Technol. 40 (2017) 45–50.
[175] H. Lian, Z. Meng, Melt electrospinning of daunorubicin hydrochlorideloaded poly (e-caprolactone) fibrous membrane for tumor therapy, Bioactive Mater. 2 (2017) 96–100.
[176] Y.Y. Tseng, et al., Advanced interstitial chemotherapy combined with targeted treatment of malignant glioma in rats by using drug-loaded nanofibrous membranes, Oncotarget 7 (2016) 59902–59916.
[177] X. Zhu, et al., Anti-neoplastic cytotoxicity of SN-38-loaded PCL/gelatin electrospun composite nanofiber scaffolds against human glioblastoma cells *in vitro*, J. Pharm. Sci. 104 (2015) 4345–4354.
[178] C. Lei, et al., Development of a gene/drug dual delivery system for brain tumor therapy: potent inhibition via RNA interference and synergistic effects, Biomaterials 34 (2013) 7483–7494.
[179] J.R. Bagó, et al., Electrospun nanofibrous scaffolds increase the efficacy of stem cell-mediated therapy of surgically resected glioblastoma, Biomaterials 90 (2016) 116–125.
[180] A.I. Mehta, et al., Current status of intratumoral therapy for glioblastoma, J. Neuro-Oncol. 125 (2015) 1–7.
[181] A. Bregy, et al., The role of Gliadel wafers in the treatment of high-grade gliomas, Expert. Rev. Anticancer. Ther. 13 (2013) 1453–1461.
[182] C. Lei, et al., Development of a gene/drug dual delivery system for brain tumor therapy: potent inhibition via RNA interference and synergistic effects, Biomaterials 34 (30) (2013) 7483–7494.
[183] A. Nel, et al., Toxic potential of materials at the nanolevel, Science 311 (2006) 622–627.
[184] T.M. Allen, P.R. Cullis, Drug delivery systems: entering the mainstream, Science 303 (2004) 1818–1822.
[185] P.D. Senter, C.J. Springer, Selective activation of anticancer prodrugs by monoclonal antibody–enzyme conjugates, Adv. Drug Deliv. Rev. 53 (2001) 247–264.
[186] J.L. Clarke, et al., A phase 1 trial of intravenous liposomal irinotecan in patients with recurrent high-grade glioma, Cancer Chemother. Pharmacol. 79 (3) (2017) 603–610, https://doi.org/10.1007/s00280-017-3247-3.
[187] S. Nagpal, et al., Phase II pilot study of single-agent etirinotecan pegol (NKTR-102) in bevacizumab-resistant high grade glioma, J. Neuro-Oncol. 123 (2015) 277–282, https://doi.org/10.1007/s11060-015-1795-0.

[188] D. Brandsma, et al., P08.03. 2B3-101, Glutathione pegylated liposomal doxorubicin, in patients with recurrent high grade gliomas and breast cancer brain metastases, Neuro Oncol. 16 (Suppl. 2) (2014) 50–51, https://doi.org/10.1093/neuonc/nou174.191.

[189] C.P. Beier, et al., RNOP-09: pegylated liposomal doxorubicine and prolonged temozolomide in addition to radiotherapy in newly diagnosed glioblastoma—a phase II study, BMC Cancer 2 (9) (2009) 1–10, https://doi.org/10.1186/1471-2407-9-308.

[190] C. Spick, K. Herrmann, J. Czernin, ^{18}F-FDG PET/CT and PET/MRI perform equally well in cancer: evidence from studies on more than 2,300 patients, J. Nucl. Med. 57 (3) (2016) 420–430, https://doi.org/10.2967/jnumed.115.158808.

[191] C.W. Chiu, T.K. Huang, Y.C. Wang, B.G. Alamani, J.J. Lin, Intercalation strategies in clay/polymer hybrids, Prog. Polym. Sci. 39 (3) (2014) 443–485, https://doi.org/10.1016/j.progpolymsci.2013.07.002.

[192] J.J. Decker, K.P. Meyers, D.R. Paul, D.A. Schiraldi, A. Hiltner, S. Nazarenko, Polyethylene-based nanocomposites containing organoclay: a new approach to enhance gas barrier via multilayer coextrusion and interdiffusion, Polymer 61 (2015) 42–54, https://doi.org/10.1016/j.polymer.2015.01.061.

[193] D. Gao, R. Li, B. Lv, J. Ma, F. Tian, J. Zhang, Flammability, thermal and physical-mechanical properties of cationic polymer/montmorillonite composite on cotton fabric, Compos. Part B 77 (2015) 329–337, https://doi.org/10.1016/j.compositesb.2015.03.061.

[194] A. Gooneie, H. Nazockdast, F. Shahsavan, Effect of selective localization of carbon nanotubes in PA6 dispersed phase of PP/PA6 blends on the morphology evolution with time, part 1: droplet deformation under simple shear flows, Polym. Eng. Sci. 55 (7) (2015) 1504–1519, https://doi.org/10.1002/pen.24098.

[195] D.J. Lowe, A.V. Chapman, S. Cook, J.J. Busfield, Micromechanical models of young's modulus of NR/organoclay nanocomposites, J. Polym. Sci. B Polym. Phys. 49 (22) (2011) 1621–1627, https://doi.org/10.1002/polb.22355.

[196] H. Mattausch, S. Laske, I. Duretek, J. Kreith, G. Maier, C. Holzer, Investigation of the influence of processing conditions on the thermal, rheological and mechanical behavior of polypropylene nanocomposites, Polym. Eng. Sci. 53 (5) (2013) 1001–1010, https://doi.org/10.1002/pen.23350.

[197] M. Moniruzzaman, K.I. Winey, Polymer nanocomposites containing carbon nanotubes, Macromolecules 39 (16) (2006) 5194–5205, https://doi.org/10.1021/ma060733p.

[198] R. Sepahvand, M. Adeli, B. Astinchap, R. Kabiri, New nanocomposites containing metal nanoparticles, carbon nanotube and polymer, J. Nanopart. Res. 10 (8) (2008) 1309–1318, https://doi.org/10.1007/s11051-008-9411-2.

[199] G.S. Ayton, W.G. Noid, G.A. Voth, Multiscale modeling of biomolecular systems: in serial and in parallelu8, Curr. Opin. Struct. Biol. 17 (2) (2007) 192–198, https://doi.org/10.1016/j.sbi.2007.03.004.

[200] P.S. Ayyaswamy, V. Muzykantov, D.M. Eckmann, R. Radhakrishnan, Nanocarrier hydrodynamics and binding in targeted drug delivery: challenges in numerical modeling and experimental validation, J. Nanotechnol. Eng. Med. 4 (1) (2013), https://doi.org/10.1115/1.4024004, 011001.

[201] J.A. Elliott, Novel approaches to multiscale modelling in materials science, Int. Mater. Rev. 56 (4) (2011) 207–225, https://doi.org/10.1179/1743280410y.0000000002.

[202] K. Kremer, F. Müller-Plathe, Multiscale problems in polymer science: Simulation approaches, MRS Bull. 26 (3) (2001) 205–210, https://doi.org/10.1557/mrs2001.43.

[203] T. Murtola, A. Bunker, I. Vattulainen, M. Deserno, M. Karttunen, Multiscale modeling of emergent materials: biological and soft matter, Phys. Chem. Chem. Phys. 11 (12) (2009) 1869–1892, https://doi.org/10.1039/b818051b.

[204] C. Peter, K. Kremer, Multiscale simulation of soft matter systems—from the atomistic to the coarse-grained level and back, Soft Matter 5 (22) (2009) 4357–4366, https://doi.org/10.1039/b912027k.
[205] D. Raabe, Challenges in computational materials science, Adv. Mater. 14 (9) (2002) 639–650, https://doi.org/10.1002/1521-4095(20020503)14:9<639::AID-ADMA639>3.0.CO;2-7.
[206] P. Sherwood, B.R. Brooks, M.S. Sansom, Multiscale methods for macro-molecular simulations, Curr. Opin. Struct. Biol. 18 (5) (2008) 630–640, https://doi.org/10.1016/j.sbi.2008.07.003.
[207] Q.H. Zeng, A.B. Yu, G.Q. Lu, Multiscale modeling and simulation of polymer nano-composites, Prog. Polym. Sci. 33 (2) (2008) 191–269, https://doi.org/10.1016/j.progpolymsci.2007.09.002.
[208] M. Jafari, F. Doustdar, F. Mehrnejad, Molecular self-assembly strategy for Encap- sulation of an amphipathic α-helical antimicrobial peptide into the different polymeric and copolymeric nanoparticles, J. Chem. Inf. Model. 59 (1) (2019) 550–563, https://doi.org/10.1021/acs.jcim.8b00641.
[209] A. Kyrychenko, D.A. Pasko, O.N. Kalugin, Poly(vinyl alcohol) as a water protecting agent for silver nanoparticles: the role of polymer size and structure, Phys. Chem. Chem. Phys. 19 (13) (2017) 8742–8756, https://doi.org/10.1039/c6cp05562a.
[210] S.S. Tallury, M.A. Pasquinelli, Molecular dynamics simulations of flexible poly- mer chains wrapping single-walled carbon nanotubes, J. Phys. Chem. B 114 (12) (2010) 4122–4129, https://doi.org/10.1021/jp908001d.
[211] S.S. Tallury, M.A. Pasquinelli, Molecular dynamics simulations of polymers with stiff backbones interacting with single-walled carbon nanotubes, J. Phys. Chem. B 114 (29) (2010) 9349–9355, https://doi.org/10.1021/jp101191j.
[212] L. Weng, S.L. Stott, M. Toner, Molecular dynamics at the interface between ice and poly(vinyl alcohol) and ice recrystallization inhibition, Langmuir 34 (17) (2018) 5116–5123, https://doi.org/10.1021/acs.langmuir.7b03243.
[213] A. Gooneie, S. Schuschnigg, C. Holzer, A review of multiscale computational methods in polymeric materials, Polymers 9 (1) (2017) 16, https://doi.org/10.3390/polym9010016.

CHAPTER 13

Iontophoretic drug permeation enhancement techniques

Sumana Roy and Tamalika Chakraborty
Department of Pharmaceutics, Guru Nanak Institute of Pharmaceutical Science and Technology, Kolkata, India

1 Introduction

Enhancement of drug permeation is required to be considered during the development of transdermal drug delivery systems [1]. Permeation is a technique where the active pharmaceutical ingredient is crossing all the layer of the skin like, epidermis, stratum corneum, dermis and deposited in any layer whereas in permeation active pharmaceutical ingredients are crossing all the layer and ultimately reached to the systemic circulation [2]. Main difficulty associated with the development of transdermal delivery system is the enhancement of the permeation of active pharmaceutical ingredients not the penetration [3–5].

The designing of different transdermal drug permeation approaches has currently become most thriving drug delivery research area [6, 7]. The non-invasive way of delivery of drug through transdermal path is patient friendly and can avoid first-pass metabolism of active pharmaceutical ingredients [8]. The main barrier to the transportation of drugs across the skin is stratum corneum, the uppermost dead cell layer. Percutaneous absorption takes place through different pathways, such as intercellular or paracellular pathway in-between the corneocytes along lamellar lipids, intercellular or transcellular pathway *via* cells or appendage or shunt pathway through sweat gland secretary glands and through hair follicles [9]. Permeation of a drug can be improved with the help of the permeation enhancer [2]. Various types of permeation enhancers are available, few are chemical permeation enhancing agents and some are permeation enhancing technique [10].

Iontophoresis is the processes of increasing the permeation of the drug through various layer of the skin with the help of an electrical field [11]. Non-ionised drugs are able to pass through the stratum corneum in a very small amount when its molecular size is 400 or less. But ionised drugs are unable to cross the barrier of the skin. Iontophoresis increases the

Advances and Challenges in Pharmaceutical Technology
https://doi.org/10.1016/B978-0-12-820043-8.00015-3

penetration of electrically charged drugs into vascular region of the skin by the application of an electrical field. Now a days, iontophoresis is considered as a prospective drug delivery approach where applied surface area, duration of application and the magnitude of the applied electrical current proportionally related with amount of drug delivered [12]. The processes are deals with the permeation of charged molecules with the influence of electrical current ($\leq 0.50\,\mathrm{mA\,cm^{-1}}$) through different barrier associated with the structure of skin. A continuous or pulsatile release of drug can be obtained just through modulation of electrical current [11, 12]. A continuous or pulsatile release of drug is possible by bringing about modulation in the profile of the electrical current [12, 13].

2 An overview of principle and mechanism behind iontophoretic dug permeation

"Like electrical charges repel each other" is the basic electrical principle of iontophoresis [14]. During iontophoresis by means of two electrodes an electrical field is applied across the skin. A drug or charged molecules, is placed under the electrode, where the charged molecules are repelled away from the electrode with the help of an electro repulsive force. A positively charged molecule is under anode (positive electrode) and molecule with negative charge placed under the cathode (negative electrode). Iontophoresis increase the penetration of charged molecules compared to passive diffusion through the impermeable stratum corneum [11].

At the physiological pH, the skin operates as a charged molecule, when an electrical field is applied across such membrane; there is a net transport of water molecules, from anode to cathode [15]. This electro-osmotic flow results in an enhanced the transport of dissolved polar uncharged molecules. The electrical connection among the electrodes developed due to movement of ionic drug molecules [12]. In "Cathodal Iontophoresis," the cathode is donor chamber to enhance the permeation flux of anionic drug and in "Anodal Iontophoresis," the anode is the donor compartment and enhance the permeation of cationic drugs [11].

In iontophoresis, mainly neutral and cationic therapeutic agents are placed under an anode whereas the anionic therapeutic agents under a cathode. During the application of a low current density or voltage (mainly DC current) ions are repelled through the skin due to electro repulsion. Cationic drugs are driven through the skin by the anode, which influence the movement of anions from the tissue into the anode [16]. Anions are driven into the skin from cathode. Iontophoresis helps in the delivery of additional charged drug in the cathode or a mixture of two compatible drugs in the anode, which helps in the delivery of two different charged drug molecules at a time. As normal iontophoresis is done with DC current sometimes pulsatile current is also used to increase rapid delivery of drug. Pulse wave depolarise the skin and return to initial state just before the next pulse. Pulsatile current is more beneficial with respect to DC as previous one reduces the chance of development of permanent polarisation and thus improves the efficiency of iontophoresis. Pulsatile waves are less damaging to skin with respect to DC waves and patient can tolerate higher level of current [16, 17].

The basic components of an iontophoresis device are electrode with chambers and a source of power. An iontophoretic system with an electrode of Ag/AgCl has been shown in Fig. 1 [18]. Considering a formulation (D^+A^-) where (D^+) is an ionised drug molecule which is introduced into the anodal chamber. The electrode chamber places away from the skin surface. Ag/AgCl electrode system is widely used in comparison to variety types of electrodes available for iontophoresis. Popularity of the Ag/AgCl electrode system is due to its good pH-stability. Further, it is most electro-stable electrode [18].

Inside the anode chamber an electro-chemical reaction is occurred between Ag and Cl^- anions. During transportation charged drug molecule compete with Na^+ ions as NaCl is commonly used to facilitate Cl^- ions. As a result, the efficiency of drug delivery reduces [18]. Further deposition of AgCl over the surface of the electrode occurred due to reaction of Cl^- ions with the metallic Ag. Accumulation of AgCl can be overcome with the use of product with low solubility. In order to maintain the electro-neutrality there is an exchange of cation and anion through skin from the anodal chamber. Cation released from anode chamber whereas anion from the layer of skin enter within the anode. Similarly, amount of AgCl is decreased inside the cathode and Ag^+ and Cl^- ions goes inside the solution which is generally developed as a by-product of the processes. The electro-neutrality balance should be maintained by approach of cation inside the cathode compartment from the skin or through a loss on anion.

Electro-repulsion and electro-osmosis are the two principal mechanisms used for permeation of drug molecules through iontophoresis [14]. The electro-repulsion mechanism is mainly for the transportation of ionised species (charged) whereas the electro-osmosis mechanism predominantly applicable for electrically neutral drug molecules. The ionisation of the acidic group associated with phospholipid structure of cell membrane influence

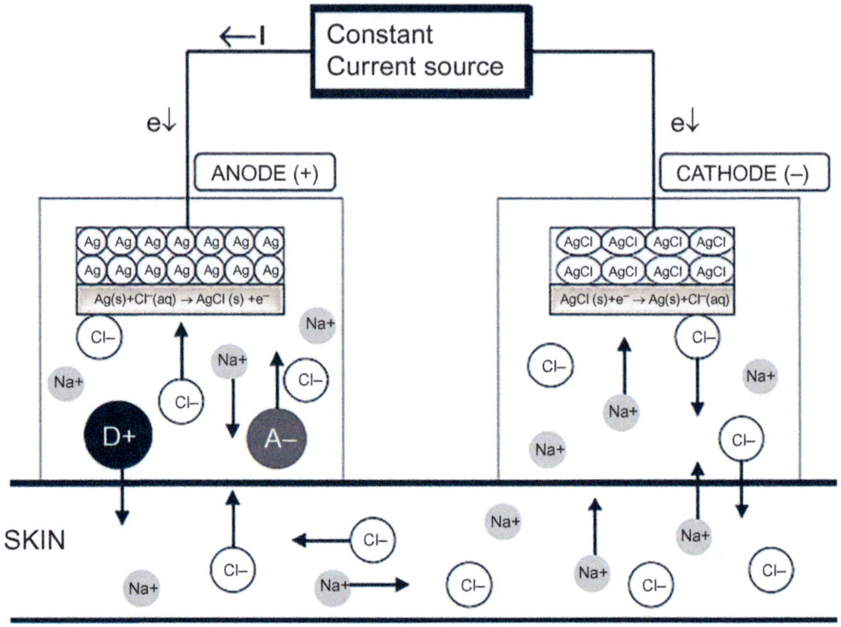

Fig. 1 A simple diagram of iontophoretic device containing an Ag/AgCl electrode. The anodal chamber comprises drug formulation (D^+A^-) containing the ionised drug molecule (D^+) with its counter ion (A^-) and Na^+Cl^-. Use of an electrical potential produces a current flowing through the electrical circuit. At the interface of the electrode solution, the Ag^+ and Cl^- react and form insoluble AgCl, which is deposited on the surface of the electrode. The electro-migration transports the cations, including the drug molecule from the anodal chamber and into the skin. At the same time, the endogenous anions (primarily Cl^-) move into the anodal chamber. In the cathodal chamber, Cl^- ions are released from the electrode and electronutrality needs that either an anion is lost from the cathodal chamber [18]. *(Copyright @ 2003, with permission from Elsevier B.V.)*

the transport of cations to the cathode, also known as electrophoresis, leads to neutralisation of the charge difference. Electro-osmosis mainly initiates anode to cathode water flow due to movement of Na^+ ions, which helps in transportation of natural drug across skin. Development of potential gradients (electrical potential) associated with iontophoresis technique significantly involved during drug transportation processes. Iontophoresis transport of drug molecules also affected by polarity, mobility and size of the active pharmaceutical ingredients [19].

3 Different applications related with iontophoresis drug permeation

3.1 Iontophoretic permeation of non-steroidal anti-inflammatory drugs (NSAIDs)

Zuo et al. established the relationship between the physicochemical properties of drugs and the transdermal enhancement effect of iontophoresis [20]. Non-steroidal anti-inflammatory drugs (NSAIDs), such as aspirin, ibuprofen and indomethacin were used as model drug. Zuo et al. measured the oil-water partition coefficients of the drug and prepared carbomer-based hydrogels [20]. They reported that strong lipophilicity of the molecules leads to high permeation and enhanced *in vitro* transdermal permeation. They noticed that the increased flowability and formation of loose stratum corneum and unordered arrangement of skin intercellular lipids by iontophoresis. The permeation of different NSAIDs has also been thoroughly investigated during iontophoresis [21]. Malinovskaja-Gomez et al. investigate permeation of polymeric nanoparticle (PLGA based) loaded with flufenamic acid using the method of iontophoresis [19]. The study emphasised the efficiency of pulsed current iontophoresis in comparison to conventional constant current, particularly for transdermal transportation of active pharmaceutical ingredients through polymeric nanoparticle. Tomoda et al. investigated enhanced both *in vivo* and *in vitro* transdermal delivery of indomethacin as a combination of nanotechnology and iontophoresis [22, 23].

3.2 Iontophoretic delivery of opoids

Yamamoto et al. reported enhanced permeation of naloxone *via* anodal iontophoretic technique with Ag/AgCl electrode. Their study is mainly based on transdermal permeation of naloxone *via* iontophoresis patches at anode applying Ag/AgCl electrodes [24]. Studies were done on the effect of the influence of drug concentration, current strength, and species variation on naloxone permeability across the skin. Rat dorsal skin and porcine ear skin were used to study *in vitro* skin permeation studies. *In vivo* absorption rate *via* the transdermal route was determined by the iontophoresis-enabled patches mainly used in the skin (rat dorsal) with a constant current supply for 24 h. They reported without significant differences between these two different skin types permeation flux of naloxone current-proportionally (0–360 $\mu A/cm^2$) increased. *In vitro* permeation of naloxone across the porcine skin independently influenced the naloxone hydrochloride dehydrate concentration over 1–10% (w/v) in the donor patch. In the *in vivo* pharmacokinetic evaluation, it was noticed that the increased sustained steady-state levels from 4 to 24 h after the commencement of current supply at a rate of 180 μA/cm^2 were 305 μg/h/cm^2, respectively. A needle free transdermal iontophoretic permeation of fentanyl was developed by Mankowitz et al. This needle free system is used mainly in the management of analgesia and postoperative pain in patient [25]. They also reported the safety and efficacy profile of fentanyl loaded iontophoretic transdermal patch to control pain associated with abdominal and pelvic surgery [26].

3.3 Iontophoretic delivery of local anaesthetics

Different techniques for delivery of local anaesthetics through topical and injectable route have been investigated [27–29]. They discovered the use of a eutectic mixture to enhance the permeation of transdermal local anaesthetic. They discussed the active drug transportation into the skin using a constantly supplied direct current of low-voltage. As topical gels, local anaesthetics have been studied by many researchers in various research endeavours [6, 28, 29]. Manjunatha et al. studied the iontophoresis associated permeation of the drug lidocaine HCl across the human skin, topically [30]. They used 2.50% and 5% lidocaine HCl for permeation study by continuous iontophoresis and modulated iontophoresis using a current density of0.5 μA cm^{-2}. Investigation emphasised the modulated iontophoresis as potential alternative to the continuous iontophoresis. Permeation of lidocaine with direct current and altering current combined iontophoretic technique was reported by Inoue et al. [31]. The study stated the rapid permeation of lidocaine ion due to alteration of voltage from direct to alternating current.

3.4 Drug targeting to central nervous system through iontophoretic delivery

Investigation on iontophoretic drug delivery to central nervous system reported improved drug permeation in most of the cases [32, 33]. *In vitro* iontophoretic permeation of ropinirole through weanling pigs skin was studied by Luzardo-Alvarez et al. [34]. Hydrochloride salt form of ropinirole enhances the delivery of drug and the influencing drug concentration in the current intensity and formulation associated with iontophoretic permeation flux in the donor chamber. From the research results, it was obviously

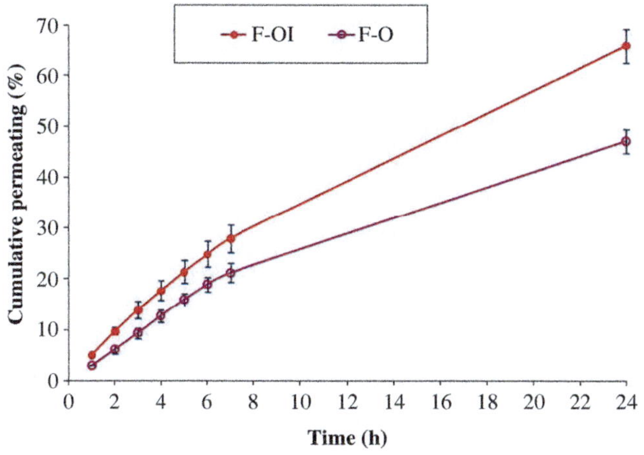

Fig. 2 A comparative *ex vivo* risperidone permeation pattern of the risperidone containing optimised transferosomal gel with (F-OI)/without (F-O) iontophoretic system under the influence of current supply, $0.5\,\text{mA}\,\text{cm}^2$ (mean$\pm$SD, $n=3$) [33]. *(Copyright @ 2017, with permission from Elsevier B.V.)*

exposed that even though the permeation flux was found to be influenced as independent of concentrations of ropinirole at current supply which is predetermined, significantly. The enhanced permeation flux of ropinirole was measured with augmented supply of current. Das et al. in the year of 2017 reported the iontophoretic permeation of risperidone transferosomal gel across the excised porcine skin at$0.5\,\mu\text{A}\,\text{cm}^{-2}$ of current supply [33]. The risperidone permeation flux forrisperidone-loaded optimised transferosomal gel *via* iontophoresis was measured as higher in comparison to that of the risperidone-loaded transferosomal gel (Fig. 2).

3.5 Iontophoretic delivery of cardiovascular drugs

Nair et al. discovered the application of different chemical permeation enhancing agents (like dimethyl formamide, PEG 400, *N*-methyl 2 pyrrolidone) in combination with iontophoresis to enhance the transportation of drug across the skin [35]. The chemical permeation enhancers used for iontophoretic delivery of metoprolol tartrate in gel formulation reported enhanced skin permeation with applied current $0.5\,\text{mA}\,\text{cm}^{-2}$. Uses of sodium lauryl sulfate as chemical enhancer with iontophoresis enhanced the permeation of metoprolol and reduce formation of depot. Keerthi et al. researched on permeation of atenolol transdermal patch [36]. This investigation projected that oleic acid can be used as a chemical enhancer in combination with iontophoresis significantly increase transdermal permeation of the active drug atenolol in comparison to passive mode of delivery of the drug atenolol. One of the research by Teong et al. studied about anti-osteoporotic actions of liposome encapsulated propranolol on rats that are overectomized by means of applying the technique of transdermal iontophoresis [37].

3.6 Proteins and peptides delivery through iontophoresis

The efficient transport of therapeutic proteins and peptides has become one of the popular areas of investigation in therapeutics [38, 39]. Different investigations have conducted successfully conducted for alternate delivery of insulin [39]. The stability of protein molecules mainly insulin was evaluated by Panchagnula et al. [40]. Improved permeation of insulin with chemical enhancer and iontophoretic technique was studied by Pillai and Panchagnula [41]. Malakar et al. studied the permeation of insulin from insulin-loaded optimised transferosomal gels though iontophoresis delivery in the porcine ear skin at a current supply of $0.5\,\mathrm{mA\,cm^{-2}}$, *in vitro* [42]. They noticed that the elevated insulin permeation flux for the insulin-loaded optimised tranferosomal gels by iontophoresis as compared the other condition, *i.e.*, without using iontophoresis (Fig. 3). La Fountaine et al. studied the

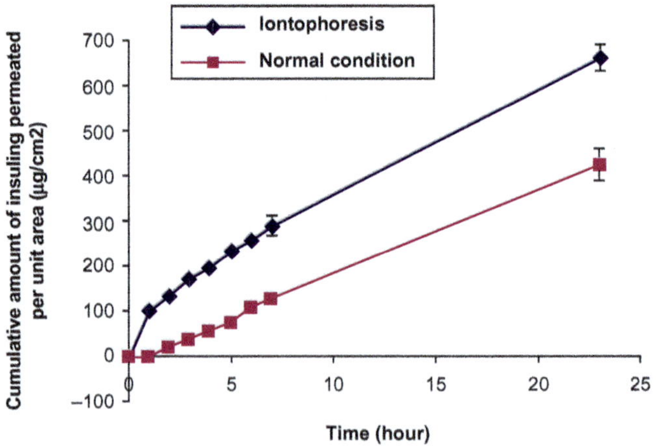

Fig. 3 Cumulative amount of *in vitro* insulin permeated through porcine skin per unit area *vs.* time profile of optimised gels containing insulin loaded transferosomes in the iontophoretic condition with the influence of current supply, $0.5\,\mathrm{mA\,cm^{-2}}$ and the normal condition (without using iontophoresis) (mean ± SE, $n = 3$) [42]. *(Copyright @ 2012, with permission from Elsevier B.V.)*

effect of microvascular perfusion which is cutaneous to the iontophoretic permeation of insulin are effected differentially by the insulin resistance due to spinal cord-injury [43].

3.7 Different modern development in iontophoretic delivery system

Kim et al. demonstrated a tattoo-based wearable alcohol biosensor system with three important functions like with amperometric enzyme biosensing direct sweat -inducing direct, non-invasive alcohol monitoring from human sweat, a thin flexible printed electronic circuitry for controlling the entire operation and a wireless real time data collection [44]. These single-use printed tattoos could be used for identifying intoxicated and preventing drunk driving. The new iontophoresis-sensing wearable platform due to its design and performance offers a promising field for non-invasive monitoring of additional target analytes.

Iontophoretic drug delivery is a novel method for *in vivo* cancer treatment with optimised delivery of drug to tumour cells with the use of electric-fields (with very low intensity) in cell culture. Charles et al. developed iontophoresis-on-chip (IOC) platform with variable voltage and current as an anti-cancer agents. The investigation was focused on heparin-based microfluidic hydrogel model targeting the passage of drug through an triple-negative breast cancer cell (MDA-MB 231) and extracellular matrix [45]. Movement of the drug molecules is based on electrophoresis and diffusion opposite to the charged electrode through the hydrogel. A finite element modelling was used to measure the drug concentration of drug in the extracellular matrix of tumour cell. IOC platform was used to validate the model prediction through a comparison with the predetermined cationic fluorescent dye with particular concentration in the microfluidic device. The result focused with an electric field (50mVDC) and an electric current (3 mA) increased the delivery of drug and responsible for tumour cell death by $48.12 \pm 14.33\%$ and $39.13 \pm 12.86\%$, respectively.

Ouyang et al. developed a triboelectric nanogenerator driven self-powdered, on-demand transdermal drug delivery system [46]. This system was designed to trigger the electric-responsive controlled drug carrier and activated iontophoresis with enhanced drug delivery. Study mainly demonstrated that manual rotation of TENG (30–40 rpm) for 1.5 min can release $3\,\mu g/cm^2$ dosage of drug. Through the alteration of duration of charging or through the power management circuit rate can be tuned from $0.05–0.25\,\mu g/cm^2$. The performance of TENG-base drug delivery was increased by almost 50% as studied through *ex vivo* experiments on porcine skin in comparison to conventional transdermal delivery system.

Wu et al. developed feedback controlled transdermal drug delivery system [47]. The study mainly focused on development of a self-powered wearable iontophoretic transdermal delivery system mainly driven by energy harvested from biochemical motions used mainly for closed -loop motion detection and therapy. Study focused on conversion of biomechanical motions into electricity for iontophoresis through a wearable triboelectric nanogenerator (TENG) without power source. A hydrogel-based soft patch with side-by-side electrodes also designed for non-invasive iontophoretic transdermal delivery system. An experiment on pig skin with different dyes have used as model drugs demonstrate the feasibility of the proposed system successfully. This work provides a cost-effective solution for non-invasive, electrically assisted transdermal delivery system with closed-loop sensing and treatment.

An et al. designed a delivery system through transdermal route for efficient and successful delivery of electrically mobile nanocarrier through iontophoresis. [48]. The system consists of a drug reservoir and a RED (portable and disposable reverse electrodialysis). The poly vinyl alcohol incorporated with poly pyrrolein the form of a hydrogel is responsible for increased electron transfer from the RED battery. The 'movement and release of electrically charged molecules through nanocarriers is accelerated from PYP hydrogel. The investigation established that DNs loaded with rosiglitazone or fluconazole could be functionalized and facilitated transdermal transport *via* repulsive RED-driven iontophoresis. RED-driven topical iontophoresis resulted in reduced body weight through controlling glucose level and conversion of adipose tissue from white to brown, *in vivo*.

Silva et al. discovered improved antioxident activity of a gel based topical formulation [49]. They developed different plant derived phenolic acids alone or in combination as well as in the form of a gel based topical formulations with hydroxyethyl cellulose. Five different methods are used by the investigator to measure *in vitro* antioxidant activity. The concentration applied in the formulation is 0.6% (w/w) both for derived isolated and mixture compound and for each phenolic acid respectively. The study mainly focused on the antioxidant property of the gallic acid applied in the form of the gel into a procine skin. Gallic acid retention is maximum when applied through iontophoresis and after 20 min of iontophoresis increased antioxidant activity was detected by the investigators.

Fukuta et al. demonstrated the use of iontophoresis with 0.3 to 0.5 mA/cm^2 current can enhance permeation of small interfering RNA and nanoparticles into the skin [50]. They hypothesised that non-invasive delivery through intra dermal route is achieved by iontophoresis particularly for biological macromolecules. The observed the fluorescence of FITC-labelled IgG antibody was penetrate up to dermis layer (for rat without hair). No passive diffusion of antibody was observed. The group also delivered antibody into inflamed skin tissue through iontophoresis in imiquimod-induced psoriasis rat model. They also discovered that iontophoresis can inhibit the pathogenic progression of psoriasis through upregulation of interleukin-6 mRNA levels which is not possible through subcutaneous injection for anti-tumour necrosis factor-α drug etanercept. Their study is mainly focused that iontophoresis is a non-invasive and efficient intradermal delivery technology for biological macromolecules.

Molokhia et al. investigated the effect of transscleral *in vivo* and *in vitro* iontophoresis of macromolecules and evaluate the efficacy in a choroidal neovascularization animal model [51]. They studied *in vivo* real time distribution of IgG and bevacizumab into the eye through magnetic resonance imaging (MRI). Study discovered 600 times improved transscleral iontophoresis of IgG at 4 mA with respect to passive diffusion and 14 times with respect to a conventional formulation. Iontophoresis enhanced the permeation to posterior tissues as shown by MRI for Galbumin. Research reported the delivery of macromolecules through the conjunctiva and sclera through transscleral iontophoresis and eventually exposing the retina to the dug molecules.

4 Conclusions

Iontophoresis is an advanced and promising Technique as a permeation enhancer. In improve the permeation of salt from of drug through the skin. Drugs with poor bioavailability profile can be introduce very easily with iontophoresis delivery. In comparison to adverse effect of this technique it is a promising technique to enhance permeation of drug through skin and is an alternative technique of drug delivery.

In this chapter, different approaches related with iontophoretic delivery with recent discovery area are presented. This discussion demonstrates recent and extensive research endeavours exploring and exploiting the importance in terms of practicability of the iontophoretic delivery approach and treatment platform in the treatment and therapeutic approaches of different category of drug with their diverse physico-chemical characteristics like, opoids, NSAIDs. local anaesthetics, drugs acting on CNS, steroids, cardiovascular drugs, antiviral drugs, antibiotics, proteins, peptides, antidiabetic drugs, *etc.* Delivery approaches through iontophoresis offer a rational alternative in therapeutic and pharmacokinetic ground particularly for molecules with low oral bioavailability shorter half-life and for multiple dosing. Taking pharmacodynamics in consideration, the approach associated with iontophoretic drug delivery presents a therapeutic advantage over the conventional drug delivery routes. Particularly for certain delivery of peptide molecules extensive care must be given in terms of physicochemical characteristics and pharmacological effectiveness. The prospect of iontophoretic transport systems of drug for customising therapeutics like feedback controlled transdermal delivery, delivery of nanocarriers to reduce glucose level and conversion of white to brown adipose tissue. Deliver through iontophoresis helps to reach to the posterior segment and is a promising field for further therapeutic efficacy.

References

[1] H. Marwah, T. Garg, A.K. Goyal, G. Rath, Permeation enhancer strategies in transdermal drug delivery, Drug Deliv. 23 (2) (2016) 564–578.

[2] A.K. Nayak, B. Mohanty, K.K. Sen, Comparative evaluation of *in vitro* diclofenac sodium permeability across excised mouse skin from different common pharmaceutical vehicles, Int. J. PharmTech. Res. 2 (1) (2010) 920–930.

[3] J. Malakar, A. Basu, A.K. Nayak, Candesartan cilexetil microemulsions for transdermal delivery: formulation, *in-vitro* skin permeation and stability assessment, Curr. Drug Deliv. 11 (2014) 313–321.

[4] S. Jana, S. Manna, A.K. Nayak, K.K. Sen, S.K. Basu, Carbopol gel containing chitosan-egg albumin nanoparticles for transdermal aceclofenac delivery, Colloids Surf. B Biointerfaces 114 (2014) 34–44.

[5] S. Jana, S.A. Ali, A.K. Nayak, K.K. Sen, S.K. Basu, Development and optimization of topical gel containing aceclofenac-crospovidone solid dispersion by "quality by design" approach, Chem. Eng. Res. Des. 92 (2014) 2095–2105.

[6] B. Das, A.K. Nayak, U. Nanda, Topical gels of lidocaine HCl using cashew gum and Carbopol 940: preparation and *in vitro* skin permeation, Int. J. Biol. Macromol. 62 (2013) 514–517.

[7] S. Lobo, G. Yan, Improving the direct penetration into tissues underneath the skin with iontophoresis delivery of a ketoprofen cationic prodrug, Int. J. Pharm. 535 (2018) 228–236.

[8] N. Akhtar, V. Singh, M. Yusuf, R.A. Khan, Non-invasive drug delivery technology: development and current status of transdermal drug delivery devices, techniques and biomedical applications, Biomed Tech (Berl) 65 (3) (2020) 243–272.

[9] E. Schnetz, M. Fartasch, Microdialysis for the evaluation of penetration through the human skin barrier: a promising tool for future research, Eur. J. Pharm. Sci. 12 (1991) 165–174.

[10] R. Neupane, S.H.S. Boddu, J. Renukuntla, R.J. Babu, A.K. Tiwari, Alternatives to biological skin in permeation studies: current trends and possibilities, Pharmaceutics 12 (2) (2020) 152.

[11] A.K. Nayak, S. Dey, K. Pal, I. Banerjee, Iontophoretic drug delivery systems, in: K. Pal, H.-B. Kraatz, C. Li, A. Khasnobish, S. Bag, I. Banerjee, U. Kuruganti (Eds.), Bioelectronics and Medical Devices, From Materials to Devices—Fabrication, Applications and Reliability, Woodhead Publishing Series in Electronic and Optical Materials, Elsevier Inc., 2019, pp. 393–420.

[12] R.G. Lohe, S.S. Jadkar, S.D. Modekar, K.M. Bhusare, Iontophoresis drug delivery system, World J. Pharm. Pharm. Sci. 5 (2016) 748–762.

[13] T.M. Karpiński, Selected medicines used in iontophoresis, Pharmaceutics 10 (4) (2018) 204.

[14] K. Pal, S.S. Sagiri, P. Pattnaik, S.S. Ray, Drug delivery, iontophoretic, in: M. Mishra (Ed.), Encyclopedia of Biomedical Polymers and Polymeric Biomaterials, vol. XI, Taylor & Francis Group, New York, NY, 2014, pp. 1–15.

[15] E. Tesselaar, F. Sjoberg, Transdermal iontophoresis as an *in-vivo* technique for studying microvascular physiology, Microvasc. Res. 81 (2011) 88–96.

[16] P. Panzade, P. Puranik, Iontophoresis: a functional approach for enhancement of transdermal drug delivery, Asian J. Biomed. Pharm. Sci. 2 (2012) 11,1–8.

[17] N.K. Jain, Controlled Drug Delivery, first ed., CBS Publishers and Distributors, New Delhi, 1997.

[18] Y.N. Kalia, A. Naik, J. Garrison, R.H. Guy, Iontophoretic drug delivery, Adv. Drug Deliv. Rev. 56 (2004) 619–658.

[19] K. Malinovskaja-Gomez, H. Labouta, M. Schneider, J. Hirvonen, T. Laaksonen, Transdermal iontophoresis of flufenamic acid loaded PLGA nanoparticles, Eur. J. Pharm. Sci. 89 (2016) 154–162.

[20] J. Zuo, L. Du, M. Li, B. Liu, W. Zhu, Y. Jin, Transdermal enhancement effect and mechanism of iontophoresis for non-steroidal anti-inflammatory drugs, Int. J. Pharm. 466 (1–2) (2014) 76–82.

[21] R. Crevenna, A. Burian, Z. Oesterreicher, E. Lackner, W. Jager, G. Rezcicek, M. Keilani, M. Zeitlinger, Iontophoresis driven concentrations of topically administered diclofenac in skeletal muscle and blood of healthy subjects, Eur. J. Clin. Pharmacol. 71 (2015) 1359–1364.

[22] K. Tomoda, H. Terashima, K. Suzuki, T. Inagi, H. Terada, K. Makino, Enhanced transdermal delivery of indomethacin-loaded PLGA nanoparticles by iontophoresis, Colloids Surf. B Biointerfaces 88 (2) (2011) 706–710.

[23] K. Tomoda, H. Terashima, K. Suzuki, T. Inagi, H. Terada, K. Makino, Enhanced transdermal delivery of indomethacin using combination of PLGA nanoparticles and iontophoresis *in vivo*, Colloids Surf. B Biointerfaces 92 (2012) 50–54.

[24] R. Yamamoto, S. Takasuga, Y. Yoshida, S. Mafune, K. Kominami, C. Sutoh, Y. Kato, M. Yamauchi, M. Ito, K. Kanamura, M. Kinoshita, In vitro and in vivo transdermal iontophoretic delivery of naloxone, an opioid antagonist, Int. J. Pharm. 422 (1–2) (2012) 132–138.

[25] H.S. Minkowitz, H. Danesi, L. Ding, J.B. Jones, Development of the fentanyl iontophoretic transdermal system (ITS) for patient-controlled analgesia of postoperative pain management, Pain Manag. 5 (5) (2015) 327–337.

[26] H.S. Minkowitz, J.P. Rathmell, S. Vallow, K. Gargiulo, C.V. Damaraju, D.J. Hewitt, Efficacy and safety of the fentanyl iontophoretic transdermal system (ITS) and intravenous patient-controlled analgesia (IV PCA) with morphine for pain management following abdominal or pelvic surgery, Pain Med. 8 (2007) 657–668.

[27] E.A. Shipton, New delivery systems for local anaesthetics-part 2, Anesthesiol. Res. Pract. 2012 (2012), 289373.

[28] M.S. Hasnain, P. Rishishwar, S. Ali, Use of cashew bark exudate gum in the preparation of 4% lidocaine HCL topical gels, Int. J. Pharm. Pharm. Sci. 9 (8) (2017) 146–150.

[29] M.S. Hasnain, P. Rishishwar, S. Ali, S. Alkahtani, M. Tabish, M. Milivojevic, M.T. Ansari, A.K. Nayak, Formulation and *ex vivo* skin permeation of lidocaine HCl topical gels using dillenia (*Dilleniaindica* L.) fruit gum, Rev. Mex. Ing. Quim. 19 (2020) 1465–1476.

[30] R.G. Manjunatha, S. Sharma, R.P. Narayan, V. Koul, Effective permeation of 2.5 and 5% lidocaine hydrochloride in human skin using iontophoresis technique, Int. J. Dermatol. 57 (11) (2018) 1335–1343.

[31] T. Inoue, T. Sugiyama, T. Ikoma, H. Shimazu, R. Wakita, H. Fukayama, Drug delivery and transmission of lidocaine using iontophoresis in combination with direct and alternating currents, J. Med. Dent. Sci. 63 (2016) 71–77.

[32] V.M. Meidan, M. Al-Khalili, B.B. Michniak, Enhanced iontophoretic delivery of buspirone hydrochloride across human skin using chemical enhancers, Int. J. Pharm. 264 (2003) 73–83.

[33] B. Das, S.O. Sen, R. Maji, A.K. Nayak, K.K. Sen, Transferosomal gel for transdermal delivery of risperidone: formulation optimization and *ex vivo* permeation, J. Drug. Deliv. Sci. Technol. 38 (2017) 59–71.

[34] A. Luzardo-Alvarez, M.B. Delgado-Charro, J. Blanco-Mendez, Iontophoretic delivery of ropinirole hydrochloride: effect of current density and vehicle formulation, Pharm. Res. 18 (2001) 1714–1720.

[35] A. Nair, H. Vyas, J. Shah, A. Kumar, Effect of permeation enhancers on the iontophoretic transport of metoprolol tartrate and the drug retention in skin, Drug Deliv. 18 (2011) 19–25.

[36] H. Keerthi, P.K. Panakanti, M.R. Yamsani, Design and characterization of atenolol transdermal therapeutic systems: enhancement of permeability via iontophoresis, PDA J. Pharm. Sci. Technol. 66 (2012) 318–332.

[37] B. Teong, S.M. Kuo, W.H. Tsai, M.L. Ho, C.H. Chen, H.H. Huang, Liposomal encapsulation for systemic delivery of propranolol via transdermal iontophoresis improves bone microarchitecture in ovariectomized rats, Int. J. Mol. Sci. 18 (2017) E822.

[38] A.K. Nayak, Advances in therapeutic protein production and delivery, Int. J. Pharm. Pharm. Sci. 2 (2) (2010) 1–5.

[39] J. Malakar, S.O. Sen, A.K. Nayak, K.K. Sen, Development and evaluation of microemulsion for transdermal delivery of insulin, ISRN Pharm. 2011 (2011) 780150.

[40] R. Panchagnula, P. Bindra, N. Kumar, C.S. Dey, O. Pillai, Stability of insulin under iontophoretic conditions, Pharmazie 61 (2006) 1014–1018.

[41] O. Pillai, R. Panchagnula, Transdermal delivery of insulin from poloxamer gel: *ex vivo* and *in vivo* skin permeation studies in rat using iontophoresis and chemical enhancers, J. Control. Release 89 (2003) 127–140.

[42] J. Malakar, S.O. Sen, A.K. Nayak, K.K. Sen, Preparation, optimization and evaluation of transferosomal gel for transdermal insulin delivery, Saudi Pharm. J. 20 (2012) 355–363.

[43] M.F. La Fountaine, C.M. Cirnigliaro, F. Azarelo 3rd., J.C. Hobson, O. Tascione, K.N. Swonger, T. Dyson-Hudson, W.A. Bauman, Cutaneous microvascular perfusion responses to insulin iontophoresis are differentially affected by insulin resistance after spinal cord injury, Exp. Physiol. 102 (2017) 1234–1244.

[44] J. Kim, I. Jeerapan, S. Imani, T.N. Cho, A.J. Bandodkar, S. Cinti, P.P. Mercier, J. Wang, Noninvasive alcohol monitoring using a wearable tattoo-based iontophoretic-biosensing system, ACS Sensors 1 (2016) 1011–1019.

[45] S. Charles, et al., Lab on a chip lab on a chip, Lab Chip 15 (2) (2013) 4277–4285, https://doi.org/10.1039/C4LC01069H.

[46] Z. Liu, J. Nie, B. Miao, J. Li, Y. Cui, S. Wang, X. Zhang, G. Zhao, Y. Deng, Y. Wu, Z. Li, L. Li, Z.L. Wang, Self-powered intracellular drug delivery by a biomechanical energy-driven triboelectric nanogenerator, Adv. Mater. 31 (12) (2019), e1807795.
[47] C. Wu, P. Jiang, W. Li, H. Guo, J. Wang, J. Chen, M.R. Prausnitz, Z.L. Wang, Self-powered iontophoretic transdermal drug delivery system driven and regulated by biomechanical motions, Adv. Funct. Mater. 30 (2020) 1–7.
[48] Y.H. An, J. Lee, D.U. Son, D.H. Kang, M.J. Park, K.W. Cho, S. Kim, S.H. Kim, J. Ko, M.H. Jang, J.Y. Lee, D.H. Kim, N.S. Hwang, Facilitated transdermal drug delivery using nanocarriers-embedded electroconductive hydrogel coupled with reverse electrodialysis-driven iontophoresis, ACS Nano 14 (4) (2020) 4523–4535.
[49] E. Monteiro, S.A. Silva, J.G. de Souza, P. Siqueira Melo, I.A.M. Moreno, S.M. Alencar, R.F.V. Lopez, G. Ricci Leonardi, Target-action of antioxidants using iontophoresis, J. Cosmet. Dermatol. (2020), https://doi.org/10.1111/jocd.13560.
[50] T. Fukuta, Y. Oshima, K. Michiue, D. Tanaka, K. Kogure, Non-invasive delivery of biological macromolecular drugs into the skin by iontophoresis and its application to psoriasis treatment, J. Control. Release 323 (2020) 323–332.
[51] S. Molokhia, K. Papangkorn, C. Butler, J.W. Higuchi, B. Brar, B. Ambati, S.K. Li, W.I. Higuchi, Transscleral iontophoresis for noninvasive ocular drug delivery of macromolecules, J. Ocul. Pharmacol. Ther. 36 (4) (2020) 247–256.

CHAPTER 14

Advances in MEMS micropumps and their emerging drug delivery and biomedical applications

Ranjit Barua, Sudipto Datta, Pavel Sengupta, Amit Roy Chowdhury, and Pallab Datta
Centre for Healthcare Science and Technology, Indian Institute of Engineering Science and Technology, Shibpur, Howrah, India

1 Introduction

Advances in microfabrication techniques have enabled the fabrication of microelectromechanical systems (MEMS) technology and their application in creating miniaturized microfluidic devices for several biomedical applications. A microfluidic device is expected to perform several functions such as sample transportation, micro-mixing of the sample with reagents, sample concentration, sorting, and analysis. One of most important factors, which determine the success of microfluidic devices is complete comprehension of the fluid flow phenomenon and consequent efficiency of fluid handling. This are implemented with the aid of micro-pumps and thus micro-pumps form an important component of MEMS-based devices. Though their initial application areas were inkjet printing, drug delivery devices and blood analyzers, applications areas in Micro Total Analysis Systems, transportation of blood in ventricular assist devices, artificial heart and heart valves and more recently, microfluidic lab-on-a-chip devices for point-of-care (POC) detection and analysis, have gained momentum. Lately, advancements in wearable, wireless devices are also emerging as important biomedical application areas of micro-pumps.

Micro-pumps are miniaturized (10^{-6} m) macroscopic pumps (water/oil pumps) used to transport a fluid from one spatial location to another. At this length scale, most prominent difference from the macro-pump counterparts are observed due to very high pressures required to transport of fluids at requisite flow rates. This occurs as the surface/volume ratio becomes higher at micro-ranges increasing the frictional effects. Typically, the pressure

Advances and Challenges in Pharmaceutical Technology
https://doi.org/10.1016/B978-0-12-820043-8.00002-5

increases by a power of 4 for every reduction in dimension of device. Micro-pumps usually operate below the range of 15 μL/min and are 100 μm or smaller in size. The extensive application demands of the micro-pumps has thus resulted in intensive research on the topic. Micro-pumps have been broadly classified into two categories—mechanical (involving moving part) and non-mechanical. While mechanical micro-pumps resemble the macro-pumps in principle differing only in size, non-mechanical micro-pumps have emerged as a consequence of satisfying the rigid requirements of fluid flow at this scales.

Reflecting the rapid growth in MEMS micropumps in the past decade, several topical reviews have been published, and most recently in though mostly limited to appraising their applications in drug delivery devices. Many of the earlier accounts have also limited their discussion to principles of the micro-pumps without informing the comparative performance advantages/disadvantages of each micro-pumps with respect to important performance standards such as flow rate and other operating conditions required. The present review comprehensively examines the performance capabilities of the conventional and emerging micro-pumps in perspective of their emerging applications in microfluidic lab-on-chip devices. Further, current advances in their integration with several technologies for future applications like wearable devices, wireless-powered devices are also discussed, highlighting the gaps and future research directions.

2 Micropumps

Micropumps are the driving devices which assist in dispensing of fluids into the desired location of human body or reaction stage. The ideal requirements of micropumps are to possess desired volumetric flow rate with high resolution over dispensing control. To obtain clinically acceptable products based on micropump technology, they should also be biocompatible, reliable and have high performance accuracy. Over the past three decades, the miniature dimension and good correctness of these micropumps were useful in applications such as chemotherapy for cancer treatment and delivery of insulin for patients suffering from diabetes. As listed in Table 1, recently, with the advent of LOC (lab-on-chip) devices for POC (point-of-care) diagnostics, micropump assume higher technological relevance. The two broad categories of micropumps are mechanical micropumps and non-mechanical micropumps [1].

Table 1 Current classification of micropumps adapted from Abhari et al. [2].

Mechanical displacement		
Actuation method	**Micropumping technique**	**Non-mechanical displacement**
Electrostatic	Vibrating diaphragm	Magnetohydrodynamic
Piezoelectric	• Vibrating diaphragm • Peristaltic • Flexural plate waves	Electrohydrodynamic
Thermo-pneumatic	• Vibrating diaphragm • Peristaltic	Electro-osmotic • AC • DC
Shape memory alloy (SMA)	Vibrating diaphragm	Electrowetting
Bimetallic		Bubble type
Ion conductive polymer film		Electrochemical
Electromagnetic	Vibrating diaphragm	

2.1 Mechanical micropumps

Mechanical pumps need mechanical or physical actuation for performing the dispensing function. The systems of mechanical micro-dosing are additionally classified into dynamic type and displacement type [3]. Micro-dosing systems of displacement type increase the pressure inside the fluid compartment at fixed time periods. To release fluids to external environment, pressure is applied to the outlet valve for overcoming the mechanical resistance. The mechanical micro dosing systems continuously increase the pressure to the fluid reservoir which helps to release the fluid to the external environment [4]. The most common categories of mechanical micropumps for biomedical application are electrostatic, thermo-pneumatic, shape memory alloy (SMA), piezoelectric, and bimetallic.

2.2 Electrostatic micropumps

Coulomb's law is the main principle behind working of electrostatic actuators [5]. A silicon membrane undergoes bending when suitable voltage is applied on parallel plate electrodes containing these silicon membranes. These silicon membranes regain their original position/shape as the voltage across the electrodes is annulled. Fig. 1 represents schematic of an

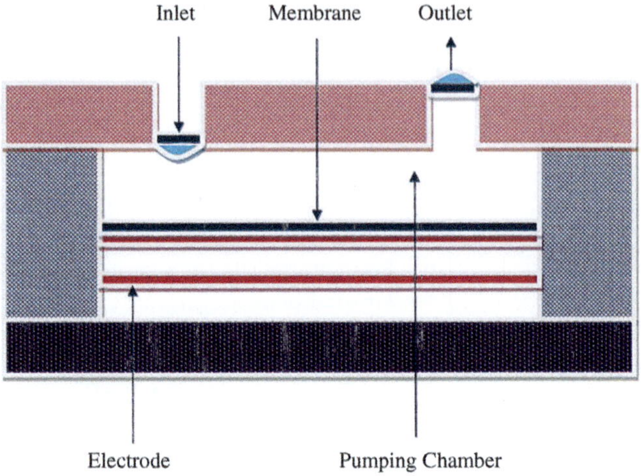

Fig. 1 Diagrammatic representation of an electrostatic micropump.

electrostatic micropump. From the pump compartment, the fluid is driven out when the pressure in the compartment changes via deformation the silicon membranes. Deforming silicon membrane creates the pumping mechanism the fluid by imparting momentum on it [6, 7]. The positive pressure which is generated by silicon membrane deformation drives the fluid out from the outlet valve. For creating the negative pressure within the compartment, applied voltage is removed and silicon membrane returns to its initial position. Further, for opening the inlet valve, the negative pressure is needed which allows the contained fluid to enter the pump compartment [8, 9]. Again the cycle repeats and as the membrane deforms creating positive pressure, fluid is released through the outlet valve. Using surface micromachining technology, Judy [10] had constructed the first electrostatic micropump. This micropump consisted of three parts chamber, active check valve and active outlet valve. Electrostatic micropump with dual diaphragm using injection molding technique was later demonstrated by Cabuz et al. [11]. The bidirectional operation was possible by this micropump but it was only gases uses. A rate of flow of about 30 μL/min was achieved in operating conditions of 160 V, 30 Hz and 8 mW power. Machauf et al. have also [12] fabricated a membrane based on electrostatically actuated micropump by exploiting the high or low permittivity of working fluids (conductive fluids) and achieved 1 μL/min flow at 50 V Similarly, dispensation of gases by electrostatic actuator has been demonstrated for gas chromatography [13]. A rate of flow of 3 μL/min at 7 kPa of backpressure at 14 kHz was observed.

Principles of fluidic resonance for maximum flow rate and multistage peristaltic arrangement is another example of micropump of their category [6]. The micropump presented the pressure variation in range of 7.3–3.3 kPa, flow rate of 0.29–0.07 cm^3/min and 0.05 plus 0.35 cycles for the valves to open. The use of reduction order model of the film in electrostatic micropump has been demonstrated [7]. Investigation of various factors like pull-in voltage, and pull-in position residual stress, thickness, initial gap and radius were also performed. Lil et al. [8] have described a electrostatic micropump where MATLAB software was used for modeling and demonstrated that for silicon electrostatic actuating membrane, 635 Hz was the calculated resonance frequency. In additional progress, by utilizing Finite Element Method, frequency of 680 Hz was seen. Zengerle et al. have described an electrostatic micropump having check valves [9]. And silicon membrane constituent micropumps. The highest pressure and volumetric rate of flow was 31 kPa and 850 μL/min was observed. Recent electrostatic micropumps have made further improvements over their previous counterparts. For example, Kim et al. [14] have reported a fully integrated 2, 4, and 18 stage peristaltic micropumps also containing integrated microvalves. The pump could achieve high flow rates, at low power for gases. Flow rates of 2.7, 3.0, and 4.0 cm^3/min were achieved for 2, 4, and 18 staged pumps respectively and the corresponding maximum pressure differentials were 17.5, 7.0, and 2.5 kPa. The total power consumptions were 57, 15.1, and 9.1 mW respectively, which is significantly low. In other examples the peristaltically and electrostatically driven four-electrode micropump for gas chromatography applications has been reported [15]. The proposed micropump had a single chamber of smaller cell and controlled embedded electrodes to generate the signals. The maximum flow rate using a basic actuating signal was approximately 38 μL/min at 90 V and 15 Hz; in the optimized signal it was approximately 136 μL/min at 90 V and 15 Hz, which meant a 4-fold increase. Sandoughsaz et al. [16] proposed a two-stage micropump having 2 kPa per stage pressure and 1.3 cm^3/min flow rate. Different innovative types of electrostatic micropumps have also already started being introduced by researchers having wide features and applications. Among them is the electrostatic traveling wave (ETW) micropump, in which flow is provided by a three-phase electro traveling wave [17]. They have achieved a flow rate of 135 μL/s. Another novel concept was given [18] where a novel miniaturized soft dielectric elastomer actuator (DEA) micropump was proposed. This microfluidic device has a simple fabrication process and achieves a large strain. It was shown to be possible to get high reliability at the operating conditions of 0.1 Hz and 0.25 Hz frequencies while a maximum flow rate 42 μL/s at an optimum operating frequency of 3 Hz. One advantage of this micropump is that the working fluid electric permittivity is directly proportional to the electrostatic force which is acting on the diaphragm. Thus, the higher the permittivity the higher the force and the pumping rate, for Thus for a fixed actuation voltage higher the permittivity higher is the force along with the pumping rate. Because of this advantage, the generated force for pumping is high enough at a relatively large gap between the two electrodes [12]. Another main advantage of this micropumps are power consumption is low with a faster response. The main disadvantage of this micropump is that the actuator has only lesser stroke and the diaphragm deflection. The disadvantage of this micropump is the electrostatic actuator has little stroke and the diaphragm deflection can be made under control without any effort by controlling the given voltage [19].

2.3 Thermo-pneumatic micropumps

Movement of fluids during heating and cooling associated with gaseous expansion and compression is the main working principle of thermo-pneumatic actuators [20]. Sealing gases make the actuator (along with aspired thermal expansion properties) inside a closed compartment in which one face is the membrane. Heating element is also present inside the compartment. When air is filled inside the compartment, the pair of heater and cooler expands and compresses the air periodically as depicted in Fig. 2.

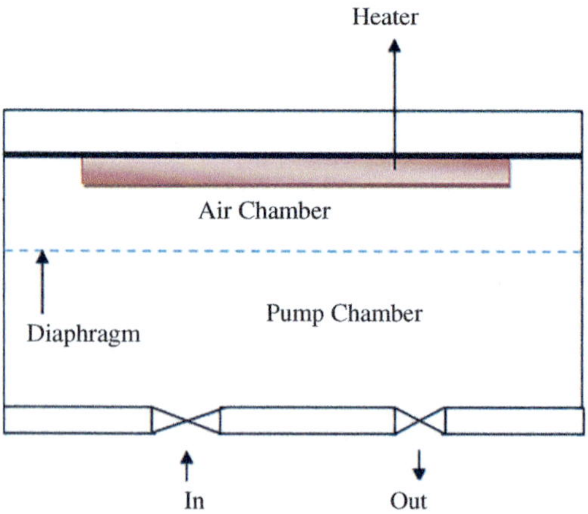

Fig. 2 Diagrammatic representation of thermo-pneumatic micropump.

This repeated change volume of the air inside the pump compartment gives the membrane of the pump momentum resulting in fluid to flow out of the pump. Pressures generated and the displacement of the membrane obtained with this micropump [21] are relatively high. However, driving power at certain minimum level is required for the pumping action. Other disadvantage of this type of pumps is the slower reaction or response time. Van De Pol et al. [22] demonstrated first the microfabrication based thermo-pneumatic micropump. A rate of flow of around 34 μL/min was obtained at 6 V at 30 °C temperature. A thermo-pneumatic micropump with corrugated diaphragm was described by Jeong et al. [23]. They found the flow rate to be about 14 μL/min at 8 V along and 4 Hz frequency. Cooney and Towe [24] demonstrated a thermo-pneumatic micropump with thin film heater and multiple reservoirs as its main constituents. The highest rate of flow of around 1.4 μL/min for about 4.5 h was demonstrated by average power of 200 mW. Similar micropump using glass layer, indium tin oxide heating element, poly-dimethyl siloxane (PDMS) chamber, PDMS membrane and PDMS cavity was fabricated by Kim et al. [25]. The rate of flow was 0.078 μL/min at 55 V and 6 Hz frequency. Three cascade thermo-pneumatic micropump and inlet/outlet port connecting microfluidic channel type peristaltic micropump has also been reported [26] for de-ionized water a flow-rate of 73.9 μL/min at zero backpressure. Micropump with two separate areas for heating of air and squeezing of fluid was demonstrated by Chia et al. [27]. On the area of fluid pumping, highest temperature reached was 2.0 K. A peristaltic micropump [28] through connecting a PDMS section containing micro-channels with PDMS/PMMA section is also reported in which a PDMS/adhesive film operated similar to a actuator of pneumatic type. An extreme rate of flow obtained was 96 μL/min. Jeong and Yang have shown a micropump [23] with a wave shape diaphragm in which maximum of 2.5 kPa pressure was reached. The highest rate of flow was around 14 μL/min with frequency, voltage input and duty ratio of 4 Hz, 8 V and 40% respectively. This type of micropumps requires air cooling and cannot work under high frequency conditions because it has low efficiency and slow response. However, fast response and low power consumption are the two major advantages of such micropumps reported by Abhari et al. [19]. Thermo-pneumatic micropumps have now been widely extended in advanced lab-on-chip devices. A disposable thermo-pneumatic micropump [29] based on thermopneumatic principle of micro scale having check valves. This ensured easy incorporation into an integrated microfluid biochip.

This device included glass and PDMS, which are biocompatible materials. The chip consisted of a glass substrate and three PDMS layers. In the biochip, the heating element Cr/Au is designed on a glass substrate by micromachining. The flow channel, pump chamber, and valves are made by replica molding using layers of PDMS. At 33% of the applied voltage, the pump achieved the most superior performance. Further, Chee et al. [30] reported a novel thermo pneumatic micropump powered by wirelessly a frequency-controlled resonant heater was used. The developed micropump has minimal moving parts that improve device performance. Output power of 0.22 W is developed through pumping the total volume of 4 mL in a single stroke under resonant frequency conditions. A variation of 2.8% for volume between each stroke was reported. Flow rate performance was between 1.01 μL/min to 5.24 μL/min with the corresponding heating power ranging from 0.07 to 0.89 W respectively. Comprehensive thermal behavior analysis of a wireless thermo-pneumatic micropump has been reported [31] Using numerical methods they investigated the temporal conduction of the planar inductor-capacitor (LC) wireless heater and the heating chamber. Their study used FEM and COMSOL analyses to maximize the pumping performance.

Wirelessly controlled thermo-pneumatic micropumps have steadily found increased application as micromixers. Nafea et al. [32] recently modeled and simulated a thermo-pneumatic zigzag type micromixer. Selectively activation of two passive wireless heaters in an external magnetic field having different resonant frequency is employed. A maximum efficiency of 96.1% after 65 s of mixing is reached after the activation switching time of 10 s, which serves as the optimum time. The micromixer provided controlled mixing-ratios. Maximum rate of flow of 3.4 μL/min and drop of pressure of 385.22 Pa was attained. An earlier work from the group [33] reported a micromixer providing a controlled mixing-ratio and having a flow rate of 0.28 μL/min. The biggest advantage was its biocompatibility, for cell culture applications.

Modern thermo-pneumatic micropumps have been used for a number of applications among which the screening devices for diseases is gaining more importance. For example, Lei et al. [34] presented a work for rapid detection of influenza A (H1N1) and (H3N2) viruses through a reagent-ready-on-chip immunological device. Four single-stroke actuators and a reaction chamber to detect the viruses based on their nucleoprotein (NP) were part of the device. The whole detection was accomplished within 1 h with minimal sample (20 μL) and reagent (50 μL) volumes.

2.4 Piezoelectric micropumps

Deformation of membranes due to piezoelectric effect is the main principle of piezoelectric micropumps [35]. As piezo-crystals like quartz are compressed or stretched, they create an electric signal [2]. Conversely, the length of the crystal changes when there is an application of voltage across it. The level of the dislocation is comparative to the quantum of given voltage [36]. The crystal returns to its initial location/shape as the applied voltage across it is removed. By coating a piezoelectric material on the pump membrane, the induced voltage generates certain deformation which acts similar to a plate which can be pushed to drive the liquid from the compartment of the micropump as depicted in Fig. 3.

Using micromachining technology, Van Lintel et al. [37] created a piezoelectric micropump. The micropump confined a chamber of pump, passive silicon check valve, and a tinny glass film piezo-disk actuated. Highest rate of flow of around 8 μL/min with 9.8 kPa back pressure were seen at 1 Hz, 125 V. Esashi et al. A three layered micropump with flow rate upto 15 μL/min and 6.4 kPa backpressure at 30 Hz, 90 V has been reported [38]. Olsson et al. [39] demonstrated a two compartment micropump to further performance improvement. On a Si membrane Lead Zirconate Titanate (PZT) was screen-printed to fabricate a piezoelectric micropump. A rate of flow of 120 μL/min along with 2 kPa backpressure was observed at 600 V with frequency of 200 Hz. Passive valves containing piezoelectric micropump was studied by Schabmueller et al. [40] who obtained 1500 μL/min flow rate along with 1 kPa of backpressure with ethanol as fluid. One-way parylene valves containing piezoelectric micropumps [41] provided a flow rate of 3.2 μL/min along with 0.2 kPa of backpressure at

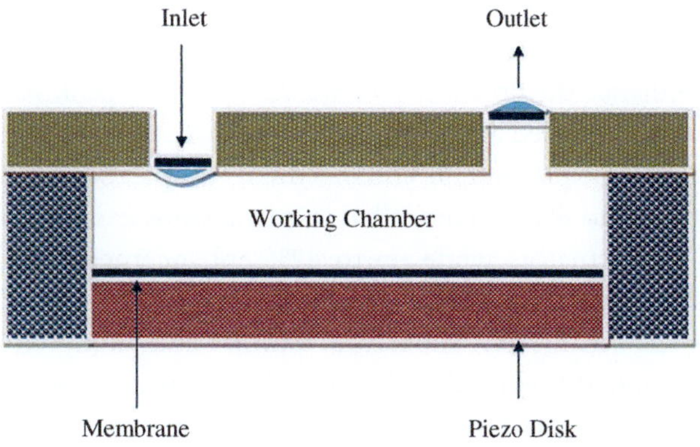

Fig. 3 Diagrammatic representation of piezoelectric micropump.

80 V and 3 mW (a significantly low power consumption). A micropump in which the pressure of backflow was independent of the rate of flow has been reported [42]. At low frequency, 20 kPa is the maximum independent back-pressure. Piezostack actuated peristaltic micropumps have also been reported [43], which achieved flow rate of 40 μL/min at 28.6 Hz by using water as the fluid. Up to 7 kPa, it was observed that the rate of flow was independent of the back pressure with a highest 45 kPa backpressure at 140 V. Fabrication using two optical mask was reported by Johari et al. [36]. The asymmetry result of the flow of longitudinal on the pumping potential by means of a piezoelectric element placed inside the passage of a simple pumping system has been examined [44]. The dynamic piezoelectric micropump process has been comprehensibly studied by Ali et al. [42]. The quantitative dimensions of the applied field, pressure generation, actuator length and frequency were investigated. A disposable piezoelectric pump containing four compartments connected in series was explored for close-loop insulin therapy [45]. The pump showed 6.23×10^{-5} mL/pulse outward flow resolution. Highest back-pressure of 22 kPa was observed at 36 V with frequency of 200 Hz. Simple construction and design, rapid response and high force of actuation makes them suitable for many applications. However due to inherent complexities of piezoelectric materials processing, fabrication of these types of pumps is more complicated. Moreover, for pump flow in valveless micropump the stroke volume and the piezoelectric membrane thickness are the two important factors [46]. To improve the performance of piezoelectric micropumps, dual frequency drives has been proposed [38]. A dual-frequency piezoelectric micropump operated on voltage of 400 V, 445.5 Hz was shown to cause the improvement of over 155% compared to single-frequency drive operating with the same driving voltage. Another area of research is to improve the flow rate. Singh et al. [47] fabricated a planar type valveless PDMS micropump and analyzed it using 3-D electro-fluid simulations. A modified chamber design was used to enhance the flow rate by 28%. At a peak-to-peak voltage of 30 V, the micropump delivered a maximum flow rate of 20 μL/min at a back pressure of 220 Pa. Displacement enhancement of piezoelectric membrane microactuator up to 6% by controlled in-plane stress can be achieved using PZT thick film actuators while up to 42% enhancement can be obtained using an actuator under an appropriate bias voltage [48].

A piezoelectric-driven miniature pump containing rib structures specifically designed for biological applications was developed with a high flow rate [49]. A high flow rate of 197 μL/min was achieved through 0.2 mm piezoelectric-actuator with a 0.05 mm rib structure oriented to the flow-field inside the pumping chamber. Another novel separable piezoelectric micropump with a boss diaphragm design for medical applications was also developed [50], where the pump can be separated into single-use part to achieve a cheaper and reusable actuator and purified liquid delivery. Flow rate was modulated from 1.58 to 6.21 mL/min below 20 Hz frequency. Valveless mechanisms are also explored at present. Numerical and experimental studies of a valveless microblower (PZT) diaphragm as actuator were reported by Dau and Dinh [44]. Flow rate of the device was found to be 0.7 L/min with back pressure 300 Pa at peak-to-peak voltage of 20 V. Their study has potential applications in health science as integrated fluidic systems. Cazorla et al. [51] made an innovative low voltage silicon based micro-pump having piezoelectric thin films. It is the first functional self-priming micro-pump based on PZT thin films and silicon substrate which is capable of working both for water and in air. A typical value obtained with water was 3.5 μL/min flow rate at 1 Hz, nil back pressure and actuation voltage of 24 V. It can normally withstand 3.2 kPa downstream pressure.

In the advancement towards developing a condensed and simplistic microfluidic droplet device based on a PZT diaphragm micropump with controller, Okura et al. [52] have recently developed a novel device for digital (Polymerase Chain Reaction) platforms. The device used a PDMS microfluidic chip with T-junction channel, Water-in-oil droplets at 1000 droplets/sec was successfully formed with negligible sample loss. Single molecule PCR amplification in the droplets was successfully demonstrated.

2.5 Bimetallic micropumps

Bimetallic micropumps contain two different metals which deform at different degrees when heated as shown in Fig. 4 [53]. The diaphragm is deflected against thermal fluctuations because the two metals have sufficiently different coefficients of thermal expansion. For fabricating micropumps using this principle, a silicon-based bimetallic membrane has been often selected [54]. An aluminum layer was deposited on the substrate of the silicon with thickness 10 μm to form a diaphragm for micro-driving. This micropump dimension was $6 \times 6 \times 1$ mm. The rate of flow and highest pressure at the back were noted near about 12 kPa and 45 μL/min correspondingly, at 5.5 V driving voltage and 0.5 Hz. Voltage needed for this micropump is relatively lower with respect to other types of micropumps but at high frequencies, the operation of this micropump is not possible. Zou et al. [55] fabricated

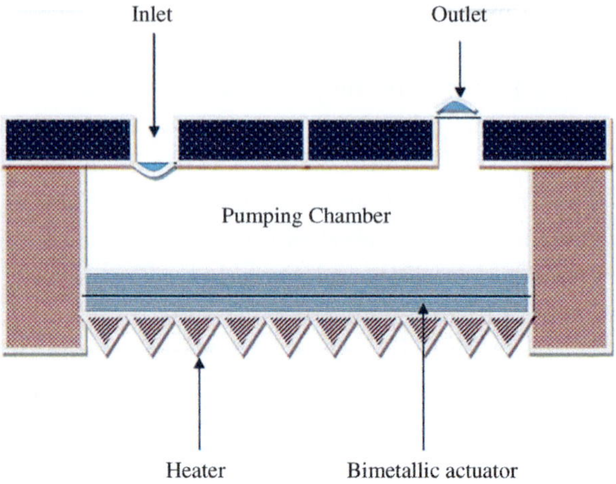

Fig. 4 Diagrammatic representation of bimetallic micropumps.

a micropump with combination of bimetallic and thermal pneumatic actuation. When heat is applied on the Al/Si layer, deformation occurs at a downward direction. Simultaneously, in the compartment the gas expands, and which effects the bimetallic actuation. Rate of flow of 336 μL/min was obtained and 0.5 kPa was the open pressure. Similarly, Pang et al. [56] designed a novel micropump which worked on the principles of both bimetallic- and electrostatic-actuation. Bimetallic micropump containing two check valves, two chips, two bimetallic actuators and a pump compartment is also reported [57]. A highest rate of around 43 μL/min was reached at 16 V with a 0.9 Hz frequency. Bimetallic actuation forces are quite high compared to the other micropumps and are simple for implementation. Generally, the, material coefficient of thermal expansion is less for bimetallic micropumps which enables attaining very small deformations. One of the main advantages of these micropumps is the ability to be operated under small voltage than other micropumps types while their principal limitation lies in their inability to be operated at high frequencies [58].

A recent advancement in bimetallic micropump segment has been the addition of a fiber-optic surface plasmon resonance (SPR) sensor along with a PCR chip in an integrated compressed device for DNA analysis [59]. The SPR sensor was made of bimetallic (Ag/Al) SPR coating with multimode optical fiber, whereas the PCR chip comprised of a PMMA substrate in microchannel form. They showed the integration of PCR micro-device and optical fiber SPR sensor into an aligned single instrument for fluorophore-free finding in PCR. The instrument contained the microfluidic PCR reactor that uses channel position-dependent changes in the temperature. The uniqueness of the work lied in the demonstration of reusability under a variety of conditions and addition in the lab-on-a-chip-platform.

2.6 Shape memory alloys (SMA)

Return to an initial shape from the deformed shape when heated or cooled is known as shape-memory effect [60]. Ni-Ti alloy is the main diaphragm material used for fabrication of these micropumps [61]. After heating, martensitic phase is converted to austenitic phase creating deformation and providing the actuating force on the SMA diaphragm in the micropump as shown in Fig. 5. Makino et al. reported a distinctive SMA micropump structure [62] with total dimension of 10 × 20 × 1.4 mm, volumetric flow rate of 340 μL/min as well as 100 kPa pump back pressure. This micropumps have several advantages like bio-compatibility, capability to regain huge alteration in stresses and strain on heating or cooling, large force-to-volume ratio, capacity of high damping and chemical resistance. Thin film SMA micropumps with actuation configurations of different types can be fabricated [63]. Electrical driving signal obtained from the Ti/Ni thin film results in Joule heating, which promotes the phase changes and ultimately SMA effect in the micropump. A highest rate of flow of 50 μL/min was found at frequency of 0.9 Hz. Application of SMA micropumps for analysis and micro-dispensing systems have been demonstrated in which highest rate of flow 0.4 μL/cycle at 100 kPa bias pressure was obtained [64].

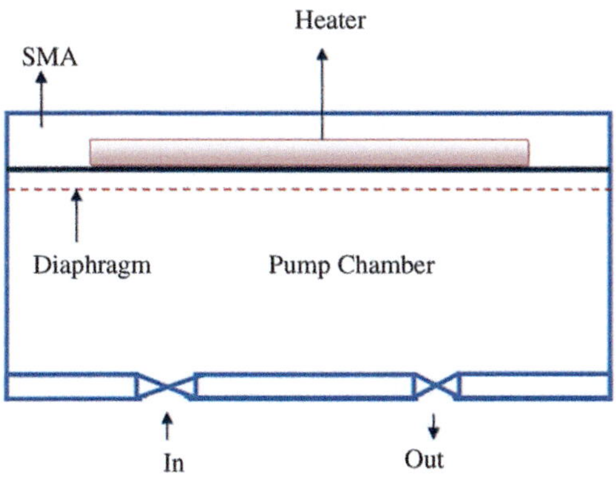

Fig. 5 Diagrammatic representation of shape memory alloy (SMA) micropumps.

A micropump consisted of membrane of NiTi/Si, two inlet—as well as the outlet-check valves and pump compartment was reported [65]. Another micropump composed of Ti/Ni/Cu SMA thin film driving film, compartment of the pump and two outlet and inlet check valves was reported by Zhang et al. [66]. For gripping manipulation, modification of TiNi SMA spring actuator of diameter 50 mm wire and 350 g hanging mass was reported by Setiawan [67]. However, due to temperature sensitive nature of actuation the temperature sensitivity, it is not possible to control precisely the SMA deformation. By designing multiple degrees of freedom and compact structures, fabrication of TiNi membrane device with practical, effective and complex characteristics can be performed. SMA micropumps are limited by availability of materials, high power consumption and uncontrolled SMA deformation due to reduced temperature sensitivity, such micropump containing several degrees of freedom and compact structures. The fabrication of TiNi membrane device with useful, realistic and intricate characteristics can be performed by [68]. SMA micropumps are limited by availability of materials, high power consumption and uncontrolled SMA deformation due to reduced temperature sensitivity. Vyawahare et al. [69] have employed multi-layer soft-lithography with SMA to fabricate electronically-activated devises such as pumps, latches and valves multiplexers assembled on printed circuit boards (PCBs). Their decisive work made electronically-controlled microfluidic chips to be integrated with other opto-electronic parts of PCB. This development opens possibilities for fabricating microfluidic circuits resembling electronic ICs. Ali and Takahata [63] reported a wireless-controlled microfluidic device with operative control given by multiple SMA actuators under external radiofrequency magnetic fields. A wireless resonant heater was used to drive each SMA actuator, which heats only the field frequency resonant with heater frequency in the range of 135–295 MHz. The heater circuits are both simultaneously and selectively controlled. A 5 μL reservoir is squeezed by 5 mm long cantilever-type actuator to eject controlled amount of liquid from the reservoir. Hilber and Jakoby [70] described that with the omission of polymer hydrogels, which make use of the normal swelling performance, actuation through shape memory effect require temperature stimulus. However, heat stimulation using surrounding energy is not easy to control while incorporated heaters additionally an outer power supply. Moreover, integration and fabrication of these materials into microfluidic structures needs substantial technological complexity. The device is compostable and biocompatible. The actuation principle is uniquely founded on ethanol fermentation. Microfluidic pumps have gained in popularity as LOC devices for their ability to manipulate fluids in very small quantities. Pawar et al. [71] simulated the diffuser/nozzle type of micropump using fluid structure interactions. Shape optimization design of geometric parameters was performed for the nozzle/diffuser type micropump. Parametric study was also carried out for improving the pump performance. The micropump delivered a maximum rate of flow of 145 μL/min for 20 Hz excitation frequency with no back pressure and 0.020 N actuating force. A new SMA micro-valve which can be switched on/off by surface acoustic wave was investigated by Liu et al. [72]. Two transducers fabricated on a piezoelectric substrate were used to excite surface acoustic waves. These waves heated a paraffin oil droplet to raise the temperature of the SMA wire through conduction. The spread distance of the two ends of the SMA wire is then increased, which squeezed the PDMS microchannel. Removing the electrical signal recovers the SMA and the fluid in the microchannel is then transported by a syringe.

3 Ionic conductive polymeric films (ICPF)

The ICPF micropump contains a core layer which contains perfluorosonic acid polymer or sulfonated tetrafluoroethylene (Nafion)/silica. The core layer consists of two metal electrodes. Normally ICPF is like a "sandwich" diaphragm because on the two sides of the polymer two thin films are deposited. An excellent electrically conductive film (like gold) deposited on the polymer acts like metal electrodes. In this type of micropump one side is fixed, the diaphragm of the ICPF can be managed to turn either upward or downward as long as a suitable pair of voltage is given to the electrodes. Major advantages of these micropumps are the ability to operate under moist environment and the driving voltage being low. Fig. 6 shows the

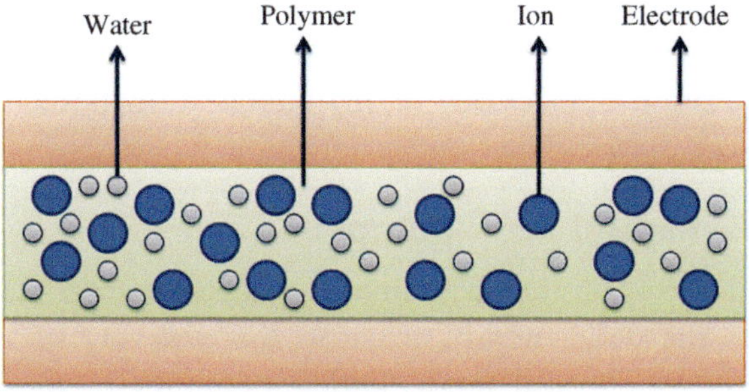

Fig. 6 Diagrammatic representation of ICPF micropump.

diagrammatic representation of ICPF micropump. The total size of the pump was 23 mm long and 13 mm in diameter. The highest rate of flow the micropump was 37.8 μL/min at 15 Hz frequency, operated at 1.5 V. Conductive polymers were stacked on each other by epoxy glues. ICPF applications are found in fine micro-robots and micro-manipulators by which surgical operations can be performed. A new design of micro-catheter having two bending degrees of freedom with active guide wire using ICPF actuator was reported by Guo et al. [73]. A micro-robot with three degrees of freedom, resembling underwater fish has also been reported [73]. A micropump of ionic metal/polymer flap valve consist of a supported by diaphragm of a material of flexible type is another example of ICPF application [74]. A highest rate of flow of 760 μL/min and 1.5 kPa backpressure was obtained with 3 V and frequency of 3 Hz. The design of an integrated sensory actuator has also been used [75]. Simultaneous bending feedback and diaphragm actuator output was measured by polyvinyldiene fluoride (PVDF) membrane. To predict the diaphragm deformation and the flow rate, the controlled oriented model was reported [54, 76]. Micropump actuated with polypyrrole (PPy) showed the highest rate of flow about 1260 μL/min at 4 V. Advantages of actuators like ICPF are low driving voltage, quick response, biocompatibility and amenability to be operated in moist environments. Low repeatability in batch manufacturing is the main disadvantage of these micropumps.

Nash and Fritsch [77] have proposed a novel micropump with poly 3,4-ethylenedioxythiophene (PEDOT) electrodes for redox-magnetohydrodynamics (MHD) application. This reduced interfaces in detection by eliminating the requirement of redox reaction in the solution, for LOC applications. PEDOT, a conducting polymer with low cellular toxicity, was polymerized electrochemically assay on electrodes (25 mm × 103 μm). Their work shows it is possible to retain the unique properties of redox-MHD along with higher sustainable voltage range and currents compared to previous reports. The system was shown to work without degradation of electrode or any bubble formation.

3.1 Non-mechanical micropumps

Design of non-mechanical micropumps is much uncomplicated, which can be ascribed to the deficiency of actuators that are very necessary for the micropumps of mechanical types. Different pumps in this category are described as follows.

3.2 Electro-chemical micropumps

In this type of micropump, bubbles generated by electrolysis process are principally utilized. The electrolysis reaction can be represented as follows.

$$\text{Anode}: 2H_2O - 4e^- \rightarrow 4H^+ + O^2 \text{ (oxidation)}$$

$$\text{Cathode}: 2H_2O + 2e^- \rightarrow 2OH^- + H_2 \text{ (reduction)}$$

Applications of this type of micropump in drug delivery (Fig. 7), with bubble generation rate of $0.02\,\mu L/s$ has been demonstrated [78]. The simple structures of this category of micropump along with ease of integration with other microfluidic devices are the main advantages.

An electrochemical micropump was manufactured [79] by micromachining to work with less voltage and consumption of power. The microfluidics device was created by combining micropump on-chip along with valves and is operated by a H_2 bubble generated electrochemically. Platinum black working electrode along with thin film electrodes was used. The flow channel and the reservoir of the electrolyte solution were composed of the PDMS substrate. In another interesting work, an artificial synapse has been created. Using electrochemical micropump [80]. This micropump contained a glass nozzle along with two electrodes of blacked platinum loaded by an active solution. A voltage of 3 V was applied on the electrodes for driving the solution towards the neuron. Similarly, an electrochemically actuated PPy-PDMS film connected to the check valves has been shown to achieve a highest rate of flow of $52\,\mu L/min$, 55 mW input power and a voltage of $\pm 1.5\,V$ [81].

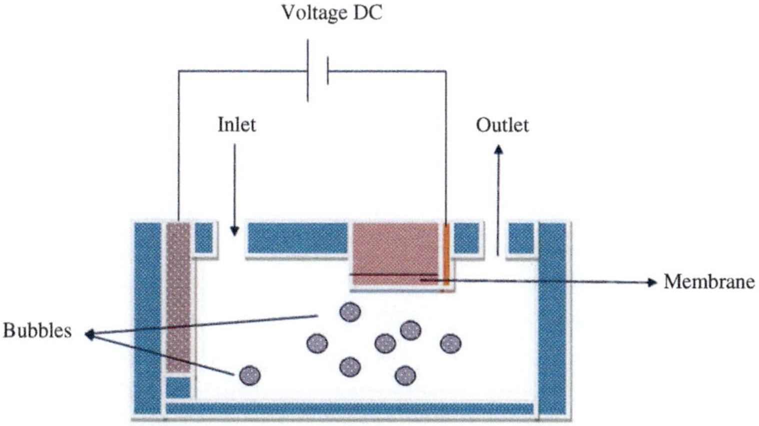

Fig. 7 Electro chemical pump micropump.

Biochemical sensing and biochemical analyses are set to be hugely impacted by the development of efficient bidirectional microfluidic transport systems, the shrinkage of hydrogen bubbles through the utilization of proper electrode potentials and under reversible electrochemical production using a three-electrode system was reported by [82]. Effect of control over electrode and flow channel configuration on the overall operation of the microfluidic system has been studied in detail. Device performance was found to be dependent on working electrode structure, paving the way for further optimization. These devices can extract solution through minimally invasive needle and process liquid plugs. Progress in the LOC segment has also taken place in the miniaturization and advancement of biosensor technology as demonstrated [83] in form of an integrated and automated LOC based biosensor device prototype (MiSens). The purpose of this device was to achieve sensitive detection of cancer biomarkers using a real-time amperometric monitoring during substrate flow. The device was designed for effortless usage exploiting sensors with unique plug-and-play docking station chip. Their prototype was able to detect prostate-specific antigen (PSA) levels in serum. Clinical results have indicated its potential for point-of-care cancer testing. Further roll of electric field relations in microfluidic channels has been explored [84] and identification of challenges in single microfluidic channel devices has been performed. Such model enable the study of electric field at low applied potentials in a microchannel and appears to be valuable for the analysis of future microchip electrochemistry methods.

LOC devices based on electrochemical analysis have also come a long way and have evolved according to a large number of applications. Ben-Yoav et al. [85] fabricated a system of arrayed electrochemical sensors for the study of DNA hybridization. Using an arrayed LOC and redox couple, the electrochemical activity of the microfluidic device containing valves was characterized and validated. A theoretical detection limit of DNA concentrations of 1 nm was achieved by device sensors. The effect of DNA hybridization on the calculated charge transfer resistance and diffusion resistance components was evaluated and demonstrated. Several microfluidic or LOC devices use microelectrodes for electrochemical detection. It has been found that electrode sensitivity is reduced in miniaturizing of electrodes, a challenge remaining to be overcome in microfluidic device applications. Though, the surface modification of microelectrode can increase the sensitivity and the surface area, a feature which has been exploited by Lee et al. [86]. Their work reports significant improvement in performance of platinum electrodes and sensitivity by carbon coating. The primary objective was to calculate the electrical resistances of RBC suspensions in a microchannel for hematocrit purpose. Their results indicated substantial improvement of platinum electrode electrocatalytic activity for RBC current capacity as carbonize the platinum electrode surfaces and erythrocytes adsorb. Other interesting areas of application include developing a new LOC platform as a smart system for detection and remediation of various pesticides and contaminants [87]. Boron doped diamond electrodes were fabricated as an LOC platform for detection and degradation of the pesticide atrazine through magneto-enzyme immunoassay. Flexible electrochemical LOC for detection of interleukin-10 (IL-10) in patients suffering from heart failure has been reported [88]. The work reported a lower time and cost of fabrication for this micro-LOC manufactured by soft lithography tools. Such simple devices pave the way for other devices for numerous other biomedical applications. An electrochemical immunosensing method for finding melanoma cells based on specific antigen and antibody combinations was used by Seenivasan et al. [89]. They were able to reach selective, highly sensitive and rapid finding of circulating tumor cells (CTCs) in blood samples. Narang et al. [90] described an electrochemical microfluidic paper based device to detect a sedative diazepam. Diazepam is a well-known anxiety-relieving and muscle-relaxing drug. The stability of electrochemical responses for diazepam in a specific concentration range was the key to its performance. Pollutant detection has also taken precedence over the years and electrochemical sensors of various types are being used in this field. For example recently, Kudr et al. [82] reviewed electrochemical microfluidic devices for in-situ pollution analysis.

3.3 Electro-wetting (EW) micropumps

The electro wetting micropumps works on the principle that under an applied electric field, tailoring of surface tension of the liquid, the solid or both takes place in dispersible liquids. Electrowetting micropumps are classified into two types on the basis of functional mechanisms, i.e., (a) pulse pump and (b) continuous pump. In case of pulse type electro-wetting micropumps, electro-potential is used for wetting of the fluid on the electrode. The purpose of the electric-potential is to level the fluid on the surface of the electrode. Another type, the continuous electro-wetting micropump employs pair of immiscible liquids, like liquid metal (mercury) with a drug solution. The liquid-liquid system is positioned in between the capillary and electric potential is applied on two capillary ends. An EW micropump of continuous type is depicted in Fig. 8. The volumetric rate of flow obtained was 70 μL/min at 2.3 V (driving voltage) as well as 170 μW consumption of power. At 2.3 V 25 Hz, the maximum pressure was about 800 Pa. At present, the demand of the EW-based micropump is increasing for LOC operations. Here, an array of single electrodes are used to direct the droplet in any position on the plane allowing new droplets for chemical and/or mixing reactions [91]. A test EW pumps at a pressure of 0.001 MPa inside 10 μm radius small channel has been demonstrated [92]. An EW-on-dielectric device [93] has been fabricated with aluminum oxide (Al_2O_3) and gained 2 μL of water droplets.

3D microfluidics with an alternating current electrowetting-on-dielectric AC-(EWOD) principle has also been described [94]. A circular coplanar electrode array was used to exert vertical as well as horizontal actuation of droplets between two parallel plates. Their study optimized the parameters of actuation frequency, switching time, and the gap between the top and bottom plates to transport droplets vertically. Horizontally, actuation of a droplet was performed at 1 kHz and 70 V. The study illustrated in droplet manipulation in 3D through switching of applied frequency for either horizontal or vertical transport performance enhancement of EW microfluidic devices has been at the centre of research in recent years. For example [95], a nanocomposite thin film made of barium strontium titanate (BST) and teflon®AF for its potential application in low-voltage EWOD device. Deposited films were shown to be uniform, smooth and crack-free hydrophobic surfaces with high dielectric constant. A single layer of coating provides simple and cost-effective method of fabrication of EWOD devices. It was able to actuate and move 5 μL of liquid droplets at 30 V, with excellent result reproducibility. A second train of thought

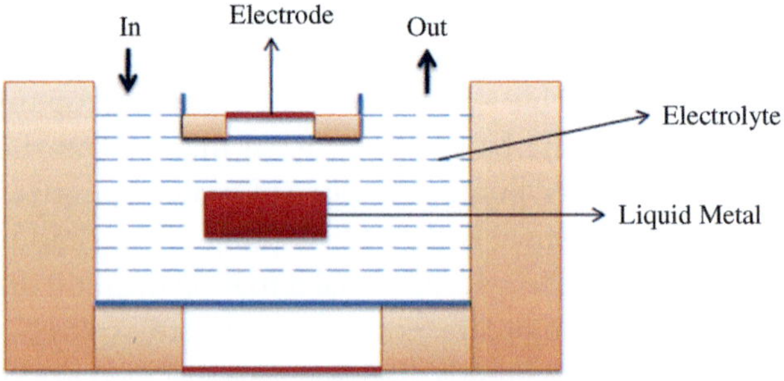

Fig. 8 Diagrammatic representation of EW micropump.

has been to decrease the saturated contact angle for droplets in microfluidics to improve the performance of EWOD devices. Li et al. [96] proposed this hypothesis and revealed contact angle decreases rapidly before reaching saturation value. Experimentally, their study showed that charges can be trapped and released repeatedly resulting in periodic decrease of contact angle. This control is a significant step in developing economic processes for mass fabrication of large-area flexible microlens arrays with controllable curvatures and wide field of view. A low-cost fabrication method for EWOD structures was demonstrated [97] by making EWOD stacks by screen printing of Ag particles, the EW effect allowed droplet movement through control of surface tension. ITO slides with a top-coated teflon polymer were used as electrical counter electrodes. The study investigated screen-printed silver particles has been used in order to reveal a new approach for quick and affordable fabrication of EWOD structures. Jain et al. [98] addressed the challenge of tracking multiple droplets on single substrate in real time using an affordable EWOD-based dynamic sensing system on (PCB). Using PCB with PDMS-based dielectric and a hydrophobic layer makes the system scalable, portable, robust, highly sensitive and inexpensive.

Interesting applications of EW microfluidic devices have come up in recent years. Applications in PCR assays with feedback on a 2D electrowetting array using BioCoder programming language have been described [99]. They created a new version of BioCoder and used it specifically for EW arrays with integrated sensors and then used this language to state the performance of two feedback-driven PCR assays for real time decision making. Through this work, biochemical assays can be described using the new version of BioCoder. A droplet based microfluidic LOC platform for the automated detection of trinitrotoluene (TNT) using EW principles is also demonstrated [100]. The reaction for detection is carried out in a programmable chip which includes a colorimetric detection. In recent times, the application of EW in electronic displays, electro-optical switches and LOC microfluidic devices was studied by Lu et al. [101]. The study addressed a key concern in theoretical as well as numerical methods to model behaviors of EW-actuated droplets under DC and AC signal. A dynamic contact angle based model and CFD models using VOF (Volume of Fluid)—CSF (Continuous Surface Force) method were used. Evolution of droplet shapes under different DC and AC were explored. This methods were able to precisely model the key parameters of EW-dynamics which is a significant step forward.

3.4 Electro-hydrodynamic (EHD) micropumps

For deposition of fluids containing charged particles, electro-hydrodynamic pumps are widely utilized. The dilating medium carries fluid. In response to an applied DC electric field across emitter and collector electrodes, electrostrictive forces drag the particles causing the fluid to move from one direction to another.

In EHD micropumps (Fig. 9), DC charge injection uses coulomb forces which are exerted in between two permeable electrodes through contact with fluid, disposing it. Ions are injected into the fluid with the help of electrochemical reactions in one or both the electrodes. Generally, EHD micropumps are induction, injection, polarization, or ion-drag, each based on the process of bringing in charged particles in the fluid. In one reported configuration, two electrically isolated grids were present in DC charged injection EHD micropump [102]. A rate of flow 15,000 μL/min with 1.72 kPa pressure head was obtained at given 800 V. A micropump [103] based on traveling wave-induced electro convection reported rate of flow in the range 0.05–5 μL/min. Electronic cooling along with EHD ion drag pump and EHD polarization micropump can also be manufactured. In this pump device a highest cooling rate of around 65 W/cm^2 along with 250 Pa pumping head was obtained. In order to address the aging problem in an ejection type EHD micropump (ITO) planar electrodes were used [104]. Ethyl alcohol was driven by planar electrodes with a rate of flow of around 356 μL/min with a voltage of 61 V DC. A micropump [105] to deliver the power for pumping using the force of the dipole moment which is produced

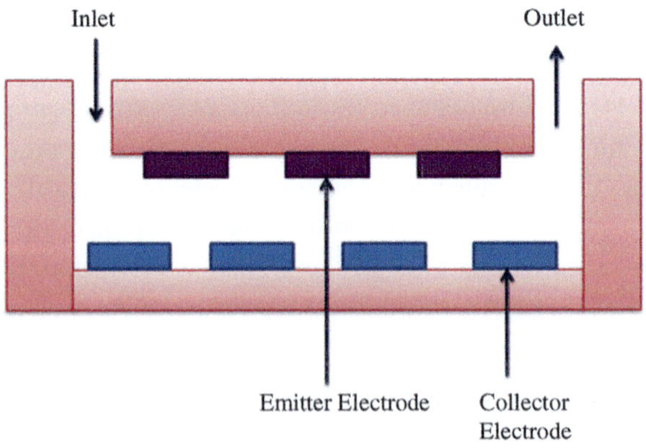

Fig. 9 Diagrammatic representation of EHD micropump.

from polarizing fluid molecules has also been described. The pump was operated at 13 kPa of pressure head and 10 W/cm^2 of wall heat flux at 500 V along with 500 μm pitch for parallel electrodes. EHD micropump having computational fluid model of ion drag is also available [106]. Interdigitated electrodes were present in this micropump with top and bottom channel. EHD micropump with an induction base for maximum heat flux cooling process [107] along with its numerical model considering 3D transient fluid flow is also reported. The numerical solution of the transportation of the charges is a problem due to the EHD simultaneous actuation as well as the diaphragm vibration is also available. The main advantage of this micropump is lower cost which requires standard silicon micromachining techniques and with microchannels it can be easily integrated. Disadvantage of this micropump is the aging problem which was improved by using ITO (Indium Tin Oxide) electrodes [104]. Several advances have been made in EHD devices to transport the dielectric liquid medium effectively. For example, a bi-directional EHD pump for dielectric liquid transporting with non-symmetric voltages applied to symmetric circular-cylindrical electrodes is described [108]. It is based on the Onsager effect. Multi-physics code COMSOL software executes numerical simulation for the fluid flow, the electric potential, and the transport of ion concentrations for the two kinds of electrode patterns. The study conducted a flow-visualization experiment to verify the physical models based on the numerical methods employed. It also reveals an optimum distance between two large grounded electrodes in order to produce higher pumping flow rate, with fixed electrode diameter. Increased size of large electrodes enhances the pumping effect. The study also focused on the comparison of two different electrode patterns on flow rate performance. Recently some greater insight into the functioning and performance of EHD micropumps has been made available. In such work, Russel et al. [109] described the doping effect of ferrocene, a redox dopant, on the working fluid of EHD micropumps under the application of DC electric field has been described. After diffusing ferrocene homogeneously using a certain HFE-7100 working liquid, the results indicated that the current and pressure generated with the doped working fluid were significantly higher than pure HFE-7100 under an applied DC field. A maximum static pressure of 6.7 kPa and flow rate of 0.47 mL/min was achieved at 700 V and no back pressure. Improving the efficiency of an EHD micropump with working fluid was shown to further improve using a micropump with 100 planar electrode pairs surrounded plus the lowest wall around 100 μm in height microchannel [110]. Performance of the micropump under the influence of pulsed voltage was enhanced by 75% compared to DC voltages. This study determined an optimum pulse repetition rate for very good micropump efficiency.

3.5 Osmotic-type micropumps

In this micropump, an osmotic agent, solvent and drug are the three main parts. The osmotic agent inside the compartment creates a concentration gradient for the solvent across the diaphragm. Propelling them to travel to the zone of lower density. Here the force of inertia is directly converted to the kinetic momentum on the actuation diaphragm. Fig 10 shows the force of osmotic is produced into the fluid which carries the drug molecules and also diffuses into the osmotic diaphragm. External powers (mechanical or electrical) are not needed in this type of micropump [111], however this micropump shows a low rate of 0.2 μL/h flow and lower response with longer delay times.

Electro-osmotic micropumps are also applied for LOC applications and especially as point-of-care (POC) devices. A study by Manshadi et al. [112] showed application of parallel electro-osmotic micropumps for different inner walls, flow rate increased due to volume force with the decrease of pump inlet area. It was seen that type of fluidic channel wall and ratio of two parallel micropump inlets has negligible effect on the flow rate. Their overall results also revealed a higher flow rate if parallel micropumps are applied and that pseudoplastic liquids have generally higher flow rate than the dilatant liquids. Using blood, enhancement in flow rate efficiency of 30% was evidenced for parallel micropumps employed as POC device. Electro-osmotic micropump analysis has also been used for water quality monitoring [113]. In this application, a novel design of electro-osmotic micropump was proposed and investigated with low voltage. Using theoretical analysis and numerical

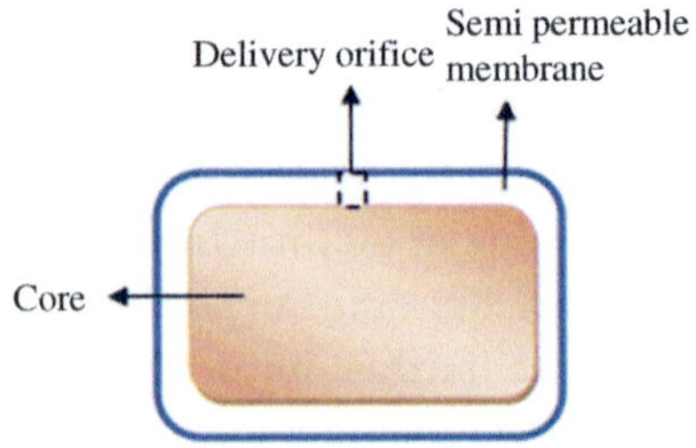

Fig. 10 Diagrammatic representation of Osmotic actuator.

simulations, effect of flow rate and the microchannels pumping efficiency were predicted. The effect of various microchannels dimension on the performance of the micropump was also elucidated.

3.6 Capillary-type micropumps

Capillary phenomenon is the basic driving force behind micro fluidic accumulators (MFA) [114], as depicted in Fig. 11. It contains many hydrophobic micro-capillaries, which are regulated by the valves, which makes the results of inlet and the outlet exposed to the outside atmosphere [115]. When the pressure is applied into the main micro-channel, fluid moves in all the capillaries which acts together like a reservoir and the fluid is released by alteration in resistance of the outlet fluid. A highest rate of flow noted was 7.2 μL/min, which is significantly below the limit needed in bio-chip applications.

Low-cost design of microfluidic capillary systems has been at the forefront of research, like all the other types of micropumps. Caprini et al. [116] have proposed a fast and cost-effective method of designing autonomous capillary systems for implementing bio-analytical protocols using saline solutions. This was achieved by rapid prototyping of glass microfluidic chips having a simple microfluidic network without breaking the glass slides. Strong capillary force and successful mixing of colored physiological solutions was observed. The chip was also shown to be reusable. Moreover, an autonomous integrated microsystem having capillary microfluidics and impedimetric sensors for detecting nanoprobe assembly and antibodies is also reported [117]. Using open-channel microfluidics, the functionalized impedance sensor detected up to 1.2 μg/mL target antibody within 10 min after functionalization. The results are a significant step in having an

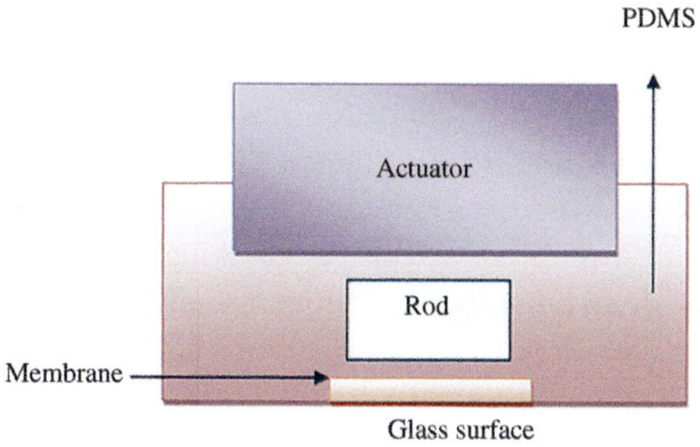

Fig. 11 Diagrammatic representation of MFA structure.

integrated system for rapid transducer functioning and biosensor utility. Tachibana et al. [118] presented a self-propelled continuous-flow polymerase chain reaction (CF-PCR) in a microfluidic device for flow control without any external pump. After mathematical formulation of the capillary flow during temperature switching, the study examined the viscosities and capillary pressures in PCR microchannels. This overcame the pervious difficulty in periodical switching of PCR solution. The simulation and experimental data of the capillary flow were promising. Other recent applications also include capillary microfluidic devices working as bionanoreceptors of virus-like particles as reported by Zang et al. [119]. They developed a capillary microfluidics-integrated sensor system using TMV virus-like particles (VLPs) as nanoceptors for antibody sensing. The VLP-functionalized impedance sensor can do label-free detection of target antibodies of 55 pm concentration within 5 min. This paves the way for nanoscale VLP-based sensors in fast, accurate, and decentralized pathogen control and detection. Temiz and Delamarche [120] addressed the challenge of fabricating microfluidic chips on a wafer level. Their novel "chip-olate" process designed microfluidic structures on a silicon wafer, and integrated several separate processes into a continuous one. Integration of reagents followed by lamination of a sealing cover was shown, creating ready-to-use individual chips. A PCB-based capillary pump for affordable POC diagnostics, adding to the promising and immense area of research in POC diagnostics is also recent achievement in this area [121]. Capillary pumps were combined with microfluidic channels of various hydraulic resistances and characterized. The capillary pump flow rate 138 μL/min is much higher than comparable devices based on glass, silicon or polymers, which is a significant advancement. Table 2 shows different types of micropums and their characteristics.

4 Microneedles

Microneedles are used to deliver drugs in transdermal and intradermal or delivery of vaccine without pain. They can be fabricated in the form of patches and syringes. Microfabrication techniques, allows needles to be precisely positioned in the stratum corneum lacking stimulating the essential nerve cells. The drugs enter the circulatory systems quickly from the dermis where the drugs are delivered [123]. The idea originated in 1970 but gained experimental acceptance in 1990 when microfabrication tools were made available by microelectronics industry. For intracellular administration of drugs silicon wafer microneedles were the first ones to be explored [124].

Table 2 An overview of typical performance parameters obtained from different micropump the comparison of different micropumps.

Reference	Name of micropump	Pumping medium	Size	Voltage (V)	Pressure	Flow rate	Frequency	Applications
Lee et al. [6]	Electrostatic	Gas	Not given	Not reported	7.3–3.3 kPa	0.29–0.07 SCCM	2.2–2.8 kHz	Drug delivery applications
Chia et al. [27]	Thermo-pneumatic	Not reported	$16 \times 18 \times 5$ mm	9	490 Pa	20.01 μL/min	1.2 Hz	Biomedical applications
Yin et al. [28]	Thermo-pneumatic	Compressed air	Not reported	Not reported	138 kPa	96 μL/min	10 Hz	Microfluidic devices
Liu et al. [45]	Piezoelectric	Insulin	15×8 mm	36	22 kPa	6.23×10^{-5} mL/min	200 Hz	Insulin therapy system
Zou et al. [55]	Bimetallic	Gas/water	$13 \times 7 \times 2$ mm	15	0.5 kPa	5.6 μL/s	Not reported	Not reported
Zhang et al. [66]	SMA	DI water	$8 \times 8 \times 1.8$ mm	Not reported	Not reported	235 μL/min	80 Hz	Not reported
Fang et al. [76]	ICPF	Water	$25 \times 25 \times 10$ mm	4	1.3 kPa	1260 μL/min	0.5 Hz	Biomedical devices
Jung et al [81]	Electrochemical	Water	$5.6 \times 16 \times 26$ mm	±1.5	11 mbar	52 μL/min	Not reported	Microfluidic applications
Yun et al. [122]	Electro-wetting	Mercury	Not reported	2.3	800 Pa	70 μL/min	25 Hz	Biomedical devices
Lim et al. [111]	MHD	PBS solution	$40 \times 25 \times 1$ mm	Not reported	100000N/m^2	0.3 μL/min	Not reported	Drug delivery

Later numerous methods for fabricating and developing microneedles were proposed by various researchers across the world. One such case of micro-fabricating hollow metallic microneedles in single and multi-needle arrays with various features like microrivets, barbs and varying tip geometries have been proposed by Chandrasekaran et al. [125]. For biological detection, extraction of fluid or system delivery microneedles and microneedles array can be used as a standalone microfluidic instrument. Microneedles are capable of being integrated with biosensors, micro-pumps, microfluidic chips and microelectronic instruments. Microneedles can be classified into many types like tissue pretreatment solid microneedles, drug coated microneedles, dissolving type microneedles and microneedles of hollow types. Solid microneedles are manufactured from Si, Ti, stainless steel and some specific polymers like polycarbonate (PC), PMMA, etc. The permeability of the skin can be increased by solid microneedles proceeding to relevant function of a drug. Water soluble drugs are coated on the solid microneedles to melt beneath the skin preceding the ejection of the microneedle. After the insertion of the microneedles beneath the skin sugar and polymer based microneedles gets dissolved. The drugs which are put in a nutshell in the microneedles are ejected after it is placed in the skin. The microneedles of hollow types are the forth class of microneedles which gives a mentioned channel for drug administration inside the skin and various tissues [126] The large compartment size can be eliminated by using transdermal microneedle which have surface mounted device which can be simply changed and fill up. Microneedle penetration will remain up to skin [127] it is not suitable for site specific drug administration. For medical and cosmetic use various microneedles based instruments have been created [128], for example, coated [129] and dissolving [130] type of microneedles are used in vaccine for influenza [131]. For injection of insulin for type 1 diabetes hollow microneedle of 1 mm depth are used [132]. For injection of drugs to eyes and nasal mucosa microneedles are used.

Two types of microneedles are differentiated on the basis of manufacturing type they are (a) in-plane microneedles and (b) microneedles of out-of-plane types. Another major advantages of in-plane microneedles is that the needle size, i.e., the length can be controlled very easily during manufacturing process [133]. But it is very challenging for creating the microneedle array in 2D geometry this is one of the main limitation of in-plane microneedle. For 1D and 2D arrays the out-of-plane microneedle is easier to manufacture because the microneedle length stick out of the surface of the substrate. Though, manufacturing of microneedles of out-of-plane type with high and length aspect ratio puzzling structure. Based on the structures needles are categorized as solid or hollow. Hollow type needles were manufactured on 1844 [134] and enlarged the implication in the field of biomedical there was no other active ways to conveyance the human body fluid. These needles become much essential after microneedles invention, the flow of fluid or drug through the microneedles was possible by the needles bore or lumen. A variety of micro-machining techniques of surface and bulk has been used to manufacture hollow in-plane microneedles of hollow in-plane type having length of 1–6 mm and silicon nitride entirely surrounded channels [135]. The strength of the solid microneedles is more than the hollow microneedles due to the presence of solid lumens. Two types of solid microneedles are dissolving and coated. In microneedles of coated type the drugs are coated on the needle lumen after the dissolution of the drug in the patient body the needle is removed. In the dissolving microneedle the microneedle dissolves but the base is non-dissolving so it is removed from the skin. Various solid coated and dissolving microneedles have been reported. Zosano Pharma, Inc. (formerly Macroflux®, ALZA Corp.) reported injection of parathyroid hormone (1–34) for the treatment of osteoporosis by coated Ti microneedles arrays having length 190 μm [136]. The proper drug delivery depends on the process of microneedle coating. Selection of material is very important to manufacture and design of the microneedles. Many researchers use silicon for fabrication of microneedles which is very harmful for human body and is brittle. Other researchers understood this situation so they started using polymeric microneedles because most polymers are biocompatible. Polymers give very good chemical and mechanical properties which are essential for microneedle production. For microneedles production many polymers like PC, (poly-L-lactide acid) PLLA, PMMA, (polyglycolic acid) PGA, PDMS, are used. Various researchers reported polymeric microneedles fabrications some researchers used other materials like glass, metal, alloy, etc. for microneedle manufacturing. Ti microneedles of in-plane type have been manufactured by bulk micromachining for drug administration. Tungsten microneedles are used in penetration of nerves is used. A microneedle is represented in Fig. 12.

4.1 Applications of microneedles

Microneedles are used in drug administration, extraction of blood, sampling of fluid, cancer treatment, dialysis micro type, printing of ink-jet type and

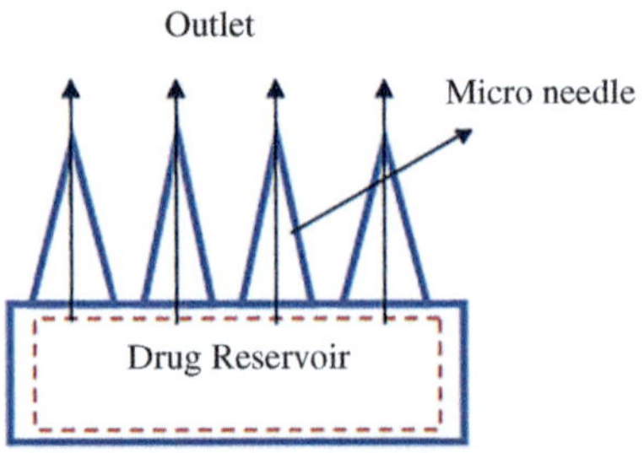

Fig. 12 Diagrammatic representation of a microneedle drug delivery apparatus.

electrodes sensing. Normally the length of microneedles 100 μm to 300 μm are used for drug delivery process but for extraction of blood the exact microneedles length is 1100 μm to 1600 μm [127]. For cell surgery solid micro needles are used and the hollow Ti microneedles are used for extraction of blood by methods like sputtering and deposition. Square hollow microneedles of SiO_2 types are represented by electrochemical etching process for flow delivery systems and for surgery of cells SiO_2 microneedles out-of-plane have been manufactured by lithography. In dermal diphtheria vaccination and influenza vaccination for surface micromachining and etching techniques the stainless-steel hollow and solid microneedles are relatively utilized and for TDD, the Si microneedles of hollow out-of-plane are manufactured. A piezoelectric micropump integrated with an array of microneedles for insulin drug delivery was reported [137]. These pumps are basically compact and wearable so integration of microneedles can be done easily with these micropumps. The characterization of infusion flow of a hollow micro-needle array driven by microneedle was reported [138]. Podder et al. [139] designed an electrostatic micropump with microneedle array to control the fluid flow with the help of COMSOL multi-physics design. Rao et.al [140] designed a piezoelectric micropump for high flow rate and optimization with microneedles. They observed that by incorporating microneedle with the micropump the flow rate was increased at low working voltages and frequencies. Zahn et.al [141] worked on microneedles integrated with on-chips MEMS positive displacement micropump for continuous drug delivery applications.

4.2 Embedded drug delivery system

MEMS based drug delivery systems have improved quickly. The feed-back specially closed loop system can help the real time observing of pump performance. The sensor can help to deliver the information of flow rate, pressure, doze site and state of pump. If the control system is incompetent for performing the work, it will create a serious health care issue even resulting in fatality.

4.3 Bio-compatibility

MEMS based drug delivery systems, the devices are either implanted or placed under the skin, and the degraded product can cause malfunction and behaves like [142]. The bio-compatibility index Foreign Body Giant Cells (FBGC) is commonly employed to assess the possibility of the instrument plus the material. Recently the polymer based MEMS systems have become more popular due to the minimal stiffness and naturally presents sufficient flexibility. Items like PDMS, PMMA, Parylene C. Table 3 lists the bio-compatibility materials for MEMS.

4.4 Microcontroller in health care systems

Microcontrollers are extensively used these days in massive variety of applications such as mobile phones, auto mobiles, CD/DVD players, Pacemakers, ATM machines, cameras, washing machines, microwave ovens, EVG machines and the list goes on and on. The single integrated circuit of the microcontroller contains a memory, a core processor and input/output (I/O) peripherals which are programmable. The memory of the

Table 3 MEMS bio-compatibility materials.

Materials	Bio-compatibility	Characteristic	Application
PMMA	Nice	Simple fabrication	Microfluidic structure
PDMS	Nice	Active layer ultra-thin	Osmotic membrane
PDDA	Nice	Porous material	Reduce toxicity of nano level particle
TiNi	Excellent	Shape memory	Actuator
Au	Excellent	Simple construction	Electrode assembly
Al_2O_3	Very nice	Brittle	Microfluidic structure
Ti	Excellent	Simple construction	Heater
Pt	Excellent	Simple construction	Temperature sensor heater

program is frequently in the form of NOR flash otherwise OTP ROM which is incorporated on the chip. Microcontrollers can behave like small computers [143]. By using peripherals on the chip it reduces the power consumption, decreases the size and reduces the cost. Based on necessity and working different microcontrollers are developed. The connection between the environment and the exterior device like serial communication interface (SCI), synchronous serial communication (SSC) etc. are done by embedded systems. To achieve a mentioned therapeutic reaction, the appropriate quantity of the drugs which are active should be fascinated and transferred to the location of working at the precise time and the input rate then can be accustomed to develop the intensities needed to keep the point of the result as lengthy as needed [144].

4.5 Wearable drug delivery systems

Wearable drug delivery technology is commonly used in medical systems and especially in life monitoring and disease detection. The technology denotes to the positive like apparatus that can be sprightly wore or joined in the accessories and cloths [145]. Currently wearable devices used detects physical index like motion tracking, heart rate, body temperature photoelectric detection method steps and bioelectricity signal. Currently the development of microfluidics is extensively applied in the medical researches and life science with main accomplishments. It is used for single synthesis of bio-macromolecules, cell screening and cell capture [146]. Also some microfluidic chips have used to find out the drug delivery, bioscience and disease detection. The arrangement of wearable microfluidics and technology of wearable devices openly associates with human skin, also it can assists the gathering and examination of the fluids in the body along with function of wireless data transfer integration. So far the application of wearable microfluidics in life monitoring is:

i. Glucose and lactic acid testing.
ii. The body fluid pH value measurement.
iii. Na^+, K^+, and Ca^{2+} testing.

Wearable microfluidic proposals an analysis, sensing, comprehensive body fluid collection, data transfer solution and control. Gowers et al. [147] established a 3D printed microfluidic system which is wearable and which can be tested online i.e. lactic acid and glucose simultaneously. For analysis of lactic acid and glucose FDS (Food and Drug Administration) accepted a micro clinical dialysis probe which is a nominally invasive method. For real-time monitoring of glucose Yu et al. [148] proposed a method of non-invasive PDMS five layers system of microfluidic for collection of subcutaneous interstitial liquid. For measuring lactic acid in the sweat microfluidic system of paper-based type was developed [149]. They manufactured hydrophobic and hydrophilic barrier by SU-8 for micro channel fabrication. Oxidation of H_2O_2 is done by horseradish peroxidase (POX) in the existence of 2, 20-azino-bis (3-ethylbenzthiazoline-6-sulfonic acid) diammonium salt (ABTS) that causes color production which ultimately gives the lactic acid concentration. A microfluidic chip of flexible type having three polyethylene terephthalate (PET) layers combined with pH sensor of silicon-based was developed by Nie et al. [150]. A microfluidic wearable system which consisted the sensors series for sweat analysis having pH sensor and for fluid transportation textile-based microfluidics was developed by Coyle et al. For drug-delivery through skin a wearable microfluidic is equipped having a collector having drugs along with micro-needle array of micro-needle [151]. The rapid and simple method of drug delivery of wearable microfluidics was developed by Di et al. [152]. In Fig. 13 the demonstration the process of wearable drug delivery, a device which is wearable containing drug is attached with the skin, when the stretching of the skin star occurs the release of drug from the device happens through the skin by the micro depots. In the microneedle array layer on the elastomer substract and then the elastomer substract the drug loaded nanoparticle with micro gel depot is attached. The tensile pressure in the elastomer film which is applied to the microneedle which punches through the skin and releases the nanoparticles with the drugs. The procedure claimed that it could be used for releasing pain killer drugs of anti-infective means.

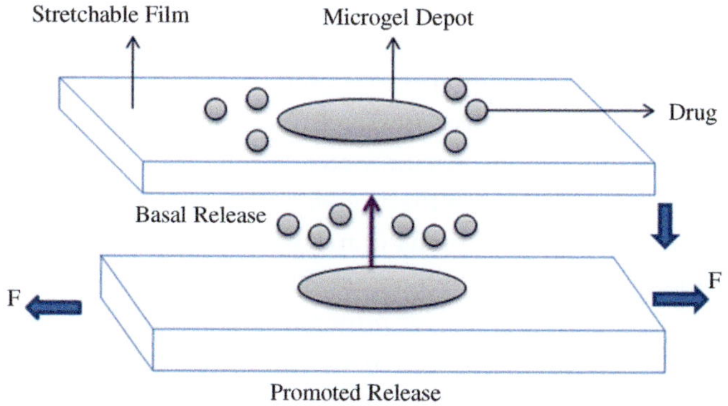

Fig. 13 Tensile strain-prompted drug release.

This method more appropriate for more complicated medical measurement like blood glucose control.

The microfluidic chips wearable type can monitor lactate, glucose, Ca^{2+}, pH of sweat and Na^{+}/K^{+} with the variety and examination of interstitial fluid, tears, saliva and sweat. The method of non-collection of blood for measuring lactate and glucose are done by wearable microfluidic technology which is somehow minimize some patient pain as it is non-invasive collection method. Other advantages of microfluidics of wearable type are (i) the body fluid required for the test is very less compared to the standard method; (ii) the requirement of the component needed for the test is also very less in rance of micro or nano liter scale which significantly decreases the cost pf every examination; (iii) in-vivo monitoring of the bodily fluids as well as wireless transfer of data [153] in real-time is possible by this process which is almost impossible in the standard measurement process; (iv) this technology of wearable microfluidics can be combined with the various other types of sensors where the microfluidics development technology is hereditary from the MEMS. Various sensors like electrochemical sensor, optical sensor, biosensor and piezoelectric sensor can be joined with the microfluidics wearable chips to fulfilling the different necessities of body fluid testing whit very less technical problem.

5 Conclusions and future directions

The first work on micropumps was introduced in 1975. But in 1980s, research and development on micropumps expending with microfabrication technology was started. Around 1990 this work start introduced to MEMS area. MEMS technologies successfully used to the requirements of biotechnology and to design of different micropump types ideas, manufacturing, various instruments and their functions in various fields like several types of biomedical applications such as glucose injection in patients suffering from diabetes, transdermal insulin delivery, anti-thrombogenic micropumps for blood transportation etc. MEMS-based micropumps biocompatibility is most vital issues for drug delivery systems, the biocompatible polymers, e.g., PMMA and PDMS, are used for biomedical equipments. Electrochemical, electro-wetting and ICPF actuator micropumps perform in low applying voltage and deliver most satisfactory flow rates. Electrostatic micropumps and piezoelectric micropumps need elevated applying voltage. ICPF micropumps seem to be the highest capable mechanical micropumps it delivers satisfactory rates of flow at extremely minimal voltages.

Non-mechanical electro-osmotic micropumps perform in huge functional voltages and it can deliver low rates of flow. This micropump is used in micro-analysis systems. Though much advancement in micropump research has been made and development is gradually increasing, still there require making different micropumps types in useful drug administration systems along with biomedical application instruments development in further.

References

[1] N. Tsai, C. Sue, Review of MEMS-based drug delivery and dosing systems, Sensors Actuators A 134 (1) (2007) 555–564.

[2] A. Geipel, A. Doll, F. Goldschmidtboing, P. Jantscheff, N. Esser, U. Massing, P. Woias, Pressure independent micropump with piezoelectric valves for low flow drug delivery systems, in: Micro Electro Mechanical Systems, 2006, January, pp. 786–789.

[3] P. Woias, Micropumps—summarizing the first two decades, microfluids and BioMEMS, Proc. SPIE 4560 (2001) 39–52.

[4] M. Staples, K. Daniel, M.J. Cima, R. Langer, Application of micro- and nano-electromechanical devices to drug delivery, Pharm. Res. 23 (5) (2006) 847–863.

[5] M. Koch, N. Harris, A.G.R. Evans, N.M. White, A. Brunnschweiler, A novel micro-machined pump based on thick-film piezoelectric actuation, Sensors Actuators A 70 (98) (1998) 98–103.

[6] S. Lee, S.Y. Yee, A. Besharatian, H. Kim, L.R. Bernal, K. Najafi, Adaptive gas pumping by controlled timing of active microvalves in peristaltic micropumps, in: TRANSDUCERS 2009–2009 International Solid-State Sensors, Actuators and Microsystems Conference, 21–25 June, 2009, pp. 2294–2297.

[7] W. Liu, Research on electrostatic micropump pull-in phenomena, in: International Conference on Intelligent Computation Technology and Automation Research, 2010, pp. 1154–1157.

[8] L. Lil, R. Zhu, Z. Zhou, J. Ren, Modeling of a micropump membrane with electrostatic actuator, in: 2nd International Conference on Advanced Computer Control, China, 27–29 March, 2010, pp. 630–632.

[9] R. Zengerle, A. Richter, A micro membrane pump with electrostatic actuation, in: IEEE Micro Electro Mechanical Systems, Germany, 4–7 February, 1992, pp. 19–24.

[10] J.W. Judy, Biomedical applications of MEMS, in: In Proceedings of the Measurement and Science Technology Conference, Anaheim, CA, USA (31 January–1 February), 1991, pp. 403–414.

[11] C. Cabuz, W.R. Herb, E.I. Cabuz, S.T. Lu, The dual diaphragm pump, in: IEEE International Conference on Microelectromechanical Systems, 2001, pp. 519–522.

[12] A. Machauf, Y. Nemirovsky, U. Dinnar, A membrane micropump electrostatically actuated across the working fluid, J. Micromech. Microeng. 15 (12) (2005) 2309–2316.

[13] A.A. Astle, H.S. Kim, L.P. Bernal, K. Najafi, P.D. Washabaugh, Theoretical and experimental performance of a high frequency gas micropump, Sensors Actuators A 134 (2007) 245–256.

[14] H. Kim, A.A. Astle, K. Najafi, L.P. Bernal, P.D. Washabaugh, An integrated electrostatic peristaltic 18-stage gas micropump with active microvalves, J. Microelectromech. Syst. 24 (1) (2014) 1–15.

[15] I. Lee, P. Hong, C. Cho, B. Lee, K. Chun, B. Kim, Four-electrode micropump with peristaltic motion, Sensors Actuators A Phys. 245 (2016) 19–25.

[16] A. Sandoughsaz, K. Najafi, A. Arbor, L.P. Bernal, A 2kPa per stage and 1.3sccm flow rate modular two-stage electrostatic gas micropump with stiffened drive electrodes, IEEE Sens. (2016) 1–3.

[17] M. Shirkosh, Y. Hojjat, H. Sadeghian, A new design of Electrostatic Traveling Wave (ETW) micropump and the effect of parameters on the flow rate, Flow Meas. Instrum. 48 (2016) 8–14.

[18] K.H. Hq, et al., Soft dielectric elastomer actuator for micropump application, in: IEEE Micro Electro Mechanical Systems, Shanghai, China,24–28 January, 2016, pp. 561–564.

[19] F. Abhari, H. Jaafar, N.A. Yunus, A comprehensive study of micropumps technologies, Int. J. Electrochem. Sci. 7 (2012) 9765–9780.

[20] S. Zimmermann, J.A.F.D. Liepmann, A.P. Pismo, A planar micropump utilizing thermopneumatic actuation, in: 17th IEEE International Conference on MEMS, 2004, pp. 462–465.

[21] W.K. Schomburg, J. Vollmer, B. Bustgens, J. Fahrenberg, H. Hein, W. Menz, Microfluidic components in LIGA technique, J. Micromech. Microeng. 4 (4) (1994) 186.

[22] F.C.M. Van de Pol, H.T.G. Van Lintel, M. Elwenspoek, J.H.J. Fluitman, A thermopneumatic micropump based on microengineering techniques, Sensors Actuators A 23 (1990) 198–202.

[23] O.C. Jeong, S.S. Yang, Fabrication and test of a thermopneumatic micropump with a corrugated p+ diaphragm, Sensors Actuators A 83 (1–3) (2000) 249–255.

[24] C.G. Cooney, B.C. Towe, A thermopneumatic dispensing micropump, Sensors Actuators A 116 (August 2003) (2004) 519–524.

[25] J. Kim, K. Na, C.J. Kang, Y. Kim, A disposable thermopneumatic-actuated micropump stacked with PDMS layers and ITO-coated glass, Sensors Actuators A Phys. 120 (2005) 365–369.

[26] O. Chan, S. Wook, S. Sik, J. Jungho, Fabrication of a peristaltic PDMS micropump, Sensors Actuators A 124 (2005) 453–458.

[27] B.T. Chia, H. Liao, Y. Yang, A novel thermo-pneumatic peristaltic micropump with low temperature elevation on working fluid, Sensors Actuators A Phys. 165 (1) (2011) 86–93.

[28] H. Yin, W. Keong, N. Nguyen, A reliable method for bonding polydimethylsiloxane (PDMS) to polymethylmethacrylate (PMMA) and its application in micropumps, Sensors Actuators B. Chem. 151 (1) (2010) 133–139.

[29] Y.A. Seung-Mo Ha, W. Cho, Disposable thermo-pneumatic micropump for bio lab-on-a-chip application, Microelectron. Eng. 86 (4–6) (2009) 1337–1339.

[30] M.S. Mohamed Ali, P.S. Chee, M.N. Minjal, P.L. Leow, Wireless powered thermopneumatic micropump using frequency-controlled heater, Sensors Actuators A 233 (2015) 1–8.

[31] P.S. Chee, M. Nafea, P.L. Leow, M. Sultan, M. Ali, Thermal analysis of wirelessly powered thermo-pneumatic micropump based on planar LC circuit, J. Mech. Sci. Technol. 30 (6) (2016) 2659–2665.

[32] M. Nafea, N. Ahmad, A.R. Wahap, M.S. Mohamed Ali, Modeling and simulation of a wireless passive thermopneumatic micromixer, in: Asian Simulation Conference, AsiaSim 2017 Modeling, Design and Simulation of Systems, 2017, pp. 312–322.

[33] M. Nafea, N. Ahmad, A.R. Wahap, M.S. Mohamed Ali, Selective wireless control of a passive thermopneumatic micromixer, in: Micro Electro Mechanical Systems (MEMS), 2016, pp. 792–795.

[34] K.F. Lei, J.-L. Liu, C.-H. Huang, R.-L. Kuo, N.-M. Tsang, A reagent-ready-on-chip microfluidic immunoassay system for rapid detection of influenza A H1N1 and H3N2 viruses no title, Biochip J. 10 (1) (2015) 34–41.

[35] Y.H. Mu, N.P. Hung, K.A. Ngoi, Optimisation design of a piezoelectric micropump, Adv. Manuf. Technol. 15 (8) (1999) 573–576.

[36] J. Johari, J. Yunas, B. Majlis, Piezoeletric micropump for drug delivery system fabricated using two optical masks, Adv. Mat. Res. 74 (2009) 279–282.
[37] H.T.G. Van Lintel, F.C.M. Van de Pol, S. Bouwstra, A piezoelectric micropump based on micromachining of silicon, Sens. Actuators A 15 (1988) 153–167.
[38] Q. Sheng, L. Guo, F. Sheng, X. Yan, Z. Hua, Piezoelectric micropump using dual-frequency drive, Sensors Actuators A Phys. 229 (2015) 86–93.
[39] A. Olsson, G. Stemme, E. Stemme, A valve-less planar fluid pump with two pump chambers, Sens. Actuators A 47 (46–47) (1995) 549–556.
[40] C.G.J. Schabmueller, M. Koch, M.E. Mokhtari, A.G.R. Evans, A. Brunnschweiler, H. Sehr, Self-aligning gas/liquid micropump, J. Micromech. Microeng. 12 (4) (2002) 420.
[41] G. Feng, E.S. Kim, S. Member, Piezoelectrically actuated dome-shaped diaphragm micropump, J. Microelectromech. Syst. 14 (2) (2005) 192–199.
[42] M.Y. Ali, C. Kuang, J. Khan, G. Wang, A dynamic piezoelectric micropumping phenomenon, Microfluid. Nanofluid. 9 (2010) 385–396.
[43] F. Trenkle, S. Haeberle, R. Zengerle, Normally-closed peristaltic micropump with re-usable actuator and disposable fluidic chip, in: Proceedings of the Eurosensors XXIII Conference, vol. 1 (1), 2009, pp. 1515–1518.
[44] V.T. Dau, T.X. Dinh, Numerical study and experimental validation of a valveless piezoelectric air blower for fluidic applications, Sensors Actuators B. Chem. 221 (December 2015) (2018) 1077–1083.
[45] G. Liu, C. Shen, Z. Yang, X. Cai, H. Zhang, A disposable piezoelectric micropump with high performance for closed-loop insulin therapy system, Sensors Actuators A Phys. 163 (2010) 291–296.
[46] Q. Cui, C. Liu, X.F. Zha, Study on a piezoelectric micropump for the controlled drug delivery system, Microfluid. Nanofluid. 3 (2007) 377–390.
[47] S. Singh, N. Kumar, D. George, A.K. Sen, Analytical modeling, simulations and experimental studies of a PZT actuated planar valveless PDMS micropump, Sensors Actuators A Phys. 225 (2015) 81–94.
[48] J. Ma, Y. Hu, J. Chen, Y. Cui, X. Xu, J. Chu, Displacement improvement of piezoelectric membrane microactuator by controllable in-plane stress, Sensors Actuators A Phys. 230 (2015) 45–51.
[49] H.K. Ma, R.H. Chen, Y.H. Hsu, Development of a piezoelectric-driven miniature pump for biomedical applications, Sensors Actuators A Phys. 234 (2015) 23–33.
[50] H. Ma, W. Luo, J. Lin, Physical development of a piezoelectric micropump with novel separable design for medical applications, Sensors Actuators A Phys. 236 (2015) 57–66.
[51] P.-H. Cazorla, et al., A low voltage silicon micro-pump based on piezoelectric thin films, Sensors Actuators A Phys. 250 (2016) 35–39.
[52] N. Okura, Y. Nakashoji, T. Koshirogane, M. Kondo, Y. Tanaka, K. Inoue, M. Hashimoto, A compact and facile microfluidic droplet creation device using a piezoelectric diaphragm micropump for droplet digital PCR platforms, Electrophoresis 38 (2017) 2666–2672. Wiley Online Library.
[53] A. Nisar, N. Afzulpurkar, B. Mahaisavariya, A. Tuantranont, MEMS-based micropumps in drug delivery and biomedical applications, Sens. Actuators B 130 (2008) 917–942.
[54] C. Zhan, T. Lo, L. Liu, T. Peihsin, L. Liu, T. Peihsin, A silicon membrane micropump with integrated bimetallic actuator, Chin. J. Electron. 5 (2) (1996) 33–35.
[55] J.X. Zou, Y.Z. Ye, Y. Zhou, Y. Yang, A novel thermally actuated silicon micropump, in: International Symposium on Micromechatronics and Human Science, Japan, 1997.
[56] J. Pang, Q. Zou, Z. Tan, X. Qian, L. Liu, Z. Li, The study of single chip integrated microfluidic system, in: 5th International Conference on Solid State and Integrated Circuit Technology, Beijing, China, 1998, pp. 895–898.

[57] Y. Yang, Z. Zhaoying, Y. Xiongying, J. Xiaoning, Bimetallic thermally actuated micropump, in: International Mechnical Engineering Congross and Exposition (ASME 1996), 1996.
[58] S. Tadokoro, S. Yamagami, M. Ozawa, Soft micromanipulation device with multiple degrees of freedom consisting of high polymer gel actuators, in: IEEE International Conference on Microelectromechanical Systems, 1999, pp. 37–42.
[59] T. Tai, K. The, L. Trinh, W. Jung, N. Yoon, Integration of a microfluidic polymerase chain reaction device and surface plasmon resonance fiber sensor into an inline all-in-one platform for pathogenic bacteria detection, Sens. Actuators B 242 (2017) 1–8.
[60] W.L. Benard, H. Kahn, A.H. Heuer, M.A. Huff, Thin-film shape-memory alloy actuated micropumps, J. Microelectromech. Syst. 7 (2) (1998) 245–251.
[61] L.G. Machado, M.A. Savi, R. De Janeiro, R. De Janeiro, Medical applications of shape memory alloys, Braz. J. Med. Biol. Res. 36 (2003) 683–691.
[62] E. Makino, T. Mitsuya, T. Shibata, Micromachining of TiNi shape memory thin film for fabrication of micropump, Sensors Actuators A Phys. 79 (2000) 251–259.
[63] M.S.M. Ali, K. Takahata, Wireless microfluidic control with integrated shape-memory-alloy actuators operated by field frequency modulation, J. Micromech. Microeng. 21 (7) (2011) 1–10.
[64] E. Makino, T. Mitsuya, T. Shibata, Fabrication of TiNi shape memory micropump, Sensors Actuators A Phys. 88 (October 2000) (2001) 256–262.
[65] S. Guo, T. Fukuda, SMA actuator-based novel type of micropump for biomedical application, in: International Conference on Robotics and Automation, New Oeleans, USA, April, 2004, pp. 1616–1621.
[66] H.J. Zhang, C.J. Qiu, A TiNiCu thin film micropump made by magnetron co-sputtered method, Mater. Trans. 47 (3) (2006) 532–535.
[67] M.A. Setiawan, The performance evaluation of SMA Spring as actuator for gripping manipulation, J. Tek. Elektro 7 (2) (2007) 110–120.
[68] S. Guo, K. Asaka, Polymer-based new type of micropump for bio-medical application, in: International Conference on Robotics and Automation, Taipei. Taiwan, 14–19 September, 2003, pp. 1830–1835.
[69] S. Vyawahare, S. Sitaula, S. Martin, D. Adalian, A. Scherer, Electronic control of elastomeric microfluidic circuits with shape memory actuators, RSC Lab Chip 8 (2008) 1530–1535.
[70] W. Hilber, B. Jakoby, Ethanol fermentation bioreactor for microfluidic actuation, Sensors Actuators A Phys. 240 (2016) 145–152.
[71] P.M. Pawar, R.R. Gidde, B.P. Ronge, Shape optimization of micro fluidic pump using fluid-structure interaction approach, in: International Conference on Advanced Technologies for Societal Applications, 2016, pp. 471–477.
[72] W. Liu, X. Fu, X. Zhang, W. Hu, A new shape memory alloy microvalve based on surface acoustic wave, Ferroelectrics 504 (1) (2016) 22–30.
[73] S. Guo, T. Fukuda, T. Nakamura, K. Oguro, Design and experiments of micropump using ICPF actuator, in: 7th International Symposium on Micro Machine and Human Science, 1996, pp. 235–240.
[74] T.T. Nguyen, N.S. Goo, V.K. Nguyen, Y. Yoo, S. Park, Design, fabrication, and experimental characterization of a flap valve IPMC micropump with a flexibly supported diaphragm, Sensors Actuators A Phys. 141 (2008) 640–648.
[75] Z. Chen, K. Kwon, X. Tan, Integrated IPMC/PVDF sensory actuator and its validation in feedback control, Sensors Actuators A Phys. 144 (2008) 231–241.
[76] Y. Fang, X. Tan, A novel diaphragm micropump actuated by conjugated polymer petals: fabrication, modeling, and experimental results, Sensors Actuators A Phys. 158 (2010) 121–131.

[77] C.K. Nash, I. Fritsch, Poly(3,4-ethylenedioxythiophene)-modified electrodes for microfluidics pumping with redox-magnetohydrodynamics: improving compatibility for broader applications by eliminating addition of redox species to solution, ACS Anal. Chem. 88 (3) (2016) 1601–1609.
[78] S. Böhm, B. Timmer, W. Olthuis, P. Bergveld, A closed-loop controlled electrochemically actuated micro-dosing system, J. Micromech. Microeng. 10 (4) (2000) 498.
[79] H. Suzuki, R. Yoneyama, A reversible electrochemical nanosyringe pump and some considerations to realize low-power consumption, Sens. Actuators B 86 (2002) 242–250.
[80] Y. Yoshimi, K. Shinoda, M. Mishima, B.S.K. Nakao, Development of an artificial synapse using an electrochemical micropump, J. Artif. Organs 7 (4) (2004) 210–215.
[81] J. Ho, K. Tong, R. Shepherd, Y. Wu, G. Wallace, D. Diamond, Performance characteristics of a polypyrrole modified polydimethylsiloxane (PDMS) membrane based microfluidic pump, Sensors Actuators A Phys. 148 (2008) 239–244.
[82] A.J. Kudr, O. Zitka, M. Klimanek, Microfluidic electrochemical devices for pollution analysis—a review, Sensors Actuators B. Chem. 246 (2017) 578–590.
[83] Y. Uludag, et al., An integrated lab-on-a-chip-based electrochemical biosensor for rapid and sensitive detection of cancer biomarkers, Anal. Bioanal. Chem. 408 (27) (2016) 7775–7783.
[84] N.M. Contento, P.W. Bohn, Electric field effects on current—voltage relationships in microfluidic channels presenting multiple working electrodes in the weak-coupling limit, Microfluid. Nanofluid. 18 (1) (2014) 131–140.
[85] H. Ben-yoav, P.H. Dykstra, W.E. Bentley, R. Ghodssi, A controlled microfluidic electrochemical lab-on-a-chip for label-free diffusion-restricted DNA hybridization analysis, Biosens. Bioelectron. 64 (2015) 579–585.
[86] Y.L. Hwi, B. Cedrick, M. Adrienne, Platinum electrode modification: unique surface carbonization approach to improve performance and sensetivity, J. Electrophor. 36 (2015) 1666–1673.
[87] M. Medina-Sánchez, C.C. Mayorga-Martinez, T. Watanabe, A. Tribidasari, An electrochemical immunosensing method for detecting melanoma cells, Biosens. Bioelectron. 68 (2015) 508–515.
[88] A. Baraket, M. Lee, N. Zine, N. Yaakoubi, J. Bausells, A. Errachid, A flexible electrochemical micro lab-on-chip: application to the detection of interleukin-10, Microchim. Acta 183 (2016) 2155–2162.
[89] R. Seenivasan, N. Maddodi, V. Setaluri, S. Gunasekaran, An electrochemical immunosensing method for detecting melanoma cells, Biosens. Bioelectron. 68 (2015) 508–515.
[90] J. Narang, et al., Lab on paper chip integrated with Si @ GNRs for electroanalysis of diazepam, Anal. Chim. Acta 980 (2017) 50–57.
[91] H. Obata, et al., Electrochemical bubble-based bidirectional microfluidic transport, ACS Sens. 1 (2) (2015) 190–196.
[92] E. Colgate, H. Matsumoto, An investigation of electrowettingbased microactuation, J. Vac. Sci. Technol. 8 (1990) 3625–3633.
[93] J.J.P. Jong-hyeon Chang, D.Y. Choi, S. Han, Driving characteristics of the electrowetting-on-dielectric device using atomic-layer-deposited aluminum oxide as the dielectric, Microfluid. Nanofluid. 8 (2) (2010) 269–273.
[94] J.B. Chae, S.J. Lee, J. Yang, S.K. Chung, 3D electrowetting-on-dielectric actuation, Sensors Actuators A 234 (2015) 331–338.
[95] S. Sohail, E.A. Mystri, A. Khan, S. Banerjee, K. Biswas, Fabrication and performance study of BST/Teflon nanocomposite thin film for low voltage electrowetting devices, Sensors Actuators A Phys. 238 (2015) 122–132.

[96] X. Li, H. Tian, J. Shao, Y. Ding, X. Chen, L. Wang, Decreasing the saturated contact angle in electrowetting- on-dielectrics by controlling the charge trapping at liquid—solid interfaces, Adv. Funct. Mater. 26 (2016) 2994–3002.
[97] A. Tröls, H. Enser, B. Jakoby, Low-cost silver screen-printed electrowetting on dielectrics structure for optofluidic switches, Procedia Eng. 168 (2016) 1061–1065.
[98] V. Jain, A. Hole, R. Deshmukh, R. Patrikar, Design, fabrication and characterization of low cost printed circuit board based EWOD device for digital microfluidics applications, Sensors Actuators A Phys. 23 (2) (2017) 389–397.
[99] C. Curtis, P. Brisk, Simulation of feedback-driven PCR assays on a 2D electrowetting array using a domain-speci fi c high-level biological programming language, Microelectron. Eng. 148 (2015) 110–116.
[100] V.K. Pamula, A digital microfluidics platform for multiplexed explosive detection, in: Electronics Noses and Sensors for the Detection of Explosives, 2004, pp. 279–288.
[101] Y. Lu, A. Sur, C. Pascente, S.R. Annapragada, P. Ruchhoeft, D. Liu, Dynamics of droplet motion induced by electrowetting, Int. J. Heat Mass Transf. 106 (2017) 920–931, https://doi.org/10.1016/j.ijheatmasstransfer.2016.10.040.
[102] J. Darabi, M.M. Ohadi, D. DeVoe, An electrohydrodynamic polarization micropump for electronic cooling, J. Microelectromech. Syst. 10 (1) (2001) 98–106.
[103] M. Electro, et al., Pumping of water solutions in microfabricated electrohydrodynamic systems, in: IEEE Micro Electro Mechanical Systems, Germany, 4–7 February, 1992, pp. 25–30.
[104] L. Yang, J. Wang, Y. Huang, The micro ion drag pump using indium-tin-oxide (ITO) electrodes to resist aging, Sensors Actuators A Phys. 111 (2004) 118–122.
[105] C. Lin, J. Jang, 3D numerical micro-cooling analysis for an electrohydrodynamic micro-pump, Sensors Actuators A Phys. 122 (2005) 167–176.
[106] E.I. Micropump, J. Darabi, M. Rada, M. Ohadi, J. Lawler, Design, fabrication, and testing of an electrohydrodynamic ion-drag micropump, J. Microelectromech. Syst. 11 (6) (2002) 684–690.
[107] V. Singhal, S.V. Garimella, Induction electrohydrodynamics micropump for high heat flux cooling, Sensors Actuators A Phys 134 (2007) 650–659.
[108] D.S. Cho, S. Kang, Y.K. Suh, Development of a bi-directional electrohydrodynamic pump: parametric study with numerical simulation and flow visualization, Adv. Mech. Eng. 8 (6) (2016) 1–15.
[109] M.K. Russel, S.M. Hasnain, P.R. Selvaganapathy, C.Y. Ching, Effect of doping ferrocene in the working fluid of electrohydrodynamic (EHD) micropumps, Microfluid. Nanofluid. 112 (2016) 1–9.
[110] M.K. Russel, P.R. Selvaganapathy, C.Y. Ching, Ion drag electrohydrodynamic (EHD) micro-pumps under a pulsed voltage, J. Electrostat. 82 (2016) 48–54.
[111] S. Lim, B. Choi, A study on the MHD (magnetohydrodynamic) micropump, J. Mech. Sci. Technol. 23 (2009) 739–749.
[112] M. Karim, D. Manshadi, D. Khojasteh, M. Mohammadi, Electroosmotic micropump for lab-on-a-chip biomedical applications, Int. J. Numer. Modell. Electron. Networks Devices Fields 29 (5) (2016) 845–858.
[113] N. Gallah, K. Besbes, Electroosmotic micropump analysis for lab On chip water quality monitoring, in: 13th International Multi-Conference on Systems, Signal and Devices, 2016, pp. 330–335.
[114] D.W. Bodhale, A. Nisar, N. Afzulpurkar, Structural and microfluidic analysis of hollow side-open polymeric microneedles for transdermal drug delivery applications, Microfluid. Nanofluid. 8 (2010) 373–392. Springer.
[115] Y. Ma, H.S. Gill, Coating solid dispersions on microneedles via a molten dip-coating method: development and in vitro evaluation for transdermal delivery of a water-insoluble drug, J. Pharm. Sci. 103 (2014) 3621–3630.

[116] D. Caprini, A. Nascetti, G. Petrucci, D. Caputo, G. De Cesare, Rapid prototyping of glass microfluidic chips based on autonomous capillary networks for physiological solutions, in: XVIII AISEM Annual Conference, IEEE, 2015, pp. 1–4.
[117] F. Zang, K. Gerasopoulos, K. Mckinzie, J.N. Culver, R. Ghodssi, C. Park, Autonomous capillary microfluidics for rapid nanoreceptor assembly and biosensing, in: IEEE Transducer, Alaska, USA, 21–25 June, 2015, pp. 548–551.
[118] H. Tachibana, M. Saito, K. Tsuji, K. Yamanaka, Self-propelled continuous-flow PCR in capillary-driven microfluidic device: microfluidic behavior and DNA amplification, Sensors Actuators B. Chem. 206 (2015) 303–310.
[119] F. Zang, K. Gerasopoulos, A.D. Brown, J.N. Culver, R. Ghodssi, Capillary microfluidics-assembled virus-like particle bio-nano-receptor interfaces for label-free biosensing, Appl. Mater. Interfaces 9 (10) (2017) 8471–8479.
[120] Y. Temiz, R.D. Lovchik, E. Delamarche, Capillary-driven microfluidic chips for miniaturized immunoassays: patterning capture antibodies using microcontact printing and dry-film resists, Methods Mol. Biol. 1547 (2017) 37–47, https://doi.org/10.1007/978-1-4939-6734-6_3.
[121] N. Vasilakis, K.I. Papadimitriou, H. Morgan, T. Prodromakis, High-performance PCB-based capillary pumps for affordable point-of-care diagnostics, Microfluid. Nanofluid. 21 (6) (2017) 1–11.
[122] K. Yun, I. Cho, J. Bu, C.C.J. Kim, E. Yoon, A surface-tension driven micropump for low-voltage and low-power operations, J. Microelectromech. Syst. 11 (5) (2002) 454–461.
[123] S. Liu, M. Jin, Y. Quan, F. Kamiyama, K. Kusamori, Transdermal delivery of relatively high molecular weight drugs using novel self-dissolving microneedle arrays fabricated from hyaluronic acid and their characteristics and safety after application to the skin, Eur. J. Pharm. Biopharm. 86 (2) (2014) 267–276.
[124] K. Oka, S. Aoyagi, Y. Arai, Y. Isono, Fabrication of a micro needle for a trace blood test, Sensors Actuators A 98 (2002) 478–485.
[125] S. Chandrasekaran, J.D. Brazzle, A.B. Frazier, Surface micromachined metallic microneedles, J. Microelectromech. Syst. 12 (3) (2003) 281–288.
[126] S.A. Coulman, et al., Microneedle mediated delivery of nanoparticles into human skin, Int. J. Pharm. 366 (2009) 190–200.
[127] E. Caffarel-salvador, J.C. Mcelnay, H.O. Mccarthy, K. Mooney, A.D. Woolfson, R.F. Donnelly, Potential of hydrogel-forming and dissolving microneedles for use in paediatric populations, Int. J. Pharm. 489 (1–2) (2015) 158–169.
[128] R.F. Donnelly, et al., Hydrogel-forming microneedles increase in volume during swelling in skin, but skin barrier function recovery is unaffected, J. Pharm. Sci. 103 (5) (2014) 1478–1486.
[129] S. Paik, et al., In-plane single-crystal-silicon microneedles for minimally invasive microfluid systems, Sensors Actuators A Phys. 114 (2004) 276–284.
[130] H. Khan, P. Mehta, H. Msallam, D. Armitage, Z. Ahmad, Smart microneedle coatings for controlled delivery and biomedical analysis, J. Drug Target. 22 (9) (2014) 1–6.
[131] Q. Bai, K.D. Wise, D.J. Anderson, A high-yield microassembly structure for three-dimensional microelectrode arrays, I.E.E.E. Trans. Biomed. Eng. 47 (3) (2000) 281–289.
[132] T.C. Mccrudden, et al., Hydrogel-forming microneedles prepared from "super swelling" polymers combined with lyophilised wafers for transdermal drug delivery, J. PLoS 9 (10) (2014) 1–12.
[133] K. Ita, Transdermal delivery of drugs with microneedles: strategies and outcomes, J. Drug Delivery Sci. Technol. 29 (2015) 16–23.
[134] S.P. Sullivan, et al., Dissolving polymer microneedle patches for influenza vaccination, Nat. Med. 16 (8) (2010) 915–920.

[135] S. Henry, D.V. McAlister, M.G. Alen, M.R. Prausnitz, Microfabricated microneedles: a novel approach to transdermal drug delivery, J. Pharm. Sci. 87 (8) (1998) 922–926.
[136] M. Pearton, et al., Microneedle delivery of plasmid DNA to living human skin: formulation coating, skin insertion and gene expression, J. Control. Release 160 (3) (2012) 561–569.
[137] M.P. Suresh, A. Vimala Juliet, Design of a valveless micro-pump for variable rate of insulin delivery, Indian J. Sci. Technol. 9 (44) (2016) 1–5.
[138] K. Chen, M. Pan, Z.-G. Feng, Modeling of drug delivery by a pump driven microneedle array system, Open Biomed. Eng. J. 10 (2016) 19–33.
[139] P.K. Podder, D. Mallick, D.P. Samajdar, A. Bhattacharyya, Design, Simulation and Study of MEMS Based Micro-Needles and Micro-Pump for Biomedical Applications, 2011, November, pp. 1–3.
[140] K.S. Rao, J. Sateesh, K. Guha, K.L. Baishnab, P. Ashok, K.G. Sravani, Design and analysis of MEMS based piezoelectric micro pump integrated with micro needle, Microsyst. Technol. 4 (2018) 1–7.
[141] J.D. Zahn, A. Deshmukh, A.P. Pisano, D. Liepmann, Continuous on-chip micropumping through a microneedle, in: Tech. Dig. MEMS 2001. 14th IEEE Int. Conf. Micro Electro Mech. Syst. (Cat. No.01CH37090), 2001, pp. 503–506.
[142] G. Kotzar, et al., Evaluation of MEMS materials of construction for implantable medical devices, Biomaterials 23 (2002) 2737–2750.
[143] J.P. Abenstein, W.J. Tompkins, A new data-reduction algorithm for real time ECG analysis, I.E.E.E. Trans. Biomed. Eng. 29 (1) (1982) 43–48.
[144] U.V. Eswaran, V.B. Sudharshan, Human like biosensor disease simulator, disease analyzer and drug delivery system, in: IEEE Conference on Information and Communication Technologies (ICT 2013), 2013, pp. 1033–1038.
[145] M. Thevis, et al., Measuring xenon in human plasma and blood by gas chromatography/mass spectrometry, Rapid Common. Mass Spectrom 28 (2014) 1501–1506.
[146] V.F. Curto, et al., Chemical Real-time sweat pH monitoring based on a wearable chemical barcode micro-fluidic platform incorporating ionic liquids, Sens. Actuators B 171–172 (2012) 1327–1334.
[147] S.A.N. Gowers, et al., A 3D printed microfluidic device with integrated biosensors for online analysis of subcutaneous human microdialysate, ACS Anal. Chem. 87 (15) (2015) 7763–7770.
[148] H. Yu, D. Li, R.C. Roberts, K. Xu, N.C. Tien, An interstitial fluid transdermal extraction system for continuous glucose monitoring, J. Microelectromech. Syst. 21 (4) (2012) 917–925.
[149] L. Trabzon, F. Nes, Microfluidic device on a nonwoven fabric: a potential biosensor for lactate detection, Text. Res. J. 84 (16) (2014) 1729–1741.
[150] C. Nie, A. Frijns, M. Zevenbergen, J. Den Toonder, Chemical an integrated flex-microfluidic-Si chip device towards sweat sensing applications, Sens. Actuators B 227 (2016) 427–437.
[151] D. Granger, M. Marsolais, J. Burry, R. Laprade, Na^+/H^+ exchangers in the human eccrine sweat duct, Am. J. Philol. 7 (2018) 1047–1058.
[152] J. Di, S. Yao, Y. Ye, Z. Cui, J. Yu, T.K. Ghosh, Y. Zhu, Z. Gu, Stretch-triggered drug delivery from wearable elastomer films containing therapeutic depots, ACS Nano 9 (9) (2015) 9407–9415.
[153] B.P.C. Demuth, X. Su, R.E. Samuel, P.T. Hammond, D.J. Irvine, Nano-layered microneedles for transcutaneous delivery of polymer nanoparticles and plasmid DNA, Adv. Mater. 22 (2010) 4851–4856.

CHAPTER 15

Ionic liquids in biomedical application

Sourav Chakraborty
Scottish Church College, Kolkata, India

1 Introduction

Ionic liquid (IL) belongs to the class of compound which consists of ions with melting point generally below 100 °C. It was in the year 1888 when Weiner and Gabriel [1] reported an organic salt, ethanolammonium nitrate with melting point (m.p.) ~53 °C which is considered as the first-ever report on organic salt, later termed as 'Ionic Liquid'. Later in 1914, Paul Walden [2] reported, for the first time, a room temperature ionic liquid (RTIL), ethylammonium nitrate with m.p. of 12 °C, which got some attention in the research fraternity. Still it was rather a slow progress of the IL-based research work which is clearly realized from the number of published articles before 1980s. But with exploration of various structural possibilities of cations and anions and discovery of 2nd and 3rd generation ILs, scientists and technologists could realize the potential of IL for different applications. In the last two decades, the publishing houses have been flooded with articles with 'Ionic liquid' as keyword showing constantly growing interest in this material. Especially the 'green' status of the material made it more demanding to the scientists to explore its utility as a replacement to the volatile organic compounds (VOCs) and to unleash its benefit for several application purposes ranging from engineering to biomedical applications.

2 Structural features of ionic liquid and its evolution

Ionic liquid is a molten salt, which is liquid at low to moderate temperature. The presence of asymmetric or large-sized cations and anions of poor coordinating nature in its structural unit possibly hinders the packing of the ionic motives in crystal lattice thereby making the compound liquid at moderate to low temperature range [3, 4]. From the last few decades, an intense investigation on ILs regarding their structure and application prospect proved that

Advances and Challenges in Pharmaceutical Technology
https://doi.org/10.1016/B978-0-12-820043-8.00014-1

these materials possess a huge and diverse potential for different purposes. This is attributed to the amazing structural diversity of these materials. The number of articles presenting various structural combinations (i.e., between cations and anions of an IL) and showing their mutual exchange is getting higher and higher every year which indicates the demand and potential of IL in the scientific domain. The popularity of IL to the researchers is driven by the fact that, the physicochemical properties of an IL can be easily tuned by exchanging its cation or anion by other respective analog. Some of the popular ionic moieties forming IL are shown in Fig. 1.

After the discovery of RTIL by Walden in 1914, few but not so many researchers really were attracted to these materials. Nearly two decades later, a patent was published showing dissolution of cellulose in a pyridinium-based molten salt [5]. This was of high value as polysaccharides such as cellulose could not be dissolved in many conventional solvents. However, the importance of the work was not recognized and the outcome was overlooked at that time. The real interest on the structural features and properties of IL was actually observed from 1960 onwards. In the next 30 years, the 1st generation ILs were analyzed both in terms of their chemical structure as well as application. Typical cations for 1st generation IL were dialkylimidazolium and alkylpyridinium derivatives whereas chloroaluminate and metal halides were used as anions [6]. Unfortunately, those were highly sensitive to moisture and air thereby limiting their application for practical purposes. Thus efforts were made to find new ionic species which can impart moisture resistive property. In the 1990s, the 2nd generation ILs started to emerge where anions from the 1st generation IL were replaced by weakly coordinating hydrophobic anions such as fluoroborate or hexafluorophosphate. The presence of fluorides made them enough hydrophobic to render stability toward moisture and air. These ILs showed strong potential for technological purposes, owing to their moisture resistivity. From the last decade, the urgency for biocompatible materials prompted the appearance of 3rd generation ILs which consist of readily available ions that is biodegradable in nature such as natural bases (e.g., choline), naturally occurring carboxylic acids, amino acids [7, 8] etc.

The 2nd generation IL was explored in several applications in various engineering fields. Still their use at commercial level is restricted to a few cases, presumably due to economic side related to IL synthesis or processability. However, along with the synthesis of 3rd generation IL, which is more biocompatible and less toxic, it is anticipated that the IL-based applications will reach the commercial level in the upcoming years. Quite a lot of

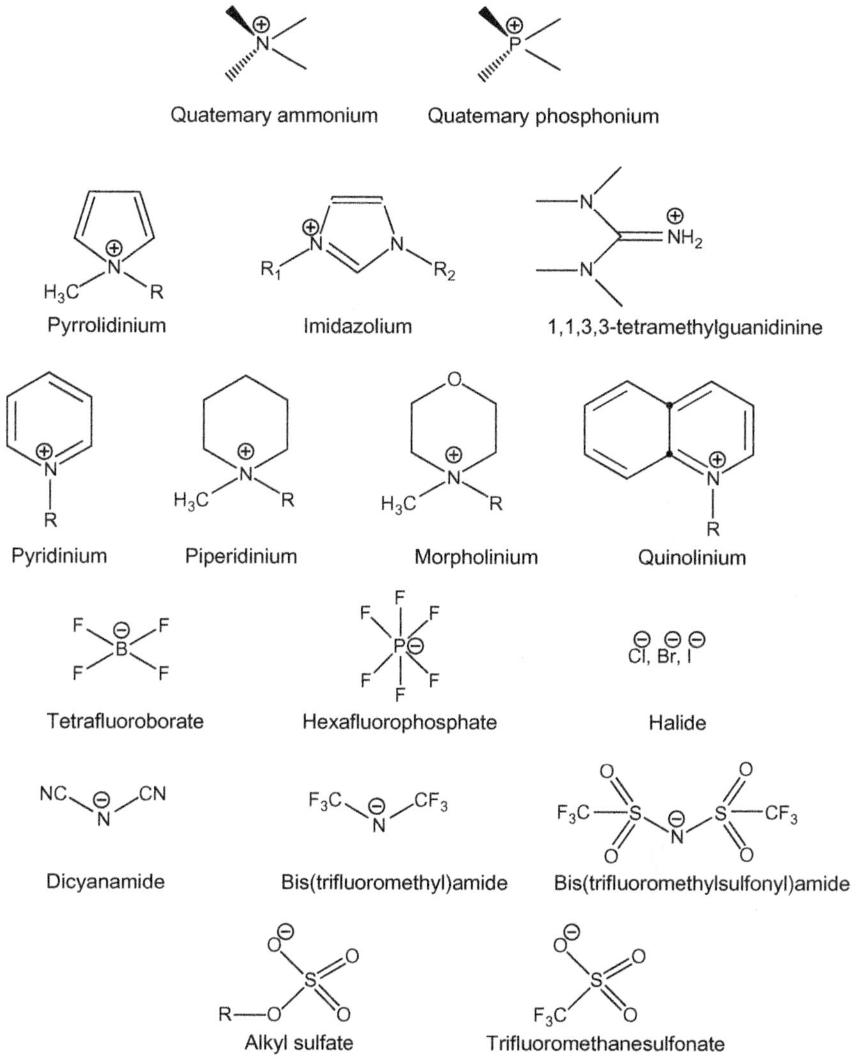

Fig. 1 Chemical structure of some cations and anions which are commonly used to synthesize ionic liquid.

funding has already been spent and are in 'execution mode' to synthesize and investigate novel IL materials of newer generation specially in order to apply ILs in industrial and biomedical fields. The advancement in research methodology and technologies like modern spectroscopic studies [9–11], study of the physical properties [12], analysis of the bonding pattern [13] and study of the molecular dynamics [14, 15] has unleashed various structural and characteristic features of ILs which are quite helpful for progress in IL-based research work.

Possibly the most exciting and effective characteristic of an IL molecule is its ability to tune the chemical structure by ion-exchange capability. The anions of the ILs can easily be interchanged by other anions [16] of variable nature and hence a huge possibility arises to tune the properties of the IL molecules. Selecting the appropriate cation or anion can produce IL of hydrophilic to hydrophobic [17] or even of amphiphilic nature [18]. The amphiphilic ILs were also successfully employed to replace conventional surfactant to execute various heterogeneous polymerization process like emulsion [19], miniemulsion [20] or microemulsion [21] polymerization. Moreover, IL molecules were shown to form self-organized structure at nano or micro-scale level [22]. The ability of ions to self-assemble in solution made IL capable of forming amphiphilic nanostructures. Such a property of nanoscale and micro-scale ordering in connection with availability of numerous cations and anions, make IL a material of choice for high-valued purposes [23].

The latest development on the IL based research work was the discovery of 3rd generation ILs. These are ILs having biologically active constituent ions. The tunable characteristics through ion-exchange capability made it possible to synthesize such ILs by replacing simple anions like halide by some bio-active anions. The examples include IL containing salicylate anion which showed excellent potential to be used in pharmaceutical field [24]. Either cation or anion or both moieties of an IL molecule can offer a biologically active site. Acetylcholine, a vital neurotransmitter of living organism, was incorporated in an IL structure as cationic part in association with halide as anion. Adamovich et al. [25] synthesized the first acetylcholine based IL, known as 'cholinomimetic IL' from the reaction between (2-hydroxyethyl)-dimethylamine and acetyl chloride. Replacing the alkyl group of acetyl chloride by other alkyl groups, it was possible to synthesize several cholinomimetic ILs which showed the ease of incorporating bioactive substances into the 3rd generation IL structure. The synthesis of the cholinomimetic IL was illustrated schematically in Fig. 2.

The discovery of such biocompatible 3rd generation ILs provided a huge opportunity for the biochemists to utilize these in several biomedical applications. Here in this chapter, we focus on the 3rd generation ILs showing different biomedical applications. Before entering into the detail of the biomedical benefits of such ILs, let us have a look on the interaction of ILs with different biologically important molecules.

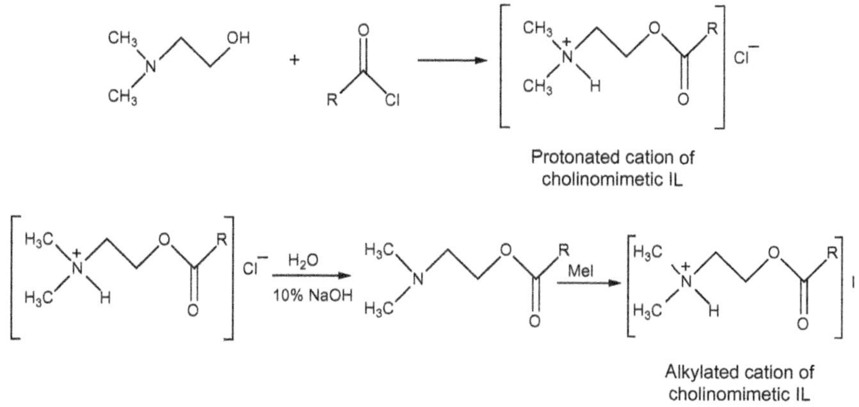

Fig. 2 Synthesis of cholinomimetic ionic liquid [25].

3 Interactions between ionic liquid and biological macromolecules

Although IL has got the status of 'green chemical' due to its thermal and chemical stability in a specified range of temperature, thorough investigation found quite a significant level of toxicity in IL toward biomolecules. This, in turn, confirmed a significant interaction of IL with biological molecules as the extent of toxicity actually provides a measure of affinity between IL and biomolecules [26]. Throughout the last decade, several experiments were executed to explore the microscopic events (at the molecular level) in the IL-biomolecule interaction [27–29]. Different types of biological system and IL have been considered for understanding the possible interactions. It was observed that ILs were able to

(1) diffuse through the bio-membranes and destroy them at moderate to high extent;
(2) provide stability to protein and enzyme molecules and influence their activity;
(3) stabilize complex DNA structures;
(4) act as solvent for the dissolution of polysaccharides;
(5) formulate drug delivery system (DDS) for targeted drug delivery without destroying non-affected cells.

The above observations [4] made IL an exciting and potential material to be used in different biomedical applications. Before going into the detail about biological applications, let's visualize the microscopic mechanism in the IL-biomolecule interaction as that may ease the understanding of the application in a large spectrum. Here, we limit our discussion within the range of four broad classes of biomolecules, namely (i) membrane (ii) nucleic acids (DNA and RNA) (iii) protein and (iv) carbohydrate, as these are the most important molecules in the biological system, thereby relevant to the scope of this chapter.

3.1 Interaction of ionic liquid with biomembrane

Biomembranes are of prime importance as any kind of interaction with the cell mostly gets initiated with penetration through cell membrane. Thus, the nature of interaction of IL with biomembrane was investigated in detail. A large interest was put on the interaction of IL with phospholipid which is one of the main constituent of biomembrane. The aqueous environment, in the surroundings of biomembrane, consists of different ions which significantly influence the functional behavior of the membrane. Thus the interaction of organic salt, like IL with their huge structural variety, was thought to be quite interesting. One of the first instances of investigating IL-biomembrane interaction was reported by Evans [30–32]. The experiments were performed with RTILs, mainly consist of pyrrolidinium and alkyl imidazolium ions. It was reported that in all cases the interaction led to complete disruption of the lipid bilayers. It was observed that at low concentration and short carbon chain length of alkyl imidazolium cation $[C_n\text{mim}]^+$ (where n is the number of carbon atoms and 'mim' stands for methyl imidazolium unit), the phospholipid bilayer was damaged slowly through creation of holes in small areas within bilayer. On the other hand, high concentration and long alkyl chain resulted in immediate penetration through bilayer. The influence of the length of cationic tail confirmed the huge impact of the alkyl groups of imidazolium unit on the destruction of lipid bilayer. Moreover, since the alkyl chain (C_n) itself is a hydrophobic one, the attraction of alkyl imidazolium unit toward phospholipid molecules indicates the presence of dispersion forces as a significant parameter apart from the Coulomb forces which is common for all cations. Lee et al. [33] examined the influence of imidazolium IL on the stability of lipid bilayers by simulating Gramicidin A dimer, IL and phospholipid, at different salt (sodium chloride) concentration. It was observed that with increase in IL concentration, the thickness of the bilayer was reduced leading to reduction in membrane curvature, which is shown in Fig. 3.

On the other hand, the role of anionic part of IL is not so clear although the dispersion forces along with Coulomb and steric interaction were supposed to play a key role in determining the fate of biomembrane in aqueous

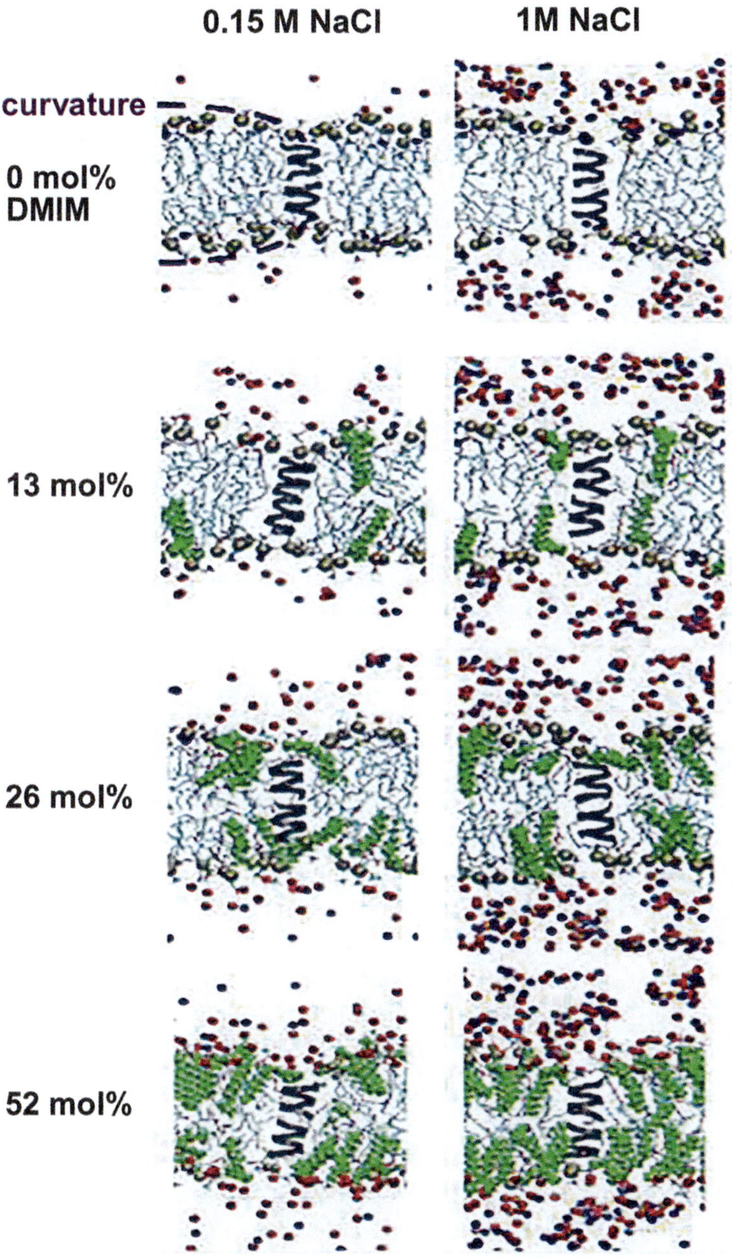

Fig. 3 Influence of ionic liquid concentration on the curvature of biomembrane at different salt concentration. *(Reprinted from H. Lee, S.M. Kim, et al., Effects of imidazolium-based ionic liquids on the stability and dynamics of gramicidin A and lipid bilayers at different salt concentrations, J. Mol. Graph. Model. 61 (2015) 53–60, with permission from Elsevier.)*

IL solution. The absorption of cations by phospholipid bilayer was also observed by computer simulations complementing the facts established by the different chemical and physical experiments to predict the character of IL-phospholipid systems [34–36]. Bhattacharya et al. [37] investigated thermodynamic parameters related to the interaction between IL and lipid monolayer. The lipid layer was used as a model system for the experimental purposes. The influence and the mechanism of penetration of the imidazolium-based IL (1-decyl-3-methylimidazolium tetrafluoroborate) into a lipid monolayer system of zwitterionic nature, was investigated in detail. The calculation of thermodynamic parameters like change in enthalpy (ΔH) and change in entropy (ΔS) showed disordering effect in the lipid layer through the interaction of IL molecules. Drücker et al. [38] investigated the interaction of alkyl imidazolium IL with lipid bilayer membrane model which represents a replica of a cell membrane with almost similar curvature as that of cell membrane. It was observed that the alkyl chain length play a significant role to determine the extent of interaction of the IL with lipid bilayers. IL with a longer alkyl chain was able to penetrate through bilayers without destroying the bilayer structure, whereas that of shorter alkyl chain resulted in complete disintegration of the bilayers. The model bilayer system was varied from neutral to negatively charged head-groups in order to observe the influence of head-groups of lipid layer on the charged IL molecules. It was found that the impact of charged IL moiety was highest for charged (negative) bilayer system as compared to the neutral one which might be attributed to the electrostatic interaction.

Thus, it can be concluded that prior to the penetration through the membrane, electrostatic interactions between IL and lipid head-group plays a significant role. On the other hand, the mechanism and extent of membrane penetration is highly controlled by the alkyl chain length or hydrophobicity of the organic part of IL moiety.

3.2 Interaction of ionic liquid with protein

The word 'Protein' originates from a Greek word 'prōteios' meaning 'primary'. Proteins provide the structural framework of cells and tissues. They are essential for transport of nutrients and to repair cells and form new ones. The building block of protein unit is the amino acid molecules. Linking of amino acids leads to polypeptide chain which ultimately folds into a defined three-dimensional structure of the protein. Thus any interaction with a protein molecule means interacting with its constituent amino acids and associated functionalities.

Simple protein is composed only of amino acid residues whereas conjugated protein, in addition to the polypeptide chain, contains other non-amino acid components known as prosthetic group (metal ions, lipids, carbohydrates etc.). The complexity in structural features and behavior is thus quite large for conjugated protein. With lots of article predicting IL as next generation solvent to replace conventional aqueous and organic solvent, the protein-IL system is expected to offer exciting outcome owing to the interaction of protein molecule with cation and anion of IL moiety. However, it is difficult to form a unique frame to explain the IL/protein interaction due to the variety of protein structures as well as various kinds of IL which are already used in experiments and keep coming through new publications. There are quite a few reports on RTIL-protein interaction [39, 40] although prediction of the nature of interaction between a specific RTIL and a given protein is not so easy. Apparently from protein point of view, it is the sequence of amino acids and secondary structure of the protein, which play an important role to decide the nature and extent of interaction with IL. As the net charge of the protein molecule depends on the pH of the medium and isoelectric point of the protein itself, both the cation and anion of an IL can play dominant role during IL-protein interaction. Molecular simulation of protein in aqueous IL solution clearly suggested that it is the higher charge density of the IL molecule that is preferably attracted to the opposite charge of the amino acid [41]. The attraction of different protein surfaces, ranging from non-polar to highly polar, was nicely explained by Schröder et al. [29], which described that the hydrophobic tail of IL unit interacts with non-polar protein surface whereas the charged surface of the protein molecules is attracted by the respective opposite charge of the constituent ions of the IL.

Several studies have been published on the solubility of protein in RTILs and their application in non-aqueous bio-catalytic system [42, 43]. In most of the cases, the results showed fairly low solubility of the protein in RTIL [44], especially in case of hydrophobic RTILs. For the hydrophilic RTILs, they mostly remain in their hydrated form which actually promotes the dissolution of protein molecules [45]. It was also found that ILs can influence the phenomenon of protein folding and crystallization as well [29].

3.3 Interaction of ionic liquid with carbohydrate

Carbohydrate is a biomolecule with molecular formulae $C_x(H_2O)_y$. The word 'saccharide' came from a Greek word, meaning 'sugar'. The simplest carbohydrate is called monosaccharide, which possesses one sugar unit in its formulae. The monosaccharides are capable of polymerization to different degree resulting oligosaccharides and polysaccharides. These molecules are of intense biological interest in living organisms due to their key roles in conversion of energy from one form to another as well as in the energy-storage purpose. Monosaccharides and oligosaccharides (low molecular weight compounds) are mostly soluble in water and remain insoluble in most of the common organic solvents. On the other hand, polysaccharides (high molecular weight compounds) remain insoluble in water as well as in most of the organic solvents. The reason for such poor solubility results from the presence of strong intra-molecular hydrogen bonds, forming a network structure which imparts crystallinity in its structure and inhibits solvent penetration. The inherent structural stability of the large polysaccharide compound made them important structural materials for different applications [46] ranging from pharmaceutical and food applications to the area of nanotechnology and nanofabrication.

The interest in carbohydrate-IL interaction received great attention from the work of Swatloski et al. [47, 48] on the dissolution of cellulose in RTIL and from the work on chitin-IL conjugates by Sakurai et al. [49]. The dissolution of cellulose in RTILs was proved to be very useful as it provided a significant method for cellulose processing, prior to its conversion into simple carbohydrates through hydrolysis. It was observed that the polarizable anions (chloride, dimethylphosphate, acetate, etc.), if present in the IL structure, were able to form hydrogen bonds with hydrogens of carbohydrates. This, in turn, reduced the dominant fraction of intra-molecular hydrogen bonding [47], thereby able to dissolve cellulose whereas RTIL with anions such as tetrafluoroborate or hexamethylphosphate, which are incapable of hydrogen-bond formation, was found to be unable to dissolve cellulose. The cationic moieties of the IL used to provide a supporting role [50] by interacting with hydrocarbon backbone of the polysaccharide unit via dispersion forces and sometimes also through hydrogen bonds. The mechanism of cellulose dissolution in IL, in presence of dimethyl sulfoxide was nicely described by Li et al. [51]. The anion from IL got involved in the formation of hydrogen bond with hydroxyl groups of cellulose thereby destructing the intra-molecular hydrogen bonds. In consequence of that, the negatively charged cellulose got surrounded by IL-cations resulting in increase in steric crowding via alkyl-chain repulsion within the IL cationic moieties. The overall effect is the extreme reduction of solvophobic interaction within cellulose structure which resulted in excellent solubility of cellulose in IL-DMSO medium.

It is worth mentioning that the nature of chemical structure of carbohydrate compounds and their conjugates are quite complicated in nature. The influence of IL on carbohydrate-conjugated systems and ternary systems like carbohydrate/RTIL/water is well under investigation and it is obvious that presence of IL in these systems can provide improved outcome in terms of their application in different areas such as separation, food science, pharmaceutical sectors [27] etc.

3.4 Interaction of ionic liquid with nucleic acid

Nucleic acids are basically polymeric nucleotides (or polynucleotides) whose monomer units (nucleotides) consist of nitrogen containing heterocyclic base, a pentose sugar with an attached phosphate group at a specific carbon. These are found in plenty in all living organism, where they used to create and encode the information of every living cell on earth. Their ultimate role is to transmit and express the information that has already been coded from the parent generation, to the next generation of each living organism. Due to the presence of anionic moiety (phosphate group) along the polymer chain of polynucleotide, nucleic acid such as DNA (deoxyribonucleic acid) or RNA (ribonucleic acid) can be considered as anionic polyelectrolytes with reasonably high linear charge. Hence, from both structural and functional point of view, nucleic acids are considered as most complicated of all biomolecules. Also, due to the presence of conformational polymorphism, nucleic acids are quite demanding molecules in the field of nanotechnology. However, their stability for long duration has been a matter of concern due to lack of proper solvent. In this context, owing to the structural characteristics and variety of constituent ions, IL was found to be quite an interesting compound to conjugate or interact with DNA or RNA. In fact, it was observed that hydrated ILs, based on choline cation or deep eutectic solvents (DES) were quite effective to serve the purpose of a potential solvent for nucleic acids [52]. Until now, most of the published articles on the interaction between nucleic acid and IL are based on DNA as the outcome fits equally well with RNA and also because the RNA is observed to be intrinsically less stable as compared to DNA thereby experiments with DNA seems to be facile. In general, the extent of interaction between DNA and simple inorganic salts is determined by the ionic interaction between phosphate groups of DNA and nature of the cation. But for the IL based

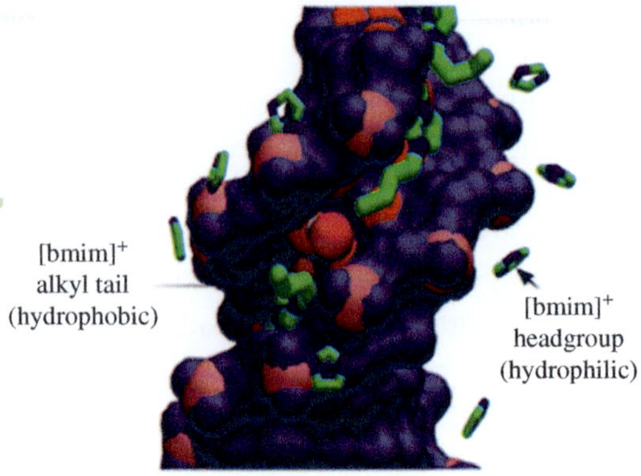

Fig. 4 Snapshot of solvation of DNA in 80 wt% [bmim][lactate] solution. It shows the interaction of $[bmim]^+$ cation with different regions of DNA. Both the polar head and alkyl tails are shown to penetrate into the DNA grooves. *(Reprinted from A. Garai, D. Ghoshdastidar, et al., Ionic liquids make DNA rigid, J. Chem. Phys. 149 (4) (2018) 045104, with the permission of AIP Publishing.)*

system, it is the binding of the ions of IL with the major or minor groove of nucleic acid molecules, which determines the extent of interaction [53]. Garai et al. [54] nicely illustrated the interaction of DNA with ionic liquid which is shown in Fig. 4.

On dealing with the nucleic acid-IL interaction, one of the major concerns was the stability of nucleic acids in IL medium. Vijayaraghavan et al. [55] investigated the stability and chemical integrity of DNA double helix in choline based RTIL-water system. Increase of DNA lifetime along with the thermal stability was reported. The attachment of RTIL cation to the DNA molecule (which is anionic in nature due to the presence of phosphate group) was found to be controlled by dispersion forces, size of cation and by the ability to form hydrogen bonding. Attachment of organic cations to DNA was observed to be effective to screen the DNA molecule from hydrolytic attack [55].

Presence of IL could also influence the stability of base pairs of DNA molecule. In presence of IL cationic moieties such as tetraalkylammonium [56] or choline [57] in physiological buffered solution, it was found that the relative stability of the base pairs of DNA got reversed which might provide opportunities in the manipulation of nucleic acids and their conjugates for various applications.

It is obvious from the above discussion that IL imparts a huge impact on the structural as well as on the physicochemical properties of biomolecules.

Till now, the highest number of studies includes proteins whereas quite a significant research work has been done on nucleic acids and phospholipids. However, in the recent scenario, the strongest practical impact can be observed from the dissolution of polysaccharides like cellulose or chitin those are insoluble in conventional solvents. These polysaccharides belong to a major portion (in mass) of biomass and hence dissolution of these materials in IL solvents might provide a facile method for the processing of polysaccharides and production of bio-fuels and biodegradable plastic materials. On the other hand, recent progress in the field of biocompatible ILs [58] might be effective to prepare potential biocompatible IL/biomolecule systems which may find applications in the field of biomedicine, food science or in drug delivery systems. In the following sections, we are going to discuss the applications of ILs in the different areas of biomedical field.

4 Biological activity of ionic liquid

Before the elaborate discussion on the biomedical application of IL, one should have a look on the bioavailability of ILs and their solubility or interaction with water molecules as most of the bio-systems reside in aqueous environment. A detail study on the influence of IL structure (cations with isobutyl or butyl substituent and anion with high fluorine content) on water solubility of the ILs showed that among the different cationic moieties, imidazolium based cations had the maximum solubility in water whereas piperidinium based ILs had the lowest. The pyrrolidinium and pyridinium based ILs had the intermediate effects. The size of the ions as well as the aromaticity of the ionic moiety had a strong influence on the extent of solubility of ILs in water [59]. A theoretical study on the influence of anionic part of IL molecule on the interaction of IL with water was published by Wang et al. [60]. The study suggested that both the cation (here imidazolium cation) and small anion such as chloride, bromide or tetrafluoroborate had a strong interaction with water molecules. On the other hand, in case of fluorine containing hydrophobic anion like hexafluorophosphate, no such significant interaction was observed with the water molecules. Hydrogen-bond between IL-anion and hydrogens of water molecule also had a significant role as observed in the aqueous alkylimidazolium acetate system [61, 62].

An excellent work was published by Ohno et al. [45] on the hydrated IL system and its influence on the solubility of protein molecules. It was observed that the biological activity of ILs depends on the state of hydration of the IL under consideration. The author proposed a number which they named as 'critical hydration number'. In most of the cases, this number was found to be 7, i.e., seven molecules of water for each IL molecule. This happened to be the threshold value at which a dramatic change in biological activity of the IL was observed which was not, at all, dependent on the type of ions constituting the IL salt. The experiments were performed with aqueous solution of 23 different ILs and the hydration number for specific application like microbial growth medium, enzyme activity etc. was noted. Interestingly, in each application, there was a threshold value of water molecule surrounding the IL for the maximum activity and the value is nearly seven. It was an important finding as tuning IL ions might influence its interaction with water which in turn, may influence the application of the hydrated IL in the concerned application.

In the following sections, we are going to focus on the utilization of ILs on some highly important biomedical applications.

4.1 Ionic liquid in antimicrobial activity

Potential of IL in antimicrobial activity came into consideration after the publication of articles showing inhibition of bacterial and fungal growth by ILs. All of a sudden, the ILs became quite an attentive material to the biotechnologists. Experiments were performed with various types of IL such as imidazolium, pyridinium, piperidinium, pyrrolidinium, ammonium etc. in order to observe the extent of inhibition of pathogenic and nonpathogenic bacterial and fungal growth [63–66] as well as to find demerits of the new IL-based antimicrobial agent, if any. The results were positive and quite advantageous as the IL offered numerous possibilities of structural variation and showed high activity toward resistant pathogenic bacterial strains. Fig. 5 shows the toxicity of imidazolium and pyridinium ILs with different alkyl chain length [63] to a broad range of micro-organism. It clearly suggested that the increase in length of the alkyl chain resulted in an increase in toxicity.

It was also obvious that the application of ILs in biotechnology might get hindered due to such significant toxicity toward microorganisms. However, this can be overcome by finding any suitable IL-resistant bacterium [67].

IL as antimicrobial agents tuned out to be an important finding from medicinal aspects. Drugs, in conjugation with antimicrobial IL, was expected to provide better outcome compared to the conventional drug formulation. Attempts have been made to synthesize antimicrobial IL and to compare the activity of the IL-based drug with the conventional antimicrobial drug. Carson et al. [68] evaluated the activity of a alkylimidazolium chloride IL as anti-biofilm against a microbial pathogens. A comparative

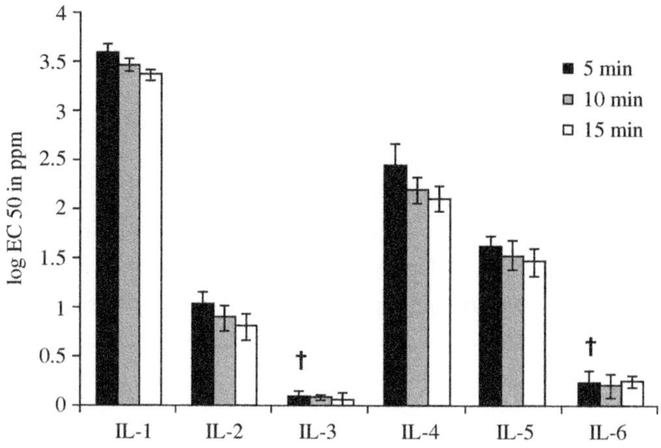

IL-1: 1-butyl-3-methyl imidazolium bromide
IL-2: 1-hexyl-3-methyl imidazolium bromide
IL-3: 1-octyl-3-methyl imidazolium bromide
IL-4: 1-butyl-3-methyl pyridinium bromide
IL-5: 1-hexyl-3-methyl pyridinium bromide
IL-6: 1-octyl-3-methyl pyridinium bromide

Fig. 5 Toxicity scale of imidazolium and pyridinium ionic liquids, at different time periods, expressed in terms of average logarithm of EC-50 value (effective concentration at 50%). † denotes insignificant difference of toxicity level for IL-3 and IL-6. *(Reprinted from K.M. Docherty, J.C.F. Kulpa, Toxicity and antimicrobial activity of imidazolium and pyridinium ionic liquids, Green Chem. 7 (4) (2005) 185–189, with permission from Royal Society of Chemistry.)*

analysis of the efficiency of antimicrobial IL was made against normal biofilms and planktonic bacteria. It was found that these ILs could offer a strong and broad spectrum anti-biofilm activity. Cole et al. [69] synthesized and investigated antibacterial nature of biocidal ILs where both imidazolium and pyridinium cation were used (separately) as the cationic part and an β-lactam unit was employed as anion. It was targeted to be used as IL-based aminopenicillin class of drugs (AMP-IL) and to make a comparative study with the efficiency of quaternary halide-based ILs. The result demonstrated that the extent of antibacterial effect on the pathogen was higher in case of Amp-ILs in majority of experiments. Moreover, the performance of Amp-ILs was even better than commercially available sodium ampicillin salt, in more than 90% of the experimental cases. About 43 times improved antibacterial activity was reported for Amp-IL compared to that of sodium ampicillin. Quite a few articles were published [70–72], utilizing new class of IL as active antimicrobial agent, prepared by exploring the tunable characteristics of IL ions. The hydroxide ion of an alkylimidazolium hydroxide IL was exchanged by equimolar amount of fumarate anion. The fumarate based IL showed good antimicrobial activity and was reported to be used in medicinal field [70]. Several other studies were reported in this context like IL based on ammonium cation as antibacterial and antifungal agent [72], hydroxylammonium cation against human pathogens [71], phosphonium cation to treat ocular pathogens [73], N-cinnamyl imidazole with different alkyl chain length against bacteria, responsible for skin and soft tissue infections (SSTIs) [74] etc. However, increase in IL tolerance in microorganism is always a matter of concern for the medicinal chemists. Moreover, likewise the conventional antimicrobial agents, appearance of resistant strains is obvious. In fact, it has been found in some cases where an IL-based antimicrobial agent induced an expression of genes having characteristic of antibiotic resistance [75]. Nevertheless, the antibacterial activity of the ILs was quite promising in many of the studies so far which is exciting for medicinal aspects of new generation ILs. It was demonstrated that there is a strong molecular interaction between cell membrane and ILs which can be attributed to the high antimicrobial impact of ILs within the cell [76]. Further experiment with a series of different anions, attached with imidazolium IL with short alkyl substituent, suggested that the toxicity toward cell membrane usually depends on the chaotropicity of the anionic part of ILs [77]. Apart from the active protection against human pathogens, the antimicrobial effect of IL has been utilized for protecting paper [78] or any metal coated surfaces [79] as well to inhibit microbial growth on their surface.

4.2 Ionic liquid as anticancer agent

Cancer has been the most concerned disease in human life for the last few decades and it has become a major cause of mortality. As per as the standard approach of treatment is concerned till now, chemotherapy, radiation and immunotherapy are the main pathways to treat the cancer cells despite their toxicity [80]. Thus, the development of new anti-tumor or anti-cancer agents with effective tumor-resistant mechanism and nearly zero toxicity to healthy cells, are the most important objectives in medicinal or pharmaceutical field [81]. From the last few decades, especially after the development of 3rd generation ILs with potential in biomedical applications, IL has been tested as a potential anti-cancer agent. With an advantage of controlling toxicity by tuning their structural features, IL was supposed to be the material of interest, with possibility of tailoring their characteristics as per demand of the specific treatment. Numerous studies on cytotoxicity of ILs on human cancer cell lines and human health cell lines have been reported [81–83]. Dias et al. [84] analyzed the toxicity data of ILs, which confirmed the cytotoxicity of ILs toward various cell line. It was observed that the factors such as alkyl side chain of the cationic moiety of IL, functionality present on the side chain attached to IL ions and overall nature of the constituent ions play a crucial role in determining the toxicity of IL toward a specific cell [85].

Experiments were performed with different cation head groups like ammonium, imidazolium, phosphonium and pyridinium with short to long side chains, associated with various anions. Toxicity was found to be increased with increase in length of the alkyl chain, attached with IL cation, thereby decreasing cell viability [86,87]. Malhotra et al. demonstrated toxicity and anti-tumor activity of imidazolium based ILs [83] with the help of structure-activity relationship. It was observed that the length of the alkyl substituent of imidazolium unit had a crucial role in the overall activity of ILs under investigation. A similar kind of experiment was executed with IL, based on substituted ammonium and phosphonium cation [81]. A similar trend was observed regarding the influence of alkyl substituent and length of the alkyl chain on anti-tumor activity. The phosphonium-based IL was found to be more active as compared to ammonium based IL.

Presence of functional groups in the side chain of IL moieties also had a profound effect on the toxicity of IL compounds. Generally, the polar functionalities, e.g., hydroxyl group results in decrease in toxicity of ILs and it has been confirmed by Frade et al. [88]. Thus, any non-polar functionality in the alkyl side chain of IL moieties is expected to be effective to impart significant anti-tumor activity.

The polarity of IL cations also has significant effect, as observed in several studies. Wang et al. investigated the influence of various cations (ammonium, imidazolium, pyridinium, phosphonium, choline etc.) on the IL-toxicity. The study reported that the toxicity of pyridinium and imidazolium cations, which are aromatic in nature, were much higher compared to that of choline or ammonium compounds which is non-aromatic and polar in nature [89]. Kumar et al. [90] also supported the observation made by Wang et al. [89] regarding the higher toxicity of the aromatic cation-based IL like imidazolium or pyridinium toward cancer cell, as compared to the non-aromatic and hydrophilic IL compounds (pyrrolidinium and piperidinium).

On the other hand, the influence of anionic part of IL compounds is quite diverse in nature regarding their effect on the toxicity against human cancer cell lines [86]. Thus, a general trend is rather difficult to find. Moreover, the detail mechanism displayed by IL within the affected cells and healthy cells is still unknown and under investigation. Exploring the microscopic events within the cancer cells along with finding suitably designed biocompatible IL may ease the development of IL based anti-cancer drugs in future.

4.2.1 Drug formulation: Challenge and way to overcome

The scientists encounter difficulties in synthesizing new drug molecules, mainly due to their low solubility, lack of biocompatibility, instability and polymorphic conversion. The efficiency of any drug molecule primarily depends on its solubility in body liquid. Poor solubility often leads to lower dissolution and absorption rate, which in turn, demands higher doses for targeting the affected cell [91,92]. On the other hand, most of the drugs are available in crystal form and hence the existence of different polymorphic form of the drug cannot be avoided. In such cases, apart from the variation in physical characteristics, the polymorphs can undergo unpredictable interchanges [93]. These make the DDS formulation a challenging task for the biochemists.

Several articles and reviews have been published on various DDS formulation techniques. One of the popular methods includes the formation of drug salts and derivatives or prodrugs [94]. An effective prodrug should have high water solubility and stability. The prodrugs easily permeate through cell membranes and remain stable in hydrolytic environment during absorption [95]. Another well-known method to inhibit the phenomenon of polymorphism is the solution shearing method [96]. Investigation on different inorganic compounds, amino acids and pharmaceutical compounds suggested that the controlling factor for controlled crystallization techniques depends on system under investigation (i.e., compound, solvent etc.) and method of crystallization (i.e., cooling, reactive or non-solvent crystallization) [97]. Besides, attachment of any functional group to the drug molecule, which can help the drug derivative for complete dissolution in body liquid, also provides an efficient pathway for increasing solubility of a drug molecule. Examples of such functional groups include amines or cyclic amines. This could also be used for salt formation of the concerned drug molecule. Amides or hydroxyl groups can be used in case of requirement of neutral solubilizing groups [98]. Apart from above methods, micellization of drug molecules via different heterogeneous emulsion system [99] or selection of proper solvents [100] were also found to be suitable in different types of DDS formulation.

Until now, majority of the drugs are available in solid state in salt form [101] except few liquid formulations [102]. Hence, the development of active 'liquid salts' is of high necessity as it can eliminate the phenomenon of polymorphism as well as increase the solubility of the drug itself. In this context, due to amazing feature like tunable structure, biocompatibility (3rd generation IL) and availability of the precursors materials, ionic liquid was expected to be an excellent material to be used in DDS formulation. Through the huge structural variation of ILs, It was possible to synthesize IL with excellent aqueous solubility, enhanced rate of absorption and rate of dissolution and ability to target the specific cells without affecting the healthy cells. Especially after the development of RTIL with medicinal potential, the experiment with IL as DDS or as a component for DDS formulation has observed a spark in the scientific community. Few of such systems will be discussed in the following section.

4.3 Ionic liquid in drug delivery system

A very important criteria for formulating DDS is that the drug molecules must be completely soluble in solvent and the choice of solvent should be based on its efficiency to prevent any precipitation or aggregation of the active pharmaceutical ingredient (API). In general, various modes of utilization of IL have been reported [103] such as IL as solvent or non-solvent for synthesis of API in crystalline form, IL as dispersing medium, IL as emulsifiers or IL-conjugates for DDS formulation. Investigation of the influence of IL structure on specific API showed that both the polarity as well as the alkyl chain attached with imidazolium or phosphonium cation had a significant impact on the solubility of API in ILs [104,105]. Cations like imidazolium, ammonium, phosphonium have been employed to synthesize IL for DDS formulation. It was observed that IL can be utilized in the formulation of broad classes of DDS.

The synthesis of non-steroidal anti-inflammatory drug (NSAID) in IL solvent was reported by Earle et al. [106] in 2000. In this work, pravadoline drug was synthesized using [C_4mim][PF_6] as solvent. It opened up the era of utilizing IL as efficient medium for the formulation of DDS. Imidazolium based IL was used most frequently compared to other kinds of IL. Several antiviral [82,107], antimicrobial [108–110], anticancer [111] drugs, contrast agents for molecular imaging and therapeutic agents [112] have been efficiently synthesized using IL as solvent.

Dispersion of nanomaterials with high biological interest has always been a major challenge due to natural tendency of the nanomaterials to form agglomerates and also due to non-availability of green and biocompatible dispersion medium. In this context, IL was expected to be an excellent dispersing medium to fulfill the concerned requirements. Fileti and Chaban [113] used aqueous solution of an imidazolium cation based IL with fluoroborate anion for dispersing fullerene homogeneously. Presence of fluorine in IL structure imparted enough hydrophobicity to disperse the hydrophobic material (C_{60} fullerene) effectively. The IL-fullerene composite showed good potential to be used in biomedical application. Soni et al. [114] used the same IL to observe the behavior of enzyme in presence of IL. It was observed that the enzyme (phytase) was self-assembled into active enzymatic nanoparticles in the IL. These nanoparticles were utilized in the synthesis of a protein based DDS for multipurpose activity.

Beside the use of IL as solvent or dispersing medium for DDS formulation, the amphiphilic IL-stabilized emulsions were also found to act as potential drug carriers [115,116]. Efficient delivery of the drug lidocaine hydrochloride was reported with the help of amphiphilic IL, synthesized by imidazolium halide with long alkyl chain (C_n where $n = 12–16$). The amphiphilic ILs were self-assembled to form nanoparticles of IL in water, which adsorbed the drug molecules onto its surface [117,118]. Both direct and inverse emulsions, in presence of amphiphilic IL, were found to be effective in DDS formulation [119]. Cholinium IL-based micro-emulsion [118] showed stability for long duration at ambient temperature and was found to possess antimicrobial activity in addition to play the role of a permeation enhancer. Furthermore, the cholinium-based ILs performed quite well in the extraction and partitioning of drug molecules, which was reported after examining with several popular drugs [120].

4.4 Ionic liquid in transdermal drug delivery system

Transdermal drug delivery (TDD) is relatively a trouble-free process of drug administration through skin of the body [121–123]. In this process, the penetration of drug molecules occurs through the outermost layer of the skin which is known as stratum corneum (Fig. 6). It is followed by crossing the epidermis and dermis layer where no accumulation of drug molecules

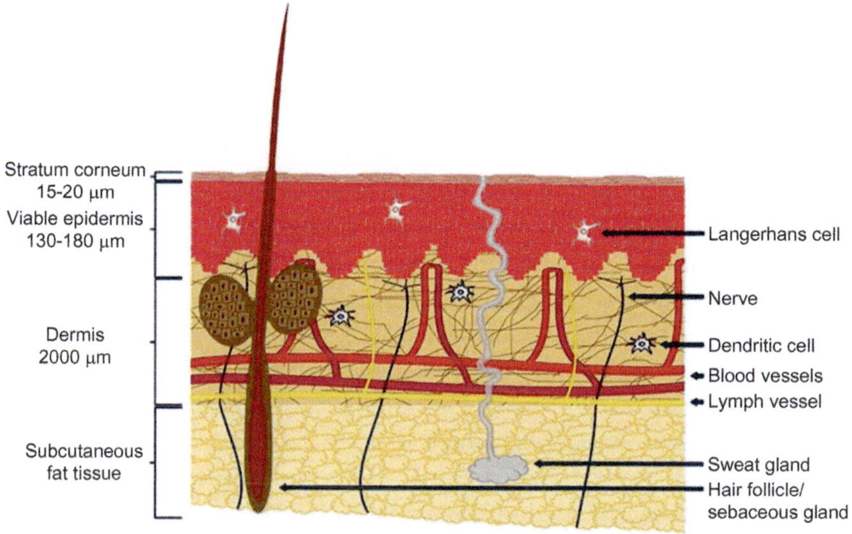

Fig. 6 Structural layers of skin. *(Reprinted from K. van der Maaden, W. Jiskoot, et al., Microneedle technologies for (trans)dermal drug and vaccine delivery, J. Control. Release 161 (2) (2012) 645–655, with permission from Elsevier.)*

usually happens. Finally the drug molecules are absorbed by the targeted cell via dermal micro-circulation [124].

TDD offers many advantages compared to the conventional drug delivery methods [125,126]. The non-invasive alternative method of drug injection removes quite a number of discomfort areas of the treatment arising from most of the drug delivery methods [122]. Due to large open surface area of the skin there are many available places to deliver drug via TDD method [123]. Moreover, reduction of dosing frequencies makes it compliance for patient and is also suitable for unconscious patients [127]. Due to the necessity for an low-cost and non-invasive pathway of vaccination, mostly in the downtrodden society of the world [128,129], it has been emphasized to develop easy, needle-free delivery method like TDD. However, the major challenge associated with TDD is the nearly impermeable barrier created by the stratum corneum having 79–90% protein content and 5–15% lipid content with variances in thickness in different parts of the body [130]. Thus, Permeation of API is historically the biggest challenge in transdermal drug delivery system (TDDS). Some strategies have already been taken with success to enhance the permeability of the membrane and thereby penetrate the barrier and deliver the drug through TDD pathway

[130]. In this context, inspired by the synthesis of novel biodegradable ILs and favorable cytotoxic characteristics, ILs were expected to be a possible solution to overcome these challenges. Already quite a number of articles and reviews have been published showing utilization of IL in TDDS as chemical permeation enhancers, drug modifiers, AIP etc. Still there are enough scope in this field to develop an IL based TDDS. Here we only highlight few of them to briefly introduce this emerging field to the readers.

Moniruzzaman et al. [131] published an article showing drug delivery through TDD method using IL. A microemulsion system assisted by IL molecules was prepared and was stabilized by combination of two surfactants namely, sorbitan laurate and polyoxyethylene sorbitan monooleate, both of which are non-toxic nature. The solubility of the drug (Fig. 7A) was tested in different ILs among which dimethylimidazolium dimethylphosphate (Fig. 7B) was selected due to its high drug dissolution efficiency. The IL-assisted microemulsion was applied successfully for the delivery of acyclovir (drug which is a sparingly soluble or insoluble in water or conventional organic solvents) using TDD method.

Zakrewsky et al. [132] used choline-geranate as IL component to enhance the delivery of antibiotic across the skin layers for treating a biofilm pathogen. The antimicrobial activity was found to be quite satisfactory and the extent of toxicity toward skin and epithelial cells was very low as well. It was also observed that the presence of IL could enhance the permeation capability of the drug molecule delivered via TDD method. The chemical structure of IL showed significant influence on the degree of permeation [133,134]. It was demonstrated that the hydrophilic ILs could open the rigid junctions within the stratum corneum which promotes para-cellular transport via fluidization of protein and lipid regions. On the other hand, the

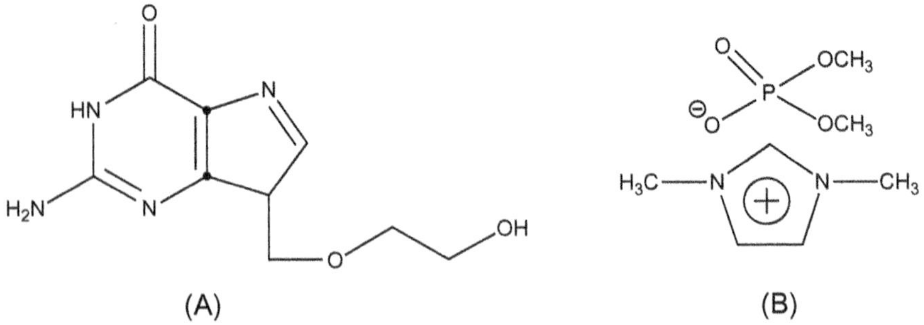

Fig. 7 Structure of (A) drug molecule, acyclovir and (B) dimethylimidazolium dimethylphosphate, ionic liquid used in the article [131].

hydrophobic ILs could create channels into the epithelial membrane which helps trans-cellular transport in the lipid regions [135]. Throughout the last few years, several articles [136] were published in this domain which further support the ability of suitably designed IL for the formulation of TDDS. One of the exciting outcome from the research on IL based TDDS was the transdermal delivery of some unlikely drug moieties such as protein [137], acyclovir [138] or nicardipine [139], those were difficult to deliver through conventional TDDS formulation. Thus, IL based TDDS offers a huge opening for an alternative drug molecules to be delivered effectively to the target cell.

Combination of IL with other permeation enhancer(s) was also successful to incorporate different kinds of API via co-solvency, which was beneficial for drug solubility. Chemical permeation enhancers such as lipid vehicles, alcohol, amines, terpenes or surfactants were successfully combined with IL to overcome the difficulties of drug penetration using TDDS [140,141]. Moreover, through the incorporation of API in IL, not only the permeation was enhanced but also the kinetics of transport could also be increased [142].

4.5 Ionic liquid as active pharmaceutical ingredient

The pharmaceutical industry usually faces two major challenges. One of them is related to its business aspect and the other is linked with the continuous need for new drugs. Hence, there are demands for novel scientific work, satisfying commercial requirements. Throughout the clinical trial process of new drugs, only <10% of the drugs used to qualify to be released in market for sell. Thus, the availability of necessary drugs is not as per requirements in the medicinal fields, raising the cost of the drugs which is a serious problem for the downtrodden people [143].

The structural features of the available drug show that a vast majority of them are administrated as solid salts as that increases the solubility of the drug in the drug carrier medium [144]. The activity or efficiency of a drug mainly depends on its chemical structure. The ion pair constituting the drug molecules, are highly responsible for the pharmacokinetic behavior of the specific drug. The processability and bioavailability of a drug mostly depend on the solubility of the salt form of the drug [145]. Even if the salts are shown to be effective enough as DDS, still the problem of transformation into other crystalline form through polymorphism exists at high extent and this not only impose problems for scientists but also can transform an effective drug

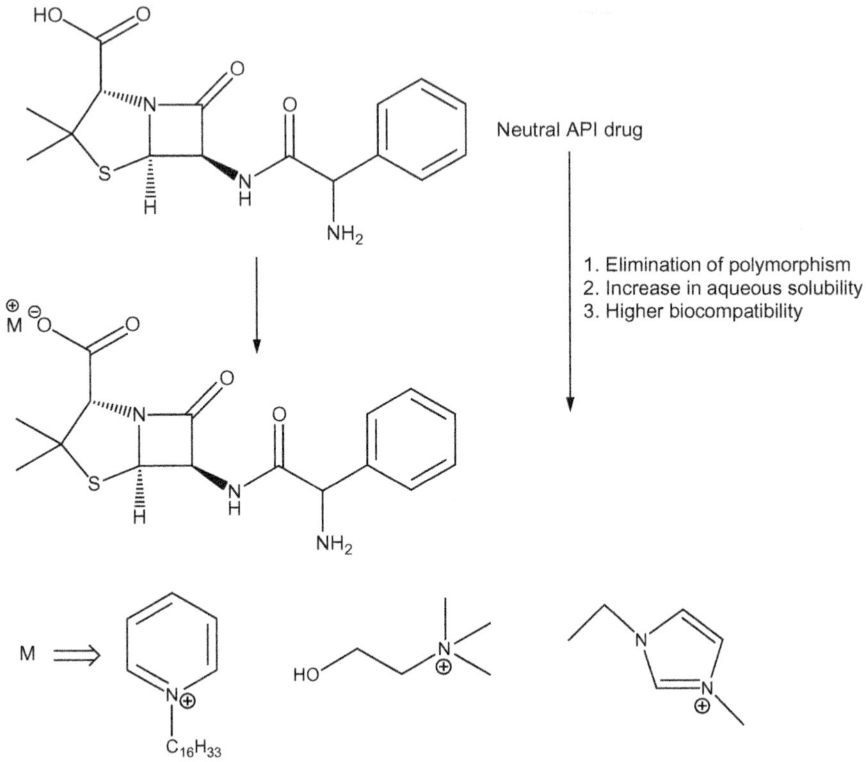

Fig. 8 Schematic representation of turning an active pharmaceutical ingredient into an ionic liquid molecule with associated advantage [149].

into a harmful one [143]. All drugs of crystalline form can suffer such polymorphism which demands a scientific innovation to overcome such drawback.

Now, turning a drug into an IL can provide an opportunity to increase the bioavailability of the drug and also to reduce the phenomenon of polymorphism [4]. Furthermore, the tunable characteristic of IL offer a possibility to modify the physicochemical and biological properties of the drug, which was reported as expected by several researchers [146–148]. Marrucho et al. [149] published a review article on IL in pharmaceutical applications where the utilization of IL as solvent for synthesis of API and advantage of converting API in the form of an IL (API-IL) have been discussed. The perspective of API-IL has been nicely represented, schematically by Marrucho et al. [149] (Fig. 8) which clearly suggests the impact of API-IL on the pharmaceutical field.

Also, API-IL with self-assembling ability was found to form API-vesicles where IL played the role of a carrier as well as drug molecule itself [150]. Such an API-vesicle was found to work efficiently as DDS. These API containing ILs belong to the class of 'third generation IL' [151]. Egorova et al. [4] summarized three possible pathways to incorporate API in IL system:

(1) Pathway I: using ionic interaction where either cation or anion can be API.

(2) Pathway II: using covalent interaction.

(3) Pathway III: using combination of ionic and covalent types of interaction [152].

Mostly it is the pathway I that is used to synthesize API-ILs. Davis et al. incorporated cation derived from miconazole, an antifungal drug [153], which was possibly the first API-IL that has been synthesized. Till then, quite a number of articles have been published regarding API-IL synthesis using above pathways. Apart from ionic or covalent interactions, hydrogen bonds can also influence the activity of a drug, as observed through the work of Wang et al. [154]. It was demonstrated that a complex formation between two drugs via hydrogen bonded interaction actually helped the trans-membrane transfer of drugs.

As a whole, API-IL offers a huge possibility of active drug formulation in liquid salt form which is expected to be an effective pharmaceutical strategy, considering all commercial aspects. Although one needs to be careful about the demerits of the strategy, as reported in few cases. The problem of polymorphism was found to exist in case of few API-IL systems [155]. Most importantly, due to the hygroscopicity of many ILs, unnecessary hydration was obvious in few cases which was undesired for an end product [156]. Hence, complete commercialization of API-IL as DDS may take longer than expected.

5 Biological activity of ionic liquid-protein conjugate

The interaction of protein molecules with IL has already been discussed in this chapter (see Section 3.2). There are huge numbers of article regarding IL-protein system directing various possible applications. Apart from the enzymatic function, proteins play the role of scaffolds for living cells or capable of forming drugs through conjugation with various biomolecules. Due to the biocompatible nature, proteins such as keratin, collagen etc. are quite useful in medical industries as well [157]. Generally, the dissolution of protein molecule, stability of its structure and folding/unfolding pattern are what is considered most while dealing with any protein molecules. It was reported that employment of IL could provide stabilization [158] thereby reducing protein aggregation and could provide controlled folding/unfolding process of the proteins [159]. IL could also enhance the activity of enzymes [160]. Numerous studies have undergone to dissolute structural proteins like keratin, silk, collagen etc. in different ILs. It was demonstrated that compared to the conventional solvents, the dissolution and processability of these proteins not only improved but also the toxicity and volatility of the solvents could be avoided by replacing them with suitable ILs [40].

Another area of for utilization of IL in protein world is its influence on enzymatic activity and selectivity [161]. Employment of IL in biosynthetic processes led to improved outcome, as reported by several groups [162,163]. The production of glucose dehydrogenase from Bacillus subtilis or reductase from Saccharomyces cerevisiae is done by Escherichia coli. It was observed that the activity of Escherichia coli was enhanced by the presence of an imidazolium based ILs containing fluorine based anion [164]. Most of the experiments, so far, have utilized imidazole or choline based IL for such enzymatic reaction.

6 Biological activity of ionic liquid-carbohydrate conjugate

Synthesis of new types of polysaccharide conjugate and their utilization in biomedical fields are of high interest to the modern day scientists. Novel carbohydrate conjugates or derivatives are of high demand in the pharmaceutical market firstly due to their biocompatibility, biodegradability and environmentally benign characteristics and more importantly due to available resources from biomass. These materials possess a strong potential to be used in various sector in biomedical fields from drug delivery to tissue engineering [165,166]. However, the problems associated with the natural polysaccharides like cellulose or chitosan is their lack of solubility in conventional organic solvents and difficulties in processability due to the presence of large number of hydrogen bonds (both intra- and intermolecular) in their polysaccharide chains (see Section 3.3). This restricts in shaping or to bring these materials in market as value-added materials. Different types of solvents (both protic or aprotic in nature) along with few additives have been employed to overcome such problems [167]. However, most of them usually possessed high level of toxicity, high cost, low dissolving capability and recovery of the solvents after the process was difficult. Hence, an eco-friendly and commercially affordable solvent with good recyclability is always on demand. In this context, ILs were expected to offer perfect combination of less toxicity and high dissolution capability with large variety of chemical composition due to tunable character of the ionic constituents. Moreover, invention of 3rd generation ILs also encouraged the scientists to utilize ILs as component for different biomedical applications. Thus, experiments began mainly with polysaccharides like cellulose or chitin and imidazole or pyridinium based IL as that is the most frequently used IL till now.

It was Graenacher [5] in 1934, who investigated the characteristics of cellulose in molten N-ethylpyridinium chloride. It was demonstrated that it is possible to dissolve cellulose in the pyridinium IL by adding some amount of nitrogen-containing base. Although, not much experiments were followed at that time. The attention was back in 2002 when Rogers et al. [48] synthesized an alkyl imidazolium based IL, which could dissolve cellulose very well. Later on, several groups started to work on solubility and processability of carbohydrates in ILs, followed by utilization of ILs in biopolymer processing. Different types of carbohydrates were possible to dissolve in ILs and were used to fabricate biocompatible materials, membranes or films. Also, the conjugate of bio-functional polysaccharides with other polymers or other biomolecules was possible to synthesize in presence of IL [168]. This allowed ILs to be used in carbohydrate processing for several sector of biomedical applications such as in dialysis or as a component for DDS formulation [169].

There are several publications on the dissolution and processing of cellulose in ILs [167,170], demonstrating possible biomedical applications of IL-cellulose system. A copolymer of cellulose and L-lactide was synthesized in an imidazolium based IL in microcrystalline form. The copolymer formed micelles which proved to be a highly efficient material as drug carrier [171]. The composite of cellulose with nanomaterials, e.g., single-walled carbon nanotube (SWCNT) was also prepared in imidazolium based IL. The biocompatible composite showed homogeneous dispersion of SWCNT in cellulose in the IL medium [172]. Employment of imidazolium salt [C_4mim][Cl] as dispersant in cellulose-hydroxyapatite composite helped dispersion of hydroxyapatite within cellulose matrix which finally turned out to be an efficient system for drug encapsulation [173].

Likewise cellulose, other polysaccharides such as keratin, chitosan etc. also found to form composite materials in ILs for controlled drug release system [174] and stimuli responsive nano-carriers [175]. Efficient biosensor was developed using chitosan-IL composite which was successfully used to entrap Hemoglobin as well. Development of potential anticancer and antimicrobial agents was also possible through utilization of IL-carbohydrate system [27,176].

Even though a promising outcome was observed through the constant studies on IL-carbohydrate system, their commercial application is still on the darker side due to high cost of the IL-precursor materials, used in these studies. This is a serious drawback in commercial domain. However, Chen et al. [177] could synthesize two inexpensive ILs with bisulfate anion which showed that cost-effective IL can also be prepared, thereby large-scale IL-based industrial production can be commercially affordable.

7 Poly (ionic liquid) in drug delivery

Poly (ionic liquid) (PIL) is a polyelectrolyte molecule that possesses IL salt in the repeating unit of the polymeric backbone. It usually possesses characteristic IL properties added with higher molecular weight that can be controlled during polymerization of IL. Thus PIL also possess properties like ionic conductivity, thermal stability etc. Due to such polymeric backbone structure, PIL can attain chemical structures, morphologies or strength of the material those are not possible to attain through IL only. From the last few years, PILs have got huge attention in the field of materials science and large number of articles, reviews and patents have been published showing synthesis of PIL (Fig. 9) with cations like pyridinium, pyrrolidinium, imidazolium etc. and application of PILs in various domains [178,179].

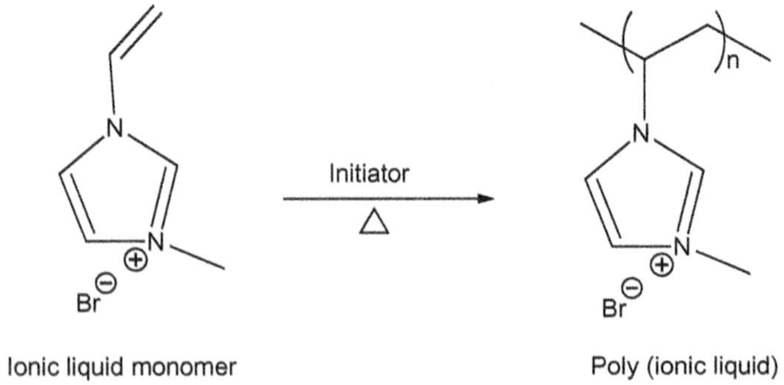

Fig. 9 Synthesis of imidazolium based poly (ionic liquid) from vinyl imidazolium monomer unit.

Inspired from the impact of IL-cations like imidazolium, pyridinium etc. on the antimicrobial activity, it was expected that PILs can also be effective in biomedical applications, offering more benefit due to the large number of IL moieties in their backbone and controlling the length of the polymer chain. Thus, PIL consisting cationic IL, like pyridinium or imidazolium based unit, was studied in detail. It was observed that due to the presence of many cationic units along the PIL molecule, the microorganisms, as soon as come in contact with PIL surface, get absorbed, leading to disruption of cell membrane, followed by bacterial cell death [180]. A broad spectrum of antimicrobial activity was found using imidazolium based PILs. Yan et al. synthesized PIL membranes having imidazolium moieties in repeating unit, followed by attachment of various amino acids unit by anion exchange method [181]. Excellent antimicrobial activity was demonstrated by the membrane. The membrane showed good biocompatibility with low cytotoxicity as well. Moreover, the membranes were recyclable without any major loss in its activity. Ye et al. [182] synthesized imidazolium based PIL-titanium dioxide composites by grafting PIL on the surface of titanium dioxide using ring-opening metathesis polymerizations. The composites showed good antibacterial activity as well as excellent bio-fouling properties. The mechanism of antibacterial activities of PIL is still not clear although analysis on the correlation between structure-antimicrobial activities suggested that a longer alkyl chain, attached with the imidazolium units and higher charge density in the repeating unit of the PIL can enhance the antibacterial properties of the PIL based materials [183]. Several quaternary ammonium based polymers were also found to possess excellent antimicrobial activities although they do not really belong to PIL category. Still the structural similarity with the PIL drives the researchers to utilize ammonium based polymers as antibacterial agents. Yuan et al. [184] prepared poly (*p*-phenylene vinylene) with quaternary ammonium unit to treat gram-positive and gram-negative bacteria. It was reported that such cationic polymers exhibit excellent potential to be used as a diagnostic materials for detection of pathogens. Not only cationic moieties of the IL monomers but the anionic part also had a tremendous effect on the overall antimicrobial efficiency of the polymeric materials. A significant influence on the antimicrobial and hemolytic activities was observed through modification of anionic unit of PIL molecules. A systematic analysis on a series of polycarbonates (with quaternary ammonium cation) having different anions such as chloride, acetate, trifluoroacetate, malonate etc. demonstrated different extent of antimicrobial activities which clearly indicated a strong influence of the counter ions in such antimicrobial aspects [185].

PILs were also found to be effective in formulating DDS. PIL nanoparticles were used as drug carrier, with doxorubicin as a model drug entity [186]. A pH-sensitive ionogel was synthesized using choline based PIL which was successfully used to formulate DDS for treating cancer-cells [187]. Temperature responsive DDS was possible to synthesize using PIL copolymer, synthesized using vinylimidazolium and N-isopropyl acrylamide as monomer unit. The PIL copolymer was successfully employed as DDS with excellent drug release capability [188]. The cytotoxicity of the material was found to be low as well. PIL composite with nanomaterials such as silica [189,190], gold [191] etc. was demonstrated to be quite effective as targeted DDS. The above-mentioned literature shows that modification or synthesis of designed PIL can offer multiple options for synthesizing antimicrobial agents or DDS. Thus, quite a large number of articles regarding biomedical applications of PIL are expected to be published in immediate future with high level of advancement in the DDS formulation.

8 Conclusion

Formulation of DDS with an aim for targeted drug delivery has always been a challenging task. Throughout the last two decades, extensive research work in the relevant field resulted opening of various pathways to overcome challenges in the biomedical field. Apart from the conventional method of oral consumption or transdermal delivery of drugs, newer pathways are flourishing, based on efficient drug carriers, nano-emulsion based drug delivery or micellar DDS etc. Thus, new generation smart materials are of high demand to meet the requirements for such new methodological developments in the pharmaceutical field.

To meet the demands of the scientists, working on development of modern DDS or API, new generation ILs may offer an exciting option to be used in drug formulation. The major advantage of using ILs is the possibility of facile and cost-effective synthesis of biocompatible ILs with a large number of ionic combinations that makes it a good component for formulating new DDS. The concept of utilizing IL itself as API offers great opportunities as well, specially to reduce the phenomenon of polymorphism. Moreover, incorporation of suitable functionalities by chemical modification of IL structure or tuning polarity of IL molecule by ion exchange can turn an IL into highly effective API.

The self-organizing characteristic of IL structure at nanoscale or microscale level provides nice opportunities to be utilized in the formulation of

DDS. IL with a long hydrophobic chain shows micellar structure in water above a specified concentration. IL monomers also showed self-organization property to form PIL particles with hydrophobic core in aqueous medium, stabilized by polar head of IL. These micro-structures may prove to be efficient drug carriers and a lots of research is already under progress to explore various possibilities of nanostructures IL as efficient drug carrier. The upcoming decade may observe some remarkable outcomes in the pharmaceutical and drug delivery domain, with ionic liquid as one of the main weapon to overcome challenges in the concerned field of study.

References

[1] S. Gabriel, J. Weiner, Ueber einige Abkömmlinge des Propylamins, Ber. Dtsch. Chem. Ges. 21 (2) (1888) 2669–2679.
[2] P. Walden, Molecular weights and electrical conductivity of several fused salts, Bull. Acad. Imper. Sci. (St. Petersburg) 1800 (1914).
[3] P.M. Dean, J.M. Pringle, et al., Structural analysis of low melting organic salts: perspectives on ionic liquids, Phys. Chem. Chem. Phys. 12 (32) (2010) 9144–9153.
[4] K.S. Egorova, E.G. Gordeev, et al., Biological activity of ionic liquids and their application in pharmaceutics and medicine, Chem. Rev. 117 (10) (2017) 7132–7189.
[5] C. Graenacher, Cellulose Solution, U.S.P. Office, Switzerland, 1934, pp. 1–4. 1943176.
[6] P.D.D. María, Ionic liquids: definition, applications, and context for biotransformations and organocatalysis, in: P.D.D. María (Ed.), Ionic Liquids in Biotransformations and Organocatalysis, John Wiley & Sons, Inc, 2012, pp. 1–14.
[7] J. Gorke, F. Srienc, et al., Toward advanced ionic liquids. Polar, enzyme-friendly solvents for biocatalysis, Biotechnol. Bioprocess Eng. 15 (1) (2010) 40–53.
[8] J.S. Wilkes, A short history of ionic liquids—from molten salts to neoteric solvents, Green Chem. 4 (2) (2002) 73–80.
[9] V.P. Ananikov, Characterization of molecular systems and monitoring of chemical reactions in ionic liquids by nuclear magnetic resonance spectroscopy, Chem. Rev. 111 (2) (2011) 418–454.
[10] V.H. Paschoal, L.F.O. Faria, et al., Vibrational spectroscopy of ionic liquids, Chem. Rev. 117 (10) (2017) 7053–7112.
[11] M.M. Seitkalieva, A.A. Grachev, et al., Nanoscale organization of ionic liquids and their interaction with peptides probed by 13C NMR spectroscopy, Tetrahedron 70 (36) (2014) 6075–6081.
[12] M. Tariq, M.G. Freire, et al., Surface tension of ionic liquids and ionic liquid solutions, Chem. Soc. Rev. 41 (2) (2012) 829–868.
[13] K. Dong, S. Zhang, et al., Understanding the hydrogen bonds in ionic liquids and their roles in properties and reactions, Chem. Commun. 52 (41) (2016) 6744–6764.
[14] M. Bühl, A. Chaumont, et al., Ab initio molecular dynamics of liquid 1,3-dimethylimidazolium chloride, J. Phys. Chem. B. 109 (39) (2005) 18591–18599.
[15] B. Docampo-Álvarez, V. Gómez-González, et al., The effect of alkyl chain length on the structure and thermodynamics of protic–aprotic ionic liquid mixtures: a molecular dynamics study, Phys. Chem. Chem. Phys. 20 (15) (2018) 9938–9949.
[16] I. Dinarès, C. Garcia de Miguel, et al., Imidazolium ionic liquids: a simple anion exchange protocol, Green Chem. 11 (10) (2009) 1507–1510.

[17] P. Yee, J.K. Shah, et al., State of hydrophobic and hydrophilic ionic liquids in aqueous solutions: are the ions fully dissociated? J. Phys. Chem. B. 117 (41) (2013) 12556–12566.
[18] J. Shen, W. He, et al., Multifunctional amphiphilic ionic liquid pathway to create water-based magnetic fluids and magnetically-driven mesoporous silica, RSC Adv. 9 (7) (2019) 3504–3513.
[19] L.F. Calheiros, B.G. Soares, et al., Ionic liquid-assisted emulsion polymerization of aniline in organic medium, Mater. Chem. Phys. 179 (2016) 194–203.
[20] S. Chakraborty, K. Jähnichen, et al., Synthesis of magnetic polystyrene nanoparticles using amphiphilic ionic liquid stabilized RAFT mediated miniemulsion polymerization, Macromolecules 47 (13) (2014) 4186–4198.
[21] C. Yuan, J. Guo, et al., Polymerization in ionic liquid-based microemulsions, Polym. Chem. 6 (22) (2015) 4059–4066.
[22] J.N.A. Canongia Lopes, A.A.H. Pádua, Nanostructural organization in ionic liquids, J. Phys. Chem. B 110 (7) (2006) 3330–3335.
[23] R. Hayes, G.G. Warr, et al., Structure and nanostructure in ionic liquids, Chem. Rev. 115 (13) (2015) 6357–6426.
[24] P.C.A.G. Pinto, D.M.G.P. Ribeiro, et al., Active pharmaceutical ingredients based on salicylate ionic liquids: insights into the evaluation of pharmaceutical profiles, New J. Chem. 37 (12) (2013) 4095–4102.
[25] S.N. Adamovich, A.N. Mirskova, et al., Biologically active ionic liquids. New analogs of acetylcholine, Russ. Chem. Bull. 61 (11) (2012) 2192–2193.
[26] A. Benedetto, P. Ballone, Room-temperature ionic liquids and biomembranes: setting the stage for applications in pharmacology, biomedicine, and bionanotechnology, Langmuir 34 (33) (2018) 9579–9597.
[27] O.A. El Seoud, A. Koschella, et al., Applications of ionic liquids in carbohydrate chemistry: a window of opportunities, Biomacromolecules 8 (9) (2007) 2629–2647.
[28] K. Jumbri, M.B. Abdul Rahman, et al., An insight into structure and stability of DNA in ionic liquids from molecular dynamics simulation and experimental studies, Phys. Chem. Chem. Phys. 16 (27) (2014) 14036–14046.
[29] C. Schröder, Proteins in ionic liquids: current status of experiments and simulations, Top. Curr. Chem. 375 (2) (2017) 25.
[30] K.O. Evans, Room-temperature ionic liquid cations act as short-chain surfactants and disintegrate a phospholipid bilayer, Colloids Surf. A Physicochem. Eng. Asp. 274 (1) (2006) 11–17.
[31] K.O. Evans, Supported phospholipid bilayer interaction with components found in typical room-temperature ionic liquids - a QCM-D and AFM study, Int. J. Mol. Sci. 9 (4) (2008) 498–511.
[32] K.O. Evans, Supported phospholipid membrane interactions with 1-butyl-3-methylimidazolium chloride, J. Phys. Chem. B. 112 (29) (2008) 8558–8562.
[33] H. Lee, S.M. Kim, et al., Effects of imidazolium-based ionic liquids on the stability and dynamics of gramicidin A and lipid bilayers at different salt concentrations, J. Mol. Graph. Model. 61 (2015) 53–60.
[34] R.J. Bingham, P. Ballone, Computational study of room-temperature ionic liquids interacting with a POPC phospholipid bilayer, J. Phys. Chem. B. 116 (36) (2012) 11205–11216.
[35] H. Lee, T.-J. Jeon, The binding and insertion of imidazolium-based ionic surfactants into lipid bilayers: the effects of the surfactant size and salt concentration, Phys. Chem. Chem. Phys. 17 (8) (2015) 5725–5733.
[36] B. Yoo, J.K. Shah, et al., Amphiphilic interactions of ionic liquids with lipid biomembranes: a molecular simulation study, Soft Matter 10 (43) (2014) 8641–8651.

[37] G. Bhattacharya, S. Mitra, et al., Thermodynamics of interaction of ionic liquids with lipid monolayer, Biophys. Rev. 10 (3) (2018) 709–719.
[38] P. Drücker, A. Rühling, et al., Imidazolium salts mimicking the structure of natural lipids exploit remarkable properties forming lamellar phases and giant vesicles, Langmuir 33 (6) (2017) 1333–1342.
[39] A. Benedetto, P. Ballone, Room temperature ionic liquids meet biomolecules: a microscopic view of structure and dynamics, ACS Sustain. Chem. Eng. 4 (2) (2016) 392–412.
[40] A. Schindl, M.L. Hagen, et al., Proteins in ionic liquids: reactions, applications, and futures, Front. Chem. 7 (347) (2019).
[41] M. Haberler, C. Schröder, et al., Hydrated ionic liquids with and without solute: the influence of water content and protein solutes, J. Chem. Theory Comput. 8 (10) (2012) 3911–3928.
[42] K. Fujita, D.R. MacFarlane, et al., Protein solubilising and stabilising ionic liquids, Chem. Commun. 38 (2005) 4804–4806.
[43] A.M. Klibanov, Improving enzymes by using them in organic solvents, Nature 409 (6817) (2001) 241–246.
[44] H. Weingärtner, C. Cabrele, et al., How ionic liquids can help to stabilize native proteins, Phys. Chem. Chem. Phys. 14 (2) (2012) 415–426.
[45] H. Ohno, K. Fujita, et al., Is seven the minimum number of water molecules per ion pair for assured biological activity in ionic liquid–water mixtures? Phys. Chem. Chem. Phys. 17 (22) (2015) 14454–14460.
[46] W.K. Czaja, D.J. Young, et al., The future prospects of microbial cellulose in biomedical applications, Biomacromolecules 8 (1) (2007) 1–12.
[47] R.C. Remsing, R.P. Swatloski, et al., Mechanism of cellulose dissolution in the ionic liquid 1-n-butyl-3-methylimidazolium chloride: a 13C and 35/37Cl NMR relaxation study on model systems, Chem. Commun. 12 (2006) 1271–1273.
[48] R.P. Swatloski, S.K. Spear, et al., Dissolution of cellulose with ionic liquids, J. Am. Chem. Soc. 124 (18) (2002) 4974–4975.
[49] Y. Wu, T. Sasaki, et al., A novel biomass-ionic liquid platform for the utilization of native chitin, Polymer 49 (9) (2008) 2321–2327.
[50] T.G.A. Youngs, C. Hardacre, et al., Glucose solvation by the ionic liquid 1,3-dimethylimidazolium chloride: a simulation study, J. Phys. Chem. B. 111 (49) (2007) 13765–13774.
[51] X. Li, Y. Zhang, et al., Efficient preparation of high concentration cellulose solution with complex DMSO/ILs solvent, J. Polym. Res. 23 (2) (2016) 32.
[52] H. Tateishi-Karimata, N. Sugimoto, Structure, stability and behaviour of nucleic acids in ionic liquids, Nucleic Acids Res. 42 (14) (2014) 8831–8844.
[53] A. Pabbathi, A. Samanta, Spectroscopic and molecular docking study of the interaction of DNA with a morpholinium ionic liquid, J. Phys. Chem. B. 119 (34) (2015) 11099–11105.
[54] A. Garai, D. Ghoshdastidar, et al., Ionic liquids make DNA rigid, J. Chem. Phys. 149 (4) (2018), 045104.
[55] R. Vijayaraghavan, A. Izgorodin, et al., Long-term structural and chemical stability of DNA in hydrated ionic liquids, Angew. Chem. Int. Ed. 49 (9) (2010) 1631–1633.
[56] W.B. Melchior, P.H. Von Hippel, Alteration of the relative stability of dA-dT and dG-dC base pairs in DNA, Proc. Natl. Acad. Sci. U. S. A. 70 (2) (1973) 298–302.
[57] H. Tateishi-Karimata, N. Sugimoto, A–T base pairs are more stable than G–C base pairs in a hydrated ionic liquid, Angew. Chem. Int. Ed. 51 (6) (2012) 1416–1419.
[58] J.M. Gomes, S.S. Silva, et al., Biocompatible ionic liquids: fundamental behaviours and applications, Chem. Soc. Rev. 48 (15) (2019) 4317–4335.
[59] K.A. Kurnia, T.E. Sintra, et al., The effect of the cation alkyl chain branching on mutual solubilities with water and toxicities, Phys. Chem. Chem. Phys. 16 (37) (2014) 19952–19963.

[60] Y. Wang, H. Li, et al., A theoretical investigation of the interactions between water molecules and ionic liquids, J. Phys. Chem. B. 110 (48) (2006) 24646–24651.
[61] Z.-D. Ding, Z. Chi, et al., Theoretical and experimental investigation of the interactions between [emim]Ac and water molecules, J. Mol. Struct. 1015 (2012) 147–155.
[62] B.A. Marekha, M. Bria, et al., Intermolecular interactions in mixtures of 1-n-butyl-3-methylimidazolium acetate and water: insights from IR, Raman, NMR spectroscopy and quantum chemistry calculations, J. Mol. Liq. 210 (2015) 227–237.
[63] K.M. Docherty, J.C.F. Kulpa, Toxicity and antimicrobial activity of imidazolium and pyridinium ionic liquids, Green Chem. 7 (4) (2005) 185–189.
[64] K.S. Egorova, V.P. Ananikov, Toxicity of ionic liquids: eco(cyto)activity as complicated, but unavoidable parameter for task-specific optimization, ChemSusChem 7 (2) (2014) 336–360.
[65] W.L. Hough-Troutman, M. Smiglak, et al., Ionic liquids with dual biological function: sweet and anti-microbial, hydrophobic quaternary ammonium-based salts, New J. Chem. 33 (1) (2009) 26–33.
[66] N. Iwai, K. Nakayama, et al., Antibacterial activities of imidazolium, pyrrolidinium and piperidinium salts, Bioorg. Med. Chem. Lett. 21 (6) (2011) 1728–1730.
[67] S.W. Singer, A.P. Reddy, et al., Enrichment, isolation and characterization of fungi tolerant to 1-ethyl-3-methylimidazolium acetate, J. Appl. Microbiol. 110 (4) (2011) 1023–1031.
[68] L. Carson, P.K.W. Chau, et al., Antibiofilm activities of 1-alkyl-3-methylimidazolium chloride ionic liquids, Green Chem. 11 (4) (2009) 492–497.
[69] M.R. Cole, M. Li, et al., Design, synthesis, and biological evaluation of β-lactam antibiotic-based imidazolium- and pyridinium-type ionic liquids, Chem. Biol. Drug Des. 78 (1) (2011) 33–41.
[70] B. He, G. Ou, et al., Antimicrobial ionic liquids with fumarate anion, J. Chem. 2013 (2013) 473153.
[71] M. Ismail Hossain, M. El-Harbawi, et al., Synthesis and anti-microbial activity of hydroxylammonium ionic liquids, Chemosphere 84 (1) (2011) 101–104.
[72] F. Walkiewicz, K. Materna, et al., Multifunctional long-alkyl-chain quaternary ammonium azolate based ionic liquids, New J. Chem. 34 (2010) 2281–2289.
[73] G.A. O'Toole, M. Wathier, et al., Diphosphonium ionic liquids as broad-spectrum antimicrobial agents, Cornea 31 (7) (2012) 810–816.
[74] O. Forero Doria, R. Castro, et al., Novel alkylimidazolium ionic liquids as an antibacterial alternative to pathogens of the skin and soft tissue infections, Molecules 23 (9) (2018) 2354.
[75] Y. Luo, Q. Wang, et al., An ionic liquid facilitates the proliferation of antibiotic resistance genes mediated by class I integrons, Environ. Sci. Technol. Lett. 1 (5) (2014) 266–270.
[76] S. Jeong, S.H. Ha, et al., Elucidation of molecular interactions between lipid membranes and ionic liquids using model cell membranes, Soft Matter 8 (20) (2012) 5501–5506.
[77] P. Mester, M. Wagner, et al., Antimicrobial effects of short chained imidazolium-based ionic liquids—influence of anion chaotropicity, Ecotoxicol. Environ. Saf. 111 (2015) 96–101.
[78] A. Koziróg, A. Wysocka-Robak, et al., Antifungal activity of paper modified with ionic liquids, Fibres Text. East. Eur. 23 (2015) 134–137.
[79] I.M. Gindri, K.L. Palmer, et al., Evaluation of mammalian and bacterial cell activity on titanium surface coated with dicationic imidazolium-based ionic liquids, RSC Adv. 6 (43) (2016) 36475–36483.
[80] R. Ali, Z. Mirza, G.M.D. Ashraf, M.A. Kamal, S.A. Ansari, G.A. Damanhouri, A.M. Abuzenadah, A.G. Chaudhary, I.A. Sheikh, New anticancer agents: recent developments in tumor therapy, Anticancer Res 32 (7) (2012) 2999–3005.

[81] V. Kumar, S.V. Malhotra, Study on the potential anti-cancer activity of phosphonium and ammonium-based ionic liquids, Bioorg. Med. Chem. Lett. 19 (16) (2009) 4643–4646.
[82] V. Kumar, S.V. Malhotra, Ionic liquid mediated synthesis of 5-halouracil nucleosides: key precursors for potential antiviral drugs, Nucleosides Nucleotides Nucleic Acids 28 (9) (2009) 821–834.
[83] S.V. Malhotra, V. Kumar, A profile of the in vitro anti-tumor activity of imidazolium-based ionic liquids, Bioorg. Med. Chem. Lett. 20 (2) (2010) 581–585.
[84] A.R. Dias, J. Costa-Rodrigues, et al., The anticancer potential of ionic liquids, Chem. Med. Chem. 12 (1) (2017) 11–18.
[85] T.P. Thuy Pham, C.-W. Cho, et al., Environmental fate and toxicity of ionic liquids: a review, Water Res. 44 (2) (2010) 352–372.
[86] R.F. Frade, C.A. Afonso, Impact of ionic liquids in environment and humans: an overview, Hum. Exp. Toxicol. 29 (12) (2010) 1038–1054.
[87] A. Jurgen, M. Marianne, Toxicity of ionic liquids towards mammalian cell lines, Curr. Org. Chem. 15 (12) (2011) 1905–1917.
[88] R.F.M. Frade, A. Matias, et al., Effect of ionic liquids on human colon carcinoma HT-29 and CaCo-2 cell lines, Green Chem. 9 (8) (2007) 873–877.
[89] X. Wang, C.A. Ohlin, et al., Cytotoxicity of ionic liquids and precursor compounds towards human cell line HeLa, Green Chem. 9 (11) (2007) 1191–1197.
[90] R.A. Kumar, N. Papaïconomou, et al., In vitro cytotoxicities of ionic liquids: effect of cation rings, functional groups, and anions, Environ. Toxicol. 24 (4) (2009) 388–395.
[91] S. Jain, N. Patel, et al., Solubility and dissolution enhancement strategies: current understanding and recent trends, Drug Dev. Ind. Pharm. 41 (6) (2015) 875–887.
[92] K.T. Savjani, A.K. Gajjar, et al., Drug solubility: importance and enhancement techniques, ISRN Pharm. 2012 (2012) 195727.
[93] D. Singhal, W. Curatolo, Drug polymorphism and dosage form design: a practical perspective, Adv. Drug Deliv. Rev. 56 (3) (2004) 335–347.
[94] D.H. Jornada, G.F. dos Santos Fernandes, et al., The prodrug approach: a successful tool for improving drug solubility, Molecules (Basel, Switzerland) 21 (2015) 42.
[95] J. Rautio, H. Kumpulainen, et al., Prodrugs: design and clinical applications, Nat. Rev. Drug Discov. 7 (3) (2008) 255–270.
[96] S.M. Guthrie, D.-M. Smilgies, et al., Controlling polymorphism in pharmaceutical compounds using solution shearing, Cryst. Growth Des. 18 (2) (2018) 602–606.
[97] M. Kitamura, Strategy for control of crystallization of polymorphs, CrystEngComm 11 (6) (2009) 949–964.
[98] M.A. Walker, Novel tactics for designing water-soluble molecules in drug discovery, Expert Opin. Drug Discovery 9 (12) (2014) 1421–1433.
[99] W. Xu, P. Ling, et al., Polymeric micelles, a promising drug delivery system to enhance bioavailability of poorly water-soluble drugs, J. Drug Deliv. 2013 (2013) 340315.
[100] A.M. Healy, Z.A. Worku, et al., Pharmaceutical solvates, hydrates and amorphous forms: a special emphasis on cocrystals, Adv. Drug Deliv. Rev. 117 (2017) 25–46.
[101] T.S. Wiedmann, A. Naqwi, Pharmaceutical salts: theory, use in solid dosage forms and in situ preparation in an aerosol, Asian J. Pharm. Sci. 11 (6) (2016) 722–734.
[102] F. Liu, S. Ranmal, et al., patient-centered pharmaceutical design to improve acceptability of medicines: similarities and differences in paediatric and geriatric populations, Drugs 74 (16) (2014) 1871–1889.
[103] N. Adawiyah, M. Moniruzzaman, et al., Ionic liquids as a potential tool for drug delivery systems, Med. Chem. Commun. 7 (10) (2016) 1881–1897.
[104] A.D. dos Santos, A.R.C. Morais, et al., Solubility of pharmaceutical compounds in ionic liquids, Fluid Phase Equilib. 356 (2013) 18–29.
[105] H. Mizuuchi, V. Jaitely, et al., Room temperature ionic liquids and their mixtures: Potential pharmaceutical solvents, Eur. J. Pharm. Sci. 33 (4) (2008) 326–331.

[106] M.J. Earle, P.B. McCormac, et al., The first high yield green route to a pharmaceutical in a room temperature ionic liquid, Green Chem. 2 (6) (2000) 261–262.
[107] V. Kumar, S.V. Malhotra, Synthesis of nucleoside-based antiviral drugs in ionic liquids, Bioorg. Med. Chem. Lett. 18 (20) (2008) 5640–5642.
[108] B. James, S. Viji, et al., Synthesis of novel highly functionalized biologically active polycyclic caged amides, Tetrahedron Lett. 48 (35) (2007) 6204–6208.
[109] Z.-G. Le, J.-P. Xu, et al., One-pot synthesis of 2-aminobenzothiazoles using a new reagent of [bmim]br3 in [bmim]BF4, J. Heterocyclic Chem. 43 (4) (2006) 1123–1124.
[110] S.P. Satasia, P.N. Kalaria, et al., Catalytic regioselective synthesis of pyrazole based pyrido[2,3-d]pyrimidine-diones and their biological evaluation, Org. Biomol. Chem. 12 (11) (2014) 1751–1758.
[111] V.K. Rao, R. Tiwari, et al., Copper triflate-mediated synthesis of 1,3,5-triarylpyrazoles in [bmim][PF6] ionic liquid and evaluation of their anticancer activities, RSC Adv. 3 (35) (2013) 15396–15403.
[112] H. Rajerison, D. Faye, et al., Ionic liquid supported organotin reagents to prepare molecular imaging and therapy agents, Org. Biomol. Chem. 14 (6) (2016) 2121–2126.
[113] E.E. Fileti, V.V. Chaban, Imidazolium ionic liquid helps to disperse fullerenes in water, J. Phys. Chem. Lett. 5 (11) (2014) 1795–1800.
[114] S.K. Soni, S. Sarkar, et al., Intrinsic therapeutic and biocatalytic roles of ionic liquid mediated self-assembled platinum–phytase nanospheres, RSC Adv. 5 (77) (2015) 62871–62881.
[115] J. Kuchlyan, N. Kundu, et al., Ionic liquids in microemulsions: formulation and characterization, Curr. Opin. Colloid Interface Sci. 25 (2016) 27–38.
[116] S. Shi, T. Yin, et al., Light induced micelle to vesicle transition in an aqueous solution of a surface active ionic liquid, RSC Adv. 5 (92) (2015) 75806–75809.
[117] A. Pal, A. Yadav, Binding interactions of anesthetic drug with surface active ionic liquid, J. Mol. Liq. 222 (2016) 471–479.
[118] Y. Pei, Y. Huang, et al., Phase behaviour and microstructure of the micro-emulsions composed of cholinium-based ionic liquid, Triton X-100 and water, J. Chem. Thermodyn. 74 (2014) 231–237.
[119] D. Dobler, T. Schmidts, et al., Ionic liquids as ingredients in topical drug delivery systems, Int. J. Pharm. 441 (1) (2013) 620–627.
[120] Y. Pei, J. Zhang, et al., Partition behavior of drug molecules in cholinium-based ionic liquids, Sep. Sci. Technol. 50 (11) (2015) 1641–1646.
[121] A.C. Anselmo, S. Mitragotri, An overview of clinical and commercial impact of drug delivery systems, J. Control. Release 190 (2014) 15–28.
[122] T. Han, D.B. Das, Potential of combined ultrasound and microneedles for enhanced transdermal drug permeation: a review, Eur. J. Pharm. Biopharm. 89 (2015) 312–328.
[123] C.M. Schoellhammer, D. Blankschtein, et al., Skin permeabilization for transdermal drug delivery: recent advances and future prospects, Expert Opin. Drug Deliv. 11 (3) (2014) 393–407.
[124] K. Kretsos, G.B. Kasting, A geometrical model of dermal capillary clearance, Math. Biosci. 208 (2) (2007) 430–453.
[125] A. Arora, M.R. Prausnitz, et al., Micro-scale devices for transdermal drug delivery, Int. J. Pharm. 364 (2) (2008) 227–236.
[126] R.F. Donnelly, T.R.R. Singh, et al., Hydrogel-forming microneedle arrays for enhanced transdermal drug delivery, Adv. Funct. Mater. 22 (23) (2012) 4879–4890.
[127] M.R. Prausnitz, R. Langer, Transdermal drug delivery, Nat. Biotechnol. 26 (11) (2008) 1261–1268.
[128] S.K. Peasah, E. Azziz-Baumgartner, et al., Influenza cost and cost-effectiveness studies globally—a review, Vaccine 31 (46) (2013) 5339–5348.

[129] H. Suh, J. Shin, et al., Microneedle patches for vaccine delivery, Clin. Exp. Vaccine Res. 3 (1) (2014) 42–49.
[130] H. Marwah, T. Garg, et al., Permeation enhancer strategies in transdermal drug delivery, Drug Deliv. 23 (2) (2016) 564–578.
[131] M. Moniruzzaman, Y. Tahara, et al., Ionic liquid-assisted transdermal delivery of sparingly soluble drugs, Chem. Commun. 46 (9) (2010) 1452–1454.
[132] M. Zakrewsky, K.S. Lovejoy, et al., Ionic liquids as a class of materials for transdermal delivery and pathogen neutralization, Proc. Natl. Acad. Sci. 111 (37) (2014) 13313–13318.
[133] D. Monti, E. Egiziano, et al., Ionic liquids as potential enhancers for transdermal drug delivery, Int. J. Pharm. 516 (1) (2017) 45–51.
[134] D. Zhang, H.-J. Wang, et al., Evaluations of imidazolium ionic liquids as novel skin permeation enhancers for drug transdermal delivery, Pharm. Dev. Technol. 22 (4) (2017) 511–520.
[135] Z. Sidat, T. Marimuthu, et al., Ionic liquids as potential and synergistic permeation enhancers for transdermal drug delivery, Pharmaceutics 11 (2) (2019) 96.
[136] G.S. Lim, S. Jaenicke, et al., How the spontaneous insertion of amphiphilic imidazolium-based cations changes biological membranes: a molecular simulation study, Phys. Chem. Chem. Phys. 17 (43) (2015) 29171–29183.
[137] S. Araki, R. Wakabayashi, et al., Ionic liquid-mediated transcutaneous protein delivery with solid-in-oil nanodispersions, Med. Chem. Commun. 6 (12) (2015) 2124–2128.
[138] S. Kandasamy, M. Moniruzzaman, et al., Formulation and characterization of acetate based ionic liquid in oil microemulsion as a carrier for acyclovir and methotrexate, Sep. Purif. Technol. 196 (2018) 149–156.
[139] R. Aboofazeli, H. Zia, et al., Transdermal delivery of nicardipine: an approach to in vitro permeation enhancement, Drug Deliv. 9 (4) (2002) 239–247.
[140] T. Hattori, H. Tagawa, et al., Transdermal delivery of nobiletin using ionic liquids, Sci. Rep. 9 (1) (2019) 20191.
[141] M.R. Prausnitz, S. Mitragotri, et al., Current status and future potential of transdermal drug delivery, Nat. Rev. Drug Discov. 3 (2) (2004) 115–124.
[142] H. Chopra, P. Kumar, et al., Ionic liquid-based transdermal delivery of propranolol: a patent evaluation of US2018/0169033A1, Pharm. Pat. Anal. 8 (5) (2019) 203–209.
[143] H. Rodríguez, K. Bica, R.D. Rogers, Ionic liquid technology: a potential new platform for the pharmaceutical industry, Trop J Pharm Res 7 (3) (2008) 1011–1012.
[144] K.C. Waterman, R.C. Adami, et al., Hydrolysis in pharmaceutical formulations, Pharm. Dev. Technol. 7 (2) (2002) 113–146.
[145] S.N. Black, E.A. Collier, et al., Structure, solubility, screening, and synthesis of molecular salts, J. Pharm. Sci. 96 (5) (2007) 1053–1068.
[146] R. Ferraz, L.C. Branco, et al., Ionic liquids as active pharmaceutical ingredients, ChemMedChem 6 (6) (2011) 975–985.
[147] W.L. Hough, R.D. Rogers, Ionic liquids then and now: from solvents to materials to active pharmaceutical ingredients, Bull. Chem. Soc. Jpn. 80 (12) (2007) 2262–2269.
[148] J. Stoimenovski, D.R. MacFarlane, et al., Crystalline vs. ionic liquid salt forms of active pharmaceutical ingredients: a position paper, Pharm. Res. 27 (4) (2010) 521–526.
[149] I.M. Marrucho, L.C. Branco, et al., Ionic liquids in pharmaceutical applications, Annu. Rev. Chem. Biomol. Eng. 5 (1) (2014) 527–546.
[150] L. Zhang, J. Liu, et al., Pharmaceutically active ionic liquid self-assembled vesicles for the application as an efficient drug delivery system, ChemPhysChem 14 (15) (2013) 3454–3457.
[151] W.L. Hough, M. Smiglak, et al., The third evolution of ionic liquids: active pharmaceutical ingredients, New J. Chem. 31 (8) (2007) 1429–1436.

[152] J.L. Shamshina, P.S. Barber, et al., Ionic liquids in drug delivery, Expert Opin. Drug Deliv. 10 (10) (2013) 1367–1381.
[153] J.H. Davis, K.J. Forrester, et al., Novel organic ionic liquids (OILs) incorporating cations derived from the antifungal drug miconazole, Tetrahedron Lett. 39 (49) (1998) 8955–8958.
[154] H. Wang, G. Gurau, et al., Simultaneous membrane transport of two active pharmaceutical ingredients by charge assisted hydrogen bond complex formation, Chem. Sci. 5 (9) (2014) 3449–3456.
[155] S. Cherukuvada, A. Nangia, Polymorphism in an API ionic liquid: ethambutol dibenzoate trimorphs, CrstEngComm 14 (23) (2012) 7840–7843.
[156] O.A. Cojocaru, S.P. Kelley, et al., Procainium acetate versus procainium acetate dihydrate: irreversible crystallization of a room-temperature active pharmaceutical-ingredient ionic liquid upon hydration, Cryst. Growth Des. 13 (8) (2013) 3290–3293.
[157] K.G. DeFrates, R. Moore, et al., Protein-based fiber materials in medicine: a review, Nanomaterials (Basel, Switzerland) 8 (7) (2018) 457.
[158] M. Reslan, V. Kayser, Ionic liquids as biocompatible stabilizers of proteins, Biophys. Rev. 10 (3) (2018) 781–793.
[159] D. Constatinescu, C. Herrmann, et al., Patterns of protein unfolding and protein aggregation in ionic liquids, Phys. Chem. Chem. Phys. 12 (8) (2010) 1756–1763.
[160] M. Naushad, Z.A. Alothman, et al., Effect of ionic liquid on activity, stability, and structure of enzymes: a review, Int. J. Biol. Macromol. 51 (4) (2012) 555–560.
[161] F.J. Hernández-Fernández, A.P. de los Ríos, et al., Enhancement of activity and selectivity in lipase-catalyzed transesterification in ionic liquids by the use of additives, J. Chem. Technol. Biotechnol. 82 (10) (2007) 882–887.
[162] H. Pfruender, M. Amidjojo, et al., Efficient whole-cell biotransformation in a biphasic ionic liquid/water system, Angew. Chem. Int. Ed. 43 (34) (2004) 4529–4531.
[163] D. Weuster-Botz, Process intensification of whole-cell biocatalysis with ionic liquids, Chem. Rec. 7 (6) (2007) 334–340.
[164] H.J. Choi, K.-N. Uhm, et al., Production of chiral compound using recombinant Escherichia coli cells co-expressing reductase and glucose dehydrogenase in an ionic liquid/water two phase system, J. Mol. Catal. B: Enzym. 70 (3) (2011) 114–118.
[165] A.K. Mohanty, M. Misra, et al., Sustainable bio-composites from renewable resources: opportunities and challenges in the green materials world, J. Polym. Environ. 10 (1) (2002) 19–26.
[166] A. Rouilly, L. Rigal, Agro-materials: a bibliographic review, J. Macromol. Sci. Polym. Rev. 42 (4) (2002) 441–479.
[167] J. Zhang, J. Wu, et al., Application of ionic liquids for dissolving cellulose and fabricating cellulose-based materials: state of the art and future trends, Mater. Chem. Front. 1 (7) (2017) 1273–1290.
[168] A.K. Mallik, M. Shahruzzaman, et al., Fabrication of polysaccharide-based materials using ionic liquids and scope for biomedical use, in: S. Maiti, S. Jana (Eds.), Functional Polysaccharides for Biomedical Applications, Woodhead Publishing, 2019, pp. 131–171 (Chapter 4).
[169] A. Takada, J.-i. Kadokawa, Fabrication and characterization of polysaccharide ion gels with ionic liquids and their further conversion into value-added sustainable materials, Biomolecules 5 (1) (2015) 244–262.
[170] F. Hermanutz, M.P. Vocht, et al., Processing of cellulose using ionic liquids, Macromol. Mater. Eng. 304 (2) (2019) 1800450.
[171] Y. Guo, X. Wang, et al., Self-assembly and paclitaxel loading capacity of cellulose-graft-poly(lactide) nanomicelles, J. Agric. Food Chem. 60 (15) (2012) 3900–3908.
[172] L. Li, L. Meng, et al., The ionic liquid-associated synthesis of a cellulose/SWCNT complex and its remarkable biocompatibility, J. Mater. Chem. 19 (22) (2009) 3612–3617.

[173] C. Tsioptsias, C. Panayiotou, Preparation of cellulose-nanohydroxyapatite composite scaffolds from ionic liquid solutions, Carbohydr. Polym. 74 (1) (2008) 99–105.
[174] C.D. Tran, T.M. Mututuvari, Cellulose, chitosan, and keratin composite materials. controlled drug release, Langmuir 31 (4) (2015) 1516–1526.
[175] D. Hua, J. Jiang, et al., Smart chitosan-based stimuli-responsive nanocarriers for the controlled delivery of hydrophobic pharmaceuticals, Macromolecules 44 (6) (2011) 1298–1302.
[176] R.F.M. Elshaarawy, A.A. Refaee, et al., Pharmacological performance of novel poly-(ionic liquid)-grafted chitosan-N-salicylidene Schiff bases and their complexes, Carbohydr. Polym. 146 (2016) 376–387.
[177] L. Chen, M. Sharifzadeh, et al., Inexpensive ionic liquids: [HSO4]−-based solvent production at bulk scale, Green Chem. 16 (6) (2014) 3098–3106.
[178] J. Claus, A. Brietzke, et al., Swelling characteristics and biocompatibility of ionic liquid based hydrogels for biomedical applications, PLoS One 15 (4) (2020), e0231421.
[179] W. Qian, J. Texter, et al., Frontiers in poly(ionic liquid)s: syntheses and applications, Chem. Soc. Rev. 46 (4) (2017) 1124–1159.
[180] S.N. Riduan, Y. Zhang, Imidazolium salts and their polymeric materials for biological applications, Chem. Soc. Rev. 42 (23) (2013) 9055–9070.
[181] J. Guo, Q. Xu, et al., Intrinsically antibacterial poly(ionic liquid) membranes: the synergistic effect of anions, ACS Macro Lett. 4 (10) (2015) 1094–1098.
[182] Q. Ye, T. Gao, et al., Grafting poly(ionic liquid) brushes for anti-bacterial and anti-biofouling applications, J. Mater. Chem. 22 (26) (2012) 13123–13131.
[183] Z. Zheng, Q. Xu, et al., Structure–antibacterial activity relationships of imidazolium-type ionic liquid monomers, poly(ionic liquids) and poly(ionic liquid) membranes: effect of alkyl chain length and cations, ACS Appl. Mater. Interfaces 8 (20) (2016) 12684–12692.
[184] H. Yuan, Z. Liu, et al., Cationic conjugated polymers for discrimination of microbial pathogens, Adv. Mater. 26 (25) (2014) 4333–4338.
[185] Y. Dai, X. Zhang, Recent development of functional aliphatic polycarbonates for the construction of amphiphilic polymers, Polym. Chem. 8 (48) (2017) 7429–7437.
[186] W. Cui, X. Lu, et al., Dual-responsive controlled drug delivery based on ionically assembled nanoparticles, Langmuir 28 (25) (2012) 9413–9420.
[187] C. Mukesh, J. Bhatt, et al., A polymerizable bioionic liquid based nanogel: a new nanocarrier for an anticancer drug, Macromol. Chem. Phys. 215 (15) (2014) 1498–1504.
[188] J.-W. Seo, J.-Y. Hwang, et al., Ionic liquid-doped and p-NIPAAm-based copolymer (p-NIBIm): extraordinary drug-entrapping and -releasing behaviors at 38–42 °C, RSC Adv. 4 (51) (2014) 26738–26747.
[189] M. Mahkam, A. Pakravan, Synthesis and characterization of pH-sensitive positive-charge silica nanoparticles for oral anionic drug delivery, J. Chil. Chem. Soc. 60 (3) (2013) 293–296.
[190] S. Rasouli, S. Davaran, et al., Positively charged functionalized silica nanoparticles as nontoxic carriers for triggered anticancer drug release, Des. Monomers Polym. 17 (3) (2014) 227–237.
[191] L. Meng, L. Niu, et al., Gold nanoparticles grown on ionic liquid-functionalized single-walled carbon nanotubes: new materials for photothermal therapy, Chem. A Eur. J. 18 (42) (2012) 13314–13319.

CHAPTER 16

Opportunities in ultrasonic drug delivery to tumor

Sarjana Raikwar[a], Shivani Saraf[a], Pritish Kumar Panda[a], Ankita Tiwari[a], Pooja Das Bidla[a], Amit Verma[a], Ankit Jain[b], and Sanjay K. Jain[a]
[a]Pharmaceutics Research Projects Laboratory, Department of Pharmaceutical Sciences, Dr. Harisingh Gour Vishwavidyalaya, Sagar, MP, India
[b]Department of Materials Engineering, Indian Institute of Science, Bangalore, Karnataka, India

1 Introduction

Recently nanotechnology has been extensively utilized as drug delivery. Nanocarriers have some benefits over conventional drug delivery systems (DDSs) such as small size; prolonged circulation time, less unwanted effects and site specific delivery. Stimuli sensitive DDS are the advanced form of nanotechnology. They selectively release their contents in regards to the particular intrinsic and extrinsic stimuli (temperature, pH, redox potential, hypoxia and ultrasound [US]). Various nanocarriers have been employed in US-based cancer treatment such as liposomes, nanoparticles (NPs), micelles, nanobubbles, nanodroplets and microbubbles (MBs) [1–3]. Polymeric NPs have been also utilized in US-based DDS for tumor treatment. The polymeric systems have several benefits like better encapsulation efficiency, passive targeting, efficient cellular uptake, sustained drug release and cytotoxicity along with safety [4]. Theranostic nanocarriers also hold abundant prospectives for US-based diagnosis and treatment. These systems facilitated the permeation of drug via capillary walls and cell membrane and improved the accumulation of drug. USDD is noninvasive, targetable and efficient approach for the effective treatment of cancer. These mechanisms induce the sonoporation into cell membrane which causes cell disruption and cell death [5]. USDD systems have improved the target specific delivery of bioactive via increasing the permeability by two mechanisms i.e. MBs and sonoporation [6]. Firstly, the MBs are formed near the target cells and pores are generated into cell membrane through cavitation. These processes facilitated the transportation of chemotherapeutic agent into target sites [1]. Focused ultrasound (FUS) mediated local hyperthermia is another mechanism of USDD system. Hyperthermia affects the penetrability of phospholipid

Advances and Challenges in Pharmaceutical Technology
https://doi.org/10.1016/B978-0-12-820043-8.00016-5

bilayer and improves the drug penetration across the membrane [5, 7]. Low-energy (0.1100 mW/cm^2) and higher-energy (10,010,000 W/cm^2) US waves are utilized for the prognosis and noninvasive therapies, respectively. The mode of USDD carriers are not properly defined but it may be because of cavitation, thermal and mechanical effects. Cavitation refers to as bubble activities under the US that occurred in liquid containing the pocket having the gas or vapor. Cavitation affects the biological activity by hyperthermia, mechanical stress and free radical formation. As US passes via tissues, some energy changed into heat and produces local hyperthermia. This hyperthermia helps in the permeation of anticancer drugs. Acoustic radiation forces (ARF) is another mechanism involved in USDD systems. It is classified into primary and secondary radiation force [8]. The primary force facilitates the extravasation of nanocarrier from the vessels in the route of wave propagation. The secondary force enhances oscillation, interaction and aggregation between micro-bubbles [8]. This book chapter discusses the mode of USDD includes hyperthermia and mechanically-induced drug release, US-responsive polymers used in chemotherapy and applications of USDD systems (nanobubbles, nanodroplets, micelles, liposomes, MBs, and NPs) in cancer therapy.

2 Mechanism of ultrasound stimulated drug delivery

Anticancer drug uptake could be enhanced by US-induced hyperthermia. It is described by rise in temperature because of FUS. When FUS is blend with DDS, it enhances the delivery of anticancer drugs [7, 9]. FUS releases acoustic energy which was focused at particular site. Deposition of acoustic energy in small area (about 110 mm) results to several therapeutic effects like tissue coagulation as thermal consequences and mechanical disruption of tissues (histotripsy). The mechanical property of US can induce the drug release, for example transient cavitation [10, 11]. Usually, drugs produce toxicities and undesirable effects because of exposure to healthy cells. Several delivery systems like liposomes, MBs, micelles, nanodroplets and nanobubbles are utilized in gene delivery, drug delivery, and macromolecular delivery for neurological disease treatment [1, 12, 13]. Generally, two types of US waves are employed in drug delivery. Low-energy US waves (0.1100 MW/cm^2) are applicable for diagnosis whereas the high-energy US waves (10,010,000 W/cm^2) are employed for noninvasive therapies.

A focused drug release can be obtained using US-based therapies. Some applications of this technique are approved for commercialization and are available for medical treatment in all over the world [14]. In USDD systems, two mechanisms are used. They include hyperthermia and mechanically-triggered drug release, which are facilitated through radiation forces or cavitation phenomena, respectively.

2.1 Hyperthermia-triggered drug release

Hyperthermia-triggered drug release from the carrier system is induced by thermal effects of HIFU. It is an immediate consequence of US due to absorption of sound waves. Because of their larger US absorption coefficient, tissues like bones experience high temperature compared to muscles [15, 16]. The localized, noninvasive and sustained release of drug is utilized by the implementation of hyperthermia induced by magnetic resonance-guided HIFU (MR-HIFU). Currently, various preclinical studies have described HIFU-triggered drug release in several animals such as rats, mice and rabbits [17]. Local heating is caused by HIFU that would induce a phase transition of polymers which cause drug release at tumor site. Two methods are involved in hyperthermia i.e. FUS and HIFU. FUS is considered as a precise and focused technique for the treatment of profound tissues through local heating phenomenon [18]. Whereas, HIFU are utilized for thermal ablation of tumor site or any diseased tissues which cause enhanced tissue temperature. This could be observed in real time imaging by magnetic resonance imaging (MRI), US imaging or interstitial thermocouple which allows to regulate the temperature for accurate therapy. Hyperthermia induced by FUS can permeates deep into the body, resulting in several clinical aspects like reduced tumor growth and produces an anticancer immune effects [18]. The main advantages of US mediated hyperthermia are localized delivery, better permeability, and less side effects. This enhances the susceptibity of anticancer drugs to tumor site [14]. US penetrate into the cells or tissues where part of energy would be absorbed and transferred into heat. This produced local heating which may induce the biological consequence on surrounding tissues [19]. US have thermal effects that could increase the tissue perfusion and capillary pore size which enhance the penetrability of drugs or gene carrier [20]. Further, FUS has improve the site specific drug delivery from temperature sensitive nanocarrier due to induction of mild hyperthermia [21, 22].

2.2 Mechanically-triggered drug release

Mechanically-triggered drug release use cavitation phenomenon for drug release. Usually, cavitation plays two functions, it disrupts the nanocarrier which causes drug release and increases the penetrability of drug via cell-membrane. The transient increase in vascular permeability that may cause increase in uptake of chemotherapeutic agents [23, 24]. At high acoustic pressure (AP) (higher than 500 kPa), an US contrast agent displays inertial cavitation, which form microjet and shockwaves [25, 26]. The transportation of drug forms transient pore in the cell membrane which helps in cytoplasmic delivery of drugs [27, 28]. When low AP is used (mechanical index <0.05–0.1), then bubble oscillates linearly and reflected frequency is analogous to the transmitted frequency. When carrier system is in the vicinity of cell membrane it leads to gentle oscillation which cause instability in membrane and increase endocytosis [25, 29]. When AP increases (0.1 < mechanical index < 0.3) led to nonlinear compression and expansion of MBs. This phenomenon is called non-linear cavitation [30]. A high AP (mechanical index >0.3–0.6) leads to expansion, compression and disruption of MBs. This is called inertial cavitation produces microjet or microsteams to form bubbles used in diagnostic reperfusion study [31]. The velocity of micro-streams is reached to peak velocity about 700 m/*sec* which led to form temporary pore in the membrane leading to enhanced permeability [32]. Acoustic cavitation consists of gas filled cavities (bubbles) that are exposed to oscillating pressure leading to oscillation of bubbles. Usually, bubbles are produced from gases dissolved in liquid or because of fall in pressure below vapor pressure of liquid. Generally cavitation is of two types i.e. inertial and stable cavitation [15, 33]. In inertial cavitation, faster progression and collapsing of bubbles takes places, which produces shock wave (pressure exceeding 10,000 atm) and temperature reaches up to 5000 K. The bubble size and abrupt collapsing ability of bubbles leads to interrupt the cells. This technique could enhance the drug delivery using sonoporation and sonophoresis [16]. The stable cavitation includes the bubbles oscillation nearly their resonant size. It generates velocities which cause variation in fluid velocities and it exerts shear strength on the adjacent tissues. This cavitation is utilized in the delivery of drug and gene into the cells [16]. Cavitation induces physical, chemical and biological effects in the tissues. In aqueous medium, the cavitation could be improved by surface-active agents. They are employed to diminish surface tension of bubbles and increase the bubble formation rate. The declination in bubbles fusion rate in the manifestation of surfactant in the water/gas interface leads to small-size cavitating bubbles in surfactant-augment media [15, 33]. The mechanism of US-triggered drug release is displayed in Fig. 1.

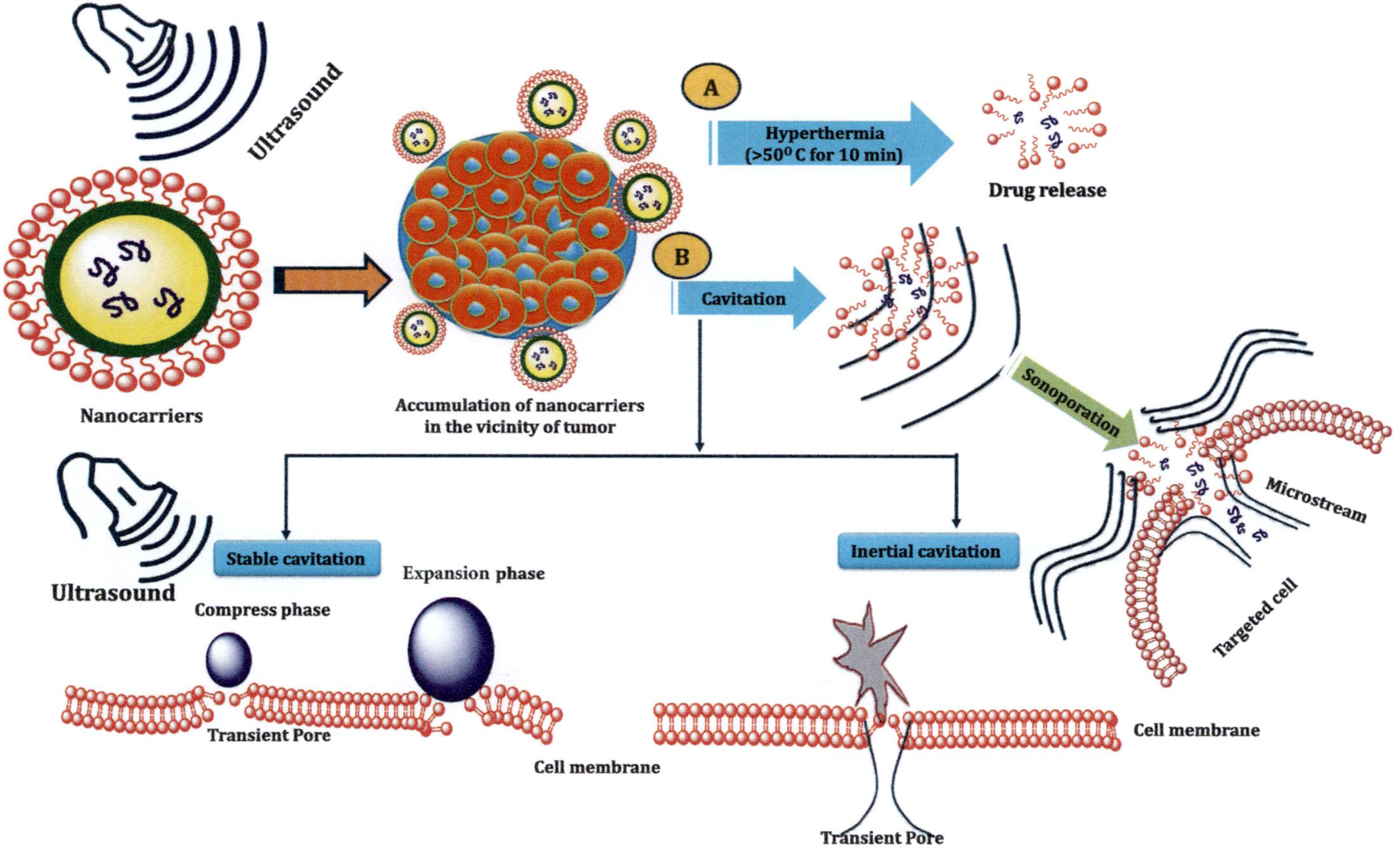

Fig. 1 Mechanism of US-triggered drug release.

3 Ultrasound-responsive polymers used in cancer therapy

US targeted DDS has been extensively used from last many years because of its simple, inexpensive, noninvasive and reproducible characteristics. Recently, it has improved by the advancement of US responsive polymers, molecular imaging, material science, and contrast agents. It helps in the targeted release of chemotherapeutic drugs/therapeutic substances to cancer sites. It will not only increase the efficacy of cancer treatment but also circumvent the undesired effects caused by the treatment [34]. Recently, various thermo-responsive polymers have been utilized in DDS and stimulated via US technology. Polymethacrylates and polyethylene oxide/polypropylene oxide *co*-block or tri-block polymers are pH-responsive and thermal-responsive polymer groups, respectively. These have been utilized for the effective treatment of cancer along with US stimulation. These polymeric systems show their physical modification after US exposure [35]. US-responsive polymers are categorized into two types such as biodegradable (polylactides, polygly*co*lides) and nonbiodegradable polymers (ethylene vinyl acetate, poly (lactide-co-glycolide)). Thermo-responsive polymers such as poly(N isopropyl acrylamide) (P(NIPAM)), Pluronic F-127, chitosan are used in USDD [36–38]. Apart from these, several polymers have been used for cancer targeting by US techniques which are depicted in Table 1. US is restricted in its capability to separate between the diseased and normal tissue. Both the healthy and diseased tissues have comparable acoustic impedances and thus reflection from their interfaces will be very less, causing the detection of diseased tissue difficult. Unique poly (lactic-*co*-glycolic) acid (PLGA) air-filled microcapsules were developed which act as US-contrast agents. Further, these microcapsules were modified by the double emulsion technique to entrapped camphor and ammonium carbonate in oil phase and aqueous phase, respectively for enhancing acoustic response [54]. Wu et al., prepared a polymeric micellar system containing pluronic P123/F127 polymer which encapsulates curcumin that directly penetrated via FUS, leading to US-triggered drug release. The micellar system showed longer circulation time and increased cellular uptake than that of free curcumin. This micellar system displayed a tumor-targeted deposition in time dependent manner with the help of US technology after systemic administration. These micellar system improved the permeability of drug into irradiated tumor cells usingUS. The in vivo site specific drug delivery was confirmed in dual-tumor models. It showed a better release of entrapped drugs after intra-tumoral injection and ensuing a significant

Polymers/lipid/major ingredients	Delivery system	agents	Application	References
Polylactic acid-glycolic acid (PLGA)	Microcapsules	Anti-HER-2 antibody	Breast cancer	[39]
Boron-containing polyanion [polyethylene glycol-b-poly ((closo-dodecaboranyl) thiomethylstyrene)	Microbubble complex	Sodium borocaptate	Glioblastoma	[40]
Polymeric gate poly-(2-(2-methoxyethoxy)ethyl methacrylate-*co*-2-tetrahydropyranyl methacrylate), p(MEO2MA-*co*-THPMA) and copolymer p(MEO2MA-co-THPMA) and Polyethylenimine	Mesoporous silica nanoparticles	Plasmid	Tumor targeting	[41]
Triblock copolymers of poly(ethylene glycol)-polylactide-poly (1H,1H,2H,2H-heptadecafluorodecyl methacrylate) (PEG-PLA-PFMA	Nanocapsules	Perfluorooctyl bromide (PFOB)	Enhanced permeability retention effect (EPR) by ultrasound	[42]
PLGA	Nanocapsules	5-Fluorouracil	Cancer therapy	[43]
PLGA	Microparticles	Doxorubicin	Tumor targeting	[44]
Hydrophilic pullulan and oxyl-alkylhydroxylamine	Nanoparticles	Doxorubicin	Tumor targeting	[45]
PLGA	Nanoparticle aggregates (NPAs)	Doxorubicin	Tumor targeting	[46]
Poly(ε-caprolactone)-poly(ethylene glycol)-poly (ε-caprolactone)	Microbubble	Honokiol	Ovarian cancer targeting	[47]
perfluoropropane (PFP), Distearoyl phosphatidylglycerol (DSPG)	Microbubble	Doxorubicin	Melanoma tumor	[48]
DSPE88 PEG2000-Biotin	Liposomes-microbubble complexes	Paclitaxel	Breast cancer	[49]
Dipalmitoylphosphatidylcholine and DSPE-PEG-biotin	Microbubbles	Nitrosourea	Glioma targeting	[50]
Dipalmitoylphosphatidylcholine and DSPE-PEG-biotin	Microbubbles	Doxorubicin	Cancer targeting	[51]
PLGA	Nanoparticles	DNA plasmid	Tumor chemo and biotherapy	[52]
Pluronic P-105	Polymeric micelles	Doxorubicin	Ovarian cancer	[53]

reduction in tumor growth by 6.5-fold [55]. Liang et al. developed a metal-supramolecular diblock copolymer (poly(propylene glycol)-PEG) with a mechano-labile bis (terpyridine)–Cu (II) complex linkage, which could self-assemble into the spherical polymeric micelles (PMs) in aqueous solution. They found the weak Cu (II)-terpyridine dynamic bond in the *co*-polymeric chain could be cleaved under the HIFU, leading to distraction of the copolymeric micelles and release the entrapped bioactive [56]. Zhang et al. prepared a polymeric micellar system via block *co*-polymer PLA-b-PEG and loaded with Nile red stain. Recently, Xuan et al. investigated the effects of various PMs with the help of polymers such as PEO-b-PTHPMA, PEO-b-PIBMA, and PEO-b-PMMA in micellar systems using HIFU. Outcomes of this study displayed that the PEO-b-PMMA micelles are shown to be more resistant to disruption. Finally, it could be concluded that the addition of tetrahydropyranyl methacrylate (THPMA) showed amplification of HIFU exposure on disassembling of micelles [57, 58]. Chen et al. prepared pullulan-*graft*-poly (carboxybetaine methacrylate) (pul PCB) *co*-polymeric hybrid NPs. It is applicable for hepatoma therapy using US technology. These NPs were stable in physiological pH of the blood. They produced echogenic CO_2 bubble under acidic tumor environment facilitated by US stimulations. Upon intravenous injection, these US sensitive polymeric hybrid NPs get accumulated efficiently into cancer cell. They exhibited six fold contrast enhancement within half an hour at cancerous site in the hepatoma-bearing mice model. So, these polymeric NPs will be a favorable system for effective imaging for hepatoma using US technique [59].

4 Applications of USDD systems in cancer therapy

4.1 Nanobubbles

Nanobubbles (NBs) are employed in USDD systems. It is an emerging strategy for noninvasive type of drug delivery. Zhou et al. [60] developed doxorubicin (DOX) encapsulated biocompatible chitosan NBs. The encapsulation efficiency and US stimulated DOX release was determined using UV spectrophotometer. Additionally, a clinical US imaging was employed for the estimation of chitosan NBs. The bio-safety profile of chitosan NBs was determined using MTT assay. Outcomes indicated that the NBs loaded with DOX displayed excellent drug-loaded ability along with US enhancement. In vitro drug release study showed the US-induced DOX release. Moreover, the efficient DOX delivery from DOX-NBs into MCF-7 (breast cancer cell lines) was accounted. So, it could be concluded that the chitosan NBs are suitable for US triggered DOX release [60]. US-targeted nanobubbles destruction (UTND) has been employed as a successful approach in drug delivery. NBs consist of core and stabilized shell. The main composition of shell consists of polymers, lipids, surfactants and other materials. Zhang et al. [57] developed US triggered PLGA nanobubbles containing methotrexate for cancer treatment. It is not only functioning as efficient US contrast agent for improving the effectiveness of HIFU ablation, but it has also a targeted anticancer drug carrier. Methotrexate (MTX) bearing NBs was prepared by double emulsion evaporation method. The surface of NBs was modified with targeting monoclonal antibodies (mAbHLA G). mAbHLA G/MTX/PLGA NBs undergo in vitro and in vivo studies that could improve the US imaging ability and targeting efficiency of NBs. Furthermore, these studies showed that NBs can increase efficiency of HIFU ablation. These NBs also promote the MTX release. The histopathology examination and immunohistological analysis showed that the NBs along with HIFU group displayed the severe coagulative necrosis with high apoptotic index and lowest proliferative index. Nanobubbles having several characteristics features, like nontoxic, long circulatory, and higher retention. NBs could load all type cytotoxic drugs to kill the residual cancer cell after HIFU ablation. These NBs reach to cancer cells which could increase the HIFU ablation efficiency and avoid the difficulty to reach residual cancer cells. The synergistic effects of NBs with HIFU ablation have provided the anticancer effect to cancer cells. These NBs could destroy cancer cells and finally reduce the tumor recurrence and undesired effects of anticancer drugs [57]. Generally, polymeric NBs are stable and modified by some functional moieties which could provide targeted delivery of drugs. The shell constituents play a significant function in the evaluation and theranostic presentation of the NBs. Shang et al. [61] prepared novel type of nanobubbles by developing PLA-PEG-NH_2 block copolymers using perfluorocarbons (PFC) gas which was stabilized by Tween 80 and span 60. The formed NBs were assessed to possess higher stability, targetability, and sustained drug release. The PLA-PEG-NH_2 polymer were utilized for the preparation of NBs in cancer chemotherapy [61]. Sonoporation is a technique which uses US-mediated gene delivery (USMD) and displays the possibilities for improving gene efficiency into tissues and cells [62–64]. The macromolecules allow entering into the cell by altering the permeability and generate pore in the membrane [64, 65]. Under the low frequency US exposure, these NBs could be displayed the sonoporation effects. It was an ideal carrier for delivery of shRNA [64, 66]. These NBs have numerous benefits in targeted gene delivery due to EPR effect, which increase the amount of drug into tumor. Deshpande et al. reported the PEI and US combination to enhance the DNA transfection activity. They used PEI-shRNA micelles which were employed in place of naked shRNA. This combined synergistic effect was utilized to enhance the shRNA transfection activity in cancer cell. This study established that the combination of PEI-shRNA NBs with US which could decrease the cell viability, stimulate apoptosis and reduce tumor growth. Histopathological examination indicated more effective destruction of cancer by NBs with US compared to other system. The mechanism involved when irradiated US to tumor area is due to the presence of NB. The cavitation in cancer cell led to enhance permeability of cell membrane [67, 68].

4.2 Nanodroplets

Nanodroplets (NDs) are nanoparticulate gaseous delivery system composed of hollow polymeric sphere of gas, with size of 1–10 nm. They are utilized for diagnosis, treatment and measurement of the disease progression. NDs are lipophilic in nature so they can be employed as a carrier for targeted DDS, enhanced site-specificity and permeability of therapeutic agent. It is an important DDS which potentiates the therapeutic effects of drugs. These NDs shows higher efficacy because of its better safety profile as compared to conventional DDSs [69, 70]. NDs have higher in vivo stability and circulation time than that of MBs, and have greater ability to extravasate into cancer cells [71–73]. Generally the stability of NDs is influenced by shell composition, size, width and boiling point of PFC [74]. Usually NDs are stable under physiological condition [75, 76]. The DDSs exposed to UV, that cause induction of vaporization due to decrease in pressure imposed on droplets through the rarefaction phase of acoustic waves [74, 77–80]. Some reports suggested that NDs can function as a DNA carrier for the gene transfection and used for effective transport of microRNA and small interfering RNA (siRNA) on US exposure [81–84]. There are significant challenges in the DDS such as incorporation of multiple drugs, mostly bioactives agents and targeting moieties. So, it requires some variation of NDs coating [85]. Ho et al. [86] developed a camptothecin-loaded fusogenic NDs (CPT-FNDs) as theranostic agents for adipose derived stem cells (ADSC). These tissues are consisting of ADSCs which is potential and ubiquitous. ADSCs provide several applications in the area of cellular therapies [87–89]. The

CPT-FNDs-loaded ADSCs were evaluated for parameters such as CPT-FNDs uptake, cell viability and migration mobility of CPT-FNDs-loaded ADSCs. Additionally, solid tumor bearing animal model was used for assessment of anticancer activity of CPT-FNDs-loaded ADSCs. The CPT-FNDs-loaded ADSCs with US stimulation led to improve the intra-tumoral drug release and enhance the contrast imaging by acoustic droplet vaporization (ADV). This report indicated that CPT-FNDs-loaded ADSCs were utilized in tumor recruitment and enhanced in vivo treatment efficiency and US imaging [86]. Enhanced US exposure of droplets led to the fusion of phospholipid coated NDs with cell membrane resulting into the drug transfer from NDs to interior of cell. The mechanism is apparently operative for lipid-coated particles, it would function for block copolymer-stabilized NDs that expose PEG chains on the surfaces [69, 90]. NDs are applicable in the field of therapeutics and anticancer drug delivery. Various advantages related with the NDs are larger effective surface area for drug loading and target specificity [90]. Mannaris et al. [85] developed US-stimulated NDs (~400 nm) stabilized with a biocompatible polymer. Polydopamine (PDA) is a polymer having several advantages like facile production, higher biocompatibility, and easily functionalized with other molecules like drug or targeting species. They demonstrate the clinical use of US via requirement of acoustic intensities to initiate the cavitation. The outcomes confirmed the in vivo biocompatibility of NDs (analyzed by hemolysis study) [85].

4.3 Micelles

Polymeric micelles are defined as the polymeric nanocarriers consisting amphiphilic block copolymer having core and shell. They are smaller in size (about 10 to 100 nm) than other nanocarriers, which is favorable for the effective extravasation into cancer cells. They provide efficient cancer targeting and enhance the entrapment of potent drugs whose application had been hindered by low aqueous solubility. It also used as solubilizing agents, lead to enhanced aqueous solubility and decrease in systematic toxicity of clinical formulations of chemotherapeutic drug like paclitaxel (PTX) and DOX etc. It also prevents the degradation of drugs under body fluids [91–93]. Surface modified micelles were developed using ligands to specific cancer cell receptors (e.g., Vitamin B12 or folic acid) [94] or using antigen specific antibodies which are overexpressed on cancer cell surface (e.g., immunomicelles) [95, 96]. Another approach used for tumor targeting is stimuli-sensitive micelles which showed the triggered release of bioactive corresponding to physical stimuli, i.e. internal (e.g., pH, hypoxia, enzyme, redox) and external (e.g., US or light) [73, 97]. Zheng et al., prepared a HIFU-triggered polymeric micelles. HA-conjugated docetaxel (DTX) micelles was actively targeting via hyaluronic acid (HA). HIFU stimulated the in vitro release of DTX from micelles [98]. Ex vivo study was performed on MDA-MB-231 and CT 26 cells. The IC_{50} of DTX-loaded micelles were found to be 1230.2 and 879.9 ng/mL for CT26 and MDA-MB-231, respectively. These IC_{50} values considerably decrease under HIFU exposure, which led to increase cell cytotoxicity. The cell uptake studies of coumarine-6 loaded micelles were performed to determine intracellular amount of drug within CT-26 (CD44 negative), MDA-MB-231 (CD44 blocked) cell lines [99, 100]. Results indicated that effective delivery of drug was found due to receptor (CD44) mediated endocytosis. Furthermore, the HIFU exposure could increase the cellular uptake by altering the penetrability of membrane. The outcomes demonstrated that the HIFU-mediated DDS can effectively enhance the site specific intracellular transport of anticancer drug to cancer cell [100]. Wu et al. [101] prepared P123/F127 curcumin (Cur) loaded micelles (M). It is a novel treatment strategy, which uses US to promote local deposition of prepared micelles into exposed area. It was observed that the curcumin loaded micelles (Cur-M) displayed longer circulation time and increases the cellular uptake as compared to free curcumin. There is an initiation of sonochemotherapy as result of cavitation induced by US. Results of in vivo study dual tumor model, demonstrated that the sustained release of drugs followed by intra-tumoral injection. The sonoporation-aided site specific chemotherapy suppressed tumor growth and reduced tumor mass (about 6.5 fold). It was concluded that the US-triggered nanomedicine offers a suitable platform for spatial and temporal drug release [101].

4.4 Liposomes

Liposomes are bilayered vesicular structures consists of phospholipid and cholesterol [102–104]. In cancer chemotherapy, anticancer drugs cause toxicity because of their nonspecific mechanism and failure to reach the target. However, liposomes are special carrier systems which decrease toxicity and improve the specificity of drug [105]. The PTX-liposomes (PTX-LIPO) were developed which was applicable for many extracranial cancers but not effective in glioblastoma due to presence of blood-brain barrier (BBB). US-triggered MBs were developed with the aim of improved the penetration across the BBB. They revealed that MBs could increase the transport of PTX in intracranial glioblastoma bearing nude mice. The liposomes with pulsed FUS penetrated effectively through BBB and BTB, respectively. Quantitative studies indicated 2-fold increase in accumulated drug in glioblastoma after FUS exposure. MBs with FUS were inhibited constantly for 4 weeks than the untreated group. It was concluded that MBs combined with low intensity FUS could be used as a potential strategy for glioblastoma therapy [106]. Awad et al. [107] prepared US-triggered PEGylated liposomes for effective treatment of tumor. Human serum albumin (HSA) conjugated to liposomes for calcein delivery to breast cancer cell. Fluorescent microscopy resulted that calcein uptake was higher for targeted liposomes than control liposomes. The uptake study was detected in MDA-MB-231 and MCF-7 breast cancer cell lines. Additionally, calcein uptake was considerably higher during the LFUS exposure than without US. It was concluded that the liposomes with US is a safe, effective and site specific treatment of breast cancer [107]. The imageable thermosensitive liposomes (iTSLs) of topotecan (Hycamtin®) were prepared. Near infrared fluorescence (NIRF) imaging have been utilize to govern the accumulation of liposomes in tumor. The hyperthermia induced by FUS (42°C for 30 min, 30 min post i.v.) was enhanced the uptake of iTSLs. It also increases florescence emission which was tested after FUS application which showed the rapid release of drug. The MRI study revealed higher uptake of liposomes after the FUS treatment. This study demonstrated that the FUS-induced hyperthermia with real time imaging applicable for tumor targeting [108]. Um et al. [109] prepared fatty acid conjugated, elastin-like peptide (FELP)-bearing thermosensitive liposomes (FTSLs) loaded with doxorubicin (DOX). Thermosensitivity of liposomes allowed the stimulation of drug release at 42°C, a mild-hyperthermic temperature with sustained drug release at 42°C. The tumor bearing mice exposed to nonthermal acoustic wave leads to higher tissue permeability without affecting vascular damage. After systematic administration of liposomes, it increases the intracellular DOX accumulation. Finally, it could be concluded that FTSLs could be an effective approach for the cancer chemotherapy [109]. DOX encapsulated liposomes are described as super-selective DDS in lung metastasis mouse model cancer treatment. It was exposed to US and delivered to lung metastasis cancer cell. This leads to increase efficiency of DOX in metastatic lung cancer cell. The outcomes indicate that the superselective DDS effective in both lungs metastasis and axillary lymph node treatment [110]. Rizzitelli et al. [111] reported MRI based therapy for enhancing the chemotherapeutic effects of liposomes (Doxil). The main objective is to attain intra-tumor release of drug and favoring its diffusion into the lesion by consequent application of low frequency pulsed US. The experiment was carrying out on mice bearing synergistic breast cancer model. The outcomes indicate that the combination of acoustic waves which permits the release of both drug and MRI agent (Gadateridol) from liposomes. The MR-T1 contrast improvement was assessed in cancer cell for the US-induced drug release. It could be concluded that developed protocol has clinical applications [111]. Salkho et al. [112] developed DOX-loaded liposomes for estrogen receptor (ER)-positive breast cancer therapy. The use of estrone was to modify the liposomes using cyanuric acid (2,4,6 tricholoro-1,3,5 triazine) as linker to attach 3-OH group of estrogen to the surface of liposomes. US-triggered mechanism at different frequencies was utilized to determine the release of drug. Cell uptake studies were assessed ER positive (MCF-7) and ER negative cell lines (MDA-MB-231). The results depicted the additive effects of liposomes due to receptor mediated targeting and US-triggered drug release. The uptake of calcein from liposomes via MCF-7 cell lines was higher compared to MDA-MB-231 cell lines. Hence, the liposomes could be a potential approach for site specific ER-positive breast cancer therapy [112].

4.5 Microbubbles

US triggered MBs produce temporary or permanent pore in the membrane via cavitation. MBs deliver the drugs in targeted organ by two ways of cavitation. (i) Stable cavitation (MBs oscillate without collapsing in an acoustic field), (ii) inertial cavitation (MBs randomly grow and collapse). Sonoporation technique is applied to deliver anticancer drugs without affecting the physiological function and defense mechanism of cells. In sonoporation, the temporary pores are created which return to their original conformation after sometime [49]. Di Ianni et al. [113] prepared miRNA loaded US triggered MB-mediated PEGylated PLGA NPs. The FUS along with MBs used to enhance the penetration of drug into cancer cell. It also enhances the permeability of drug loaded NPs. The US selection study was based on image directed therapy where SonoVue used as a contrasting agent. They evaluated the MBs cavitation dose through acoustic parameters, such as pulse length and pulse frequency. Further, in vivo study demonstrated that an increase in miRNA delivery by 1.9–3.7 folds by US treated area compared to untreated area [113]. US-induced chemotherapy was developed using

PTX liposomes MB complex (PLMC) for breast cancer therapy. MBs surface was modified by biotin–avidin linkage. Upon US-exposure the drug release from PLMC was increased. Breast cancer cells (4T1) were treated with QD-QLMC and US, (QD is the fluorescent quantum dots) which is used as reference drug that is released from QDs. In vitro and in vivo studies indicated that PLMC with US reduced the tumor growth efficiently than PLMC without US. The reduction in tumor growth was due to increased apoptosis in tumor xenografts. An increase in drug concentration to tumors site was observed on the treatment with US-based PLCM as compared to without US-based PLCM [49]. Chang et al. [114] prepared VEGFR2 targeted and cationic MBs (VCMBs) for efficient gene delivery. Tumor models like firefly luciferase gene and herpes simplex virus were used to monitor the developed DDS. The chemotherapeutic effect and transgene expression was assessed under FUS exposure. Results of in vitro study demonstrated that VCMBs possessed higher DNA entrapment efficiency and affinity for brain tumor cells [114].

4.6 Nanoparticles

US has abundant potential for improving the delivery of NPs. US-directed MRI is another approach for the theranostic application, where US contrast agent and magnetic NPs complex was used in contrast MRI. It became an effective theranostic application in biomedical field [14]. NPs used as targeted DDS for selective delivery of drug at desired site [115]. Chumakova et al. [52] developed complex NPs composed of air filled PLGA NPs and PEI (Polyethylenimine)/DNA b-gal plasmid NPs. These NPs exhibit positive and negative zeta potential, respectively. These NPs were inserted intravenously into mice (with prostate cancer cell lines DU145). The outcomes of in vivo study suggested that the ultrasonication of the NPs significantly improved tumor cell transfection. Tumor cell transfection study showed eightfold increases the cytotoxicity in irradiated tumor compared to nonirradiated tumor. The overexpression of b-gal gene was measured by histochemical analysis and western blot techniques [52]. Paris et al. [116] prepared US stimulated mesoporous NPs. The mechanism to stimulate their extravasation in flow-through-tissue mimicking agarose phantom was monitored using US-induced inertial cavitation. Cavitation was measured by two different type of US frequencies 0.5 or 1.6 MHz with pressure (0.5–4 MPa) for real time detection. These NPs improved cavitation activity reduced in situ AP to achieve extravasation (when injected with sub-micrometric polymeric cavitation nuclei). Thus, it could be concluded that the mesoporous silica NPs combined with submicrometric cavitation energy improved extravasation of nanocarrier. This leads to consequent release of NPs at cancerous tissue [116]. For tumor treatment, US-triggered ADV has been described as a theranostic strategy. But there is challenge in targeting of droplets to cancerous site. Thus, above study was employed to overcome this problem by incorporating superparamagnetic iron oxide (SPIO) NPs into acoustic droplets to permit both magnetic-assisted targeting and MR-guided US-triggered ADV. The versatile nature of droplets was enhanced by loading anticancer drug (DOX) and modified by antivascular endothelial growth factor receptor-2 antibody. Further, the potentiality of dual model imaging of SPIO (with droplets) was established by MRI technique. The animal study was demonstrated a 5.4 fold increase in targeting efficiency of SPIO embedded droplets. This led to effective destruction of cell by US-induced ADV which could be utilized for further clinical applications [117].

5 Conclusion and future perspectives

USDD using several DDS such as nanobubbles, nanodroplets, micelles, liposomes, MBs and NPs, etc. could improve drug delivery to target site. It enhances the bioavailability, therapeutic efficacy and diminishes the frequency of drug administration. Moreover, it offers controlled and targeted release of drug. USDD system is a potential strategy to target specific disease by combining the drug with diagnostic molecules. This will be an effective therapy and help to improve the disease. Ligand-targeted DDS may provide specific targeting to breast cancer, brain tumor and lung cancer etc. MRI with US is a favorable technique for therapeutic purpose, in which a US contrast agent is entrapped in nanocarrier to provide contrast in magnetic resonance imaging. Hence, the USDD is capable of delivering drug to targeted site and reduce the unwanted effects related with chemotherapeutic drug.

Consent for publication

Not applicable.

Conflict of interest

The authors confirm that this chapter contents have no conflict of interest.

Acknowledgments

Sarjana Raikwar (SRF), Shivani Saraf (SRF), Pritish Kumar Panda (SRF), Ankita Tiwari (SRF), Pooja Das Bidla (SRF) and Amit Verma (RA) are highly obliged to Indian Council of Medical Research (ICMR, New Delhi) for rendering funding assistance.

References

[1] W.G. Pitt, G.A. Husseini, B.J. Staples, Ultrasonic drug delivery—a general review, Expert Opin. Drug Deliv. 1 (1) (2004) 37–56.
[2] S.R. Sirsi, M.A. Borden, State-of-the-art materials for ultrasound-triggered drug delivery, Adv. Drug Deliv. Rev. 72 (2014) 3–14.
[3] A. Jain, S.K. Jain, Stimuli-responsive smart liposomes in cancer targeting, Curr. Drug Targets 19 (3) (2018) 259–270. 26853324 (Epub 2016/02/09. Eng).
[4] S. Padhi, M.A. Mirza, D. Verma, T. Khuroo, A.K. Panda, S. Talegaonkar, et al., Revisiting the nanoformulation design approach for effective delivery of topotecan in its stable form: an appraisal of its in vitro behavior and tumor amelioration potential, Drug Deliv. 23 (8) (2016) 2827–2837.
[5] Q.-L. Zhou, Z.-Y. Chen, Y.-X. Wang, F. Yang, Y. Lin, Y.-Y. Liao, Ultrasound-mediated local drug and gene delivery using nanocarriers, Biomed. Res. Int. 2014 (2014) 1–13.
[6] A. Jain, A. Tiwari, A. Verma, S.K. Jain, Ultrasound-based triggered drug delivery to tumors, Drug Deliv. Transl. Res. 8 (1) (2018) 150–164.
[7] D. Gourevich, O. Dogadkin, A. Volovick, L. Wang, J. Gnaim, S. Cochran, et al., Ultrasound-mediated targeted drug delivery with a novel cyclodextrin-based drug carrier by mechanical and thermal mechanisms, J. Control. Release 170 (3) (2013) 316–324.
[8] F. Yan, Z. Song, M. Du, A.L. Klibanov, Ultrasound molecular imaging for differentiation of benign and malignant tumors in patients, Quant. Imaging Med. Surg. 8 (11) (2018) 1078.
[9] A. Jain, S.K. Jain, In vitro release kinetics model fitting of liposomes: an insight, Chem. Phys. Lipids 201 (2016) 28–40.27983957.
[10] A. Schroeder, R. Honen, K. Turjeman, A. Gabizon, J. Kost, Y. Barenholz, Ultrasound triggered release of cisplatin from liposomes in murine tumors, J. Control. Release 137 (1) (2009) 63–68. 19303426 (Epub 2009/03/24. eng).
[11] J. Kost, U. Pliquett, S.S. Mitragotri, R.S. Langer, J.C. Weaver, Effect of Electric Field and Ultrasound for Transdermal Drug Delivery, Google Patents(2000).
[12] S. Saraf, A. Jain, A. Tiwari, A. Verma, P.K. Panda, S.K. Jain, Advances in liposomal drug delivery to cancer: an overview, J. Drug Delivery Sci. Technol. 56 (2020).
[13] A. Jain, P. Hurkat, S.K. Jain, Development of liposomes using formulation by design: basics to recent advances, Chem. Phys. Lipids 224 (2019) 1–45.
[14] B. Joshi, A. Joshi, Ultrasound-based drug delivery systems, in: Bioelectronics and Medical Devices, Elsevier, 2019, pp. 241–260.

[15] A. Joshi, R. Chaudhari, R.D. Jayant, On-demand controlled drug delivery, in: Advances in Personalized Nanotherapeutics, Springer, 2017, pp. 131–156.
[16] S. Mitragotri, Healing sound: the use of ultrasound in drug delivery and other therapeutic applications, Nat. Rev. Drug Discov. 4 (3) (2005) 255–260.
[17] A. Jain, A. Tiwari, A. Verma, S.K. Jain, Ultrasound-based triggered drug delivery to tumors, Drug Deliv. Transl. Res. 8 (2017) 1–15.
[18] J.M. Escoffre, A. Novell, M. de Smet, A. Bouakaz, Focused ultrasound mediated drug delivery from temperature-sensitive liposomes: in-vitro characterization and validation, Phys. Med. Biol. 58 (22) (2013) 8135–8151. 24200816 (Epub 2013/11/10. eng).
[19] R. Deckers, C.T. Moonen, Ultrasound triggered, image guided, local drug delivery, J. Control. Release 148 (1) (2010) 25–33.
[20] R. Staruch, R. Chopra, K. Hynynen, Localised drug release using MRI-controlled focused ultrasound hyperthermia, Int. J. Hyperthermia 27 (2) (2011) 156–171.
[21] A. Ranjan, G.C. Jacobs, D.L. Woods, A.H. Negussie, A. Partanen, P.S. Yarmolenko, et al., Image-guided drug delivery with magnetic resonance guided high intensity focused ultrasound and temperature sensitive liposomes in a rabbit Vx2 tumor model, J. Control. Release 158 (3) (2012) 487–494.
[22] C. Yang, Y. Li, M. Du, Z. Chen, Recent advances in ultrasound-triggered therapy, J. Drug Target. 27 (1) (2019) 33–50.
[23] S. Mura, J. Nicolas, P. Couvreur, Stimuli-responsive nanocarriers for drug delivery, Nat. Mater. 12 (11) (2013) 991.
[24] A. Schroeder, J. Kost, Y. Barenholz, Ultrasound, liposomes, and drug delivery: principles for using ultrasound to control the release of drugs from liposomes, Chem. Phys. Lipids 162 (1) (2009) 1–16.
[25] M. Afadzi, L. Davies Cde, Y.H. Hansen, T. Johansen, O.K. Standal, R. Hansen, et al., Effect of ultrasound parameters on the release of liposomal calcein, Ultrasound Med. Biol. 38 (3) (2012) 476–486. 22264409 (Epub 2012/01/24. eng).
[26] C.-D. Ohl, M. Arora, R. Ikink, N. De Jong, M. Versluis, M. Delius, et al., Sonoporation from jetting cavitation bubbles, Biophys. J. 91 (11) (2006) 4285–4295.
[27] I. Lentacker, N. Wang, R.E. Vandenbroucke, J. Demeester, S.C. De Smedt, N. N. Sanders, Ultrasound exposure of lipoplex loaded microbubbles facilitates direct cytoplasmic entry of the lipoplexes, Mol. Pharm. 6 (2) (2009) 457–467.
[28] B. Geers, I. Lentacker, A. Alonso, N.N. Sanders, J. Demeester, S. Meairs, et al., Elucidating the mechanisms behind sonoporation with adeno-associated virus-loaded microbubbles, Mol. Pharm. 8 (6) (2011) 2244–2251.
[29] B.D. Meijering, L.J. Juffermans, A. van Wamel, R.H. Henning, I.S. Zuhorn, M. Emmer, et al., Ultrasound and microbubble-targeted delivery of macromolecules is regulated by induction of endocytosis and pore formation, Circ. Res. 104 (5) (2009) 679–687.
[30] E.C. Unger, E. Hersh, M. Vannan, T.O. Matsunaga, T. McCreery, Local drug and gene delivery through microbubbles, Prog. Cardiovasc. Dis. 44 (1) (2001) 45–54.
[31] K. Kalantarinia, J.T. Belcik, J.T. Patrie, K. Wei, Real-time measurement of renal blood flow in healthy subjects using contrast-enhanced ultrasound, Am. J. Physiol. Renal Physiol. 297 (4) (2009) F1129–34.
[32] Y.-Z. Zhao, L.-N. Du, C.-T. Lu, Y.-G. Jin, S.-P. Ge, Potential and problems in ultrasound-responsive drug delivery systems, Int. J. Nanomedicine 8 (2013) 1621.
[33] B.E. Polat, D. Hart, R. Langer, D. Blankschtein, Ultrasound-mediated transdermal drug delivery: mechanisms, scope, and emerging trends, J. Control. Release 152 (3) (2011) 330–348.
[34] M. Du, Z. Chen, Y. Chen, Y. Li, Ultrasound-targeted delivery technology: a novel strategy for tumor-targeted therapy, Curr. Drug Targets 20 (2) (2019) 220–231.

[35] A.-Z. Zardad, Y.E. Choonara, L.C. Du Toit, P. Kumar, M. Mabrouk, P.P. D. Kondiah, et al., A review of thermo-and ultrasound-responsive polymeric systems for delivery of chemotherapeutic agents, Polymers 8 (10) (2016) 359.
[36] S.K. Prajapati, A. Jain, A. Jain, S. Jain, Biodegradable polymers and constructs: a novel approach in drug delivery, Eur. Polym. J. 120 (2019) WOS:000498309000001 (English).
[37] A. Jain, S.K. Jain, Environmentally responsive chitosan-based nanocarriers (CBNs), in: Handbook of Polymers for Pharmaceutical Technologies, Biodegradable Polymers, vol. 3, Scrivener Publishing, 2015, p. 105.
[38] A. Jain, A. Gulbake, S. Shilpi, A. Jain, P. Hurkat, S.K. Jain, A new horizon in modifications of chitosan: syntheses and applications, Crit. Rev. Ther. Drug Carrier Syst. 30 (2) (2013) 91–181.23510147.
[39] J. Lin, M. Stevens, J. Smith, Preparation of anti-HER-2 antibody PLGA polymer nano-ultrasound contrast agent in vitro targeting experiment, (2019)bioRxiv, 619742.
[40] C.-H. Fan, T.-W. Wang, Y.-K. Hsieh, C.-F. Wang, Z. Gao, A. Kim, et al., Enhancing boron uptake in brain glioma by a boron-polymer/microbubble complex with focused ultrasound, ACS Appl. Mater. Interfaces 11 (12) (2019) 11144–11156.
[41] J.L. Paris, P. de la Torre, M.V. Cabañas, M. Manzano, A.I. Flores, M. Vallet-Regí, Suicide-gene transfection of tumor-tropic placental stem cells employing ultrasound-responsive nanoparticles, Acta Biomater. 83 (2019) 372–378.
[42] S. Houvenagel, L. Moine, G. Picheth, C. Dejean, A. Brûlet, A. Chennevière, et al., Comb-like fluorophilic-lipophilic-hydrophilic polymers for nanocapsules as ultrasound contrast agents, Biomacromolecules 19 (8) (2018) 3244–3256.
[43] Z. Abed, S. Khoei, B. Ghalandari, J. Beik, A. Shakeri-Zadeh, H. Ghaznavi, et al., The measurement and mathematical analysis of 5-Fu release from magnetic polymeric nanocapsules, following the application of ultrasound, Anticancer Agents Med Chem. 18 (3) (2018) 438–449.
[44] K.W. Jang, D. Seol, L. Ding, D.N. Heo, S.J. Lee, J.A. Martin, et al., Ultrasound-triggered PLGA microparticle destruction and degradation for controlled delivery of local cytotoxicity and drug release, Int. J. Biol. Macromol. 106 (2017) 1211–1217.
[45] J. Xia, J. Wang, X. Wang, M. Qian, L. Zhang, Q. Chen, Ultrasound-responsive nanoparticulate for selective amplification of chemotherapeutic potency for ablation of solid tumors, Bioconjug. Chem. 29 (10) (2018) 3467–3475.
[46] A.-L. Papa, N. Korin, M. Kanapathipillai, A. Mammoto, T. Mammoto, A. Jiang, et al., Ultrasound-sensitive nanoparticle aggregates for targeted drug delivery, Biomaterials 139 (2017) 187–194.
[47] Q. Wu, M. Zhang, H. Luo, T. Yi, Self-assembled honokiol-loaded microbubbles in the treatment of ovarian cancer by ultrasound irradiation, J. Biomed. Nanotechnol. 14 (10) (2018) 1796–1805.
[48] R. Abdalkader, S. Kawakami, J. Unga, R. Suzuki, K. Maruyama, F. Yamashita, et al., Evaluation of the potential of doxorubicin loaded microbubbles as a theranostic modality using a murine tumor model, Acta Biomater. 19 (2015) 112–118.
[49] F. Yan, L. Li, Z. Deng, Q. Jin, J. Chen, W. Yang, et al., Paclitaxel-liposome–microbubble complexes as ultrasound-triggered therapeutic drug delivery carriers, J. Control. Release 166 (3) (2013) 246–255.
[50] C.-H. Fan, C.-Y. Ting, H.-L. Liu, C.-Y. Huang, H.-Y. Hsieh, T.-C. Yen, et al., Antiangiogenic-targeting drug-loaded microbubbles combined with focused ultrasound for glioma treatment, Biomaterials 34 (8) (2013) 2142–2155.
[51] I. Lentacker, B. Geers, J. Demeester, S.C. De Smedt, N.N. Sanders, Design and evaluation of doxorubicin-containing microbubbles for ultrasound-triggered doxorubicin delivery: cytotoxicity and mechanisms involved, Mol. Ther. 18 (1) (2010) 101–108.

[52] O.V. Chumakova, A.V. Liopo, V.G. Andreev, I. Cicenaite, B.M. Evers, S. Chakrabarty, et al., Composition of PLGA and PEI/DNA nanoparticles improves ultrasound-mediated gene delivery in solid tumors in vivo, Cancer Lett. 261 (2) (2008) 215–225.
[53] Z. Gao, H.D. Fain, N. Rapoport, Ultrasound-enhanced tumor targeting of polymeric micellar drug carriers, Mol. Pharm. 1 (4) (2004) 317–330.
[54] D.M. El-Sherif, M.A. Wheatley, Development of a novel method for synthesis of a polymeric ultrasound contrast agent, J. Biomed. Mater. Res. A 66 (2) (2003) 347–355.
[55] Q. Wu, X. Chen, L. Jia, Y. Wang, Y. Sun, X. Huang, et al., Ultrasonic irradiation enhanced the ability of fluorescein-DA-Fe(III) on sonodynamic and sonocatalytic damages of DNA molecules, Ultrason. Sonochem. 39 (2017) 1–11. 28732924 (Epub 2017/07/25. eng).
[56] B. Liang, R. Tong, Z. Wang, S. Guo, H. Xia, High intensity focused ultrasound responsive metallo-supramolecular block copolymer micelles, Langmuir 30 (31) (2014) 9524–9532.
[57] Z. Zhang, J. Wang, X. Nie, T. Wen, Y. Ji, X. Wu, et al., Near infrared laser-induced targeted cancer therapy using thermoresponsive polymer encapsulated gold nanorods, J. Am. Chem. Soc. 136 (20) (2014) 7317–7326.
[58] Q. Zhou, L. Zhang, T. Yang, H. Wu, Stimuli-responsive polymeric micelles for drug delivery and cancer therapy, Int. J. Nanomedicine 13 (2018) 2921.
[59] S. Chen, X.-L. Xu, B. Zhou, J. Tian, B.-M. Luo, L.-M. Zhang, Acidic pH-activated gas-generating nanoparticles with Pullulan decorating for Hepatoma-targeted ultrasound imaging, ACS Appl. Mater. Interfaces 11 (25) (2019) 22194–22205.
[60] X. Zhou, L. Guo, D. Shi, S. Duan, J. Li, Biocompatible chitosan nanobubbles for ultrasound-mediated targeted delivery of doxorubicin, Nanoscale Res. Lett. 14 (1) (2019) 24.
[61] M. Shang, K. Wang, L. Guo, S. Duan, Z. Lu, J. Li, Development of novel ST68/PLA-PEG stabilized ultrasound nanobubbles for potential tumor imaging and theranostic, Ultrasonics 99 (2019) 105947.
[62] Z. Cheng, A. Al Zaki, J.Z. Hui, V.R. Muzykantov, A. Tsourkas, Multifunctional nanoparticles: cost versus benefit of adding targeting and imaging capabilities, Science 338 (6109) (2012) 903–910.
[63] H. Liang, J. Tang, M. Halliwell, Sonoporation, drug delivery, and gene therapy, Proc. Inst. Mech. Eng. H J. Eng. Med. 224 (2) (2010) 343–361.
[64] H. Li, Z. Wang, J. Zhang, C. Yuan, H. Zhang, X. Ho, et al., Enhanced shRNA delivery by the combination of polyethylenimine, ultrasound, and nanobubbles in liver cancer, Technol. Health Care 27 (S1) (2019) 1–10.
[65] R. Cavalli, A. Bisazza, M. Trotta, M. Argenziano, A. Civra, M. Donalisio, et al., New chitosan nanobubbles for ultrasound-mediated gene delivery: preparation and in vitro characterization, Int. J. Nanomedicine 7 (2012) 3309.
[66] S. Horie, Y. Watanabe, M. Ono, S. Mori, T. Kodama, Evaluation of antitumor effects following tumor necrosis factor-α gene delivery using nanobubbles and ultrasound, Cancer Sci. 102 (11) (2011) 2082–2089.
[67] M.C. Deshpande, M.R. Prausnitz, Synergistic effect of ultrasound and PEI on DNA transfection in vitro, J. Control. Release 118 (1) (2007) 126–135.
[68] M. Figueiredo, R. Esenaliev, PLGA nanoparticles for ultrasound-mediated gene delivery to solid tumors, J. Drug Deliv. 2012 (2012) 1–20.
[69] N.Y. Rapoport, A.M. Kennedy, J.E. Shea, C.L. Scaife, K.-H. Nam, Controlled and targeted tumor chemotherapy by ultrasound-activated nanoemulsions/microbubbles, J. Control. Release 138 (3) (2009) 268–276.
[70] P. Shende, S. Jain, Polymeric nanodroplets: an emerging trend in gaseous delivery system, J. Drug Target. 27 (10) (2019) 1035–1045.

[71] N. Rapoport, K.-H. Nam, R. Gupta, Z. Gao, P. Mohan, A. Payne, et al., Ultrasound-mediated tumor imaging and nanotherapy using drug loaded, block copolymer stabilized perfluorocarbon nanoemulsions, J. Control. Release 153 (1) (2011) 4–15.
[72] M. Liu, G. Zeng, K. Wang, Q. Wan, L. Tao, X. Zhang, et al., Recent developments in polydopamine: an emerging soft matter for surface modification and biomedical applications, Nanoscale 8 (38) (2016) 16819–16840.
[73] N. Rapoport, A. Payne, C. Dillon, J. Shea, C. Scaife, R. Gupta, Focused ultrasound-mediated drug delivery to pancreatic cancer in a mouse model, J. Ther. Ultrasound 1 (1) (2013) 11.
[74] P.S. Sheeran, S.H. Luois, L.B. Mullin, Matsunaga TO, P.A. Dayton, Design of ultrasonically-activatable nanoparticles using low boiling point perfluorocarbons, Biomaterials 33 (11) (2012) 3262–3269.
[75] Matsunaga TO, P.S. Sheeran, S. Luois, J.E. Streeter, L.B. Mullin, B. Banerjee, et al., Phase-change nanoparticles using highly volatile perfluorocarbons: toward a platform for extravascular ultrasound imaging, Theranostics 2 (12) (2012) 1185.
[76] D. Hasty, J. Drapekin, T. Subramanian, T.C. Winter, J.S. Chickos, A.A. Samarov, et al., Applications of correlation gas chromatography and transpiration studies for the evaluation of the vaporization and sublimation enthalpies of some perfluorinated hydrocarbons, J. Chem. Eng. Data 57 (8) (2012) 2350–2359.
[77] P.A. Dayton, S. Zhao, S.H. Bloch, P. Schumann, K. Penrose, T.O. Matsunaga, et al., Application of ultrasound to selectively localize nanodroplets for targeted imaging and therapy, Mol. Imaging 5 (3) (2006) 7290.2006. 00019.
[78] J.Y. Lee, D. Carugo, C. Crake, J. Owen, V.M. de Saint, A. Seth, et al., Nanoparticle-loaded protein–polymer nanodroplets for improved stability and conversion efficiency in ultrasound imaging and drug delivery, Adv. Mater. 27 (37) (2015) 5484–5492.
[79] M.L. Fabiilli, K.J. Haworth, I.E. Sebastian, O.D. Kripfgans, P.L. Carson, J.B. Fowlkes, Delivery of chlorambucil using an acoustically-triggered perfluoropentane emulsion, Ultrasound Med. Biol. 36 (8) (2010) 1364–1375.
[80] Z. H-f, H.W. Chan, S.A. Wickline, G.M. Lanza, C.T. Pham, $\alpha v\beta 3$-Targeted nanotherapy suppresses inflammatory arthritis in mice, FASEB J. 23 (9) (2009) 2978–2985.
[81] D. Gao, M. Xu, Z. Cao, J. Gao, Y. Chen, Y. Li, et al., Ultrasound-triggered phase-transition cationic nanodroplets for enhanced gene delivery, ACS Appl. Mater. Interfaces 7 (24) (2015) 13524–13537.
[82] J.Y. Lee, C. Crake, B. Teo, D. Carugo, V.M. de Saint, A. Seth, et al., Ultrasound-enhanced siRNA delivery using magnetic nanoparticle-loaded chitosan-Deoxycholic acid Nanodroplets, Adv. Healthc. Mater. 6 (8) (2017) 1601246.
[83] M.T. Burgess, T.M. Porter, Acoustic cavitation-mediated delivery of small interfering ribonucleic acids with phase-shift nano-emulsions, Ultrasound Med. Biol. 41 (8) (2015) 2191–2201.
[84] R.J. Paproski, A. Forbrich, M. Hitt, R. Zemp, RNA biomarker release with ultrasound and phase-change nanodroplets, Ultrasound Med. Biol. 40 (8) (2014) 1847–1856.
[85] C. Mannaris, C. Yang, D. Carugo, J. Owen, J.Y. Lee, S. Nwokeoha, et al., Acoustically responsive polydopamine nanodroplets: a novel theranostic agent, Ultrason. Sonochem. 60 (2020) 104782.
[86] Y.-J. Ho, Y.-J. Chiang, S.-T. Kang, C.-H. Fan, C.-K. Yeh, Camptothecin-loaded fusogenic nanodroplets as ultrasound theranostic agent in stem cell-mediated drug-delivery system, J. Control. Release 278 (2018) 100–109.
[87] H. Mizuno, M. Tobita, A.C. Uysal, Concise review: adipose-derived stem cells as a novel tool for future regenerative medicine, Stem Cells 30 (5) (2012) 804–810.

[88] D.T. Josiah, D. Zhu, F. Dreher, J. Olson, G. McFadden, H. Caldas, Adipose-derived stem cells as therapeutic delivery vehicles of an oncolytic virus for glioblastoma, Mol. Ther. 18 (2) (2010) 377–385.
[89] L. Casteilla, V. Planat-Benard, P. Laharrague, B. Cousin, Adipose-derived stromal cells: their identity and uses in clinical trials, an update, World J Stem Cells 3 (4) (2011) 25.
[90] N. Rapoport, Drug-loaded perfluorocarbon nanodroplets for ultrasound-mediated drug delivery, in: Therapeutic Ultrasound, Springer, 2016, pp. 221–241.
[91] G. Gaucher, R.H. Marchessault, J.-C. Leroux, Polyester-based micelles and nanoparticles for the parenteral delivery of taxanes, J. Control. Release 143 (1) (2010) 2–12.
[92] R. Savić, A. Eisenberg, D. Maysinger, Block copolymer micelles as delivery vehicles of hydrophobic drugs: micelle–cell interactions, J. Drug Target. 14 (6) (2006) 343–355.
[93] N. Rapoport, Ultrasound-mediated micellar drug delivery, Int. J. Hyperthermia 28 (4) (2012) 374–385.
[94] X. Han, J. Liu, M. Liu, C. Xie, C. Zhan, B. Gu, et al., 9-NC-loaded folate-conjugated polymer micelles as tumor targeted drug delivery system: preparation and evaluation in vitro, Int. J. Pharm. 372 (1) (2009) 125–131.
[95] V. Torchilin, Targeted polymeric micelles for delivery of poorly soluble drugs, Cell. Mol. Life Sci. 61 (19–20) (2004) 2549–2559.
[96] V.P. Torchilin, A.N. Lukyanov, Z. Gao, B. Papahadjopoulos-Sternberg, Immunomicelles: targeted pharmaceutical carriers for poorly soluble drugs, Proc. Natl. Acad. Sci. 100 (10) (2003) 6039–6044.
[97] N. Rapoport, Physical stimuli-responsive polymeric micelles for anti-cancer drug delivery, Prog. Polym. Sci. 32 (8–9) (2007) 962–990.
[98] H. Zhang, H. Xia, J. Wang, Y. Li, High intensity focused ultrasound-responsive release behavior of PLA-b-PEG copolymer micelles, J. Control. Release 139 (1) (2009) 31–39.
[99] C. Yang, Y. Liu, Y. He, Y. Du, W. Wang, X. Shi, et al., The use of HA oligosaccharide-loaded nanoparticles to breach the endogenous hyaluronan glycocalyx for breast cancer therapy, Biomaterials 34 (28) (2013) 6829–6838.
[100] S. Zheng, Z. Jin, J. Han, S. Cho, S.Y. Ko, J.-O. Park, et al., Preparation of HIFU-triggered tumor-targeted hyaluronic acid micelles for controlled drug release and enhanced cellular uptake, Colloids Surf. B Biointerfaces 143 (2016) 27–36.
[101] P. Wu, Y. Jia, F. Qu, Y. Sun, P. Wang, K. Zhang, et al., Ultrasound-responsive polymeric micelles for sonoporation-assisted site-specific therapeutic action, ACS Appl. Mater. Interfaces 9 (31) (2017) 25706–25716.
[102] A. Jain, S.K. Jain, Chapter 9: Application potential of engineered liposomes in tumor targeting, in: A. Grumezescu (Ed.), Multifunctional Systems for Combined Delivery, Biosensing and Diagnostics, Elsevier—Health Sciences Division, 2017, pp. 171–192.
[103] A. Jain, S.K. Jain, Multipronged, strategic delivery of paclitaxel-topotecan using engineered liposomes to ovarian cancer, Drug Dev. Ind. Pharm. 42 (1) (2016) 136–149. 26006330 (Epub 2015/05/27. Eng).
[104] A. Jain, A. Gulbake, A. Jain, S. Shilpi, P. Hurkat, S.K. Jain, Dual drug delivery using "smart" liposomes for triggered release of anticancer agents, J. Nanopart. Res. 15 (7) (2013) 1–12.
[105] L. Sercombe, T. Veerati, F. Moheimani, S.Y. Wu, A.K. Sood, S. Hua, Advances and challenges of liposome assisted drug delivery, Front. Pharmacol. 6 (2015) 286. 26648870 4664963.
[106] Y. Shen, Z. Pi, F. Yan, C.-K. Yeh, X. Zeng, X. Diao, et al., Enhanced delivery of paclitaxel liposomes using focused ultrasound with microbubbles for treating nude mice bearing intracranial glioblastoma xenografts, Int. J. Nanomedicine 12 (2017) 5613.

[107] N.S. Awad, V. Paul, M.H. Al-Sayah, G.A. Husseini, Ultrasonically controlled albumin-conjugated liposomes for breast cancer therapy, Artif. Cells Nanomed. Biotechnol. 47 (1) (2019) 705–714.
[108] M.N. Centelles, M. Wright, P.-W. So, M. Amrahli, X.Y. Xu, J. Stebbing, et al., Image-guided thermosensitive liposomes for focused ultrasound drug delivery: using NIRF-labelled lipids and topotecan to visualise the effects of hyperthermia in tumours, J. Control. Release 280 (2018) 87–98.
[109] W. Um, S. Kwon, D.G. You, J.M. Cha, H.R. Kim, J.H. Park, Non-thermal acoustic treatment as a safe alternative to thermosensitive liposome-involved hyperthermia for cancer therapy, RSC Adv. 7 (47) (2017) 29618–29625.
[110] T. Ouchi, A. Sukhbaatar, S. Horie, M. Sakamoto, K. Shiga, S. Mori, et al., Superselective drug delivery using doxorubicin-encapsulated liposomes and ultrasound in a mouse model of lung metastasis activation, Ultrasound Med. Biol. 44 (8) (2018) 1818–1827.
[111] S. Rizzitelli, P. Giustetto, D. Faletto, D.D. Castelli, S. Aime, E. Terreno, The release of doxorubicin from liposomes monitored by MRI and triggered by a combination of US stimuli led to a complete tumor regression in a breast cancer mouse model, J. Control. Release 230 (2016) 57–63.
[112] N.M. Salkho, V. Paul, P. Kawak, R.F. Vitor, A.M. Martins, M. Al Sayah, et al., Ultrasonically controlled estrone-modified liposomes for estrogen-positive breast cancer therapy, Artif. Cells Nanomed. Biotechnol. 46 (sup2) (2018) 462–472.
[113] T. Di Ianni, R.J. Bose, U.K. Sukumar, S. Bachawal, H. Wang, A. Telichko, et al., Ultrasound/microbubble-mediated targeted delivery of anticancer microRNA-loaded nanoparticles to deep tissues in pigs, J. Control. Release 309 (2019) 1–10.
[114] E.-L. Chang, C.-Y. Ting, P.-H. Hsu, Y.-C. Lin, E.-C. Liao, C.-Y. Huang, et al., Angiogenesis-targeting microbubbles combined with ultrasound-mediated gene therapy in brain tumors, J. Control. Release 255 (2017) 164–175.
[115] S. Raikwar, S. Vyas, R. Sharma, N. Mody, S. Dubey, S.P. Vyas, Nanocarrier-based combination chemotherapy for resistant tumor: development, characterization, and ex vivo cytotoxicity assessment, AAPS PharmSciTech 19 (8) (2018) 3839–3849.
[116] J.L. Paris, C. Mannaris, M.V. Cabañas, R. Carlisle, M. Manzano, M. Vallet-Regí, et al., Ultrasound-mediated cavitation-enhanced extravasation of mesoporous silica nanoparticles for controlled-release drug delivery, Chem. Eng. J. 340 (2018) 2–8.
[117] C.-H. Wang, S.-T. Kang, C.-K. Yeh, Superparamagnetic iron oxide and drug complex-embedded acoustic droplets for ultrasound targeted theranosis, Biomaterials 34 (7) (2013) 1852–1861.

CHAPTER 17

A scientometric review of hydrogel-based ocular drug delivery systems

Kunal Pal[a], Arfat Anis[b], Amit Kumar Nayak[c], and Samarendra Maji[d]
[a]Department of Biotechnology and Medical Engineering, National Institute of Technology Rourkela, Rourkela, Odisha, India
[b]SABIC Polymer Research Center, Department of Chemical Engineering, King Saud University, Riyadh, Saudi Arabia
[c]Department of Pharmaceutics, Seemanta Institute of Pharmaceutical Sciences, Mayurbhanj, Odisha, India
[d]Department of Chemistry, Faculty of Engineering and Technology, SRM Institute of Science and Technology, SRM Nagar, Kattankulathur, Tamil Nadu, India

1 Introduction

The ocular anatomy and physiology present a great challenge for the ophthalmic doctors to deliver drugs to the internal structures of the eye [1]. The eyeball consists of different tissues, namely, a cornea that allows the light to gain entry within the internal structure of the eyeball, a sclera, a choroid, and a retina that is involved in the visual perception. Internally, the eyeball can be segregated into two chambers, namely, anterior chamber consisting of the cornea, iris, ciliary body, crystalline lens, and aqueous humor, and posterior chamber consisting of the vitreous humor, retina, retinal pigment epithelium, and optic disk [2]. A large population, across the globe, becomes incapacitated due to the visual impairment caused by ocular diseases (e.g., diabetic retinopathy, age-related macular degeneration, conjunctivitis, retinal vascular diseases, glaucomatous optic neuropathies, posterior uveitis, and glaucoma) [2, 3]. If proper medical intervention is not taken at the early stage, the ocular diseases may lead to the total loss of the visual activity of the patient [4]. It is important to note that every patient is unique in himself/herself. The onset and the progression of the diseases have been found to be differential in each patient. This phenomenon also creates hindrance toward not only the diagnosis but also the treatment of the patient by the ophthalmic doctors [2].

The delivery of the drugs to the posterior segment of the eye is a challenging issue [5]. Most of the ocular diseases that may impair visual activity is associated with the pathophysiological changes in the tissues of the posterior

Advances and Challenges in Pharmaceutical Technology
https://doi.org/10.1016/B978-0-12-820043-8.00011-6

segment of the eye. This has been apprehended to the fact that the drug formulations are expected to improve the bioavailability of the drugs in the posterior segment of the eye [6]. However, these formulations are either instilled in the cul-de-sac of the eye [7] or administered through systemic delivery routes [8]. When the drug formulations are instilled in the cul-de-sac of the eye, it is expected that the drugs will penetrate through the cornea and pass the anterior segment of the eye to reach the posterior segment of the eye. It is important to note that cornea acts as a natural barrier that prevents the delivery of the drugs into the anterior segment of the eye [9]. In general, the corneal permeation of the drugs is $<1\%$ and may be improved slightly using various permeation enhancers [10]. On the contrary, the administration of the drugs to the eye via the systemic circulation requires delivery of the drugs into the bloodstream [11, 12]. In either of the delivery routes, the bioavailability of the drugs in the posterior segment of the eye is poor. Hence, there is a need for the administration of a larger amount of drugs. However, this may increase the side effects associated with drugs. In this regard, various researchers have come up with novel drug delivery technologies, including micelles, liposomes, nanotubes, nanoparticles, microparticles, hydrogels, and ocular implants.

Among the various drug delivery technologies, hydrogels have attracted the attention of many of the researchers due to its ability not only to encapsulate the drug molecules but also to release the same in a controlled and sustained fashion [13]. Hydrogels are polymeric networks that can hold a large amount of water within its network structure [14]. The water-holding capacity of the hydrogels is related to the hydrophilic property of the polymers that are used for the synthesis of the hydrogels. Under normal circumstances, the water is not released by the hydrogels. Further, the hydrogels are further reported to increase the shelf-life of the drug molecules [15]. This has been reported to improve the hydration properties and biocompatibility of the hydrogels. It is important to note that the mechanical properties of the hydrogels can be easily tailored to match the properties of the surrounding tissues. Due to the aforesaid advantages, many hydrogel-based drug delivery systems are being marketed. Interestingly, the clinical use of hydrogels, apart from the non-oral route of drug administration, remains quite limited even though the hydrogel-based formulations have shown great promise as drug delivery systems, including ocular drug delivery systems [3]. There has been a surge in the use of hydrogel-based drug delivery systems for the ocular delivery of drugs in recent years.

Unfortunately, there is no systematic review of the applications of the hydrogel-based formulations as ocular drug delivery systems. A systematic literature review allows us to gain a considerable amount of knowledge in a specific domain in a relatively short time [16]. One of the quickest methods to perform a literature review of a large number of publications is bibliometric analysis [16]. The bibliometric analysis can not only help us in understanding the various relationships (e.g., author cooperation relationship, author-citation relationship, etc.) but also the current research trends in a specific field of study. Considering the above discussion, we propose to introduce the bibliometric analysis of the hydrogel-based ocular drug delivery systems so as to gain an insight on the most influential authors, institutions and countries involved in the said field of research, the most cited research articles and journals, and lastly, the recent trends in the field of the study. For the bibliometric analysis, the publication records were retrieved from Scopus.

2 Methodology

2.1 Data source

The bibliometric analysis was carried out by retrieving the publications from the Scopus database. The retrieval of the publications was done on March 25, 2020. The article searching was done using the search parameters (TITLE-ABS-KEY (hydrogels) AND TITLE-ABS-KEY ("drug delivery") AND TITLE-ABS-KEY (ocular)). No restriction on the time span was applied. This returned a total number of documents of 287. Out of the 287 documents, 284 documents were in English while one document each was in Croatian, Czech, and Spanish. Further, the documents other than research articles like review papers, book chapters, conference papers, editorials, and erratum were removed from the searched documents. This reduced the total number of documents to 205. All the 205 documents were in English. The citation information, bibliographical information, abstract and keywords, funding details, and other information was exported as CSV file.

2.2 Bibliometric analysis

The bibliometric analysis was carried out using the VOSviewer software. This software was developed by van Eck and Waltman [17]. The VOSviewer software provides a graphical representation of the bibliometric data as maps that are easy-to-interpret [17]. The distances between the nodes of the bibliometric map are related to the closeness between the nodes.

In other words, if the distance between a pair of nodes is closer than the distance between the other pair of nodes, then this information suggests that the first pair of nodes are closely related as compared to the second pair of nodes. Using the software, different bibliometric maps based on co-authorship, abstracts and keywords, co-occurrence of keywords in the titles, and co-citation of common references used in the articles were generated and analyzed [16, 17].

3 Results and discussions

3.1 Yearly distribution and growth trend

The number of publications that are published year-on-year basis provides a good estimate of the research trend in a specific field of research. The analysis of the trend in the number of publications can provide information on the probable research trend in the near future. A plot of the number of publications and cumulative publications on a year-on-year basis was plotted to analyze the research trend on the application of the hydrogels in the field of ocular drug delivery (Fig. 1). The research on the application of the hydrogels in ocular drug delivery dates back to the year 1985. However, it was not unit the year 2010 that researchers have shown great interest in the said field of research. Post-2009, except 2011, at least 10 articles have been published. In the year 2018, the highest number of articles (31) were published. This year (2020), until now, 10 articles have already been published. Furthermore, the cumulative publication graph suggests that researchers across

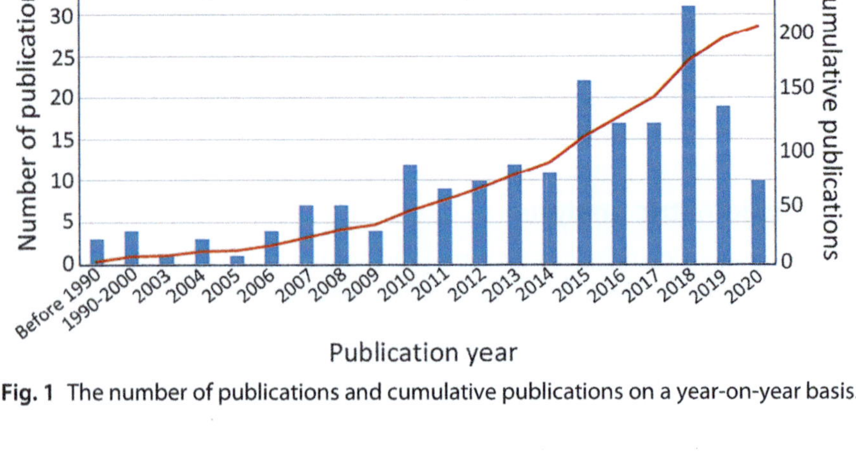

Fig. 1 The number of publications and cumulative publications on a year-on-year basis.

the globe have started paying attention to the field of hydrogel-based ocular drug delivery since the year 2003. This is evident from the fact there has been an exponential increase in the number of cumulative publications since the year 2003.

3.2 Distribution of publications on a country level

The 205 articles were published from 40 countries. The highest number of publications (41 articles, 16.94% of the total documents) was from the United States of America (USA), followed by 32 articles (13.22%) from China. India and Spain shared the 3rd rank, with each of them publishing 19 articles (7.85%). Subsequently, four countries published articles in the range of 10 and 13, five countries that published articles in the range of 5 and 9, four countries that published four articles, and two countries that published three articles. Furthermore, nine countries published two articles each and 12 countries that published one article each. The total number of publications by adding the contributions from each of the countries is 242, which is higher than 205. This suggests that there has been collaborative work between different countries. In total, there were 19 countries which published at least three articles in the field of hydrogels for ocular drug delivery applications. The statistics of these countries have been provided in Table 1. It is quite evident from Table 1 that out of 19 countries, 12 countries were having a nominal GDP rank of 15 or less. This is indicative of the fact that the economically developed countries have identified the benefits of hydrogels based drug delivery research and are exploring their feasibility for clinical applications. The United States of America has received the most number of citations from the 41 documents that were published from the country. Interestingly, the average citation of the United States of America was in the second position (43.27), just after Germany that had an average citation of 45.33 from total documents of three. The subsequent order of the countries that had an average citation of 25 or higher was Spain, Israel, Japan, Taiwan, and Canada. It is important to note that China and India are ranked second and third, respectively, in terms of the number of documents published. However, China ranks eighth, and India ranks sixteenth in terms of average citations.

The Total Link Strength (TLS) provides an estimation of the collaborative research of one country with the other (Table 1). The analysis of TLS suggested that the USA was by far the most superior country in terms of collaborative research with a TLS of 92. It was found that the USA had

Table 1 Top countries that published more than three documents.

#	Country	Quantity	Percentage (%)	Citation	Average citation per document	Nominal GDP rank[a]	Total link strength
1	United States	41	16.94	1774	43.27	1	92
2	China	32	13.22	798	24.94	2	77
3	India	19	7.85	252	13.26	5	20
4	Spain	19	7.85	657	34.58	13	64
5	Portugal	13	5.37	250	19.23	47	59
6	United Kingdom	13	5.37	202	15.54	6	32
7	Canada	10	4.13	259	25.90	10	19
8	Taiwan	10	4.13	282	28.20	–	12
9	Italy	9	3.72	180	20.00	8	25
10	Iran	7	2.89	110	15.71	25	15
11	Japan	7	2.89	201	28.71	3	13
12	Australia	5	2.07	55	11.00	14	5
13	Israel	5	2.07	168	33.60	31	0
14	Argentina	4	1.65	66	16.50	28	11
15	Egypt	4	1.65	51	12.75	40	8
16	Romania	4	1.65	89	22.25	46	10
17	Singapore	4	1.65	96	24.00	34	5
18	Germany	3	1.24	136	45.33	4	2
19	Malaysia	3	1.24	30	10.00	33	3
20	Others 21 countries	30	12.40	667	22.23	–	–

[a]Nominal GDP Rank as per the International Monetary Fund (2019 estimates), World Economic Outlook Database, October 2019.

published documents in collaboration with China, United Kingdom, Japan, Taiwan, Egypt, Italy, Romania, Australia, India, Malaysia, Canada, Spain, Portugal, Iran, and Argentina (Fig. 2). In the second position was China, with a TLS of 77. China had published research documents with the USA, Australia, Canada, India, Egypt, Taiwan, Japan, the United Kingdom, Romania, Italy, Singapore, Spain, and Portugal. With a TLS score of 64, Spain was in the third position in collaborative research. The researchers from Spain had published documents in collaboration with researchers from Portugal, the USA, the United Kingdom, China, India, Canada, Egypt, Taiwan, Italy, Iran, and Argentina. From the analysis of the TLS score and Country cooperation network map, it is evident that there is a robust research collaboration of most of the countries with the USA and China.

Further, the top organizations were identified that published more than three documents in the said field of research (Table 2). The bibliometric analysis showed that five organizations had published at least three documents across the years. It was found that Universidad de Santiago de Compostela, Spain had published eight documents, which could gather 459 citations. This resulted in the average citations of each document to 57.38. The citation metrics of Universidad de Santiago de Compostela, Spain was far superior to Hadassah Hebrew University Medical Center, Israel, and the University of Florida, US. The Hadassah Hebrew University Medical Center and the University of Florida had three documents each and

Fig. 2 Country cooperation network on hydrogel-based ocular drug delivery research. (N.B.: (1) Countries that published at least three documents were considered, and (2) Israel with no research collaboration was not considered for the mapping).

Table 2 Top organizations that published more than three documents.

#	Organization	Country	Documents	Citations	Average citation per document
1	Departamento de Farmacia y Tecnología Farmacéutica, Facultad de Farmacia, Universidad de Santiago de Compostela	Spain	8	459	57.38
2	Department of Ophthalmology, Hadassah Hebrew University Medical Center, Jerusalem	Israel	3	105	35.00
3	Department of Chemical engineering, University of Florida, Gainesville	United States	3	95	31.67
4	Institute of Biomaterials and Engineering, Wenzhou Medical University, Wenzhou	China	3	22	7.33
5	School of Pharmacy, Shenyang Pharmaceutical University, Shenyang	China	3	18	6.00

had an average citation of 35.00 and 31.67, respectively. The rest of the two organizations (Wenzhou Medical University and Shenyang Pharmaceutical University) were from China and had three documents each. However, the average citation for both organizations was lower than 10. From the analysis, it can be concluded that Universidad de Santiago de Compostela, Spain, plays a dominant role in this area of research.

3.3 Authors and co-authors relationship

The analysis of the author's and co-author's relationship helps to identify the major research groups that are working in a particular field. This can be achieved easily by mapping the author's and co-author's relationship (Fig. 3). The mapping process provides a visual representation of the relationship, which makes it easier to study not only the activity of an author but also the interconnectivity with other research groups. Fig. 3 divulged that there were five major groups (considering at least five authors in a group) that are working in the field of hydrogels-based drug delivery systems.

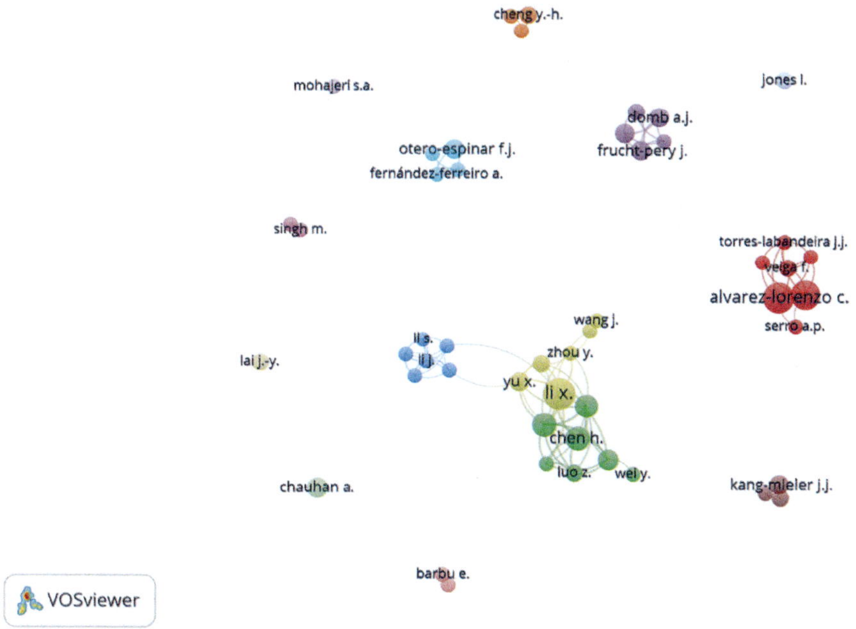

Fig. 3 Authors and co-authors relationship of the researchers who are working on hydrogel-based ocular drug delivery research. (N.B.: (1) Authors who have published at least three documents were considered).

The research groups of the authors X. Li, H. Chen, J. Li, C. Alvarez-Lorenzo, and J. Frucht-Pery were the prominent groups in the said field of research. Among all the researchers, X. Li had the highest number of documents (number of documents = 14), followed by C. Alvarez-Lorenzo, and A. Concheiro, who had 13 and 12 documents, respectively. The research groups of X. Li, H. Chen, and J. Li were highly interconnected and together formed the largest cluster of 19 researchers. X. Li had a direct relationship with H. Chen, suggesting that X. Li and H. Chen were involved in collaborative research. Though there was no direct relationship between X. Li and J. Li, both the authors shared a relationship through Y. Wang. It is highly possible that Y. Wang was trained in X. Li's laboratory, before he joined J. Li's laboratory. Both X. Li and H. Chen were connected to X. Yu, who was in turn connected with J. Li through the author X. Yang. In gist, it may be concluded that the personnel of the research groups of X. Li, H. Chen, and J. Li invariantly join each other's research group and share their research expertise. The research group of C. Alvarez-Lorenzo and A. Concheiro formed the second-largest cluster, where seven researchers were involved in collaborative research. The fifth research group of J. Frucht-Pery consisted of five researchers. Interestingly, the research groups of C. Alvarez-Lorenzo and J. Frucht-Pery were closed and did not have any collaboration with any other research group. It may be concluded from the discussion that there is a need for extensive research collaboration among the different research groups that may allow crucial information and knowledge flow among the different research groups. This, in turn, will help in improving the quality of the research in ocular drug delivery systems using hydrogels.

The number of documents published by the authors and the citation metrics received by the authors allows identifying the most active researchers in a said field of research (Table 3). As mentioned earlier, X. Li, followed by C. Alvarez-Lorenzo and A. Concheiro, had the highest number of documents. However, the number of citations of C. Alvarez-Lorenzo was highest, followed by A. Concheiro and X. Li. This indicated that these three authors are the most active researchers across the globe. However, the average citations per document provide information about the most impactful researchers. In other words, a good quality paper would receive a higher number of citations, which can be identified by calculating average citations per document. It can be observed that X. Wang (documents: 3, citations: 149), followed by A. Chauhan (documents: 6, citations: 295) and A. Concheiro (documents: 12, citations: 483), had the highest average citations per document. This is suggestive of the fact that the documents of

Table 3 Top authors who published more than three documents.

#	Author	Documents	Citations	Average citations	Total link strength
1	Li X.	14	321	22.93	45
2	Alvarez-Lorenzo C.	13	514	39.54	25
3	Concheiro A.	12	483	40.25	25
4	Chen H.	8	288	36.00	30
5	Yu J.	8	155	19.38	34
6	Zhang Z.	7	177	25.29	27
7	Chauhan A.	6	295	49.17	0
8	Shi S.	6	223	37.17	25
9	Domb A.J.	5	168	33.60	17
10	Eljarrat-Binstock E.	5	168	33.60	17
11	Frucht-Pery J.	5	168	33.60	17
12	Kang-Mieler J.J.	5	125	25.00	4
13	Otero-Espinar F.J.	5	99	19.8	9
14	Yu X.	5	35	7.00	14
15	Cheng Y.-H.	4	112	28.00	6
16	Jones L.	4	122	30.5	0
17	Lai J.-Y.	4	136	34.00	0
18	Luo Z.	4	128	32.00	21
19	Mieler W.F.	4	158	39.5	4
20	Orucov F.	4	118	29.5	15
21	Veiga F.	4	128	32.00	15
22	Wang Q.	4	25	6.25	7
23	Barbu E.	3	101	33.67	3
24	Chiou S.-H.	3	112	37.33	5
25	Fernández-Ferreiro A.	3	54	18.00	9
26	Gil-Martínez M.	3	54	18.00	9
27	Jin L.	3	119	39.67	16
28	Lamas M.J.	3	54	18.00	9
29	Li J.	3	30	10.00	10
30	Li S.	3	20	6.67	8
31	Liu C.J.-L.	3	75	25.00	5
32	Liu W.	3	80	26.67	2
33	Mohajeri S.A.	3	54	18.00	0
34	Pan W.	3	30	10.00	10
35	Pe'er J.	3	71	23.67	12
36	Ribeiro A.	3	112	37.33	15
37	Santos D.	3	112	37.33	15
38	Serro A.P.	3	16	5.33	2
39	Singh M.	3	25	8.33	3
40	Singh V.	3	25	8.33	3

Continued

Table 3 Top authors who published more than three documents—cont'd

#	Author	Documents	Citations	Average citations	Total link strength
41	Torres-Labandeira J.J.	3	112	37.33	15
42	Tsibouklis J.	3	101	33.67	3
43	Wang J.	3	101	33.67	1
44	Wang X.	3	149	49.67	3
45	Wang Y.	3	46	15.33	6
46	Wei Y.	3	90	30.00	1
47	Yang X.	3	52	17.33	6
48	Yu S.	3	30	10.00	10
49	Zhou Y.	3	47	15.67	8

X. Wang, A. Chauhan, and A. Concheiro were more impactful compared to the others. The average citation per document of C. Alvarez-Lorenzo was 39.54 and was in the fifth position. Interestingly, X. Li was at a thirty-first position with average citations per document of 22.93, indicating that even though X. Li is a prolific researcher with good networking and collaboration, somehow, his research outcomes are not cited frequently.

3.4 Distribution and co-citation relationship

The sources and citations relationship provides an indication of the important journals where the authors prefer to publish their results (Fig. 4). Table 4 enlists the important journals that published research articles related to hydrogel-based ocular drug delivery systems. It can be seen from Table 4 that the most preferred choice of the authors to publish their research on hydrogel-based ocular drug delivery was the International Journal of Pharmaceutics. In the journal, 17 articles have been published to date. These documents were able to garner 507 citations for the journal. On the second spot was the Journal of Controlled Release, where 11 articles have been published, and the journal received 813 citations. The third spot was the European Journal of Pharmaceutics and Biopharmaceutics, with nine publications. The journal received 182 citations for these research articles. It is also noteworthy that all these three journals had higher total link strength as compared to the other journals. This is suggestive of the fact that these journals were highly cited by the documents published in the other journals, which can be confirmed from the source-citation relationship map. In terms of average citations per documents, the top three journals were Journal of Controlled Release, Biomacromolecules, and Biomaterials, respectively.

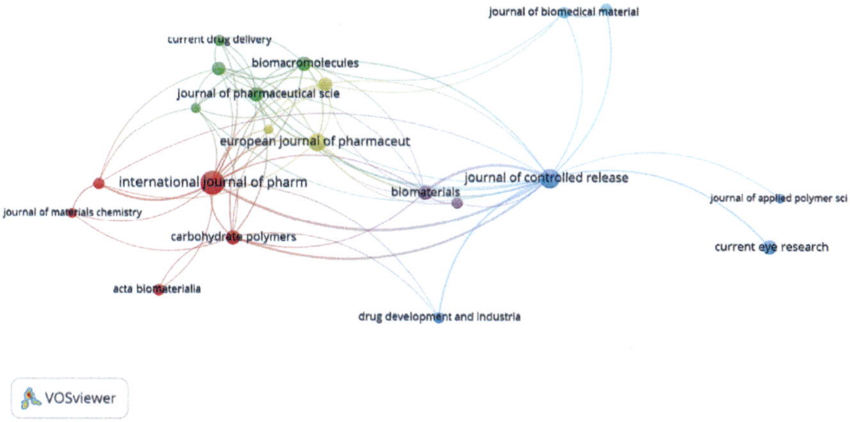

Fig. 4 Source and citations relationship of the sources that published documents on hydrogel-based ocular drug delivery research. (N.B.: (1) Sources that have published at least three documents were considered).

The average citations per document indicate the impactful articles that are published in the journals. This is indicative of the fact the journals Journal of Controlled Release, Biomacromolecules, and Biomaterials published high-quality research work. This is substantiated from the fact that the impact factor of these journals was >5.00. The impact factor of the rest of the journals was <5.00, except for Acta Biomaterialia and Carbohydrate polymers.

3.5 Documents and citations relationship

The document and citation relationship provides information on the quality of the published document. A higher citation metric suggests that the quality of the paper is very good and hence has been cited by many researchers. In our review, we have selected the documents that have been cited at least 40 times, which allowed us to narrow down the number of documents to 47 (Table 5). This allowed us to analyze the highly-cited documents. The top-cited document was Carlfors et al. (1998), which reported the evaluation of the Gelrite® in situ gels for probable use in ophthalmic applications [18]. The second most cited document described the controlled release of the drug, norfloxacin, through the soft contact lenses [19]. The third most cited document was on zero-order therapeutic release from the hydrogel-based contact lenses [20]. Among these three top-cited documents, the document by Ali et al. (2007), had been cited the most (six times) by the selected group of 47 documents [20, 23, 27, 31, 34, 36, 56]. The documents by Venkatesh et al. (2007) [23], Ribeiro et al. (2011) [34], Xie et al. (2015) [59], and Casolaro et al. (2012) [64] were cited five times by

Table 4 Top journals where more than three documents were published.

#	Source	Documents	Citations	Average citations per document	Impact factor (2018)	Total link strength
1	International Journal of Pharmaceutics	17	507	29.82	4.213	34
2	Journal of Controlled Release	11	813	73.91	7.901	51
3	European Journal of Pharmaceutics and Biopharmaceutics	9	182	20.22	4.708	20
4	Biomacromolecules	6	377	62.83	5.667	18
5	Biomaterials	6	355	59.17	10.273	18
6	Carbohydrate Polymers	6	134	22.33	6.044	19
7	Current Eye Research	6	121	20.17	1.672	3
8	Journal of Pharmaceutical Sciences	6	134	22.33	3.197	16
9	Colloids and Surfaces B: Biointerfaces	5	103	20.60	3.973	12
10	Materials Science and Engineering C	5	38	7.60	4.959	13
11	Acta Biomaterialia	4	109	27.25	6.638	2
12	Current Drug Delivery	4	70	17.50	1.645	6
13	Drug Delivery	4	57	14.25	3.095	11
14	Drug Development and Industrial Pharmacy	4	148	37.00	2.367	6
15	European Journal of Pharmaceutical Sciences	4	231	57.75	3.532	4
16	Journal of Biomedical Materials Research—Part B Applied Biomaterials	4	145	36.25	2.674	6
17	Journal of Materials Science: Materials in Medicine	4	105	26.25	2.467	5
18	International Journal of Biological Macromolecules	3	85	28.33	4.784	9
19	Journal of Applied Polymer Science	3	21	7.00	2.188	1
20	Journal of Biomedical Materials Research—Part A	3	117	39.00	3.221	4

Table 5 Top documents that have been cited for at least 40 times.

#	Document	Citations	Links	Reference
1	Carlfors, Edsman et al. 1998	183	2	[18]
2	Alvarez-Lorenzo, Yañez et al. 2006	159	3	[19]
3	Ali, Horikawa et al. 2007	140	6	[20]
4	Le Bourlais, Treupel-Acar et al. 1995	132	0	[21]
5	Hori, Sotozono et al. 2007	126	2	[22]
6	Venkatesh, Sizemore et al. 2007	119	5	[23]
7	Lai and Li 2010	113	0	[24]
8	Cho, Chung et al. 2003	105	1	[25]
9	Jung, Abou-Jaoude et al. 2013	104	4	[26]
10	Rosa dos Santos, Alvarez-Lorenzo et al. 2009	98	4	[27]
11	Mullarney, Seery et al. 2006	97	0	[28]
12	Yang, Tyagi et al. 2012	95	1	[29]
13	Budai, Hajdú et al. 2007	83	2	[30]
14	Peng and Chauhan 2011	82	4	[31]
15	Derwent and Mieler 2008	82	1	[32]
16	Kapoor and Chauhan 2008	82	3	[33]
17	Ribeiro, Veiga et al. 2011	75	5	[34]
18	Holden, Tyagi et al. 2012	74	1	[35]
19	Winterton, Lally et al. 2007	74	1	[36]
20	Maulvi, Lakdawala et al. 2016	70	1	[37]
21	Hao, Wang et al. 2014	70	2	[38]
22	Gao, Sun et al. 2010	70	1	[39]
23	Liu, Griffith et al. 2008	68	0	[40]
24	Anumolu, Singh et al. 2009	66	4	[41]
25	Turturro, Guthrie et al. 2011	63	2	[42]
26	Xu, Liang et al. 2010	62	1	[43]
27	Cavalieri, Miano et al. 2004	59	1	[44]
28	Xu, Weng et al. 2013	55	0	[45]
29	Zhang, He et al. 2016	53	2	[46]
30	Anumolu, DeSantis et al. 2010	50	4	[47]
31	Eljarrat-Binstock, Raiskup et al. 2004	50	1	[48]
32	Yu, Xu et al. 2014	49	4	[49]
33	Sivaram, Rajitha et al. 2015	48	0	[50]
34	Li, Zhang et al. 2013	48	4	[51]
35	Eljarrat-Binstock, Orucov et al. 2008	47	1	[52]
36	Huang, Zhong et al. 2016	46	0	[53]
37	Widjaja, Bora et al. 2014	46	1	[54]
38	Hui, Boone et al. 2008	46	0	[55]
39	Glisoni, García-Fernández et al. 2013	45	4	[56]
40	Verestiuc, Nastasescu et al. 2006	45	0	[57]

Continued

Table 5 Top documents that have been cited for at least 40 times—cont'd

#	Document	Citations	Links	Reference
41	Lovett, Wang et al. 2015	44	0	[58]
42	Xie, Jin et al. 2015	44	5	[59]
43	Cheng, Hung et al. 2014	44	0	[60]
44	Gonzalez-Mira, Nikolić et al. 2012	44	1	[61]
45	Su, Tan et al. 2015	42	1	[62]
46	Yin, Gong et al. 2010	42	0	[63]
47	Casolaro, Casolaro et al. 2012	40	5	[64]

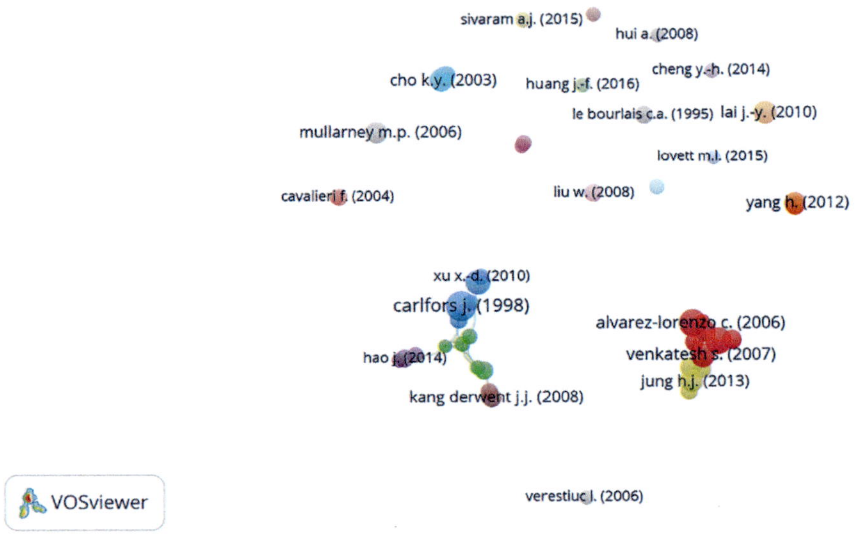

Fig. 5 Documents and citations relationship of the publications that have been cited for at least 40 times.

the selected group of 47 documents. This suggests that these five documents are followed by other high-quality documents that are in the field of hydrogel-based ocular drug delivery systems. From Fig. 5, it can be observed that there were two distinct and highly interconnected clusters. In the first group, the document from Carlfors et al. (1998) was the most cited publication [18]. In the second group, the documents from Alvarez-Lorenzo et al. (2006), and Ali et al. (2007) were among the most cited publications [19, 20]. In gist, it can be concluded that the documents from Carlfors et al. (1998), Alvarez-Lorenzo et al. (2006), and Ali et al. (2007) were the most significant publications that had a considerable impact on the research involving hydrogels for ocular drug delivery.

3.6 Identifying different hydrogel-based delivery system

The density visualization map (Fig. 6) was created by co-occurrence and all keywords analysis. The keywords were manually screened, such that only the different types of delivery systems are selected for the generation of the density visualization map. The analysis of the density visualization map suggested that among the different delivery systems, the hydrogel/ hydrogels terms were highly correlated to nanoparticle/ nanocarrier, ophthalmic solutions, and contact lens/lenses. The relationship between ophthalmic solutions and hydrogels is quite understandable. This can be explained by the fact that the need for hydrogels was envisioned to improve the residence time of the drug formulations, which the ophthalmic solutions were not able to provide. Hence, researchers must be comparing the results of ophthalmic solutions and hydrogels. Hence, it can be summarized that the ocular delivery of drugs through contact lenses and nanoparticles/nanocarriers is the most preferred method of drug delivery.

4 Conclusion

The current paper discusses the scientometric analysis of hydrogel-based ocular drug delivery systems. In gist, the information on 205 articles related to the research area mentioned above was extracted from the Scopus

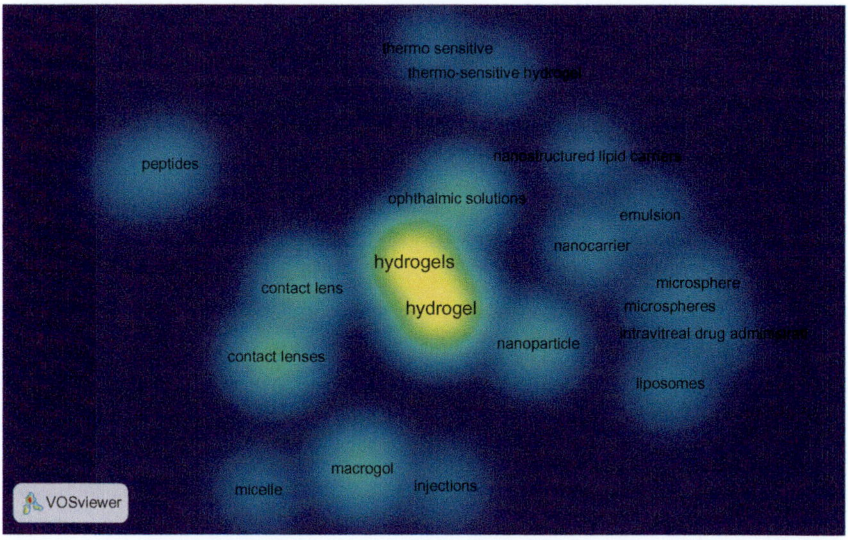

Fig. 6 Density visualization map of the different delivery systems used in ocular drug delivery.

database. The scientometric analysis was performed on the extracted information. It was observed that the research in the field started as early as in the year 1984. However, the research in the said field started receiving much attention post-2010. Since the year 1984 till the year 2009, 34 research articles were published. However, since the year 2010 to date, 170 research articles were published. It was found that most of the economically developed countries were involved in ocular drug delivery research. It was quite evident that the USA was the single largest country that published the most number of documents. The USA was followed by China, which was not far behind the USA. India and Spain shared the distant third position. The publications from Germany and the USA were most impactful. Departamento de Farmacia y Tecnología Farmacéutica, Facultad de Farmacia, Universidad de Santiago de Compostela, Spain was the most influential and impactful organization that worked in the said field of research. The publications of Department of Ophthalmology, Hadassah Hebrew University Medical Center, Jerusalem, Israel, and Department of Chemical engineering, University of Florida, Gainesville, the USA also received good attention from the researchers. Interestingly, there are only five authors groups that published research articles in collaboration with at least five other authors. Hence, there is a huge scope for research collaboration across the globe, which would help to improve the overall research quality. The most number of documents were published by X. Li, C. Alvarez-Lorenzo, and A. Concheiro. But, X. Wang was the most impactful researcher in terms of average citations per document. X. Li was the much influential researcher in terms of research collaboration. Most numbers of the documents were published by the International Journal of Pharmaceutics, followed by the Journal of Controlled Release, which published the most impactful documents in terms of average citations per documents. The document, Carlfors et al. (1998) [18], was the most cited publication, while the document, Ali et al. (2007) [20], was the most cited among the documents that have been cited at least 50 times. Interestingly, it was found that contact lenses have been explored as drug delivery systems. Such delivery systems perform a dual purpose, i.e., vision correction and sustained drug delivery. Also, nanoparticulate and nanocarriers have found profound applications in ocular controlled drug delivery systems.

Conflict of interest statement

There is no conflict of interest among the authors.

References

[1] A. Mandal, et al., Polymeric micelles for ocular drug delivery: from structural frameworks to recent preclinical studies, J. Control. Release 248 (2017) 96–116.
[2] R.C. Cooper, H. Yang, Hydrogel-based ocular drug delivery systems: emerging fabrication strategies, applications, and bench-to-bedside manufacturing considerations, J. Control. Release 306 (2019) 29–39.
[3] D. Chang, K. Park, A. Famili, Hydrogels for sustained delivery of biologics to the back of the eye, Drug Discov. Today 24 (8) (2019) 1470–1482.
[4] L. Petit, H. Khanna, C. Punzo, Advances in gene therapy for diseases of the eye, Hum. Gene Ther. 27 (8) (2016) 563–579.
[5] K. Peynshaert, et al., In vitro and ex vivo models to study drug delivery barriers in the posterior segment of the eye, Adv. Drug Deliv. Rev. 126 (2018) 44–57.
[6] R.M. Farid, et al., Lipid-based nanocarriers for ocular drug delivery, in: Nanostructures for Drug Delivery, Elsevier, 2017, pp. 495–522.
[7] C. Gupta, V. Juyal, U. Nagaich, Formulation and optimization of a thermosensitive in situ gel of moxifloxacin hydrochloride for ocular drug delivery, Int. J. Appl. Pharm. 10 (3) (2018) 123–130.
[8] J.H. Jung, J.J. Chae, M.R. Prausnitz, Targeting drug delivery within the suprachoroidal space, Drug Discov. Today (2019).
[9] J. Ban, et al., Corneal permeation properties of a charged lipid nanoparticle carrier containing dexamethasone, Int. J. Nanomedicine 12 (2017) 1329.
[10] S. Chandran, A. Roy, R.N. Saha, Effect of pH and formulation variables on in vitro transcorneal permeability of flurbiprofen: a technical note, AAPS PharmSciTech 9 (3) (2008) 1031–1037.
[11] R. Gaudana, et al., Ocular drug delivery, AAPS J. 12 (3) (2010) 348–360.
[12] M.J. Herpin, et al., Production and characterization of drug-loaded toroidal vortices from a novel ocular drug delivery device, Drug Deliv. Transl. Res. 8 (5) (2018) 1139–1151.
[13] J. Li, D.J. Mooney, Designing hydrogels for controlled drug delivery, Nat. Rev. Mater. 1 (12) (2016) 1–17.
[14] G. Muhammad, et al., Polysaccharide based superabsorbent hydrogel from Mimosa pudica: swelling–deswelling and drug release, RSC Adv. 6 (28) (2016) 23310–23317.
[15] X. Xu, et al., Prevention of corneal neovascularization by subconjunctival injection of avastin® loaded thermosensitive hydrogels in rabbit model, Int. J. Pharm. 552 (1–2) (2018) 164–170.
[16] Y. Yang, et al., A bibliometric review of laboratory safety in universities, Saf. Sci. 120 (2019) 14–24.
[17] N.J. van Eck, L. Waltman, Software survey: VOSviewer, a computer program for bibliometric mapping, Scientometrics 84 (2) (2010) 523–538.
[18] J. Carlfors, et al., Rheological evaluation of Gelrite® in situ gels for ophthalmic use, Eur. J. Pharm. Sci. 6 (2) (1998) 113–119.
[19] C. Alvarez-Lorenzo, et al., Imprinted soft contact lenses as norfloxacin delivery systems, J. Control. Release 113 (3) (2006) 236–244.
[20] M. Ali, et al., Zero-order therapeutic release from imprinted hydrogel contact lenses within in vitro physiological ocular tear flow, J. Control. Release 124 (3) (2007) 154–162.
[21] C.A. Le Bourlais, et al., New ophthalmic drug delivery systems, Drug Dev. Ind. Pharm. 21 (1) (1995) 19–59.
[22] K. Hori, et al., Controlled-release of epidermal growth factor from cationized gelatin hydrogel enhances corneal epithelial wound healing, J. Control. Release 118 (2) (2007) 169–176.

[23] S. Venkatesh, S.P. Sizemore, M.E. Byrne, Biomimetic hydrogels for enhanced loading and extended release of ocular therapeutics, Biomaterials 28 (4) (2007) 717–724.
[24] J.-Y. Lai, Y.-T. Li, Functional assessment of cross-linked porous gelatin hydrogels for bioengineered cell sheet carriers, Biomacromolecules 11 (5) (2010) 1387–1397.
[25] K.Y. Cho, et al., Release of ciprofloxacin from poloxamer-graft-hyaluronic acid hydrogels in vitro, Int. J. Pharm. 260 (1) (2003) 83–91.
[26] H.J. Jung, et al., Glaucoma therapy by extended release of timolol from nanoparticle loaded silicone-hydrogel contact lenses, J. Control. Release 165 (1) (2013) 82–89.
[27] J.-F. Rosa dos Santos, et al., Soft contact lenses functionalized with pendant cyclodextrins for controlled drug delivery, Biomaterials 30 (7) (2009) 1348–1355.
[28] M.P. Mullarney, T.A.P. Seery, R.A. Weiss, Drug diffusion in hydrophobically modified N, N-dimethylacrylamide hydrogels, Polymer 47 (11) (2006) 3845–3855.
[29] H. Yang, et al., Hybrid dendrimer hydrogel/PLGA nanoparticle platform sustains drug delivery for one week and antiglaucoma effects for four days following one-time topical administration, ACS Nano 6 (9) (2012) 7595–7606.
[30] L. Budai, et al., Gels and liposomes in optimized ocular drug delivery: studies on ciprofloxacin formulations, Int. J. Pharm. 343 (1) (2007) 34–40.
[31] C.-C. Peng, A. Chauhan, Extended cyclosporine delivery by silicone–hydrogel contact lenses, J. Control. Release 154 (3) (2011) 267–274.
[32] J.J.K. Derwent, W.F. Mieler, Thermoresponsive hydrogels as a new ocular drug delivery platform to the posterior segment of the eye, Trans. Am. Ophthalmol. Soc. 106 (2008) 206.
[33] Y. Kapoor, A. Chauhan, Ophthalmic delivery of cyclosporine a from Brij-97 microemulsion and surfactant-laden p-HEMA hydrogels, Int. J. Pharm. 361 (1) (2008) 222–229.
[34] A. Ribeiro, et al., Bioinspired imprinted PHEMA-hydrogels for ocular delivery of carbonic anhydrase inhibitor drugs, Biomacromolecules 12 (3) (2011) 701–709.
[35] C.A. Holden, et al., Polyamidoamine dendrimer hydrogel for enhanced delivery of antiglaucoma drugs, Nanomedicine 8 (5) (2012) 776–783.
[36] L.C. Winterton, et al., The elution of poly (vinyl alcohol) from a contact lens: the realization of a time release moisturizing agent/artificial tear, J. Biomed. Mater. Res. B Appl. Biomater. 80B (2) (2007) 424–432.
[37] F.A. Maulvi, et al., In vitro and in vivo evaluation of novel implantation technology in hydrogel contact lenses for controlled drug delivery, J. Control. Release 226 (2016) 47–56.
[38] J. Hao, et al., Fabrication of a composite system combining solid lipid nanoparticles and thermosensitive hydrogel for challenging ophthalmic drug delivery, Colloids Surf. B: Biointerfaces 114 (2014) 111–120.
[39] Y. Gao, et al., PLGA–PEG–PLGA hydrogel for ocular drug delivery of dexamethasone acetate, Drug Dev. Ind. Pharm. 36 (10) (2010) 1131–1138.
[40] W. Liu, M. Griffith, F. Li, Alginate microsphere-collagen composite hydrogel for ocular drug delivery and implantation, J. Mater. Sci. Mater. Med. 19 (11) (2008) 3365–3371.
[41] S.S. Anumolu, et al., Design and evaluation of novel fast forming pilocarpine-loaded ocular hydrogels for sustained pharmacological response, J. Control. Release 137 (2) (2009) 152–159.
[42] S.B. Turturro, et al., The effects of cross-linked thermo-responsive PNIPAAm-based hydrogel injection on retinal function, Biomaterials 32 (14) (2011) 3620–3626.
[43] X.-D. Xu, et al., Peptide hydrogel as an intraocular drug delivery system for inhibition of postoperative scarring formation, ACS Appl. Mater. Interfaces 2 (9) (2010) 2663–2671.

[44] F. Cavalieri, et al., Study of gelling behavior of poly(vinyl alcohol)-methacrylate for potential utilizations in tissue replacement and drug delivery, Biomacromolecules 5 (6) (2004) 2439–2446.
[45] X. Xu, et al., Sustained release of avastin® from polysaccharides cross-linked hydrogels for ocular drug delivery, Int. J. Biol. Macromol. 60 (2013) 272–276.
[46] Z. Zhang, et al., Fabrication of a micellar supramolecular hydrogel for ocular drug delivery, Biomacromolecules 17 (3) (2016) 798–807.
[47] S.S. Anumolu, et al., Doxycycline loaded poly(ethylene glycol) hydrogels for healing vesicant-induced ocular wounds, Biomaterials 31 (5) (2010) 964–974.
[48] E. Eljarrat-Binstock, et al., Delivery of gentamicin to the rabbit eye by drug-loaded hydrogel iontophoresis, Invest. Ophthalmol. Vis. Sci. 45 (8) (2004) 2543–2548.
[49] J. Yu, et al., In situ covalently cross-linked PEG hydrogel for ocular drug delivery applications, Int. J. Pharm. 470 (1) (2014) 151–157.
[50] A.J. Sivaram, et al., Nanogels for delivery, imaging and therapy, WIREs Nanomed. Nanobiotechnol. 7 (4) (2015) 509–533.
[51] X. Li, Z. Zhang, H. Chen, Development and evaluation of fast forming nanocomposite hydrogel for ocular delivery of diclofenac, Int. J. Pharm. 448 (1) (2013) 96–100.
[52] E. Eljarrat-Binstock, et al., Charged nanoparticles delivery to the eye using hydrogel iontophoresis, J. Control. Release 126 (2) (2008) 156–161.
[53] J.-F. Huang, et al., A hydrogel-based hybrid theranostic contact lens for fungal keratitis, ACS Nano 10 (7) (2016) 6464–6473.
[54] L.K. Widjaja, et al., Hyaluronic acid-based nanocomposite hydrogels for ocular drug delivery applications, J. Biomed. Mater. Res. A 102 (9) (2014) 3056–3065.
[55] A. Hui, A. Boone, L. Jones, Uptake and release of ciprofloxacin-HCl from conventional and silicone hydrogel contact lens materials, Eye Contact Lens 34 (5) (2008) 266–271.
[56] R.J. Glisoni, et al., β-Cyclodextrin hydrogels for the ocular release of antibacterial thiosemicarbazones, Carbohydr. Polym. 93 (2) (2013) 449–457.
[57] L. Verestiuc, et al., Functionalized chitosan/NIPAM (HEMA) hybrid polymer networks as inserts for ocular drug delivery: synthesis, in vitro assessment, and in vivo evaluation, J. Biomed. Mater. Res. A 77A (4) (2006) 726–735.
[58] M.L. Lovett, et al., Silk hydrogels for sustained ocular delivery of anti-vascular endothelial growth factor (anti-VEGF) therapeutics, Eur. J. Pharm. Biopharm. 95 (2015) 271–278.
[59] B. Xie, et al., An injectable thermosensitive polymeric hydrogel for sustained release of Avastin® to treat posterior segment disease, Int. J. Pharm. 490 (1) (2015) 375–383.
[60] Y.-H. Cheng, et al., Sustained delivery of latanoprost by thermosensitive chitosan–gelatin-based hydrogel for controlling ocular hypertension, Acta Biomater. 10 (10) (2014) 4360–4366.
[61] E. Gonzalez-Mira, et al., Improved and safe transcorneal delivery of flurbiprofen by NLC and NLC-based hydrogels, J. Pharm. Sci. 101 (2) (2012) 707–725.
[62] X. Su, et al., Recent progress in using biomaterials as vitreous substitutes, Biomacromolecules 16 (10) (2015) 3093–3102.
[63] H. Yin, et al., Toxicity evaluation of biodegradable and thermosensitive PEG-PCL-PEG hydrogel as a potential in situ sustained ophthalmic drug delivery system, J. Biomed. Mater. Res. B Appl. Biomater. 92B (1) (2010) 129–137.
[64] M. Casolaro, I. Casolaro, S. Lamponi, Stimuli-responsive hydrogels for controlled pilocarpine ocular delivery, Eur. J. Pharm. Biopharm. 80 (3) (2012) 553–561.

Index

Note: Page numbers followed by "*f*" indicate figures, and "*t*" indicate tables.

A

H

I

Q

R

S

T

U

V

W

X

CPI Antony Rowe
Eastbourne, UK
March 21, 2023